Third International Youth Scientific Forum with International Participation “New Materials” 2017

IOP Conference Series: Materials Science and Engineering
Volume 347

Moscow, Russia
21 – 24 November 2017

ISBN: 978-1-5108-6235-7
ISSN: 1757-8981

Printed from e-media with permission by:

Curran Associates, Inc.
57 Morehouse Lane
Red Hook, NY 12571

Some format issues inherent in the e-media version may also appear in this print version.

Printed by Curran Associates, Inc. (2018)

Institute of Physics
Dirac House, Temple Back
Bristol BS1 6BE UK

Phone: 44 1 17 929 7481
Fax: 44 1 17 920 0979

techtracking@iop.org

Additional copies of this publication are available from:

Curran Associates, Inc.
57 Morehouse Lane
Red Hook, NY 12571 USA
Phone: 845-758-0400
Fax: 845-758-2633
Email: curran@proceedings.com
Web: www.proceedings.com

Table of contents

Volume 347

The Third International Youth Scientific Forum with International Participation "New Materials"

21–24 November 2017, Moscow, Russian Federation

Accepted papers received: 9 April 2018
Published online: 27 April 2018

Preface

Papers

IOP Conf. Series: Materials Science and Engineering **347** (2018) 011001 doi:10.1088/1757-899X/347/1/011001

Introduction

From November 21 to 24, 2017, the Third Interdisciplinary Youth Scientific Forum with international participation "New Materials" was held in Moscow in the building of the Presidium of the Russian Academy of Sciences, where topical and applied problems of creating new materials for the future and promising technologies for their processing were discussed. The Forum was held for the third time. The Forum was organized by the Council of Young Scientists of the Russian Academy of Sciences and the Coordinating Council for Youth Affairs under the President of the Russian Federation. Co-organizers of the Forum were the Russian Academy of Sciences (RAS), The Russian Chemical Society D.I. Mendeleyev and Institute of Metallurgy and Materials Science. A.A. Baikova

The Ministry of Education and Science of the Russian Federation, the Federal Agency of Scientific Organizations and the Russian Foundation of Basic Research supported the Forum.

Thematic areas included in the program of the Forum, helped to gather in one place representatives of different interdisciplinary research groups in the area of interest which includes the creation of the fundamentals of materials and their application in various industries. The Forum program was expanded by discussing the problems of introducing scientific developments and interaction of scientists with consumers of scientific results, foundations and development institutions. Additionally, the role of international cooperation, public and educational organizations in the scientific process, and the strategy of popularizing scientific achievements were discussed at the Forum.

The scientific sections of the forum covered a wide range of topics on nanomaterials and nanotechnologies, experimental methods of researching materials and structures, ceramic materials, composite and functional materials, structural materials, biocompatible materials, and materials and technologies for green chemistry.

IOP Conf. Series: Materials Science and Engineering **347** (2018) 011002 doi:10.1088/1757-899X/347/1/011002

Peer review statement

All papers published in this volume of *IOP Conference Series: Materials Science and Engineering* have been peer reviewed through processes administered by the proceedings Editors. Reviews were conducted by expert referees to the professional and scientific standards expected of a proceedings journal published by IOP Publishing.

IOP Conf. Series: Materials Science and Engineering **347** (2018) 012001 doi:10.1088/1757-899X/347/1/012001

Structure and microhardness of the plasma sprayed composite coatings after combined treatment

A Yu Ivannikov[1], V I Kalita[1], D I Komlev[1], A A Radyuk[1], V P Bagmutov[2], I N Zakharov[2], S N Parshev[2] and D S Denisevich[2]

[1] Baikov Institute of Metallurgy and Material Science, Russian Academy of Sciences, 49, Leninsky av., 119334 Moscow, Russian Federation
[2] Chair of Resistance of Materials, Volgograd State Technical University, 28, Lenin ave., 400005, Volgograd, Russian Federation

ivannikov-a@mail.ru

Abstract. The principal aim of this study was to evaluate the effect of combination of electromechanical treatment (EMT) and ultrasonic treatment on structure and microhardness of air plasma sprayed composite coatings from Ni–20Cr alloy and R6M5 high speed steel (HSS). The results of the microstructural studies showed fundamental changes of the treated by the EMT plasma sprayed coating with the formation of nanostructured crystalline phases. As a consequence of the coating thus formed, the number of pores in the coating structure reduced from 10.0±1.5% to 2.0±0.5%, the surface microhardness increased from 3100±500 MPa to 7900±400 MPa. Additional ultrasonic treatment on the selected mode decreased surface waviness, which was formed on the surface of the plasma sprayed composite coatings after the EMT. The obtained results revealed the high potential of the combined treatment for post-treatment of the plasma sprayed coatings.

1. Introduction

There are various post-treatment techniques, for example, laser re-melting, surface induction heating, heat treatment, spark plasma sintering and etc., that are applied to enhance cohesive strength and to increase coating's density [1-4].

Electromechanical treatment of coatings is a modern post-treatment technology. The articles [5-11] show that rolling by the roller of the coating with synchronous heating by the electric current improve coating density, cohesive and adhesive strength, while the advantages of the metastable structure stay the same. The important feature of this method is implementation on the usual metalworking machines. The authors of the articles [5-11] use similar terms to describe the post-treatment of the thermal sprayed coatings. Electric contact strengthening (ECS) was accepted as term by the authors of the articles [5-7]. Coating density and adhesive strength were increased after the ESC of the coatings. The authors of the articles [8-11] adopted the term electromechanical treatment (EMT) for method of post-treatment of the coatings. Improvement of the cohesive, adhesive strength, increase of density were determined in the papers [5-11] after the ECS and the EMT of thermal sprayed coatings. In the articles [9, 11] formation of nanostructured crystalline phases was determined after the EMT of plasma sprayed coatings.

IOP Conf. Series: Materials Science and Engineering **347** (2018) 012001 doi:10.1088/1757-899X/347/1/012001

Therefore, in the articles [12, 13] mathematical modeling of these processes was performed. Mathematical modeling showed that the cooling rate in the process of the EMT reached the value of 10^4 K/s, for this reason, the nanostructured crystalline phases were formed in the structure of the coatings.

Ultrasonic treatment (UT) of bulk materials and coatings is widely used in industry [14, 15]. Also, the important feature of this method is implementation on the usual metalworking machines. Ultrasonic treatment is used to increase the surface quality, i.e., to reduce surface roughness, as well as to relieve residual stresses.

Therefore, the combination of these types of post-treatment of the plasma sprayed coating is a topical way to improve the properties of the plasma sprayed coatings.

The purpose of this study is to estimate the effect of combination of the EMT and the UT on morphology, structure and microhardness of plasma sprayed composition coating.

2. Materials and methods

Two types of powders were used for plasma spraying. Commercial feedstock powder of high speed steel R6M5 (chemical composition of 1.00 – 1.04 wt.% C, 4.00 – 4.07 wt.% Cr, 6.70 – 6.71 wt.% W, 5.55 – 5.57 wt.% Mo, 2.10 – 2.12 wt.% V, 0.50 – 0.55 wt.% Mn, ≤ 0.4 wt.% Si, ≤ 0.03 wt.% S, ≤ 0.03 wt.% O, Fe constituted the balance) was chosen for plasma spraying of the main coating layer. The powder was prepared using atomization method (Metsintez OOO, Tula, Russia). The powder particles had spherical form with mean particle size of 50 – 71 μm. Sublayer and protective layer were plasma sprayed with commercial feedstock powder of Ni–20Cr alloy (chemical composition of 0.04-0.06 wt.% C, 20 wt.% Cr, ≤ 0.1 wt.% Si, ≤ 0.06 wt.% S, ≤ 0.03 wt.% O, Ni constituted the balance). The powder was produced by calcium-hydride method (Metsintez OOO, Tula, Russia). The powder particles had irregular shape and size range of 32 – 50 μm.

Substrates from low-carbon steel were used for plasma spraying of coating and subsequent combination treatment. The sample for determining the effect of the EMT modes was a cylinder 20 mm in diameter and 150 mm in length.

Table 1. The modes of the combined treatment

Track	1	2	3	4	5	6
Current density, A/mm^2	400	400	700	700	1000	1000
Linear speed, m/min				3.39		
Feed rate, mm/rev				0.2		
Ultrasonic treatment *	+		+		+	

*- 22 kHz, load 3 H, 3,39 m/min, 0.1 mm/rev

Plasma spray equipment was the air plasma spray system UPU-3D (JSCo "Electromechanica", Rzhev, Russia) with the plasma spray gun PP-25. The maximum power of the plasma spray gun PP-25 is 25 kW.

Plasma spraying was carried out according to a standard protocol [11].

Ni–20Cr powder was plasma sprayed on cylindrical samples as a sublayer to reduce residual stresses. The thickness of the sublayer was 25 ± 5 μm.

Then the plasma spraying of R6M5 powder onto the sublayer as the main coating layer of the composition coating was carried out. The thickness of this layer was 250 ± 25 μm.

As the final stage, the Ni–20Cr powder was plasma sprayed on the surface of the R6M5 coating layer as a protective layer to prevent damage from oxidation. The thickness of the protective layer was 25 ± 5 μm.

The plasma sprayed composite coating thickness was 300 ± 50 μm.

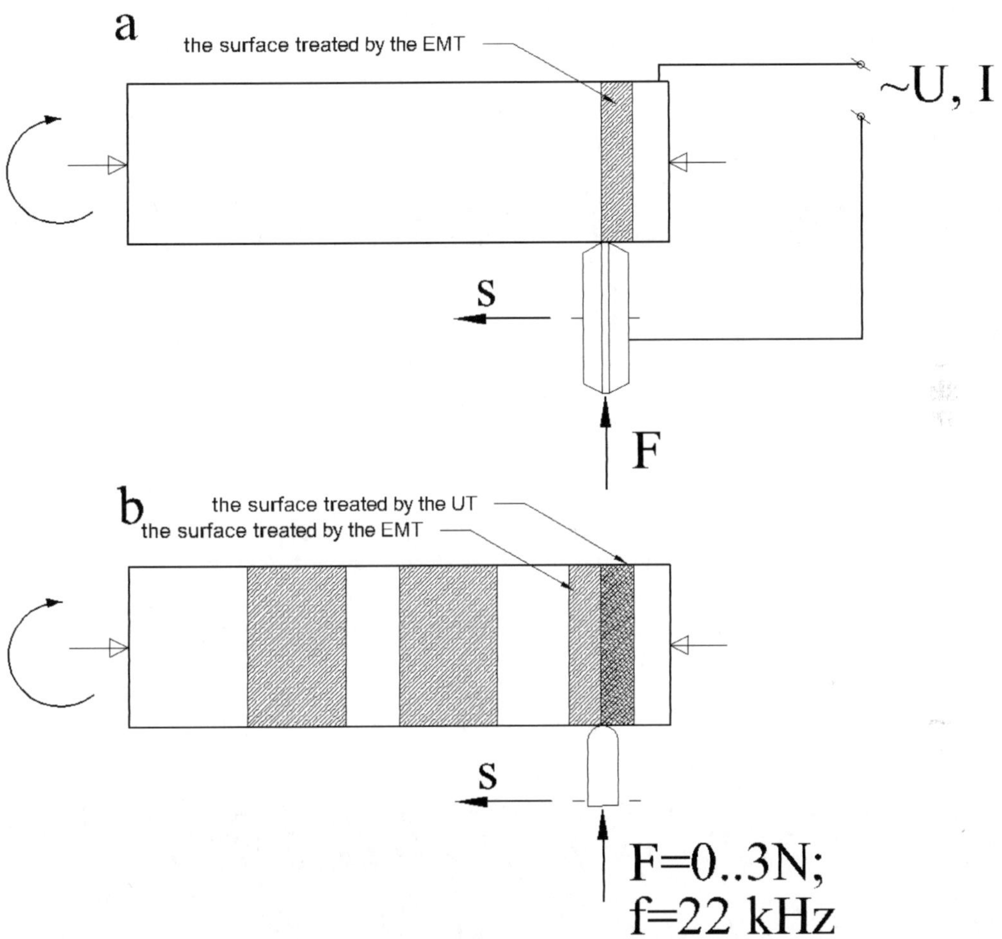

Figure 1. Schematic diagram of the EMT (a) and the UT (b)

The installation of BUFO (UTINLAB OOO, St. Petersburg, Russia) was used for the ultrasonic treatment of the plasma sprayed coating after the EMT.

Ultrasonic treatment (UT) was performed in the following mode, the frequency of 22 kHz, the load in the contact zone 3 N, the processing speed of 3.39 m/min, longitudinal feed 0.1 mm/rev, the contact area of the indenter with the coating was 2 mm^2.

Preparing of the test samples was carried out according to a standard protocol [11].

Structures of the plasma sprayed and treated by the EMT coatings were examined by means of the optical microscope (Carl Zeiss Jenavert Interphako) and the scanning electron microscope (SEM), attached with the EDS (Versa 3D). Measurements of porosity and second phases were made with the image analyzer, having software of VideoTest Structure 4.0 (VideoTest OOO, St. Petersburg, Russia). The coating microhardness was determined in cross-section using the Vickers indenter (PMT-3) at the load of 1.96 N (200 gf), for 15 s. Minimum 15 indentations were applied to each sample.

3. Results and discussion

The results of the EMT showed reduction of roughness of free surface of the plasma sprayed coating. During the EMT process, the roller was indented in the coating, compacting the coating in this way. Due to the fact that part of the material was squeezed out along the roller's edges under pressure, the wavy structure was formed on the surface of the coating. Period of the wavy structure was 200 ± 25 μm because of the constant feed rate (0.2 mm/rev).

Optical microscopy showed that the splats had different colors: white and gray (Figure 2, 3). SEM analysis detected nanocrystal phases in the structure of the gray splats [6]. They were formed because of breakdown of solid solution. And, perhaps, the nanocrystal phases were carbides of W, Mo, V, Cr.

SEM analysis showed that after the UT the nanocrystal phases in the structure of the gray splats were saved (Figure 4).

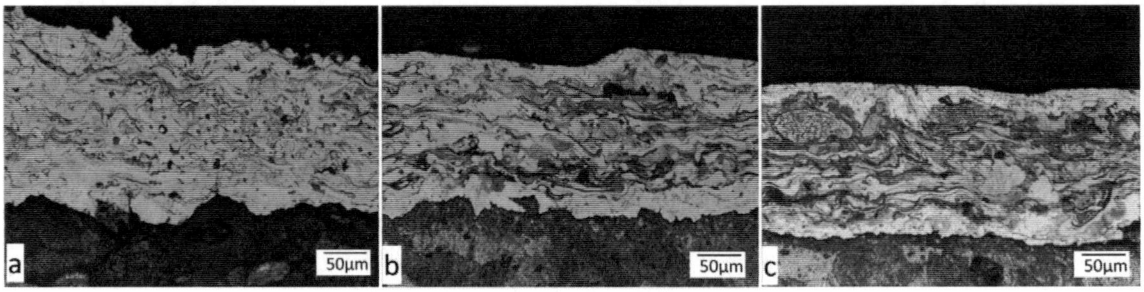

Figure 2. Macrostructure of cross-section of the plasma sprayed composition coating: a – initial coating, b – coating after the EMT, c – coating after combined treatment.

The relief surface of the coating, which was formed in the process of the EMT, was smoothed after the UT (Figure 3). The waviness of the coating was reduced after the UT (Figure 2c). Reducing the waviness was due to plastic deformation of the surface layer from the alloy Ni–20Cr (Figure 2c, Figure 3b). The plastic deformation of the surface layer was caused by high-frequency deformation by the indenter in the UT process.

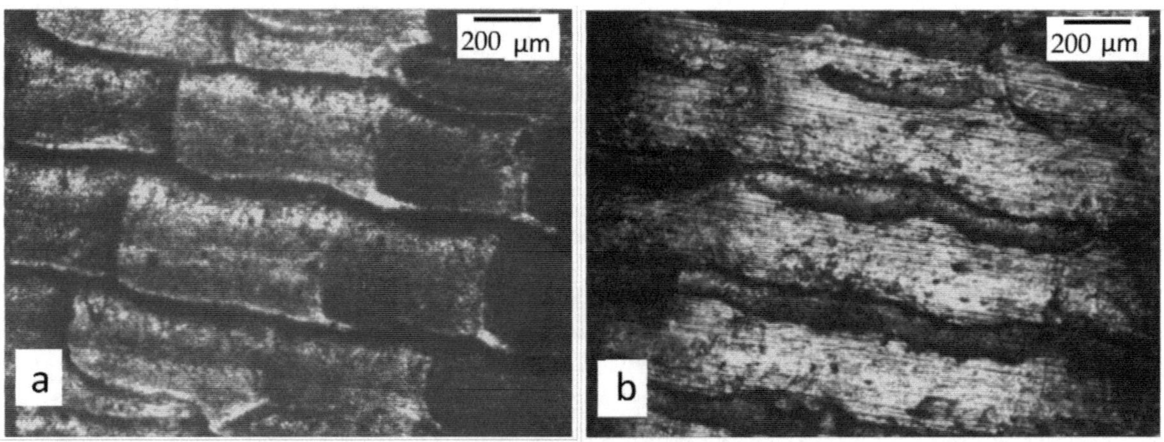

Figure 3. Macrostructure of the surface of the coating: a – after the EMT, b – after combined treatment.

The measurement results of the microhardness of the R6M5 layer are presented in Table 2.

IOP Conf. Series: Materials Science and Engineering **347** (2018) 012001 doi:10.1088/1757-899X/347/1/012001

Table 2. Microhardness of the R6M5 layer.

Track	R6M5	1	2	3	4	5	6
Microhardness, MPa	3100	5350	5400	5650	5900	7900	7900
±S, MPa	560	450	500	550	450	400	500

The microhardness of the R6M5 layer using a load of 200 gf was 3100±500 MPa. The microhardness of the plasma sprayed coating under a 200 gf load increased from 5300 ±500 MPa to 7900 ± 400 MPa after the combined treatment. Such increase of the microhardness (200 gf) is due to the compaction of the coating, the resistive welding of the boundaries of the splats, and the formation of nanostructured crystalline phases in the splats. The microhardness of the coating after the EMT and the combined treatment within measurement error was the same (Table 2). In our opinion this is due to the low energy of the mode of the UT. The force of deformation with frequency of 22 kHz of the plasma sprayed composition coating led to deformation of the surface layer in the UT process, but microhardness of the basic layer was not changed because of low pressure of the indenter.

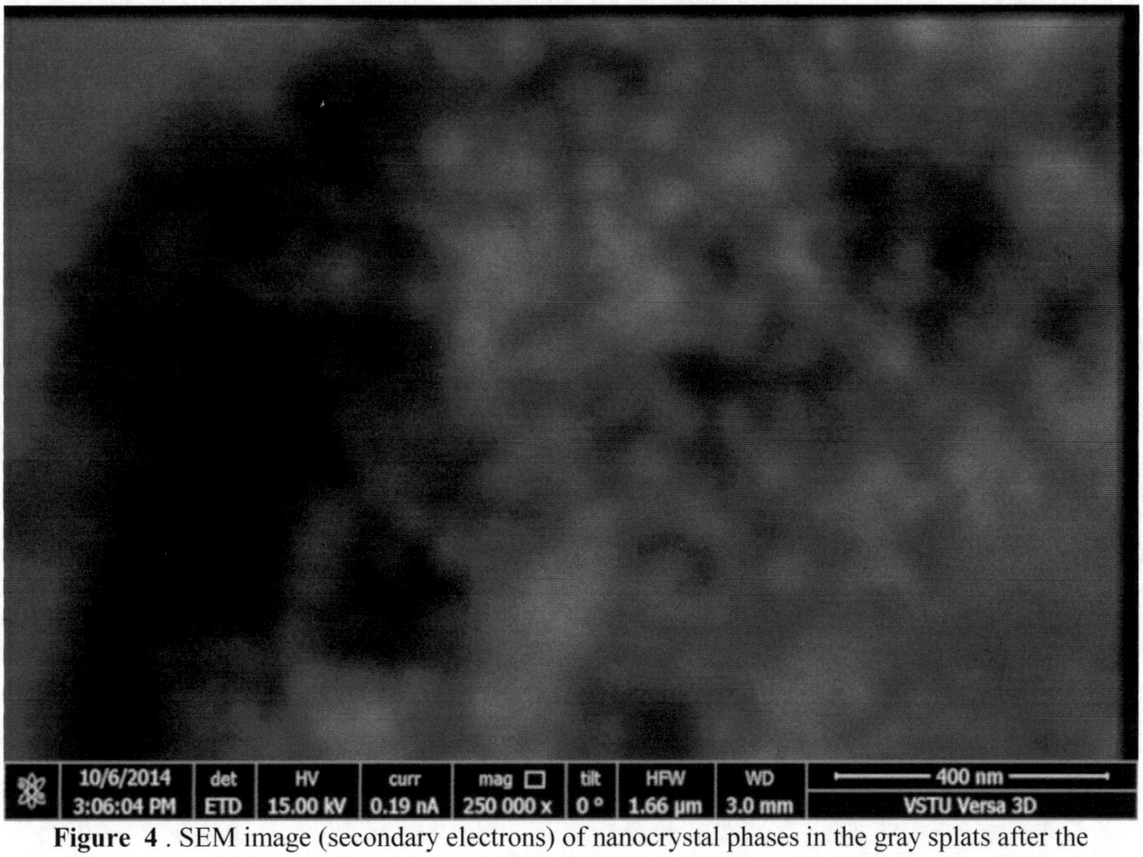

Figure 4 . SEM image (secondary electrons) of nanocrystal phases in the gray splats after the combined treatment

IOP Conf. Series: Materials Science and Engineering **347** (2018) 012001 doi:10.1088/1757-899X/347/1/012001

The UT allows reducing roughness and waviness of the coating's surface, for this reason, subsequent grinding of the coating may be omitted. Important feature of the combined treatment of plasma sprayed coating is realization of the EMT and the UT on the usual metalworking machines in one step.

4. Conclusion

1. After the combined treatment the microhardness of the plasma sprayed coating under a 200 gf load increased from 3100 ± 500 MPa to 7900 ± 400 MPa.

2. The nanocrystal phases in the structure of the gray splats were saved after the UT.

3. In this investigation, the main effect of the UT is to reduce the surface waviness that should improve operational properties of the coating

Acknowledgments
The reported study was supported by grant for young PhD № MK-1429.2017.8, ultrasonic treatment was supported by grant RFBR 18-08-00842 a, scanning electron investigation was supported by grant RFBR 17-08-01742 a, preparing of the test samples was supported by State targets № 007-00129-18-00.

References

[1] Pawlowski L 2008 *The Science and Engineering of Thermal Spray Coatings* (Chichester: Wiley &Sons,) p 627
[2] Chraska T, Pala Z, Mušálek R, Medřický J, Vilémová M 2015 *J. Therm. Spray Technol.* 24 pp 637–643
[3] Fomin A, Fomina M, Koshuro V, Voyko A, Rodionov I 2016 *Journal of Physics: Conference Series* 741(1),012170.
[4] Fomin A, Fomina M, Koshuro V, Voyko A, Rodionov I. 2017 *Journal of Physics: Conference Series* 917(8),082004
[5] Wang Y, Zhu, S., Gu, W., Qi, X. 2011 *Exp. Tech.* 35pp 66–70. [10]
[6] Xu M, Zhu S, Ding H 2015, *Surf. Coat. Technol.* 279pp 32–38.
[7] Wang Y, Zhu, S., Gu, W., Qi, X. 2010, *J. Therm. Spray Technol.* 19pp 1142–1146.
[8] Bagmutov V , Zakharova E , Zakharov I , Kosogorov A , Kalita V, Komlev D, Ivannikov A 2013 *Steel in Translation*, 43 pp 351-355
[9] Komlev D, Kalita V, Menshikov G, Vlasenko A, Radyuk A, Ivannikov A 2013, *Inorganic Materials: Applied Research*. 4(3) pp 236-246.
[10] Kalita V, Bagmutov V, Zakharov I, Komlev D, Ivannikov A 2008 *Physics and chemistry of materials processing* 1 pp 38-42
[11] Ivannikov A, Kalita V, Komlev D, Radyuk A, Bagmutov V, Zakharov I, Parshev S 2018, *Surf. Coat. Technol* 335 pp 327-333
[12] Bagmutov V, Zakharov I, Ivannikov A, Poplavsky E 2007, *Vestn. Voronezh. Gos. Tekhn. Univ.* 3 (8) pp 135-140
[13] Bagmutov V, Denisevich D, Zakharov I, Ivannikov A 2017, *PNRPU Mechanics Bulletin*, (1) pp 233-250
[14] Malaki M, Ding H 2015 *Materials & Design* 87 pp 1072-1086
[15] Han Q 2015 *Metallurgical and Materials Transactions B* 46 pp 1603–1614

IOP Conf. Series: Materials Science and Engineering **347** (2018) 012002 doi:10.1088/1757-899X/347/1/012002

Shear strength of a three-dimensional capillary-porous titanium coating for biomedical applications

V I Kalita, D I Komlev, A A Radyuk, A Yu Ivannikov, A V Alpatov, V S Komlev, V I Mamonov, M A Sevostyanov and A S Baikin.

Baikov Institute of Metallurgy and Material Science, Russian Academy of Sciences, 49, Leninsky ae., 119334 Moscow, Russian Federation

vkalita@imet.ac.ru

Abstract. The effect of pretreatment and plasma preheating of Ti-substrate on shear strength of three-dimensional capillary porous Ti-coating was studied. After sandblasting the shear strength of the plasma sprayed coating was 200 ± 2 MPa, and after additional matting it was 68 ± 4 MPa. The use of plasma preheating of the substrates for 9 seconds decreased difference between values of the shear strength to 249 ± 17 MPa and 229 ± 16 MPa, respectively. After plasma spraying the microhardness of the surface layer of the substrate was 4.34 ± 0.35 GPa, the microhardness of the boundary between the coating and the substrate was 8.08 ± 0.45 GPa, and the microhardness of the coating was 3.48 ± 0.25 GPa. High shear strength of the coating was attributed to the activation of the substrate by means of plasma preheating and hardening of the boundary between the coating and the substrate by oxides and nitrides.

1. Introduction

This work considers a complex system "implant – bone tissue". The implant surface must simultaneously satisfy many requirements in the field of mechanics, plasma spraying, and medicine [1,2]. Over the last 40 years many works have been published on the topic, which is reflected in reviews [1,2].

The authors of this study developed three-dimensional capillary porous (3DCP) Ti-coating for endosteal implants with a thickness of 1 mm or more [3]. Such coatings consist of ridges and valleys. Coating valleys form the basic volume of pore space, up to 50%, in which new bone tissue can grow freely and operate. Ridges of the 3DCP Ti-coating transmit the main mechanical load between the implants and bone tissue. Owing to this effect, the strength of the formed ridges is comparable with that of monolithic materials. Bioactive properties of the coating are provided with additional surface layers, which are formed, for example, by micro-plasma oxidation or plasma spraying [4, 5].

The purpose of this study is to estimate the effect of the plasma preheating of the substrates to achieve the maximum shear strength of the 3DCP Ti-coatings with high porosity.

2. Materials and methods

VT1-00 (chemical composition of ≤ 0.15 wt% Fe, ≤ 0.04 wt% N, ≤ 0.1 wt% O, Ti constituted the balance) was used to manufacture cylindrical substrates with diameter of 9 mm. The wire for plasma spraying was made of VT1-00, too.

Before spraying, the cylindrical substrates were divided into two groups. The surface of the substrates in the first group was roughened by grit blasting with Al2O3 powder (average grit size 700 μm). The surface of the substrates in the second group was roughened by grit blasting with Al2O3 powder (average grit size 700 μm) and then it was additionally processed by means of matting using glass beads (average grit size 120 μm).

Plasma spray equipment was the plasma spray system UPU-3D (JSCo "Electromechanica", Rzhev, Russia) with the plasma spray gun PP-25. Optimization of the plasma sprayed modes (plasma gas Ar, arc voltage 30 V, arc current 300 A) was performed in work [3]. The angle between the axis of rotation of the substrate and the spray cone was 45°.

The surface of the substrate was preheated using plasma before plasma spraying. In this experiment the preheating time was changed from 0 to 15 s.

MultiFast plastic material (Struers, Denmark) with average shear strength of 96.1 MPa was used for modeling of bone tissue. The plastic was used to press-fit on the 3DCP Ti-coatings in accordance with method of the shear test [3].

The original procedure for determining the adhesion of the coating to the substrate was used [3]. Test specimens for shear test were performed on the universal testing machine Instron 5882.

The microstructure of the substrate, the coating, and the boundary between them were examined by means of the optical microscope (Carl Zeiss Jenavert Interphako).

The samples were etched in the bath 5% HF + 25% HNO_3 + 70% H_2O for metallography. The etching time was 5-10 s. The time of deep etching was 180 s.

The coating microhardness was determined in cross-section using the Vickers indenter (PMT-3) at loads of 0.098 and 1.96 N (10 and 200 gf) for 15 s. Minimum 15 indentations were applied to each sample.

The nitrogen and oxygen contents were measured with Leco TC-436 and Leco TC-600 analyzers, respectively.

3. Results

For the specimens from the first batch the average thickness of the 3DCP Ti-coating was 1.47 ±0.15 mm, and for the specimens from the second batch the average thickness was 1.24 ±0.15 mm.

Qualitative differences in the shear deformation curves of the Ti-bulk material and of the 3DCP Ti-coating on the Ti-bulk surface were not detected (Figure 1). The shear strength of the Ti-bulk material, which was used to manufacture the substrate, was 432 ± 29 MPa.

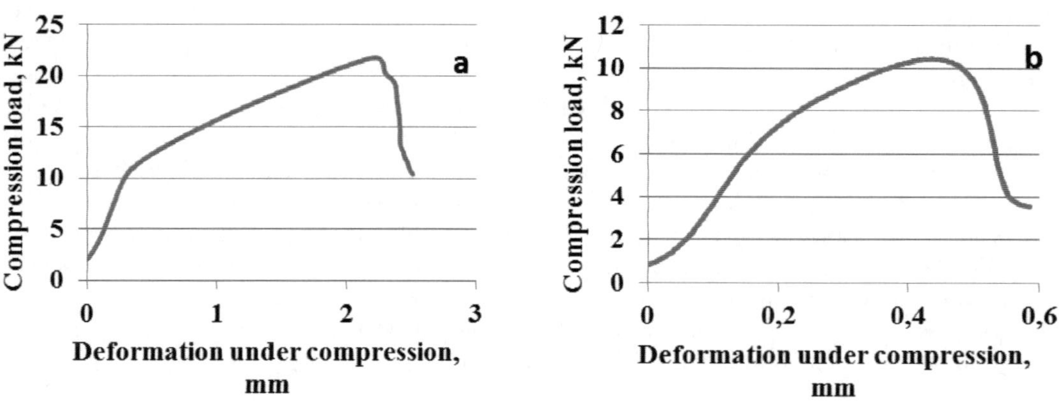

Figure 1. Diagram of shear testing: (a) the specimen from the second group with preheating for 9 s; (b) the bulk specimen.

IOP Conf. Series: Materials Science and Engineering **347** (2018) 012002 doi:10.1088/1757-899X/347/1/012002

The highest value of the shear strength was determined for the specimens from the first batch after the sandblasting process (Figure 2). The maximum shear strength was 249±17 MPa. This value was obtained for the specimens with the preheating for 9 s.

For the specimens from the second batch, the maximum value of the shear strength was 229 ±16 MPa after the plasma preheating for 9 s, and the minimum value was 68± 4 MPa without preheating.

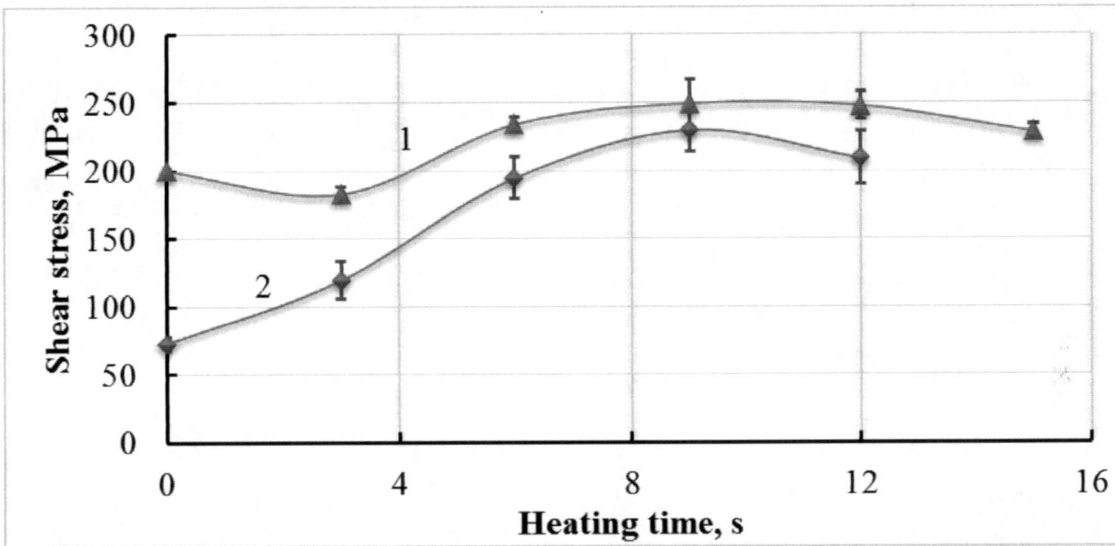

Figure 2. The shear strength of the specimens depending on the time of their preheating by the plasma jet. 1 – the specimens from the first batch, 2 – the specimens from the second batch.

Adhesive fracture of the boundary between the coating and the substrate was determined for the specimens from the second batch without the additional plasma preheating and after the additional preheating for 3 s. For all other specimens the fracture was originated in the coating.

The boundary between the coating and the substrate in the cross-section without etching was not determined by means of optical microscopy. A bright stripe with width of 10 to 20 μm appeared along the border after the etching up to 10 s (Figure 3). The chain of voids appeared on the boundary with increase of the etching time, the relative length of this chain increased with the increasing of the etching time. Near the boundary, the increase of Ti-grain size was identified both in the substrate and in the coating. In certain areas, the Ti-particles grew from the substrate into the coating (the so-called epitaxy).

Table 1 represents the content of oxygen and nitrogen in the Ti-wire and the plasma sprayed coating. The oxygen content in the coating increased 2.8 times, and the nitrogen content increased 15 times in comparison with the content in the Ti-wire.

Increase of the contents of oxygen and nitrogen enhanced the microhardness of the substrate to 4.34 ± 0.35 GPa, the microhardness of the coating to 3.48± 0.25 GPa, and the boundary between them to 8.08± 0.45 GPa, which confirmed the increase of the shear strength (initial microhardness was 1.86± 0.25 GPa) (Figure 4).

4. Discussion

The positive effect of sandblasting can be explained by the activation of the inert surface by removing the part of the coating material from the surface. The depth of the removed layer is about 4 μm, it was previously defined [4].

Moreover, the sandblasting increases the surface roughness, which increases the contact area of the sprayed particles and the substrate.

IOP Conf. Series: Materials Science and Engineering **347** (2018) 012002 doi:10.1088/1757-899X/347/1/012002

Additional matting with use of the glass beads decreases the roughness parameters of the surface [7]. For this reason, the specific surface area of the substrate is reduced, consequently, the value of the shear strength is reduced, too. This is confirmed experimentally.

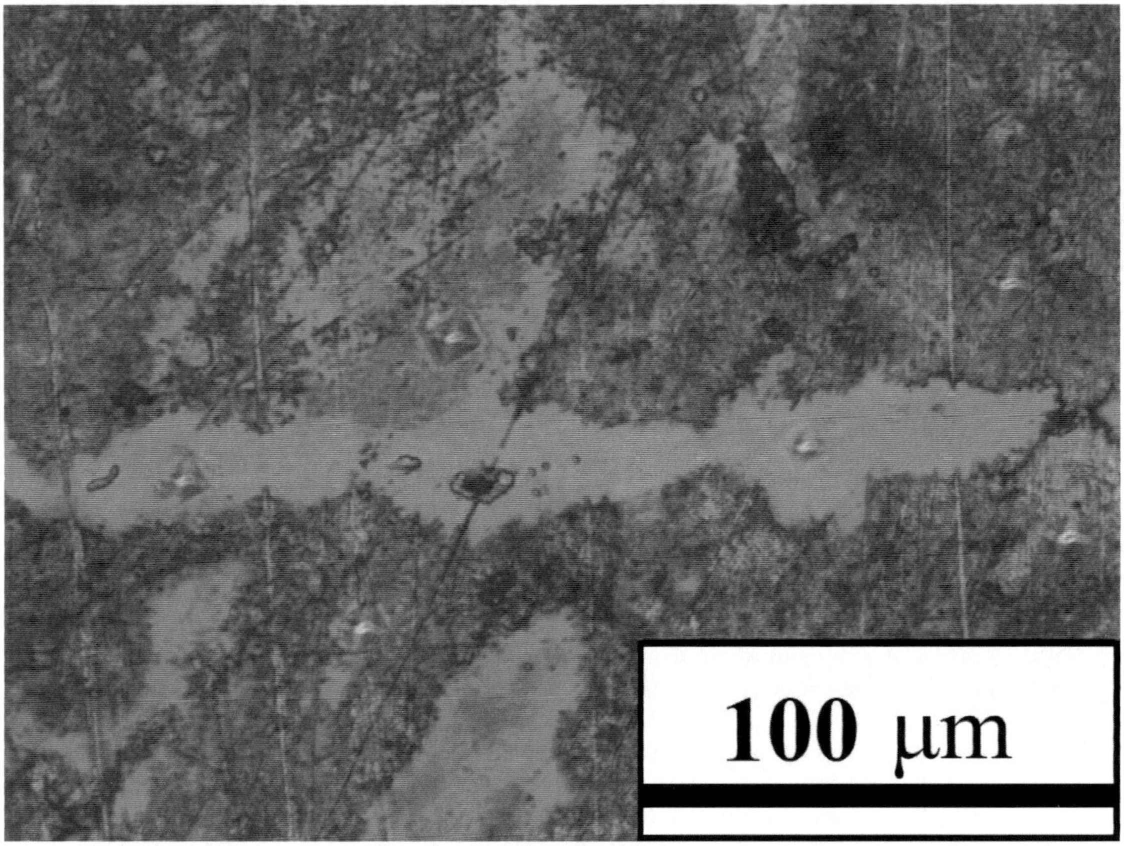

Figure 3. The boundary between the substrate and the coating in the cross-section of the specimen from the first batch, which was preheated for 12 s. Indentation of the diamond pyramid was obtained at the load of 10 gf.

Table 1. The content of the elements O and N in the wire and the coating on the substrate, which were plasma preheated for 15 s.

Sample	Content of elements in % by weight		
	O	N	Ratio O/N
Ti-wire	0.089	0.001	89
3DCP Ti-coating	0.249	0.015	16.6

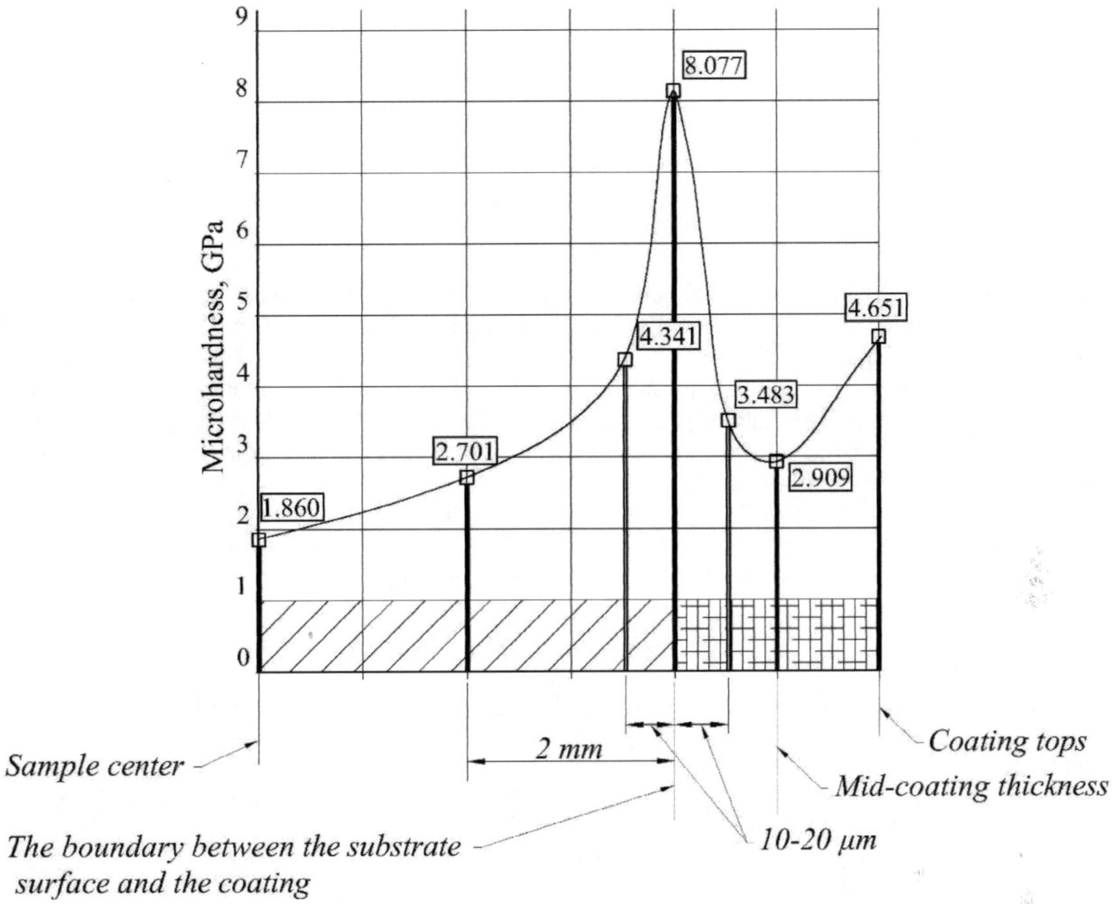

Figure 4. Distribution of values of microhardness (load 10 g) in the thickness of the shear sample: on the left Ti substrate, on the right TCP Ti coating.

The plasma preheating of the substrate for 3 s does not increase the shear strength, for this reason, adhesive fracture is determined between the coating and the surface of the substrate. That is, the surface morphology of the substrate does not provide enough mechanical adhesion with the coating and the low temperature does not allow achieving chemical interactions. There is no wetting of the surface of the substrate by the sprayed particles, which is necessary for solid connection.

Cohesive failure is determined for the coatings on the substrates after the sandblasting. The destruction occurs on the surface of a larger diameter than the diameter of the substrate, and capture valleys of the 3DCP Ti-coating, which are filled with plastic. An immediate challenge is to enhance the bond strength of the plasma sprayed porous coating with the substrate.

The influence of the substrate pretreatment on getting a strong connection with the coating is not fully studied. In our experiments the roughness factor determines the strength of the connection of the sprayed material with the substrate at relatively low substrate temperature (without additional plasma preheating). In this case, at the stage of cooling the sprayed particles can partially detach from the surface under the influence of residual stresses (thus some part of the pore volume is formed).

At temperatures of the substrate higher than 300 °C, the substrate surface becomes more active. The wetting of the substrate by the sprayed particles increases [1]. Preheating of the substrate reduces the difference between the temperature of the substrate and coating, for this reason, residual stresses reduce, and, consequently, the strength of the connection of the coating with the substrate increases.

The preheating of the substrate after sandblasting increases the shear strength of the 3DCP Ti-coatings by 30%. For the 3DCP Ti-coating with additional treatment of the substrate using matting the shear strength increases by 73%. We can assume that thermal activation is the dominant factor in relation to the factor of surface morphology.

High values of the microhardness at the surface of the coatings with the substrate can be explained by the influence of oxygen and nitrogen. Oxygen and nitrogen increase the strength and the hardness of the Ti [6].

5. Conclusion

1. The shear strength of the three-dimensional capillary porous Ti-coating without preheating of the Ti-substrate after the sandblasting was 200 ± 2 MPa, additional processing of the substrate by the glass beads reduced the shear strength to 68 ± 4 MPa.

2. After plasma preheating of the substrate for 9 s the value of the shear strength was 249 ± 17 MPa for the specimens with the substrate after the sandblasting and 229 ± 16 MPa for the specimens with additional treatment of the substrate by the glass beads. The plasma preheating decreased difference between values of the shear strength.

3. Increase of the contents of oxygen and nitrogen enhanced the microhardness of the substrate to 4.34 ± 0.35 GPa, the microhardness of the coating to 3.48± 0.25 GPa, and the boundary between them to 8.08± 0.45 GPa, which confirmed the increase of the shear strength (initial microhardness was 1.86± 0.25 GPa).

Acknowledgments

The study was supported by the Russian Scientific Foundation, project no. 15-19-00078.

References

[1] Berndt C, Hasan Md, Tietz U, and Schmitz K 2014 *Advances in Calcium Phosphate Biomaterials* (Berlin Heidelberg: Springer) pp267–329

[2] Dorozhkin S 2015 *Mater. Sci. Eng. C* 55 pp 272–326

[3] V.I. Kalita, et al., 2016 *Mater. Sci. Eng. C* 60 pp255–259

[4] Kalita V, Gnedovets A, Mamaev A, Mamaeva V, Malanin D and Pisarev V 2005 *In Abstracts and Full-Papers CD of 17th Int. Symp. on Plasma Chemistry* (Toronto: Univ. of Toronto Press) pp. 1105-1106

[5] Kalita V, Komlev D, Ivannikov A, Radyuk A, Komlev V, Mamonov V, Sevast'ianov M, Baikin A 2017 *Inorganic Materials: Applied Research* 8(2) pp 296–304

[6] Maltsev V 2013 *Metallography of industrial non-ferrous metals and alloys* (Moscow: Ripol Classic) p 367

Development of phase analysis methods of impurity elements in alloys based on iron and nickel

N A Andreeva, S N Anuchkin, V A Volchenkova, E K Kazenas, T N Penkina and A A Fomina

Baikov Institute of Metallurgy and Materials Science of the Russian Academy of Sciences (IMET RAS). Russia

E-mail: andreeva150388@mail.ru; AnuchkinSN@yandex.ru; volch.v.a@mail.ru; kazenas@ imet.ac.ru; t-penka-01@yandex.ru; fomina2402@mail.ru

Abstract. Using the method of AES with ICP, new methods have been developed for quantifying the content of various forms of existence of impurity elements: Al-Al_2O_3; Zr-ZrO_2 in alloys based on iron (Fe-Sn) and nickel (Ni-Sn). Open systems were used to dissolve Al and Zr. To translate difficult-to-open aluminum oxides (corundum) and zirconium oxide (baddeleyite) into the solution, accelerated techniques were developed using the microwave system Mars 5. To confirm the completeness of the dissolution of oxides, a classical scheme of alloy fusion with alkali metal salts was used. Optimal analytical parameters for determining the elements: Al and Zr were chosen. The influence of matrix elements (iron and nickel) and methods of its elimination were studied. This made it possible to determine the elements in a wide concentration range from $1 \cdot 10^{-3}$ to n% Al and from $1 \cdot 10^{-4}$ to n% Zr without preliminary separation of the matrix with good metrological characteristics. The relative standard deviation (S_r) does not exceed 0,2. The separate determination of the contents of aluminum and aluminium oxide in the model melt of Fe-Sn-Al_2O_3 and zirconium and zirconium oxide in the Ni-Sn-ZrO_2 model melt allowed us to estimate the number of nanoparticles participating in the heterophase interaction with tin and retired to the interface in the form of ensembles and the number of nanoparticles present in the melt and affecting the crystallization process and the structure of the metal.

In metallurgy the search of new ways of improving properties of metal, including by inoculation of it with the use of nanosize powders, presents a very actual task [1]. For the study of physicochemical principles of extraction of metals from complex ores and research of interaction of exogenous refractory nanophases with capillary-active substances in the alloyed fusions of nickel and iron with the purpose of development of technological recommendations on the affinage of metal new methodologies of quantitative determination of maintenance of different forms of being of aluminium were worked out in ferrous alloys: aluminium dissolved in the metal, and aluminium in the form of corundum Al_2O_3 nanoparticles and methodologies of separate determination of amount of zirconium dissolved in the nickel, and zirconium in the form of nanoparticles ZrO_2.

Methodologies of separate quantitative determination of maintenances of different forms of existence of elements-admixtures: Al - Al_2O_3 and Zr - ZrO_2 in the alloys of the indicated composition were not found in literature. The method of atomic-emission spectrometry with the inductively-

IOP Conf. Series: Materials Science and Engineering **347** (2018) 012003 doi:10.1088/1757-899X/347/1/012003

constrained plasma (AES with ICP) was used for researches. The method is the most perspective for determination of maintenances of aluminium and zirconium, regarding of sensitivity, selectivity and metrological characteristics [2, 3]. Work was conducted on the plasma spectrometer of ULTIMA 2 firms "HORIBA JOBIN YVON" (France-Japan).

The high resolution (5 pm) of spectrometer provides the best ratio of signal to the background, which leads to low detection limits of elements. The optimal analytical parameters of determination of aluminium and zirconium were experimentally found: the discharge power - 1,2 kW; cooling argon flow is 14 l/ min; transporting is 0,80 l/ min; plasma is 0,5 l/ min; observation height 14 mm above the top turn of the induction coil; the speed rate samples is 1,0 ml/min. Analytical lengths of waves were chosen for determination of elements, that provide the suitable detection limits, a minimum of adjustments for background and overlapping of spectral lines [4]. Optimal analytical lengths of waves for determination of Al and Zr are presented in tables 1, 2. The most sensitive and selective were λ_{Al}=309,278 nm, λ_{Zr} =339,198 nm.

A significant factor is the choice of the concentration of the matrix element in the solution, which is not violated to the stability of work of the system burner - nebulizer. For iron favourable is a concentration no more than 500 ppm and for a nickel - 1000 ppm.

The possibility of determination of low concentrations of aluminium on the background of iron and zirconium on the background of nickel was by a study of the spectral interference from the base element at the scan-out of the spectrum from 190 to 800 nm. Solutions of pure iron (500 ppm) or nickel (1000 ppm) and solutions of aluminium or zirconium (5, 10, or 100 ppb) were introduced in the plasma. The detection limits of elements were estimated by the formula [5]:

$$LOD = k \cdot BEC \cdot RSD_0$$

LOD - limit of detection; k - coefficient equal to 3 for detection limits at 3σ; BEC - the concentration of noise; RSD_0 – relative standard deviation for the zero standard.

For the evaluation of detection limits the calibration curve is built on the two points 0 and 5 ppm. The calibration is used to determine the concentration of the noise BEC. The relative standard deviation RSD_0 is determined by the results of the analysis of a zero standard during 5 minutes. As the zero calibration solution was used a standard solution of iron 500 ppm or 10000 ppm of nickel (prepared from metals of high purity). The second calibration solution was prepared on the background solution of iron 500 ppm or 10000 ppm of nickel. Concentrations of the determined elements (Al, Zr) were 5 ppm.

In tables 1, 2 are presented the calculated detection limits of aluminium and zirconium. It is seen that the detection limits of Al on the background 100 and 500 ppm of iron and Zr on a background 1000 and 10000 ppm of nickel are significantly different from those in the pure solutions and not for the better.

Table 1. Characteristics of analytical lines of aluminum.

Length wave, λ nm	Detection limit, ppb (3σ)			Interfering elements
	in pure solution	on the background		
		100 ppm Fe	500 ppm Fe	
308.215	2.0	4.5	9.0	V, Pr, Er, Ce
309.278	0.5	2.5	10.0	Nd, Tb
394.401	1.5	3.5	7.1	Pr, U, Ru, Ce
396.152	1.0	2.5	5.3	Pr, Er

Table 2. Characteristics of analytical lines of zirconium.

Length wave, λ,нм	Detection limit, ppb (3σ)			Interfering elements
	in pure solution	on the background		
		1000 ppm Ni	10000 ppm Ni	
339.198	0.02	0.05	0.10	Eu, Dy, Er, Nd, Ru, Ce
343.823	0.04	0.06	0.20	Fe, Hf, Ce, W
349.621	0.09	0.17	0.52	Nb ,Dy, Ce, Tb, Pr

The main problem of the emission analysis is to account for the influence of the matrix. On the background of iron, which is a very complex matrix, determination of trace elements is especially difficult. This is due to a temporal drift of the background, and a sophisticated form of background in area of analytical lines, and weak signal intensity. The increase in integration time reduces the detection limits, but does not decide the problem of the influence matrix. It is primarily the curvature of the calibration graphs due to significant changes of analytical signals of the elements near the limits of detection and, therefore, the deterioration of the metrological characteristics of the determination of elements. At determination of elements near-by the detection limits on their analytical signals other elements-admixtures begin to influence [6].

The influence of matrix elements (iron and nickel) and methods of their removal were studied.

On 'Figure 1' the dependences of analytical signals of different concentrations of zirconium from the concentration values of nickel are presented. The nickel content of up to 10g/l has no effect on analytical signals of solutions containing more than 2 ppb of zirconium. It does possible direct, without separation of the nickel matrix, determination of Zr beginning from $2 \cdot 10^{-4}$%. In the case of smaller concentrations of zirconium needed to make a preliminary separation of the matrix or to use methods of mathematical modeling.

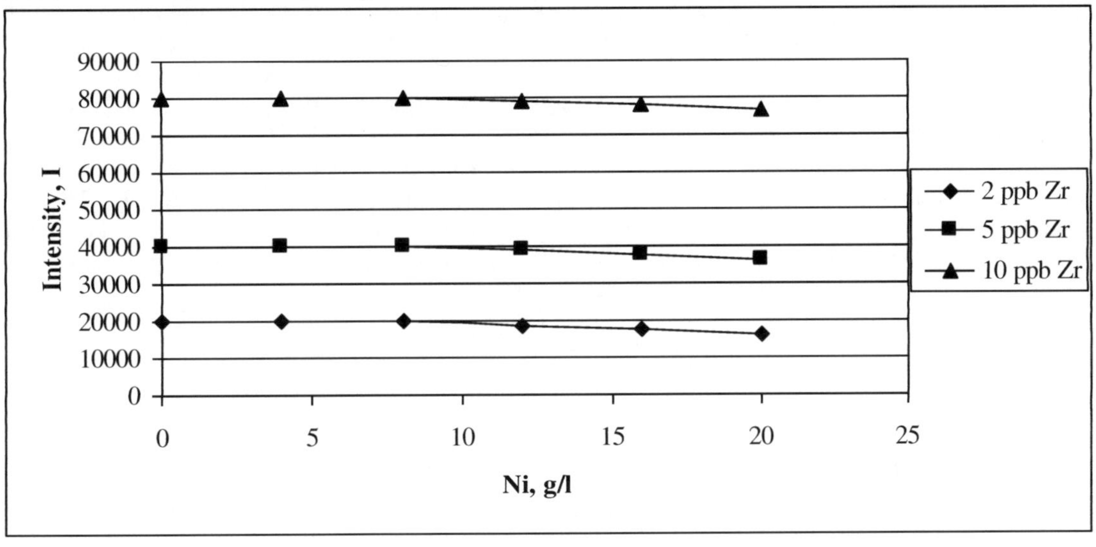

Figure 1. Matrix effect of nickel on AES with ICP determination of zirconium.

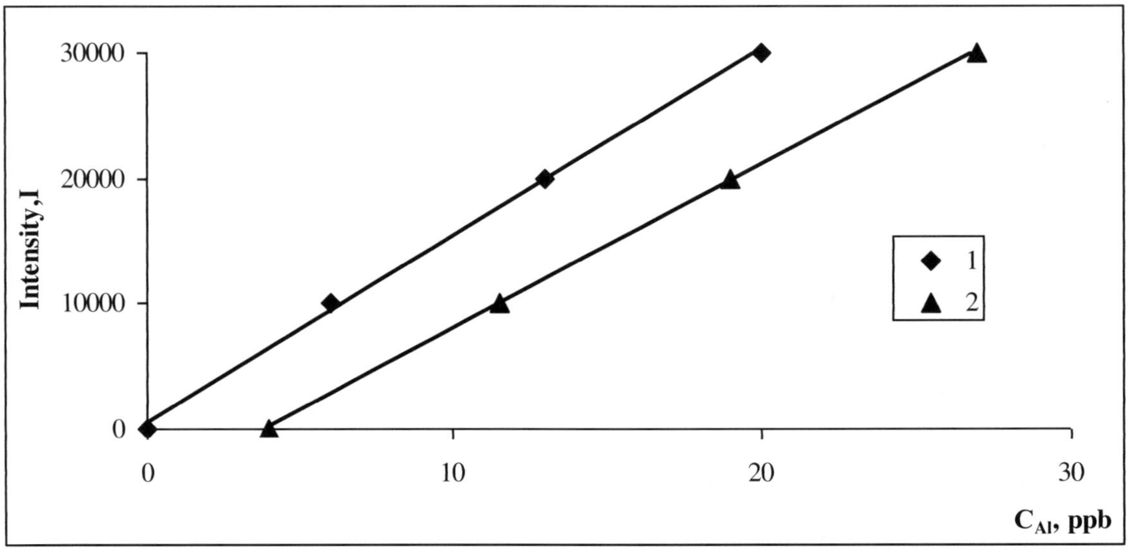

Figure 2. Calibration curves of aluminum on the background 0-100 ppm Fe (1) and on the background 500 ppm Fe (2).

On 'Figure 2' the calibration graphs for the aluminium on the background of iron from 0-100 ppm and the background 500ppm iron are shown. The calibration graphs to determine the Al for the selected analytical lines coincide for pure solutions of aluminum and containing not more than 100 ppm of iron. With the increase of iron concentration up to 500 ppm, the sensitivity of the determination of aluminium worsens and calibration curves do not coincide.

In this case it is needed to use calibration curves with the respective iron content, or to apply the method of interactive concordance of matrix [3].

The influence of the concentration of acids: HCl, HNO_3, H_2SO_4 on the analytical signals of elements were studied. On 'Figure 3' the dependences of analytical signals of Al from the nature and concentration of acid in the analyzed solution are shown. As follows from experimental data, the change in the concentration of acid in the analyzed solution leads to a change in analytical signal, and the degree of change depends on the nature and concentration of acid. So a solution of 2M HCI decreases the analytical signal Al on 3%, while the sulfuric acid of the same concentration by 11%. The increase in the concentration as HCL and HNO_3, from 0.01 M to 1 M was changed analytical signals of the elements is not more than 3%. A significant influence on the analytical signals of the elements renders the change concentrations of sulfuric acid in solution. Depressing the action of acids is due to changes in the spray system or the system of sample introduction. When you change the concentration of acid the effectiveness of spraying and, consequently, sensitivity changes. However, the practical error can be avoided by maintaining an adequate acid content in tests and standard solutions. Uncontrolled fluctuations in the concentration of acids, especially sulphuric, can lead to errors which are significantly greater than instrumental.

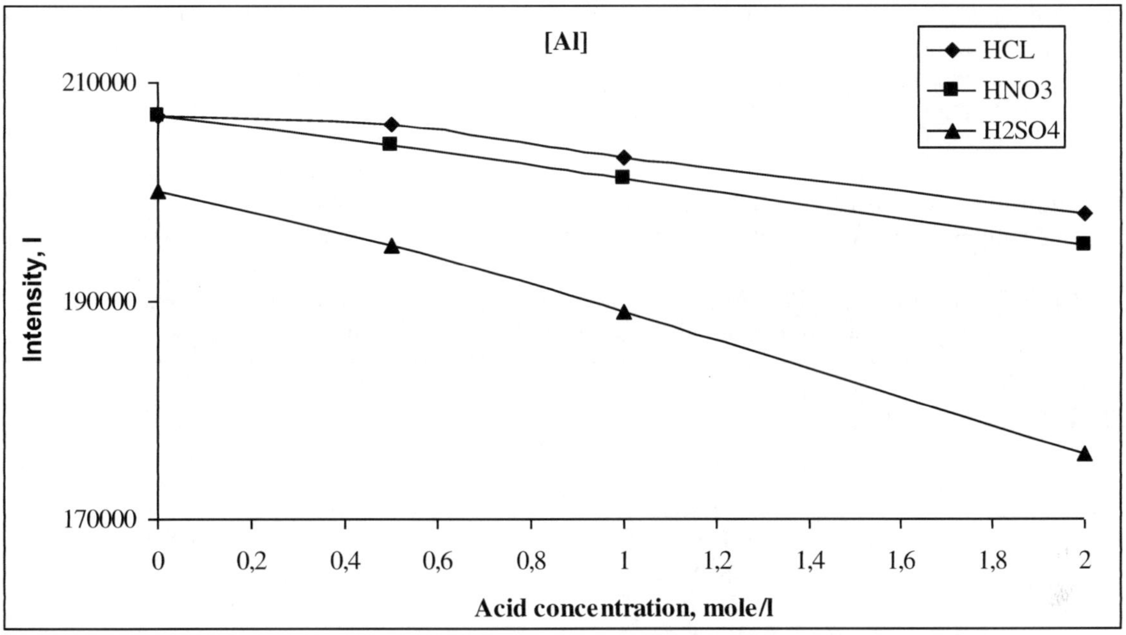

Figure 3. The dependence of the intensity of the signal Al from nature and concentration of acid.

Method AES with ICP requires translation the analyzed samples in solutions for performing analysis. All used reagents had high degree of purity. Mineral acids and their mixtures of different concentration were used for dissolution. Compositions of acids for separate translation in solution of different forms of being of elements: Al and Al_2O_3; Zr and ZrO_2 were found. In a table 3 the results of determination of aluminium are presented at the different ways of dissolution. Aluminium contained in the alloy iron in the form of Al, is easily translated into a solution with mineral acid: 10% HCl or 10 % HNO_3 at heating in glass heat-resistant open vessels. Its contents were defined at the level of 0.001%. For the conversion to a solution of Zr, dissolved in the metal, Aqua Regia (mixture concentrated HCl and HNO_3 (3:1)) was used by heating samples in open systems.

Oxides of aluminum and zirconium: corundum (Al_2O_3) and baddeleyite (ZrO_2) are some of the most difficult open connections. The dissolution of Al_2O_3 and ZrO_2 was carried out in a microwave sample preparation system MARS 5 (CEM Corporation, USA) under different conditions (P, T) of different mixtures of acids. The degree of extraction of oxides of aluminium and zirconium was studied depending on nature of solvent and time of treatment. The results are presented in tables 3, 4. Mixtures of acids for complete dissolution of samples, providing the transfer in solution of corundum (Al_2O_3) and baddeleyite (ZrO_2) were selected. Conditions of microwave exposure and temperature and time modes of decomposition of the samples were optimized. The corundum was dissolved in mixtures: HCl+ HNO_3 (1:3); HCl+ HNO_3 (3:1); HCl+ HNO_3 (1:3) + H_2O_2 and HCl+ HNO_3 (3:1) +H_2O_2 for 40 minutes at a temperature of 210^0C during realization of double cycle of autoclave microwave exposure. Baddeleyite was dissolved in a mixture of HF+ HNO_3+ HCl (4:3:1) for 40 minutes at a temperature of 210^0C when conducting a triple cycle of the autoclave microwave exposure.

The fullness of being transferred into a solution of oxides of aluminum and zirconium with found mixtures was controlled by the classical method: by converting Al_2O_3 and ZrO_2 into solution after melting of samples with salts of alkali metals [6]. Similar results for the determination of aluminium and zirconium contents were received in the solution obtained after dissolution of the samples in the MARS 5 system and after fusing. The use of modern autoclave sample preparation for dissolution with acids is preferable to the classic melting, since the time of analysis decreases, and most importantly, the metrological characteristics of the determination of elements improve. When using fusion with the oxides and salts of alkali metals the detection limits of elements deteriorate for two reasons: the value of the blank experience increases dramatically and the presence of large concentrations of alkali metals in solution adversely affects on the analytical signals of the determined elements in spectroscopic methods of analysis. According to the developed methods of sample dissolution in the microwave sample preparation system MARS 5 obtained solutions were analyzed. The total content of zirconium does not exceed 0.3%, and aluminum of 0,03% (table. 3, 4). These values are fundamentally different from the results obtained after dissolution of samples in open systems.

Table 3. The results of AES-ICP determination of the content of various forms of existence of aluminium in alloys based on iron using different methods of dissolution

Dissolution method samples	Mixtures for dissolution	Conditions of dissolution	Content of Al, %	S_r
Dissolution in open vessels	10% HCl	Heating in heat-resistant glass dish	0.0011	0.2
	10 % HNO_3		0.0012	0.2
	HCl		0.0010	0.2
	HNO_3		0.0009	0.2
	50 % HCl		0.0012	0.2
	50 % HNO_3		0.0010	0.2
	$HCl+H_2O_2$		0.0012	0.2
	$HNO_3+H_2O_2$		0.0012	0.2
	$HCl+HNO_3$ (1:3)		0.0011	0.2
	$HCl+HNO_3$ (3:1)		0.0013	0.2
	$HCl+HNO_3$ (1:3) $+H_2O_2$		0.0013	0.2
	$HCl+HNO_3$ (3:1) $+H_2O_2$		0.0012	0.2
Dissolution in closed autoclave microwave system Mars5	10% HCl	210 °C 40 min.	0.0015	0.2
	10 % HNO_3		0.0019	0.2
	HCl		0.0019	0.2
	$HNO_{3\text{конц}}$		0.0020	0.2
	50 % HCl		0.0021	0.2
	50 % HNO_3		0.0022	0.2
	$HCl+H_2O_2$		0.0073	0.2
	$HNO_3+H_2O_2$		0.0081	0.2
	$HCl+HNO_3$ (1:3)		0.028	0.04
	$HCl+HNO_3$ (3:1)		0.029	0.04
	$HCl+HNO_3$ (1:3) $+H_2O_2$		0.030	0.04
	$HCl+HNO_3$ (3:1) $+H_2O_2$		0.029	0.04
Fusion	Na_2CO_3	Pt crucible, 1000-1200 °C 10-20 min.	0.021	0.06
	$Na_2CO_3+Na_2B_4O_7$		0.030	0.05
	$LiBO_2$		0.028	0.05

Table 4. The results of AES-ICP determination of the content of various species of zirconium in alloys based on nickel using of different methods of dissolution

Dissolution method samples	Mixtures for dissolution	Conditions dissolution	Content of Zr, %	S_r
Dissolution in open vessels	HCl	Heating in heat-resistant glass dish	0.0008	0.2
	HNO_3		0.0009	0.2
	H_2SO_4		0.0008	0.2
	$HCl+HNO_3$ (3:1)		0.0030	0.2
	$HCl+HNO_3$ (3:1) + H_2O_2		0.0029	0.2
	$H_2SO_4+HNO_3$ (1:1)		0.0030	0.2
Dissolution in closed autoclave microwave system Mars5	$HCl+HNO_3+HF$ (1:3:4)	210 °C 45 min	0.29	0,04
	$H_2SO_4+HNO_3$ (1:1)		0.0030	0,2
	$HCl+HNO_3$ (3:1) + H_2O_2		0.0030	0,2
Fusion	Na_2CO_3	Pt crucible, 1000-1200 °C 10-20 min	0.22	0.06
	Na_2CO_3 + $Na_2B_4O_7$		0.30	0.05
	$LiBO_2$		0.28	0.05

Due to the lack of standard samples of composition of studied materials to confirm the obtained results for the determination of aluminium and zirconium by means of methods AES with ICP were compared with the data of the method of atomic absorption spectrometry and mass spectrometry with inductively coupled plasma (tables 5, 6). A good convergence of the results of determination by different methods confirms the correctness of determination of Al and Zr by the method of AES with ICP.

Table 5. The results of determination of aluminium by various methods.

Method of sample preparation		AES-ICP λ = 309,278 нм			AAC (flame$N_2O+C_2H_2$) λ = 309,3 нм		
		Limit of detection Al, ppb (3σ)	Al, %	S_r	Limit of detection Al, ppb (3σ)	Al, %	S_r
Acid dissolution (autoclave microwave system Mars5)	1) 100 ppm Fe	2.5	0.030	0.04	5.0	0.031	0.06
	2) 500 ppm Fe	10.0	0.028	0.12	10.0	0.028	0.10
Fusion with Na_2CO_3 + $Na_2B_4O_7$	1) 100 ppm Na	9.5	0.031	0.10	15	0.031	0.12
	2) 500 ppm Na	25.0	0.028	0.22	50	0.028	0.25

Table 6. The results of determination of various forms of elements (n=10, P=0,95)

The matrix composition of the sample	Define form elements	Methods of translation of of sample into the solution	Mass contents of the elements, %	
			The method of analysis	
			AES-ICP	AAS, MS-ICP
Fe	Al	Open system: HCl	0.0013	0.0011
	Al_2O_3 (corundum)	Microwave autoclave system MARS 5: $HCl+HNO_3$	0.031	0.033
		Fusion with Na_2CO_3	0.032	0.034
Ni	Zr	Open system: $HCl+HNO_3+HF$	0.0052	0.0056
	ZrO_2 (baddeleyite)	Fusion with Na_2O_2	0.21	0.22
		Microwave autoclave system MARS 5: $HNO_3+HF+H_2O_2$	0.22	0.23

Conclusions:

1. Using the method AES with ICP a new express methods of phase analysis of systems were developed:

- aluminium, dissolved in the doped alloy of iron and nanoparticles of Al_2O_3 in the form of corundum,
- zirconium dissolved in the doped alloy of nickel and ZrO_2 nanoparticles in the form of baddeleyite.

2. New methods for rapid dissolution of corundum and baddeleyite with using the autoclave microwave system Mars 5 were developed.

The proposed methods allowed to determine the content of various forms of aluminum, beginning from $n \cdot 10^{-3}$ % and zirconium, beginning from $n \cdot 10^{-4}$ % and above without prior separation of matrix with good metrological characteristics. The relative standard deviation (S_r) does not exceed 0,2.

3. Separate determination of the contents of aluminum and aluminum oxide in a model molten Fe-Sn-Al_2O_3 and zirconium and zirconium oxide in the model melt of the Ni-Sn-ZrO_2 made it possible to assess the number of nanoparticles involved in heterophase interaction with the tin, and retired to the phase interface in the form of bands and the number of nanoparticles present in the melt and affect the crystallization process and the structure of the metal.

Acknowledgments
The reported study was funded by RFBR, according to the research project No. 16-38-60129 mol_a_dk.

References

[1] Anuchkin S N, Burtsev V T and Samokhin A V 2016 *J. Russian Metallurgy (Metally)* **6** pp 44-51
[2] 1989 Atomic–emission analysis with inductive plasma. The results of science and technology. *Ser. Analytical chemistry.*(Moscow: VINITI) vol. 2 255 p.
[3] Thompson M P and Walsh J N 1989 *Handbook of Inductively Coupled Plasma Spectrometry* (New York: Blackie) 316 p.

IOP Conf. Series: Materials Science and Engineering **347** (2018) 012003 doi:10.1088/1757-899X/347/1/012003

[4] Boumans P W J M 1984 Coincidence tables for inductively coupled plasma. Atomic emission spectrometry vol. 1,2 (Oxford Pergamon Press) 902 p.
[5] Volchenkova V A, Kazenas E K, Dergunova N N, Degtyareva A P and Kraskov I I 2010 *J. Perspective materials* **9** pp. 128-131
[6] Bock R *A* 1979 *Handbook of Decomposition Methods in Analytical Chemistry* (New York: John Wiley and Sons) 444 p.
[7] Nygaard D D,Chase D S and Leighty D A 1983 *J.Appl. Spectrosc.* **37** vol 5 pp.432-5

IOP Conf. Series: Materials Science and Engineering **347** (2018) 012004 doi:10.1088/1757-899X/347/1/012004

Formation of composite structure of amorphous alloy after treatment of different types

N D Bakhteeva and E V Todorova
Baikov Institute of Metallurgy & Materials Science, Russia, Moscow, Leninsky av., 49

nbach@imet.ac.ru

Abstract. The dependence of structure and microhardness of aluminium amorphous alloy were investigated and determined on samples after deformation, heat treatment and irradiation with fast carbon ions.

1. Introduction

Aluminum alloys are widely utilized as light, high-strength, temperature-resistant and corrosion-resistant materials for aviation and aerospace equipment. Aluminum doping with transition and rare-earth metals allows to receive amorphous structure during highspeed quenching from liquid state. Amorphous structure is an intermediate stage for amorphous-nanocrystalline composites preparation. Different methods of such structures preparation are known: heat treatment, severe plastic deformation (SPD), pulsed photonic irradiation, laser irradiation, radiation exposure. Ample quantity of scientific work is devoted to amorphous nanocrystalline structures preparation methods consideration. In such a way high deformation rates in event of severe plastic deformation [1-5] provide fine grained structure in alloys including nanoscale multiphase structure with increased strength performance. Scientific work results issued over the last years give an evidence of intense interest both to the new aluminum alloys development and to the formation behavior research of submicro- and nanocrystalline structures in the aluminum-based alloys doped with transition and rare-earth metals [3-11]. Amorphous alloys irradiation by high-energy ions is reviewed as one of the nanocrystalline composites preparation methods. Information presented in scientific literature about irradiation influence on amorphous alloys structures is limited and commonly contradictory, irradiation parameters vary in types of particles, energies and fluence.

Amorphous alloys structure modification is considered now as a method of simultaneous mechanical properties (such as strength and ductility) enhancement. As aluminum alloys doped with transition metals are embrittled due to bulk large-sized intermetallic compounds allocation, best available compositions pulverization technology development is an actual task for research studies.

2. Material and techniques

Different types of structure are formed at the $Al_{85}Ni_7Fe_4La_4$ alloy depending on processing type. Amorphous structure was obtained in the sample pieces presented in the form of ribbons with 25 μm thickness and 1 mm width after high-rate quenching (spinning method) at a rate of 10^6 K/s. Structure of amorphous ribbons is metastable and any high-energy influence (high-temperature processing, plastic deformation, charged particles and laser irradiation, etc.) can lead to significant structural changes. $Al_{85}Ni_7Fe_4La_4$ alloy amorphous ribbons are treated according to the following scheme: 1) annealing was conducted in the temperature range from 150 to 400°C with the 50 °C interval and 15 and 30 min exposure at each temperature with further cooling at the air; 2) severe plastic deformation was conducted at the ambient temperature in the Bridgman chamber at 8 GPa of hydrostatical pressure and at the rotation angle of movable anvil $\varphi=360°\times n$ (where n=1, 3, 6); 3) irradiation with carbon $^{12}C^{+3}$ ion with 38 MeV energy and $\Phi=5\cdot 10^{16}$ particles/cm^2 fluence.

Comprehensive analysis of amorphous alloys structure in initial condition and after various types of processing with modern research techniques engagement (high-resolution electron microscopy, transmission scanning microscopy, energy dispersive analysis of X-rays, X-ray phase analysis) allows receiving of experimental findings in alloys structural and phase rating.

The following types of equipment were utilized for structural and ultimate composition research after above mentioned processing types:

1) dual-beam scanning electron-ion microscope (FEI, USA) equipped with energy-dispersive X-ray microanalysis system with film-type silicon drift detector of 40 mm diameter (EDAX, USA);

2) Titan 80-300 TEM/STEM (FEI, USA) electron microscope with illumination system spherical aberration corrector equipped with energy-dispersive X-ray microanalyzer (EDAX, USA), ring-shaped dark-field detector for wide angle scattered electrons (Fischione, USA), energy-dispersive X-ray microanalyzer system (EDAX, USA) and characteristic electronic energy loss spectrometer (Gatan, USA);

3) dual-beam electron-ion microscope Helios (FEI, USA). Utilization of reduced accelerating voltage at samples preparation allows to avoid their substantial heating which may result in structural and constitutional change in alloys not associated with the processes under consideration. Research on ionic thinning (by Ar^+ ions) influence in wide range of parameters on phase transformation in amorphous alloys [12] was conducted earlier by the authors to determine optimum polishing regime.

Alloy phase compositions were studied by X-ray diffraction method utilization: DRON-3M (Russia) diffractometer with Cu-Kα- irradiance utilization (exposure mode U = 38 kV, I = 20 mA, slot 1x2x0,5 + Soller slot system). XPertPro (PANanalitical) diffractometer in Cu-Kα- irradiance (40 kV, 15 mA, Ni-K$_\beta$-filter, step angle 0.016 degree/s). PDF-2 (sets 1-52) data were used during X-ray photogram phase analysis.

PMT-3 measuring device as well as 402 MVD – Wolpert Wilson Instruments (Germany) microhardness tester were used for samples microhardness measuring at 10 g load and 10 sec exposure time.

3. Results and its discussion

Alloy in amorphous state has single-phase structure with high oversaturation by Ni, Fe, La doping elements-halo presence at the X-ray diagrams and electron-diffraction photographs is distinctive for this case. Amorphous alloy microhardness is conventionally exceeding microhardness of polycrystalline alloy and is 405 HV. Structure of amorphous ribbons is metastable and any high-energy influence (high-temperature processing, plastic deformation, charged particles and laser irradiation, etc.) can result in significant structural alteration.

Isothermal annealing of ribbons resulted in amorphous condition stability loss: crystallization with crystalline aluminum and intermetallic phases on the basis of Al–Ni, Al–Fe, Al–La formation is developed in the alloy at a temperature of 250 °C and higher. Overall crystallization development behavior during annealing for the alloys of $Al_{85}Fe_xNi_yLa_z$ (where x=1÷4 at. %, y=7÷9 at. %, z=4÷5 at. %) system is described at [13]. Crystalline constituent amount is increasing in alloys during annealing temperature rising from 250 to 400 °C. Large amount of peak amplitudes is put in appearance at the X-ray diagrams with that and the rings with large number of reflexes are formed at the submicroscopic diffractions. Alloy phase composition at 400 °C includes crystalline aluminum solid solution with the grain sizes of about 100 nm and a large number of intermetallic compounds ($Al_{11}La_3$, $AlNi_3$, $Al_{13}Fe_4$, Al_3Ni_2, Al_3Ni, Al_9Fe_2-xNi_x, Al_8Fe_2-xNi_xLa, Al_5Fe_2) due to multicomponent doping process of an alloy. Relaxation and crystallization processes developed during annealing are reflected at the evaluations of such structure-sensitive property as microhardness which is an important characteristic value of material resistance to plastic deformation for amorphous alloys. As far as annealing temperature increases microhardness is descending as much as 10-15 % at first and then as temperature reaches the values of crystallization commencement it goes upward and reaches the value of 509 HV. Microhardness alteration behavior is not depending upon retention time.

Al_3Fe ($Al_{13}Fe_4$ analog) phase microhardness rate ranges from 800 to 1100 HV, Al_3Ni phase – from 700 to 770 HV, and Al_9Co_2 phase (Al_9Fe_2-xNi_x analog [14]) from 650 to 750 HV. Al_3Ni_2 phase has hardness of 1100 HV according to the data from [15]. Considerable accession of alloy hardness during annealing operation is determined by high hardness of intermetallic phases.

Multiple phase amorphous-nanocrystalline structure consisting of aluminum in crystalline and amorphous conditions and nanoscale intermetallic compounds ($Al_{13}Fe_4$, Al_3Ni_2, Al_3Ni, $Al_9Fe_{2-x}Ni_x$, $Al_8Fe_{2-x}Ni_xLa$) are formed in the alloys after severe plastic deformation as was shown by X-ray crystal structure analysis. At that stratification by spinodal decomposing type to two amorphous components and nanocrystallization in one of them (doped with nickel and lanthanum to a small extent) accompanying the crystallization is discovered in the structure by micro X-ray spectrum analysis. Amorphous-nanocrystalline fragmented structure of "collar" type with fragments sizes 10…20 nm with distinctive preservation of amorphous component at the center of the fragments and with the nanocrystalline welt formation with the grain sizes of 3-6 nm at the boundary zones [16] is formed after severe plastic deformation which was shown by high-resolution electron-microscope investigation. It should be noted that contrast similar to the "collar" at the transmission electronic microscopy pictures is increasing due to samples thickness variations conditioned by preferential etching of amorphous regions with various chemical composition at samples preparation for transmission electronic microscopy research. With increase of degree of deformation, the volume ratio of nanoscale crystalline phases somewhat increases without noticeable changes of phase composition and nanocrystals sizes. Crystallization is not terminated along the whole investigated interval of deformation parameters. Structure formation mechanism after severe plastic deformation is extensively reviewed in scientific work [13]. Microhardness values after severe plastic deformation with $\varphi=(360°\times3)\times2$ amounted up to 706 HV.

Research of radiation exposure to amorphous alloys structural condition and mechanical properties is of certain interest. Increase in hardness during corresponding processing and magnetic characteristic alteration are mentioned in a number of works; there exists conflicting information about radiative effect on amorphous alloys structural and thermal stability. Amorphous aluminum alloys are not without interest in terms of new radiation-proof materials development for the space hardware. New experiments on amorphous alloys irradiation with high-energy carbon ions were conducted; they required calculations for irradiation conditions, samples holder development and irradiation performance experimental testing. Residual radioactivity after irradiation is substantially extended duration of the experiment due to samples prolonged subjection to withdraw this effect. Results of that experiments are resulted as follows.

Noticeable changes in amorphous alloy structure were discovered after irradiation with carbon $^{12}C^{+3}$ ions with 38 MeV energy and $\Phi=5\cdot10^{16}$ particles/cm^2 fluence. Multiple narrow-width Bregg reflexes specific for polycrystalline condition were observed at the X-ray diffractogram after specified three-hour processing. Such diffractogram conversion is associated with multiphase nanocrystallization development at the initially amorphous single-phase structure. It has been established that crystallization in alloy after irradiation is developed with the solid solutions with face-centered cubic latitude development – Al, intermetallic compounds $Al_{11}La_3$, Al_3(Ni, Fe), $Al_8Fe_{2-x}Ni_xLa$. Besides that, peaks do appear at the diffractogram which can be associated with the graphite. Multiphase amorphous-nanocrystalline structure with the grain sizes 10-20 nm is formed in the alloy which was also determined by transmission electronic microscopy method. Detailed structure study of this sample by the high-resolution transmission electronic microscopy confirms data about amorphous alloy crystallization. Both face-centered cubic latitude Al grains and intermetallic compounds [17] were observed on transmission electronic microscopy pictures. Experiment conditions don't allow establishing of phase transformation sequence during the crystallization at irradiation exposure. Utilized experimental methodologies record only the final result of multiphase crystallization after three hours exposure of the sample. Nanocrystallization development is followed by substantial microhardness increasing up to 683 HV. Such irradiated sample hardness growth is provided (as in the case of nanocrystallization during annealing and severe plastic deformation) by the high content of the intermetallic compounds with increased strength and nanosized grains of crystalline aluminum.

4. Conclusions

Comparative analysis of the structures formed in the $Al_{85}Ni_7Fe_4La_4$ alloy shows the possibility of multiphase amorphous-nanocrystalline structures formation at the initially amorphous $Al_{85}Ni_7Fe_4La_4$ alloy under the conditions of severe plastic deformation, annealing and irradiation with carbon ions.

Based on structural parameters attestation and results of hardness measurement optimum processing conditions were defined:

- amorphous ribbon isothermal annealing at T = 400 °C, 30 min (509 HV);
- severe plastic deformation of the amorphous ribbon by rotation under pressure: P=8 GPa, φ=(360°x3)x2 (706 HV);
- irradiation with carbon $^{12}C^{+3}$ ions with 38 MeV energy and $\Phi = 5 \cdot 10^{16}$ particles/cm^2 fluence of the amorphous alloy leads to hardness increasing up to 683 HV.

It has been established that the most perspective development method of the amorphous-nanocrystalline composite material with the advanced structural performance in amorphous alloy $Al_{85}Ni_7Fe_4La_4$ is severe plastic rotational deformation under pressure, as a result of its hardness values achieve maximum magnitude.

Acknowledgments

We would like to express our special thanks to our colleagues Vasil'ev A.L. and Kolobylina N.N. for their help concerning electron microscopic investigations.

The investigations have been fulfilled according to the Federal order № 007-00129-18-00.

References

[1] Brodova I.G., Bashlykov D.V. and Shirinkina I.G. Formation of the microcrystalline states in aluminum alloys with transition metals 2005 *Materialovedenie*, **2 (95)** 48–53.

[2] Shabashov V.A., Brodova I.G., Mukoseev A.G., Sagaradze V.V. and Litvinov A.V. Structural and phase transformations in the Al-Fe system upon severe plastic deformation 2005 *The Physics of Metals and Metallography* **100 (4)** 380-391.

[3] Brodova I.G., Shirinkina I.G. and Antonova O.V. Phase and structural transformations in the Al-Cr-Zr alloy after rapid melt quenching and high-pressure torsion 2007 *The Physics of Metals and Metallography* **104 (9)** 281-288.

[4] Mazilkin A.A., Straumal B.B., Protasova S.G., Kogtenkova O.A. and Valiev R.Z. Structural changes in aluminum alloys upon severe plastic deformation 2007 *Physics of the Solid State* **49 (5)** 868-873.

[5] Valiev R.Z. and I. V. Aleksandrov 2007 *Bulk nanostructured metallic materials* (Akademkniga, Moscow)

[6] Luzgin D.V. Aluminum-base amorphous and nanocrystalline materials 2011 *Metallovedenie i termicheskaya obrabotka metallov* **10** 12-17.

[7] Perevezentsev V.M., Shcherban' M.Yu., Murashkin M.Yu. and R. Z. Valiev High-rate super-plasticity of a nanocrystalline grade 1570 aluminum alloy 2007 *Pis'ma Zh. Tekh. Fiz.* **33 (15)** 40–46.

[8] Brodova I.G., Yablonskikh T.I., Shirinkina I.G. and Dobatkin S.V. The effect of fast hardening of melt, annealing and intensive plastic strain on formation of scandium and zirconium aluminide in Al - Mg - Mn-alloys 2008 *MiTOM* **10** 39-45.

[9] Noskova N.I., Churbaev R.V., Vil'danova N.F., Elkina O.A. and Zemnukhova L.A. Structure

and microhardness of nanocrystal composite alloys based on Al and Ti 2008 *MiTOM* **10 (640)** 23-26.

[10] Sabirov I.N., Yunusova N.F., Islamgaliev R.K. and Valiev R.Z. High-strength state of a nanostructured aluminum alloy produced by severe plastic deformation 2002 *The Physics of Metals and Metallography* **93 (1)** 94-99.

[11] Rokhlin L.L., Bochvar N.R. and Dobatkina T.V. Combined Effect of Some Transition Metals on Change of Phase Composition in Aluminium and on Its Recrystallization 2009 *Texnologiya legkix splavov* **2**, 20-27.

[12] Volkov P.A., Todorova E.V., Bakhteeva N.D., Ivanova A.G. and Vasil'ev A.L. Specific features of sample preparation from amorphous aluminum alloys for transmission electron microscopy 2011 *Crystallography Reports* **56 (3)** 463-469.

[13] Bakhteeva N.D. Nanocrystallization in Al-based amorphous Al-Ni-Fe-La alloys 2010 *Nanotech. in Russia* **5 (3-4)** 235-249

[14] Vasil'ev A.L., Ivanova A.G., Bakhteeva N.D., Kolobylina N.N., Orechov A.S., Presnyakov M.Yu. and Todorova E.V. Microstructure of the Al-La-Ni-Fe system 2015 *Crystallography Reports* **60 (1)** 23-29..

[15] V. S. Zolotarevskii and N. A. Belov 2005 *Physical metallurgy of cast aluminum alloys* (MISiS, Moscow)

[16] Bakhteeva N.D., Todorova E.V., Sirotinkin V.P., Kolobylina N.N. and Vasil'ev A.L. Structure of polycrystalline Al-Ni-Fe-La alloys after quenching and plastic deformation by shear under pressure 2013 *Russian metallurgy (Metally)* **3** 206-216.

[17] Bakhteyeva N.D., Vasilyev A.L., Kolobylina N.N., Todorova Y.V., Ryazanov A.I., Latushkin S.T., Unezhev V.N. and Ivanova A.G. Structural transformations in Al85Ni7Fe4La4 alloy under carbon ion irradiation 2017 *Inorg. Materials: Applied Research* **8 (2)** 268-274.

IOP Conf. Series: Materials Science and Engineering **347** (2018) 012005 doi:10.1088/1757-899X/347/1/012005

Spectral properties of nanocomposites based on fluorine-containing polymer and gold nanoparticles

E V Barmina[1], N N Mel'nik[2], I I Rakov[1], V E Ivanov[3], A V Simakin[1], S V Gudkov[1,4,5] and G A Shafeev[1,6]

[1] Prokhorov General Physics Institute, Russian Academy of Sciences, 38 Vavilova st., Moscow, 119991 Russia
[2] Lebedev Physical Institute, Russian Academy of Sciences, Leninskiy pr. 53, Moscow, 119991 Russia
[3] Institute of Theoretical and Experimental Biophysics, Russian Academy of Sciences, Institutskaya St. 3, Pushchino, Moscow Region 142290, Russia
[4] Lobachevsky State University of Nizhni Novgorod, pr. Gagarina 23, Nizhni Novgorod, 603950 Russia
[5] All-Russia Research Institute for Phytopathology, B. Vyazyomy, Moscow Region 143050, Russia
[6] National Research Nuclear University MEPhI (Moscow Engineering Physics Institute), Kashirskoe sh. 31, Moscow, 115409 Russia

S_makariy@rambler.ru

Abstract. The optical properties of nanocomposites of gold nanoparticles and fluorine-containing polymer have been studied. Gold nanoparticles were obtained by laser ablation of gold or terbium targets in organic solvents. The thus formed colloidal solutions were used to prepare nanocomposites of gold nanoparticles in polymer matrices of transparent and colorless fluorine-containing polymer. The polymer matrix is found to promote aggregation of nanoparticles of metal under study into elongated chains. In turn, metal nanoparticles influence on the polymer matrix. Gold nanoparticles amplify the Raman signal of the polymer matrix. In addition, the Raman spectra of nanocomposites indicate aggregation of disordered carbon around the nanoparticles obtained by laser ablation in organic solvents.

1. Introduction

Nanocomposites of metal particles in a polymer matrix have a number of unusual and unexpected properties [1−6]. As was shown previously, polymers nanocomposites containing in their composition metal nanoparticles with have a high permittivity due to the high polarizability of the metal nanoparticles [7]. This function is of interest not only for electronic and microelectronic applications, but for light-conversion technologies. The optical properties of these nanocomposites, which are of value for light-conversion technologies, have been studied less thoroughly [8]. Apparently, it can be expected that plasma oscillations of electrons in metallic nanoparticles will have an extremely short reaction to the optical excitation of nanoparticles. The typical time of this response is comparable with electron−phonon relaxation time (on the order of 1 ps). Due to such short response times, these nanocomposites can find application even as optical keys in photonic elements. If we do not consider

IOP Conf. Series: Materials Science and Engineering **347** (2018) 012005 doi:10.1088/1757-899X/347/1/012005

the main advantage of the response time, then such composites can find application in biophotonics, namely in biological and agricultural research and applications [9-10].

At present, a large number of methods are used to implant nanoparticles in solid substrates. At the heart of modern nanocomposites, as a rule, there are substances capable of forming highly ordered structures with a certain geometry (alternating flat layers, filaments, point inclusions, more complex compositions). [11]. In this study, we use technique of forming nanocomposites of gold nanoparticles in a polymer matrix, based on laser ablation of solid targets in liquids [12]. A method for obtaining metal nanoparticles and their properties before and after incorporation into a fluoroplastic polymer are described.

2. Methods

Gold nanoparticles were obtained by laser ablation in a liquid. A solid target was located at the bottom of a glass cuvette under a thin layer of working fluid (usually several millimeters) and irradiated with laser radiation (90 J/cm2 at 1.06 μm, pulse width of 130 ns) [13]. The working fluids are chosen to be transparent at the laser wavelength, so the laser radiation is absorbed in the target material. In the case of volatile liquids, the cell is covered with a thin glass to slow the evaporation. This method of manufacturing nanoparticles is of a physical nature in contrast to known chemical methods. The generation rate of nanoparticles and their average size depend on a number of laser parameters. A distinctive feature of laser ablation of solids in liquids is that nanoparticles can be obtained in virtually any liquid. This feature was used in this study to form nanocomposites: metal nanoparticles embedded in polymer matrices. A distinctive feature of the laser method is the potential to generate "pure" nanoparticles that do not contain undesirable ions and surfactants [14].

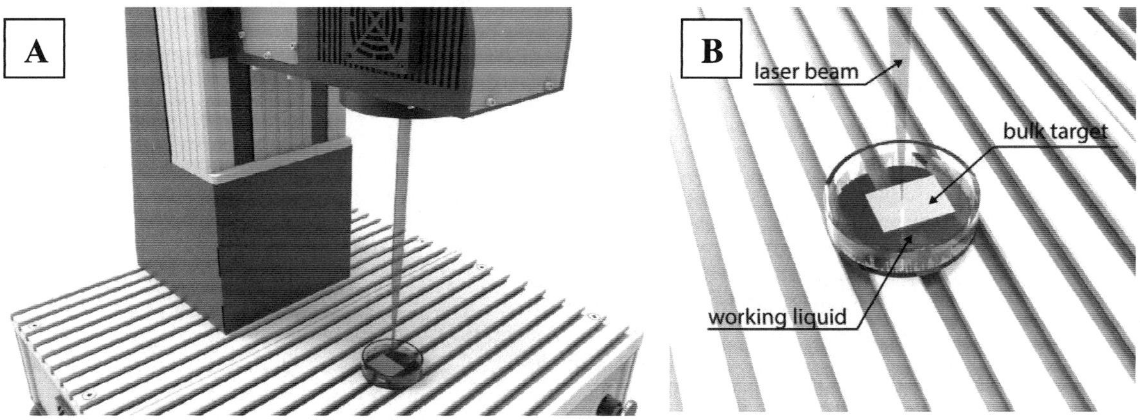

Figure 1. Experimental setup for laser ablation of solids in a liquid environment (A). An enlarged image of an experimental cell for laser ablation (B).

The size of the nanoparticles was determined using an analytical centrifuge DC24000 (CPS Instruments), a fast and sensitive particle size analyzer for colloidal solutions [XXXX]. To confirm the morphology of nanoparticles, a transmission electron microscope Carl Zeiss 200FE was used. The spectrometer USB3000T (OceanOptics) was used to study the spectral characteristics of nanoparticles. The Raman spectra of the nanocomposites were recorded with a U1000 spectrometer.

3. Results

The gold nanoparticles obtained by us are characterized by size and morphology. It is shown that in our gold nanoparticle preparation there are two types of nanoparticles (Fig. 2). All our gold nanoparticles have a size of less than 50 nm. Smaller ones with a size of 6-10 nm (70%) and larger ones 15-20 nm (30%). The distribution of these groups of nanoparticles by mass is the opposite. A

IOP Conf. Series: Materials Science and Engineering **347** (2018) 012005 doi:10.1088/1757-899X/347/1/012005

group of smaller nanoparticles with a size of 6-10 nm accounts for about 30%. A group of larger nanoparticles with a size of 15-20 nm account for about 70%.

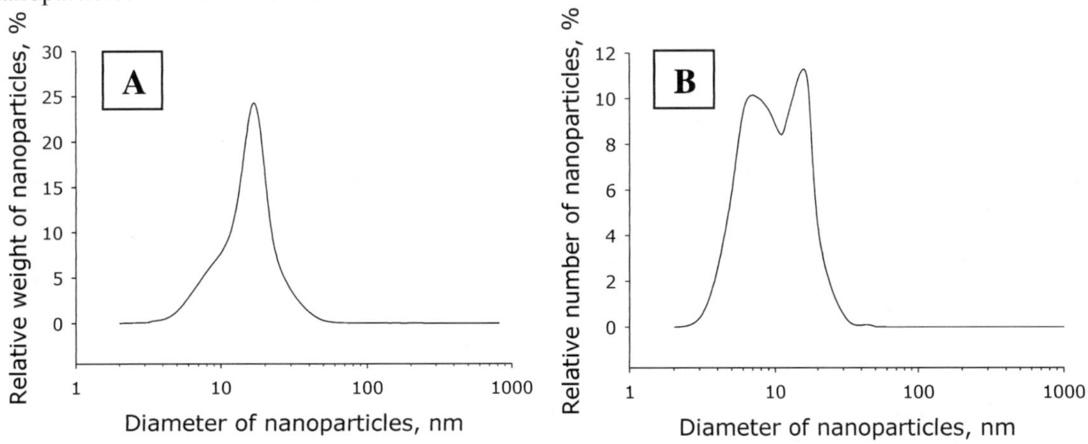

Figure 2. Distribution of nanoparticles by weight (A) and size (B).

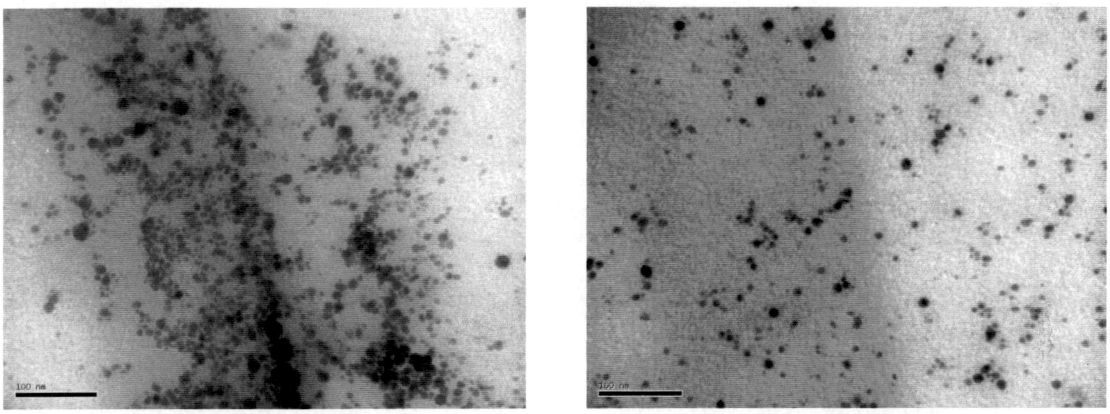

Figure 3. TEM image of gold nanoparticles.

The data obtained in the analytical centrifuge were confirmed by TEM microscopy. As can be seen in Figures 3 and 4, two nanoparticle fractions are observed. The average size corresponds to the particle size established by means of an analytical centrifuge.

The absorption spectra of the Au nanoparticles-fluorine-containing nanocomposite radically differ from the spectra of nanoparticles in liquids (Fig. 5). The spectrum of colloidal solution of gold nanoparticles in acetone exhibits a peak near 530 nm, which corresponds to the transverse plasmon resonance. One can also see a wide absorption wing in the red spectral region, which is indicative of the presence of some fraction of elongated Au nanoparticles in the colloid. The same nanoparticles in the nanocomposite with fluorine-containing polymer demonstrate the second plasmon resonance peak near 650 nm, which is due to elongated gold nanoparticles with an aspect ratio of about three.

The initial Raman peaks of polymer in nanocomposite have a larger amplitude. This can be explained by the enhancement of Raman scattering (the socalled surface-enhanced Raman scattering (SERS)) from composite nanoparticles. Gold nanoparticles enhancement some low-frequency vibrations in the fluorine-containing polymer.

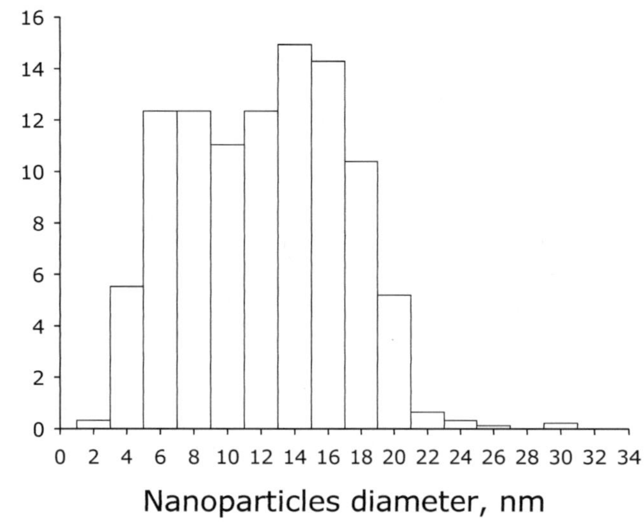

Figure 4. Size distribution averaged over 100 nanoparticles from TEM images.

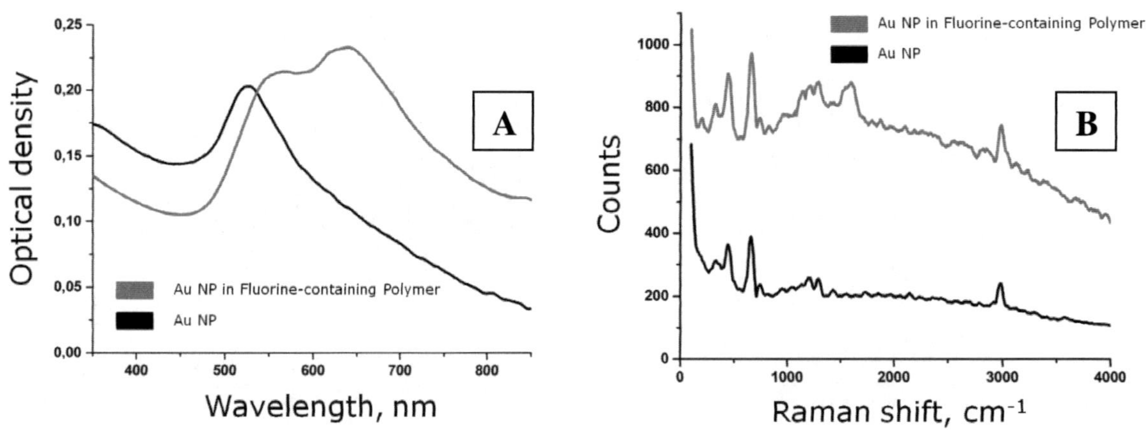

Figure 5. Extinction spectra of a colloidal solution of gold nanoparticles in acetone and Au nanoparticles (Au NP)-fluorine-containing nanocomposite (A). Raman spectra of fluorine-containing polymer in the initial state and fluorine-containing nanocomposite with gold nanoparticles (B).

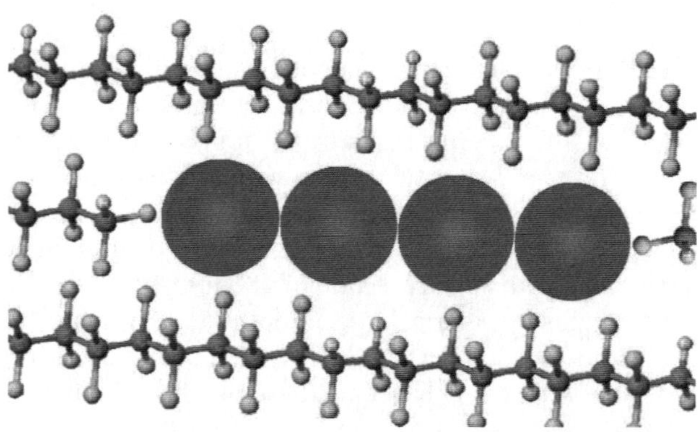

Figure 6. Polymer molecules align gold nanoparticles into chains in the nanocomposite. Carbon atoms are depicted in gray, fluorine atoms are green, gold nanoparticles are crimson.

Fluoroplast containing a polymer matrix apparently makes gold nanoparticles aggregated in a polymer. Nanoparticles cannot be incorporated directly into polymer chains, and the matrix flows around them. This process is schematically shown in Fig. 6. This aggregation is accompanied by a change in the plasmon resonance spectrum of metal nanoparticles. The newly formed absorption bands overlap almost entirely the visible spectral region and partially the IR region. This circumstance makes it possible to measure the high-speed response of nanocomposites using most diverse laser sources.

4. Conclusions

Colloidal solutions of gold nanoparticles were obtained by laser ablation of gold targets in organic liquids. Nanocomposites were prepared by mixing these solutions with solutions of fluorine-containing polymer. An analysis of the extinction spectra of the nanocomposites showed that the polymer matrix changed the morphology of the nanoparticles incorporated into it in comparison with the initialmorphology. The matrix aligns particles into elongated chains with a large aspect ratio. In turn, the Raman spectra of the nanocomposites show that nanoparticles also affect the polymer matrix. In the fluorine-containing matrix, gold nanoparticles increase some vibrational modes of the polymer in the low-frequency spectral region. The nanocomposites fabricated in this study have an acceptable optical quality and can be tested for ultrafast optical response.

Acknowledgments

The work was partially supported by the Grant MD-3811.2018.11 of the President of the Russian Federation for Governmental Support of Young Russian Scientists - Doctors of Sciences; The Grant MK-3606.2017.2 of the President of the Russian Federation for Governmental Support of Young Russian Scientists - Candidates of Sciences and RFBR Projects 18-52-70012 e-Asia_a, 16-02-01054_a.

References

[1] Grannan DM, Garland JC and Tanner DB 1981 *Phys. Rev. Lett.* **46** 375-378
[2] Badi N, Benqaoula A, Simakin AV and Shafeev GA 2013 *Mater. Lett.* **108** 225-227
[3] Kaneti YV, Tang J, Salunkhe RR, Jiang X, Yu A, Wu KCW and Yamauchi Y 2017 *Adv. Mater.* **29** 1604898
[4] Ma K, Gong L, Cai X, Huang P, Cai J, Huang D and Jiang T 2017 *Int. J. Nanomedicine.* **12** 3665–3679
[5] Toor A, So H and Pisano AP 2017 *ACS Appl. Mater. Interfaces* **9** 6369–6375
Santhosh G, Nayaka GP, Madhukar BS and Siddaramaiaha *2017 Mat. Today:Proc.* **4** 2061-12069
[6] Santhosh G, Nayaka GP, Madhukar BS and Siddaramaiaha *2017 Mat. Today:Proc.* **4** 2061-12069
[7] Berezkin AV, Kudryavtsev YV, Gorkunov MV and Osipov MA 2017 *J. Chem. Phys.* **146** 144902
[8] Ahmad SHA, Saidur R, Mahbubul IM and Al-Sulaimana FA 2017 *Renewable and Sustainable Energy Reviews* **73**, 1014-1030
[9] Gudkov SV, Andreev SN, Barmina EV, Bunkin NF, Kartabaeva BB, Nesvat AP, Stepanov EV, Taranda NI, Khramov RN and Glinushkin AP 2017 *Phys. Wave Phenomen.* **25** 207–213
[10] Ivanov VE, Usacheva AM, Chernikov AV, Bruskov VI and Gudkov SV 2017 *J. Photochem. Photobiol. B* **176** 36–43
[11] Berezkin AV, Kudryavtsev YV, Gorkunov MV and Osipov MA 2017 *J. Chem. Phys.* **146** 144902
[12] Barmina EV, Mel'nik NN, Rakov II and Shafeev GA 2017 *Phys. Wave Phenomen.* **25** 165–169
[13] Barmina EV, Gudkov SV, Simakin AV and Shafeev GA *2017 J. Laser Micro/Nanoeng.* **12**, 254-257
[14] Shafeev GA 2010 Formation of Nanoparticles under Laser Ablation of Solids in Liquids. *Lasers And Electro-Optics Research And Technology* ed SE Black (New York: Nova Science Publishers, Inc.) chapter 8 pp 191–226

IOP Conf. Series: Materials Science and Engineering **347** (2018) 012006 doi:10.1088/1757-899X/347/1/012006

Application of micro- and nanocrystalline cellulose

Yu S Sotnikova[1,2], T S Demina[1,2], A V Istomin[1], G P Goncharuk[1], Ch Grandfils[3], T A Akopova[1], A N Zelenetskii[1] and P G Babayevsky [2]

[1] Enikolopov Institute of Synthetic Polymer Materials, Russian Academy of Sciences, 70, Profsoyuznaya str., 117393, Moscow, Russia;
[2] Moscow Aviation Institute (National Research University), 4, Volokolamskoe shosse, 125993, Moscow, Russia
[3] Interfacultary Research Centre on Biomaterials (CEIB), University of Liège, Chemistry Institute, B6C, Allée du 6 août 11, B-4000 Liege (Sart-Tilman), Belgium.

Email: detans@gmail.com

Abstract. Micro- and nanocrystalline forms of cellulose were extracted from flax stalks and evaluated in terms of their applicability for various materials science tasks. It was revealed that both form of cellulose had anisometric morphology with length of 27.1 μm and 159 nm; diameter of 8.7 μm and 85 nm, respectively. They were used as reinforcing fillers for fabrication of composite films based on hydroxyethylcellulose. Film-forming and mechanical properties of the composite materials were significantly varied in dependence on filler content (0–10 wt.%) and size. As a second option of micro- and nanocrystalline cellulose application, a study of their effectiveness as stabilizing agents for oil/water Pickering emulsions was carried out. In contrast to micron-sized cellulose the nanocrystalline form appeared to be successful in the process of CH_2Cl_2/water interface stabilization and fabrication of polylactide microparticles via oil/water Pickering emulsion solvent evaporation technique.

1. Introduction

Cellulose is a natural occurring linear homopolysaccharide, which is widely applied for fabrication of paper, packing, coatings, composites, etc. Among other modifications of cellulose, its micro- and nano-sized forms attract growing attention due to the improved properties, which they could bring to composite materials [1]. Micro- and nanocrystalline cellulose are produced by extraction of crystalline regions from native cellulose using acid hydrolysis of disordered/amorphous regions, which link together assemblies of microfibrils [2]. As a function of cellulose origin and processing conditions, the extracted crystalline regions could significantly vary in size and aspect ratio, although they are normally anisometric. Micro-sized cellulose is mainly used as texturizing agent and for formulation of solid pharmaceuticals dosage forms, i.e. tablets [3], while nanocrystalline cellulose is intensively studied for fabrication of mesoporous and smart (shape-memory and optical-tunable) materials [4, 5]. Nevertheless, the main application area of micro- and nanocrystalline cellulose is a reinforcing filling of composite materials, especially films and membranes used in food packaging industry [6-8]. Another promising application of cellulose particles is emulsion stabilization, which could be even more effective than that in a case of application of molecular surfactants [9, 10]. These, so-called Pickering, emulsions stabilized by solid particles instead of surfactants possessed a high resistance to coalescence and could be successfully used for food industry as well as for fabrication of core-shell nano/microparticles for drug or cell delivery [11, 12].

Thus, the work aims to fabrication of micro- and nanocrystalline forms of flax stalks cellulose and evaluation of their applicability as reinforcing fillers in composite hydroxyethylcellulose films as well as stabilization agents for oil/water Pickering emulsions, including ones used for preparation of polylactide microparticles through oil/water solvent evaporation technique.

2. Materials and methods

2.1. Materials

Micro- and nanocrystalline cellulose was extracted from bleached flax stalks cellulose (APK "Vologodchina", Russia). Hydroxyethylcellulose marked as Natrosol 250HHR (Hercules Inc., USA) was used as received. Poly(L,L-lactide) (PLLA) with an average molecular weight of 160 kDa was purchased from Sigma-Aldrich.

2.2. Preparation and characterization of micro- and nanocrystalline cellulose

The micro- and nanocrystalline cellulose were obtained by acidic hydrolysis and marked as MCC and NCC, respectively. Briefly, MCC was extracted from flax stalks cellulose by hydrolysis in 10% HNO_3 using an ecoclave Buchi ("Buchi AG", Switzerland) at temperature of 120°C, pressure of 4 MPa for 120 min [13]. The obtained MCC was washed with distilled water on membrane filter until neutral pH was reached. Nanocrystalline cellulose (NCC) was extracted from the MCC by its additional hydrolysis in a mixture of 10% H_2SO_4 and 10% H_2O_2 using the ecoclave at temperature of 100–105°C and pressure of 5 MPa under a simultaneous mechanical stirring at 3000 rpm for 120 min. Then, the NCC was washed with distilled water on membrane filter with pores of 100 nm until pH reached to 6.5–7. As a final step, 1 wt.% suspension of the extracted NCC was treated with ultrasound at 1 kW and 44 kHz for 20 min.

Morphology and size distribution of the obtained MCC was evaluated by optical microscopy using Levenhuk D320 L ("Levenhuk", Russia) and analyzed using ImageJ (National Institute of Health, USA) software. Morphology characterization of NCC was carried out using atomic-force microscopy NtegraPrima ("NT-MDT", Russia) as published in [14].

2.3. Fabrication and characterization of hydroxyethylcellulose films filled with micro- and nanocrystalline cellulose

Hydroxyethylcellulose films filled with various amount (0–10 wt.%) of MCC or NCC were cast from 1 wt.% hydroxyethylcellulose water solutions. The MCC and NCC in a form of water dispersions were added to the polymer solutions by mixing using a four-blade propeller stirrer for 10 min at 700rpm. The obtained MCC/NCC dispersions in hydroxyethylcellulose solutions were cast on polystyrene Petri dishes and, then, dried in a dust-free chamber at RT (about 48 hours). The thickness of the films was 60 ± 20 μm.

Mechanical properties of the film samples were evaluated using a AGS-H universal tensile machine (Shimadzu, Japan) at a speed of 1 mm/min after films conditioning in a desiccator over a $(NH_4)_2SO_4$ saturated water solution.

2.4. Preparation and characterization of polylactide microparticles stabilized with micro- and nanocrystalline cellulose

A possibility to use MCC and NCC for stabilization of oil/water emulsions was evaluated for simple CH_2Cl_2/water emulsions as well as for fabrication of polylactide microparticles via oil/water solvent evaporation technique. Simple oil/water emulsions were prepared as follows: 2 mL of CH_2Cl_2 was added to 18 mL of water dispersion of NCC of MCC and mixed using a four-blade stirrer at a rotation speed of 700 rpm for 10 min. The effectiveness of oil/water interface stabilization was evaluated visually and optically. Poly(L,L-lactide) (PLLA) microparticles stabilized with NCC were prepared via Pickering oil/water emulsion solvent evaporation technique as follows: 6 wt. % solution of PLLA in CH_2Cl_2:acetone (9:1 v/v) was rapidly added to NCC water dispersion (0.1 – 1 wt.%) to achieve oil/ water phase ratio as 1/9 v/v. The mixing of the phases was carried out using a four-blade stirrer at a rotation speed of 700 rpm until a full evaporation of the organic solvents. The obtained

IOP Conf. Series: Materials Science and Engineering **347** (2018) 012006 doi:10.1088/1757-899X/347/1/012006

solid PLLA microparticles were collected, washed several times with deionized water and fractionated using sieves with apertures of 400, 315, 200 and 100 μm; then, freeze-dried and weighted.

3. Results and discussion

3.1. Preparation and characterization of micro- and nanocrystalline cellulose

Extracted micro- and nanocrystalline cellulose form suspension and colloidal opalescent water dispersion, respectively (figure 1).

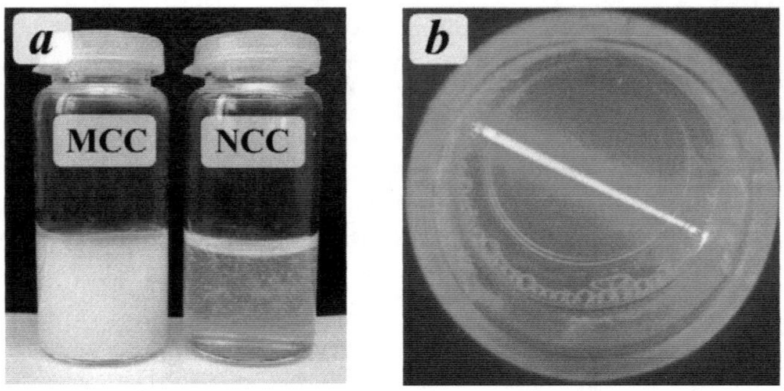

Figure 1. Photographs of MCC and NCC water dispersions (***a***) and NCC dispersion under a laser beam (***b***).

As could be seen in figure 2 the MCC had anisometric rod-like morphology with a mean length and diameter of 27.1 μm and 8.7 μm, respectively. The microcrystals significantly vary in length and diameter according to calculation of their size distribution. A study of NCC by atomic-force microscopy has also revealed anisometric morphology with a mean length and diameter of 85 ± 39 nm and 159 ± 89 nm, respectively [13]. Therefore, microcrystalline cellulose possessed higher aspect ratio than its nanocrystalline form: 3.1 and 1.9, respectively.

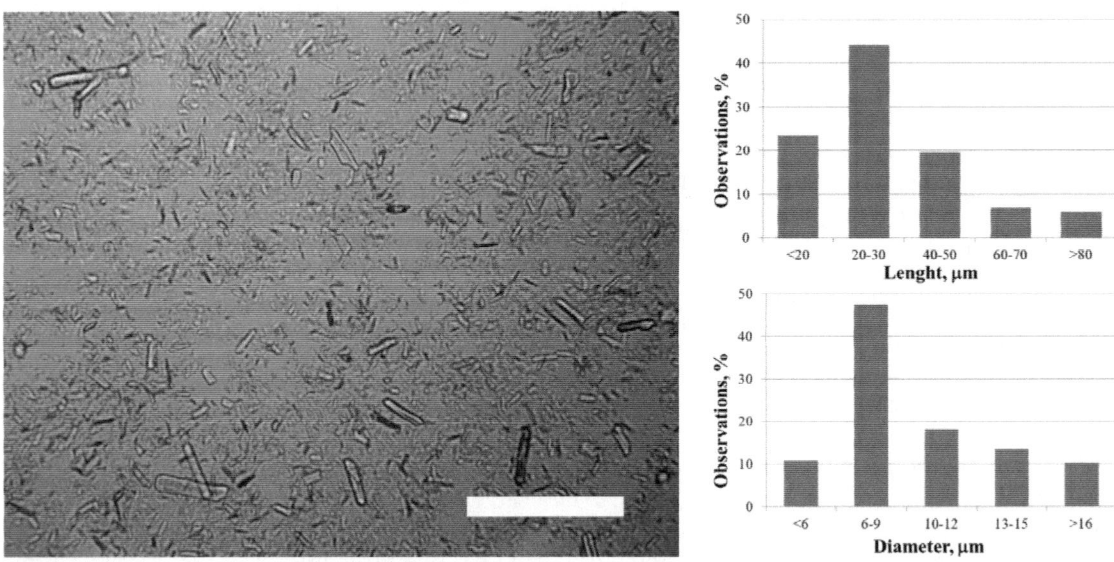

Figure 2. Optical micrographs of 0.5 wt.% MCC water dispersion and histograms of MCC length and diameter distributions. Scale bar is 500μm.

3.2 Fabrication and characterization of hydroxyethylcellulose films filled with micro- and nanocrystalline cellulose

Here we investigated an effect of filling of hydroxyethylcellulose films with MCC and NCC at various concentrations on thier mechanical properties. In contrast to the neat hydroxyethylcellulose film and one filled with NCC, the presence of MCC at relatively high concentrations (5–10 wt.%) led

IOP Conf. Series: Materials Science and Engineering **347** (2018) 012006 doi:10.1088/1757-899X/347/1/012006

to a cracking of films after the preparation, which could be caused by accumulation of internal stress on MCC/polymer interface during a solvent evaporation process and higher adhesion to Petri dish (figure 3).

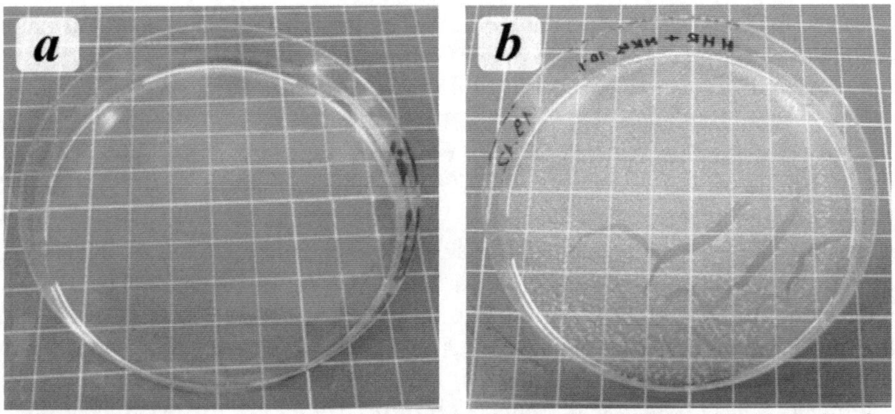

Figure 3. Photographs of hydroxyethylcellulose films filled with 10 wt.% of NCC (***a***) and MCC (***b***).

The study of the filling on mechanical properties showed that both NCC and MCC had similar positive effect on films tensile strength and modulus of elasticity. As could be seen in figure 4 this effect of filling has an optimum at 5 wt.% content of NCC or MCC, when tensile strength increased up to 41–44% and modulus of elasticity up to 41–47% in comparison with neat hydroxyethylcellulose film. Previously, the extracted NCC was also used as reinforcing filler for fabrication of guar-based films and showed an filling optimum at 0.5–1 wt.% [14]. Apparently, the optimum is determined by filler distribution effectiveness over a polymer matrix, i.e. by filler concentration, polymer nature and composite films preparation conditions, while filler size (MCC or NCC) affects film-forming properties.

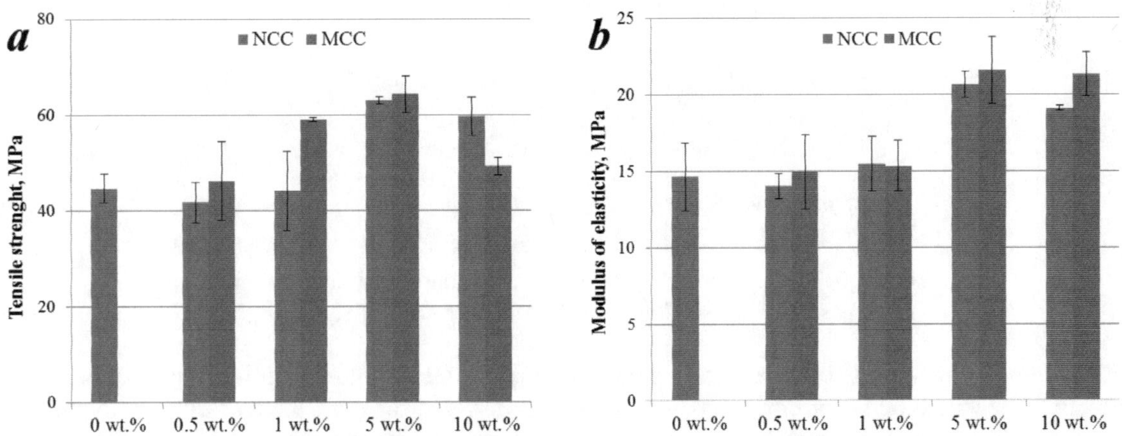

Figure 4. Tensile strength (***a***) and modulus of elasticity (***b***) of the hydroxyethylcellulose films filled with micro- and nanocrystalline cellulose at various concentrations.

Thus, both MCC and NCC could be successfully used as reinforcing fillers for fabrication of composite films. However, the presence of micron-sized MCC leads to significant decrease in film-forming abilities of hydroxyethylcellulose, which makes NCC more suitable filler for fabrication of films and membranes, for example, in packaging industry.

IOP Conf. Series: Materials Science and Engineering **347** (2018) 012006 doi:10.1088/1757-899X/347/1/012006

3.3. Preparation and characterization of polylactide microparticles stabilized with micro- and nanocrystalline cellulose

A study of applicability of the MCC and NCC for stabilization of Pickering emulsions was evaluated for simple CH_2Cl_2/water system as a first step. In comparison with nanocrystalline cellulose MCC couldn't provide the interface stabilization at the described conditions, which is rather predictable due to micron-sized particles are rarely effective for Pickering emulsions. Surprisingly enough, NCC showed pronounced ability to stabilize CH_2Cl_2/water emulsions only at lower concentration (0.1 wt. %) in aqueous phase, while its increase up to 0.5–1 wt.% led to a phase separation. We assume that this decisive decrease of stabilization effectiveness could be caused by aggregation of the nanocrystals in water dispersions. As it was previously shown using dynamic light scattering the NCC at low concentration (0.1 wt.%) had mean particle size of ~60 nm, while an increase of their concentration up to 1 wt.% led to a formation of aggregates with a size up to 6.5 μm [14]. Thus, along with aggregation the NCC lost their stabilizing ability.

An effectiveness of MCC and NCC to stabilize an interface during the fabrication of polylactide microparticles through an oil/water solvent evaporation technique was in a good agreement with results obtained using the previously discussed simple system. The polylactide microparticles were successfully fabricated only in the case of application of 0.1 wt.% NCC dispersions as aqueous phase. As could be seen in figure 5 the total yield of the produced microparticles was reached up to 54 wt.%; however, the particles had rather large mean size and their yield was drastically decreased with increase of NCC concentration in the aqueous phase. Apparently, NCC stabilization effectiveness, i.e. resistance to aggregation, could be controlled by its surface modification or adjustment of pH/ionic strength of aqueous phase.

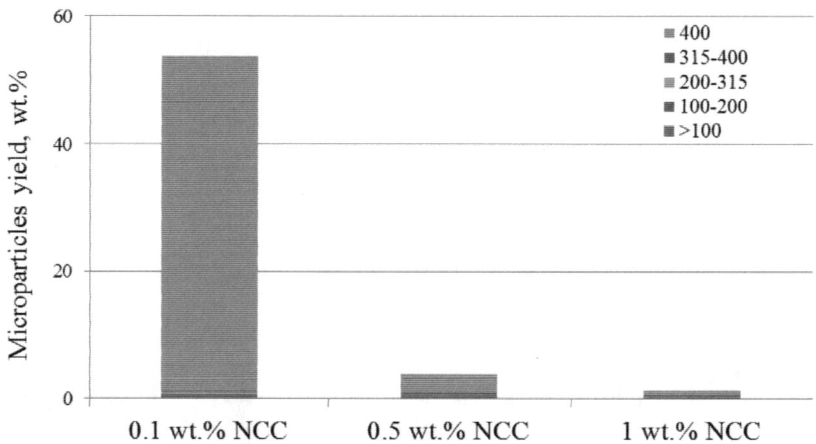

Figure 5. The total yield and size distribution of polylactide microparticles prepared using NCC at various concentration in aqueous phase (***b***).

Therefore, nano-sized form of cellulose, i.e. nanocrystalline cellulose at low concentration, could effectively stabilize oil/water emulsions, including ones tailored for fabrication of polymeric particles through solvent evaporation technique. Aggregation of the nanocrystals at higher concentrations as well as application of microcrystalline cellulose led to dramatic decrease in their stabilization effectiveness.

4. Conclusions

Micro- and nanocrtystalline flax stalks cellulose possessed anisometric particle morphology and could be used as fillers for fabrication of reinforced composite materials and stabilizing agents for oil/water Pickering emulsions. Application of both types of cellulose allowed to increase mechanical properties of composite hydroxyethylcellulose films up to 41–47% at optimal filling content of 5 wt.%. Nanocrystalline cellulose showed higher effectiveness as agents for interface stabilization than micron-sized cellulose and allowed to fabricate polylactide microparticles through oil/water Pickering emulsion solvent evaporation technique with total yield up to 54 wt.%

5. References

[1] Siro I and Plackett D 2010 *Cellulose* **17** 459-494
[2] Habibi Y et al. 2010 *Chem. Rev* **110** 3479-3500
[3] Thoorens G et al. 2014 *Int J Pharm* **473** 64-72
[4] Huang J et al. 2015 *Polysaccharide-Based Nanocrystals: Chemistry and Applications* (Wiley-VCH Verlag GmbH & Co.) p 328
[5] Lin N et al. 2012 *Nanoscale* **4** 3274-3294
[6] Samir M A S A et al. 2005 *Biomacromolecules* **6** 612-626
[7] Siqueira G et al. 2010 *Polymers* **2** 728-765
[8] Miri N E et al. 2015 *Carbohyd. Polym* **129** 156-167
[9] Chevalier Y and Bolzinger M-A 2013 *Colloids Surf A Physicochem Eng Asp* **439** 23-34
[10] Rayner M et al. 2014 *Colloids Surf A Physicochem Eng Asp* **458** 48-62
[11] Kalashnikova I et al. 2011 *Langmiur* **27** 7471-7479
[12] Rescignano N et al. 2015 *J Colloid Interface Sci.* **445** 31-39
[13] Istomin A V et al. 2016 *Fibre Chem* **48** 199-201
[14] Sotnikova Yu S et al. 2017 *Fibre Chem* **49** 188-194

IOP Conf. Series: Materials Science and Engineering **347** (2018) 012007 doi:10.1088/1757-899X/347/1/012007

Improvement of chemical composition, structure and mechanical properties of heat-resistant chromium-nickel alloy

S Varlamova[1], A Trushnikova[1], B Rumyantsev[1], V Butrim[1] and V Simonov[2]

[1]Open Joint Stock Company "Kompozit", Pionerskay, 4, Korolev city, Moscow region, Russia
[2]Bauman Moscow State Technical University, ul. Baumanskaya 2-ya, 5, Moscow, Russia

E-mail: sbvarlamova@gmail.com

Abstract. A thermodynamic analysis of a multicomponent system of the Cr-Ni alloy (Cr-32Ni-1,5W-0,25V-0,5Ti) with small additions of refractory metals was carried out. The microstructure and phase composition of the base alloy (I) and alloy with additional alloying (II) were studied. The effect of additives on the mechanical properties of the Cr-Ni alloy at 20, 900 and 1080 °C was shown. The microstructure of alloys I and II was studied in the fracture zone of samples after tensile tests at different temperatures. We studied the effect of small additives on the microstructure of alloys and changes in the morphology of the structural components (phases) as a function of temperature and degree of deformation.

1. Introduction

Development and improvement of hot-strength and heat-resistant materials is one of the key issues of metal science. The use of new modern materials is a necessary condition for achieving the required tactical, technical and operational characteristics of material for space application. Extreme operating conditions of products are forced to tighten the requirements to the level and stability of the properties for materials being developed [1].

High-temperature chromium-based alloys due to significant high-temperature strength at temperatures above 1000 °C, thermal stability and high corrosion resistance are promising substitutes for nickel alloys for aerospace and power engineering application [2].

At present, high-temperature Cr-based alloys has a limited employing in industry. The known alloy Cr- (31-35) Ni (1-3) W- (0.1-0.4) V- (0.05-0.3) Ti mass. % used to manufacture fuel decomposition chambers and nozzle of satellite's thrusters that operate in an impulse power setting.

However, from the point of view of the ongoing update of the requirements to mechanical properties and high-temperature strength, particularly, the potential of the existing alloy is almost exhausted and its further improvement is required.

The grain boundaries are the "weakest" elements of the structure in polycrystalline alloys in terms of creep resistance and a tendency to embrittlement. It is known that the properties of grain boundaries depend on their chemical composition that differs from the average chemical composition of the alloy due to segregation phenomenon [3]. Low-alloying is a well-known hardening method at which the structure and properties of alloys are changed due to local formation on borders of grains and sub-grains of the strengthening phases with limited solubility in α-γ solid solutions of alloy. The specific state of the grain boundaries facilitates and intensifies the diffusion and self-diffusion processes of various elements in polycrystalline metals that proceeds more actively in boundary zones than in grain volume. The low alloying additives introduced into the alloy can change the diffusion rate in the volume and along the grain boundaries, approximating the grain boundaries state in a thermodynamic and kinetic relation to the peculiar grain volume state [4].

The effect of low alloying on the properties of alloy can be explained by means of the theory of internal adsorption in metals according to which the defective parts in the bulk of structure are enriched with atoms of the dissolved component (with low impurity, in particular) that leads to a decrease of surplus defect energy [5]. Adsorption enrichment, occurring predominantly in the inter-crystalline sections, creates an uneven distribution of properties in different parts of the volume of metal. With low alloying additives, such unevenness becomes particularly significant. Intercrystallite boundaries form a continuous, unified and very branched netting, so low alloying additions seem to reinforce the matrix of alloy.

The high temperature strength of Cr-Ni alloy mainly depends on creep strength of the nickel phase at high temperature [2]. So it is of interest to harden this phase to improve the high temperature strength of the alloy.

In paper [6] the results of a first principles theoretical research of alloying elements effect on cohesive properties of high temperature superalloys are presented. The authors [6] showed that such low alloying elements as Zr, Hf, Ta and Nb promote increasing of cohesive strength and work of grain boundaries separation, as well as tend to segregate to the grain boundaries.

In accordance with Ref. [7] when alloying of Cr-Ni alloy by low alloying elements of refractory metals (up to 0.2% mass) is observed increase in tensile strength of alloy at simultaneous decrease in elongation.

It should be noted that current literature data confirm the increase in the high-temperature strength of the chromium-nickel alloy, however, the mechanism for strengthening the Cr-based and Cr-Ni-based alloys were studied insufficiently.

In accordance with the foregoing, the aim of this paper is to thermodynamically analyze the phase distribution of small additions of refractory metals in a chromium-nickel alloy, as well as to study the effect of these additives on the microstructure of the alloy and the properties of the material.

In this paper, we undertake the thermodynamic analysis of the phases distribution of low alloying refractory metals in a Cr-Ni-base alloy, as well as the impact of these additives on a microstructure and the mechanical properties of material.

2. Materials and experimental techniques

Base (I) and experimental with low alloying of refractory metals (II) alloys were chosen as the materials to be studied. Both alloys have been smelted in a vacuum induction furnace with the subsequent electroslag refining. The chemical composition (Table 1) was determined on a glow discharge spectrometer on a Leco GDS 850 unit.

Table 1. Chemical composition of alloys, % mass.

	Cr	Ni	W	V	Ti	Ta	Nb	Zr	Hf
Alloy I	62,0	34,9	2,0	0,28	0,4	-	-	-	-
Alloy II	63,7	32,8	2,0	0,3	0,45	0,15	0,16	<0,002	<0,002

Calculation in the program Thermo-Calc was carried out with reference to an alloy containing low alloying additives of refractory metals. The temperatures chosen for the calculation corresponded to the temperatures of the tensile test.

Tensile tests of the samples were carried out on a Zwick Z050 test machine at temperatures of 20, 900 and 1080 °C.

The microstructure was studied before and after tensile tests with use an electronic scanning microscope (SEM) JEOL JSM 6610LV, equipped with the system Oxford Instruments Advanced AZ tec

IOP Conf. Series: Materials Science and Engineering **347** (2018) 012007 doi:10.1088/1757-899X/347/1/012007

to X-ray microanalysis (MRSA). Samples for metallographic studies were prepared by mechanical polishing.

The program "Thermo-Calc" with the TCNI8 database was used for thermodynamic calculations; TCSNi-based Superalloys Database [8]. The program is based on the model of equilibrium thermodynamics that takes into account the change in the thermodynamic activity of the alloy components in phases that exist in an equilibrium state. Thus, the calculations use the temperature, pressure, chemical potential of the components to determine the chemical composition of the equilibrium phases in the alloy [9].

3. Results and discussion

By calculations it has been established that the low-alloying additives of refractory metals are mainly dissolved in the γ-Ni phase with a FCC lattice. Niobium and tantalum in minimal amounts are present in Ni_3M intermetallides.

To confirm the results of the calculations, using SEM for metallographic specimens, the microstructure of the alloys after extrusion and annealing was studied. The analysis in various areas of specimens was carried out. The specimens were selected from three different parts of the rod: the sagging end, the middle and the inlet end. Figure 1 shows typical photographs of the microstructure and the results of MRSA analysis.

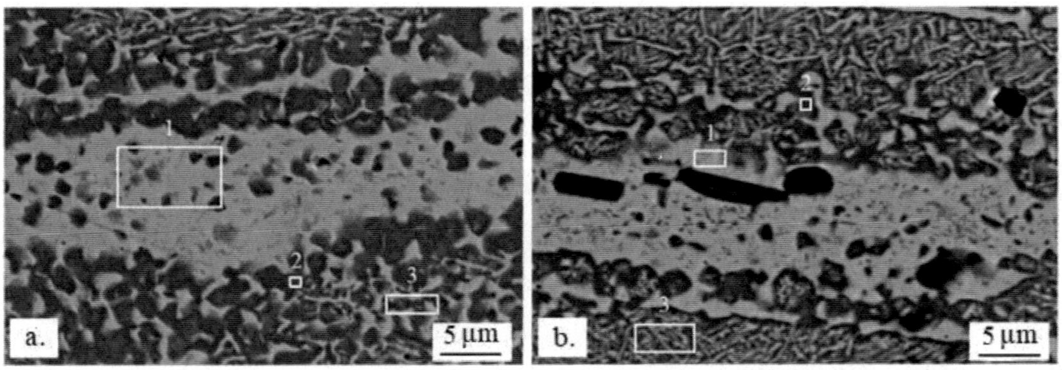

Спектр	Содержание элемента, масс. %				
	Cr	Ni	W	V	Ti
1	43,14	53,35	2,43	0,49	0,59
2	93,15	4,87	1,98	-	-
3	70,20	26,68	2,85	-	0,27

Спектр	Содержание элемента, масс. %						
	Cr	Ni	W	V	Ti	Nb	Ta
1	46,97	47,45	2,85	0,42	0,74	0,37	1,2
2	87,94	10,79	1,26	-	-	-	-
3	69,27	28,12	2,31	0,30	-	-	-

Figure 1. Typical microstructure of alloy I (a) and alloy II (b) after hot extrusion and annealing, × 3000 and characteristic MRSA spectra

As a result, studies have found that tantalum and niobium are located predominantly in the nickel γ-phase and it is consistent with the results of the calculation. Niobium partially present in the form of nitride.

Both alloy's samples were analyzed by XRF. The X-ray diffraction maximum method established increasing of lattice period of γ-phase (Ni) in alloys with low alloying, while the lattice period of α-phase (Cr) did not change. Figure 2 shows the diffractograms.

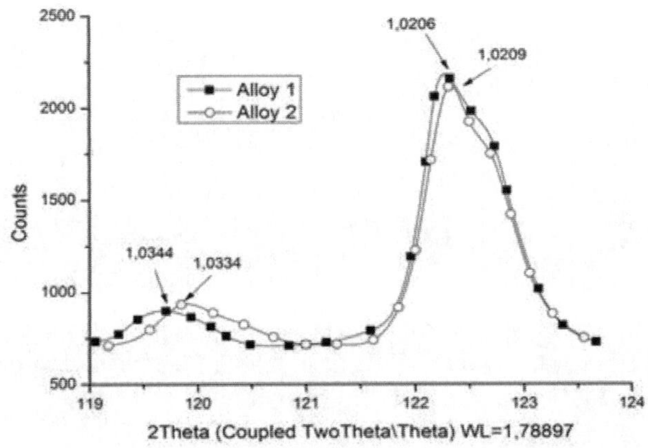

Figure 2. Diffractograms of Alloy I and Alloy II

The observed results lead us to the conclusion that low alloying refractory elements dissolve in a γ-solid solution, and their distribution over the lattice is fairly uniform.

The results of the tensile tests are presented in the Table 2. Figure 3 shows exterior of the samples after tensile tests at a temperature of 1080 °C.

Table 2. Mechanical properties of alloys I and II at different temperatures

	T = 20 °C		T = 900 °C		T = 1080 °C	
	δ	σ_B	δ	σ_B	δ	σ_B
Alloy I	20	1182	24	374	136	89
Alloy II	17,5	1224	12,3	384	20,2	137

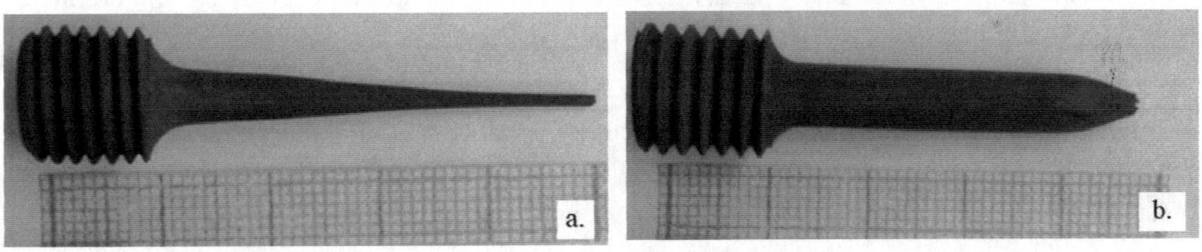

Figure 3. Exterior of the samples after tensile tests at a temperature of 1080 °C

Tensile tests showed that low alloying of refractory metals significantly increase the tensile strength (σ_B) from 89 MPa (Alloy I) to 137 MPa (Alloy II). At the same time the elongation (δ) had decreased from 136% (Alloy I, Fig 3a) to acceptable level for structural materials 20% (Alloy II, Fig.3b).

Samples for studying the microstructure after tensile tests were taken from the fracture zone (Fig. 4). The SEM showed that with temperature increase the morphology of the phases also significantly changed. Although at test temperature of 20 °C the difference in the microstructure of the alloys is insignificant, with temperature increase in alloy I take place coarsening of the secondary precipitations of γ-Ni solid solution in α-Cr solid solution due to the dynamic recrystallization [10] (Fig. 4c, e). However, in an alloy II disperse precipitates of γ-Ni are smaller in size than in alloy I.

IOP Conf. Series: Materials Science and Engineering **347** (2018) 012007 doi:10.1088/1757-899X/347/1/012007

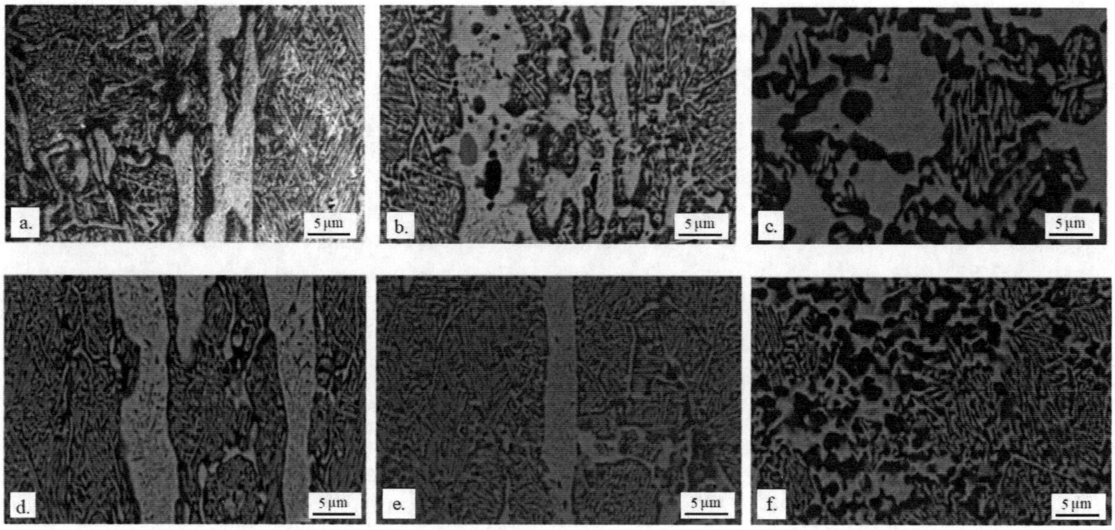

Figure 4. Microstructure of alloys after tensile tests at a temperature: a, d - 20 ° C; b, e - 900 ° C; c, f - 1080 ° C

It may be result of increasing the stability of the alloy structure due to segregation low alloying Ta, Nb, Hf and Zr at the interphase boundaries and slow down of their dissolution with increasing temperature.

4. Conclusions

Low-alloying additives of refractory metals is an effective way to improve the high temperature strength of the chromium-nickel alloy.

Thermodynamic calculations in Thermo-Calc show that low-alloying additives of refractory metals are mainly dissolved in the γ-Ni phase with a FCC lattice.

Tensile tests show that low alloying of refractory metals significantly increase the tensile strength from 89 MPa to 137 MPa and decrease the elongation from 136% to acceptable level for structural materials 20%

The X-ray diffraction maximum method established increasing of lattice period of γ- Ni phase in alloys with low alloying, while the lattice period of α- Cr phase did not change. The observed results lead us to the conclusion that low alloying refractory elements dissolve in a γ-solid solution, and their distribution over the lattice is fairly uniform.

We assumed that increasing in stability of alloy structure is connected with a segregation low alloying Ta, Nb, Hf and Zr at the interphase boundaries and slow down of their dissolution with increasing temperature.

References

[1] Butrim V N 2017 *Design from Composite Materials*. Improvement of Chromium-Nickel Alloys for Space Application **2** (134) pp. 26-38.

[2] Ault G M 1969 *Requirement to the High-Temperature Strength Materials*. High-Temperture Materials for Spaceships, M.: Metalurgy pp. 41-65.

[3] Goldstein M I, Litvinov V S and Bronfin B M 1986 *Metalphysics of high-strength alloys*. M: Metallurgy, pp. 312.

[4] Goldstein Y E and Mizin V G 1986 *Modifying and a cast pig iron and steel low alloying*. M: Metallurgy, pp 272.

[5] Gorelik S S 1978 *Rekristallization of metals and alloys*.- M: Metallurgy, -pp. 568.
[6] Beresnev A G, Razumovsky V I, Lozovoy A Y, Logachova A I and Razumovski I M 2012 *Technology of light alloys* Development of an alloying theory for creation of new generation of the high temperature strength nickel alloys produced by powder metallurgy **2**.
[7] Butrim V N, Razumovskiy V I, Beresnev A G, Trushnikova A S and Razumovskii I M 2017 *Materials Science Forum* Effect of a Number of Transition Metals on the Cohesive Properties of Cr-Ni-Base Alloys, **879**, pp 1998-2002.
[8] TCNI8; TCS Ni-based Superalloys Database [Electronicresource]. -www.thermocalc.com.
[9] Thermo-Calc Software: Thermocalc State Variables and State Variables. – Stockholm, Sweden. – 2006.
[10] Mironenko V N, Aronin A S, Vasenev V V, Aristov I M, Shmytko I M and Trushnikova A S 2016 *Physics of Metals and Metallography* Structure, Mechanisms and Behavior Model of a Two-phasic Chromium-nickel alloy at high-temperature deformation **117** 9, pp. 969-976.

IOP Conf. Series: Materials Science and Engineering **347** (2018) 012008 doi:10.1088/1757-899X/347/1/012008

Self-propagating high-temperature synthesis and luminescent properties of ytterbium doped rare earth (Y, Sc, Lu) oxides nanopowders

D.A. Permin[1], A.V. Novikova[1], S.S. Balabanov[1], E.M. Gavrishchuk[1], S.V. Kurashkin[1] and A.P.Savikin[2]

[1] G.G. Devyatykh Institute of Chemistry of High-Purity Substances RAS, Nizhny Novgorod, Russia
[2] N.I. Lobachevsky State University of Nizhny Novgorod, Nizhny Novgorod, Russia

E-mail: permin@ihps.nnov.ru

Abstract. This paper describes a comparative study of structural and luminescent properties of 5%Yb-doped yttrium, scandium, and lutetium oxides ($Yb:RE_2O_3$) powders and ceramics fabricated by self-propagating high-temperature synthesis. According to X-ray diffractometry and electron microscopy the chosen method ensures preparation of low-agglomerated cubic C-type crystal structured powders at one step. No crucial differences in luminescence spectra were found the $Yb:RE_2O_3$ powders and ceramics. It was shown that the emission lifetimes of the $Yb:RE_2O_3$ powders are lowered by crystal structure defects, while its values for ceramics samples are compared to that of monocrystals and more influenced by rare earth impurities.

1. Introduction

In recent years, ytterbium doped rare-earth sesquioxide (RE_2O_3, RE=Y, Sc, Lu) materials have been paid attention because of interesting properties of those materials which are promising for laser applications [1-4]. At that it is very difficult to produce them by melt-growth process (conventional single crystal technology). Ceramic approach seems to be preferable to achieve highly transparent RE_2O_3 media due to several advantages, such as cost-effectiveness, large-scale production, feasibility of shape controlling and better mechanical properties.

Ceramics transparency is determined by the absence of scattering centers – pores, secondary phases, thick grain boundaries. Such an ideal microstructure can be achieved by using powders that meet a number of requirements. They include high dispersity and chemical purity, narrow size distribution and low agglomeration degree.

Respectively several methods have been developed to synthesize powders for transparent ceramics. The most commonly used method is the co-precipitation [5-8]. But there are also some alternative approaches allowing making high quality transparent ceramic material. Different scientific groups use laser sublimation method [9,10], flame-spray pyrolysis [11], sol-gel technology [12], as well as self-propagating high-temperature synthesis (SHS) [13,14]. The last one is a very promising approach, since it does not require any complicated equipment, high energy consumption and is characterized by sufficiently high productivity.

Usually in the case of high-disperse powders synthesis metal nitrate acts as an oxidant and crystalline organic substances serve as a fuel. Earlier transparent ceramics of different rare earth oxides were obtained by sintering of SHS-powders derived with the metal acetate [15], citric acid [16] and glycine [14,17].

The aim of this study was to establish basic structural and luminescent properties of the ytterbium doped yttrium, lutetium and scandium oxide powders and ceramics prepared by self-propagating high-temperature synthesis.

2. Experimental

Among all types of fuel, glycine is the most commonly used for the synthesis of highly disperse powders by the SHS method. Moreover, only with the use of glycine it was possible to obtain optical ceramics based on all three target RE oxides. In this connection yttrium oxide powder (Y_2O_3, 99.99% purity, Polirit (Russia)), scandium oxide powder (Sc_2O_3, 99.99% purity, Lanhit (Russia)), lutetium oxide powder (Lu_2O_3, 99.99% purity, Lanhit (Russia)), ytterbium oxide powder (Yb_2O_3, 99.99% purity, Dalchem (Russia)) nitric acid (99.999% purity), glycine (NH_2CH_2COOH 99.9% purity) were used as the starting materials.

To prepare nitrate solutions a portion of the oxides of the respective metals was dissolved in a stoichiometric amount of dilute nitric acid at heating. Solutions of metal nitrates and glycine were mixed in a 1: 1 molar ratio in a quartz flask, which was placed in an oven preheated to 500 °C to initiate the SHS reaction. At the end of the reaction a white foam-like bulk product was formed.

The corresponding scheme of proceeding chemical reactions can be written as follows:

$$6RE(NO_3)_3 + 10NH_2CH_2COOH \rightarrow 3RE_2O_3 + 20CO_2\uparrow + 14N_2\uparrow + 25H_2O\uparrow$$

X-ray diffraction analysis was carried out on a Shimadzu XRD-6000 diffractometer with CuKα radiation (λ = 1.54178 Å) in the 2θ range of 10-60 deg to study the structural properties of the powders. The scanning step for 2θ was 0.02 deg., the scanning rate was 2 deg/min. The diffraction patterns were refined by the Rietveld method using the RIETAN-98 software. The coordinates of the atoms and the unit cell parameters of the corresponding rare earth oxides were used as the basic model. The ICSD/Retrieve and PCPDF2 databases were used in the analysis. The morphology of the powders was studied by the scanning electron microscopy method using JEOL JSM-6390 LA microscope.

Disks with a diameter of 15 mm and a thickness of 3-5 mm were produced from the calcined powders by uniaxial pressing at a pressure of about 10 MPa to measure the luminescence spectra and the decay time of luminescence. The luminescence was excited by the emission of a semiconductor laser Latus with fiber output at a wavelength of 975 nm (FWHM is about 3 nm). The laser radiation was focused on the sample using an optical condenser and a short-focus lens (F = 50 mm). Then the image was transferred to the entrance slit of the double monochromator M833 using the system of wide-aperture lenses. The radiation passed through the monochromator was detected by high-sensitivity InGaAs photodiode FGA10 operating in a synchronous detection circuit (synchronous detector Stanford Research Systems SR810).

Luminescence decay of Yb^{3+} ions was studied under excitation of a semiconductor laser diode with a center of the radiation band at a wavelength of about 975 nm, operating in a pulse generation mode with a 5 ms repetition period. The lens system focused the luminescence emission at the entrance slit of the monochromator. The radiation passed through the monochromator was converted to an electrical signal using the photoelectric multiplier FEU-62, which was then fed to the input of a digital oscilloscope.

Deagglomeration of the powders was carried out on an attrition mill LDU-3MPR (Labotex, Russia) with 2∅ mm yttria-stabilized zirconia balls in isopropyl alcohol medium for 30 minutes.

The synthesized powders were uniaxially pressed at 300 MPa to prepare ceramic samples. The resultant green compacts in the form of disks 15 mm in diameter and 2 mm in thickness were sintered in an SNVE 1.3.1 vacuum furnace with tungsten heaters at a residual pressure of 10^{-3} Pa and temperatures up to 1750°C for 3 h at a heating rate of 10°C/min. After that samples were additionally calcined in air at a temperature of 1300°C for 2 h to eliminate oxygen vacancies (forming at the vacuum sintering stage) and then polished on both sides with diamond slurry.

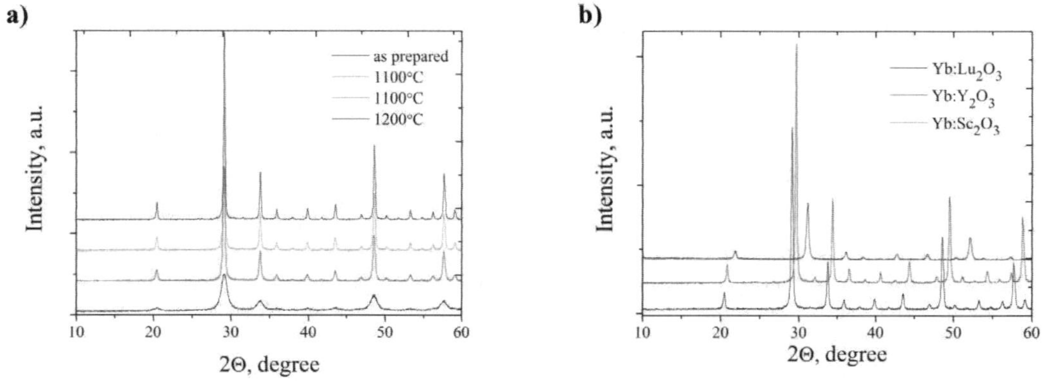

Figure 1. Diffraction patterns of (a) 5%Yb:Y_2O_3 powders as prepared and calcined at different temperatures and (b) 5%Yb:Sc_2O_3, 5%Yb:Y_2O_3 and 5%Yb:Lu_2O_3 powders calcined at 1100°C

3. Results and discussion

3.1. Structural and morphological studies of Yb:RE_2O_3 SHS-powders

Figure 1 shows the results of XRD analysis of the SHS products. In each case the synthesized powders represent respectively yttrium, scandium and lutetium oxide with a cubic C-type crystal lattice (space group Ia3¯, № 206, Z = 16). As can be seen from figure 1a powders after the synthesis are noticeably amorphized, calcination of the powders leads to a decrease in the number of defects and an increase in the crystallite size, which results in the increase of reflections intensity on the X-ray diffraction pattern.

Figure 1b shows X-ray diffraction patterns of 5%Yb:Sc_2O_3, 5%Yb:Y_2O_3 and 5%Yb:Lu_2O_3 powders calcined at 1100 °C. The highest intensity of reflections is observed for lutetium oxide. A possible explanation for this is the higher density of lutetium oxide, which provides a more intense reflection of X-radiation. According to the calculations by Rietveld method, the largest crystallite size (320 nm) was observed for lutetitium oxide, for yttrium oxide and scandium oxide powders the corresponding values amount to 165 and 45 nm.

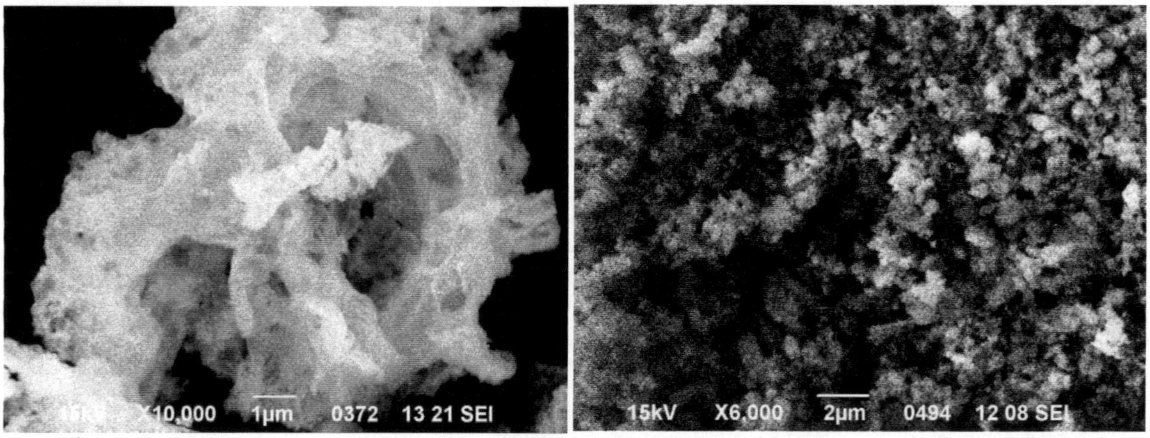

Figure 2. SEM micrograph of the SHS-derived Y_2O_3 nanopowders (a) as-prepared and (b) after attrition milling for 30 minutes

The prepared powders are similar in morphology and consist of loose agglomerates with a porous structure. For example a typical SEM micrograph of SHS-derived Y_2O_3 powder is shown on the figure 2a. The microstructure of the powders is determined by the mechanism of the particle formation upon combustion. The propagation of the reaction front first causes foaming of the precursor, followed by the onset of the combustion reaction, which is accompanied by the evolution of a large amount of gaseous products. As can be seen from figure 2b, during attrition milling large particles are destroyed to submicron particles of rounded shape, which proves a low agglomeration degree.

Vacuum sintering of these high-disperse low-agglomerated powders make it possible to achieve a material with 97-99% density of theoretical value at temperatures even as low as 1650-1700°C, which is ~ 700 °C below the melting points of the corresponding materials. However, obtaining highly transparent materials based on synthesized powders requires the use of sintering additives and special temperature regimes to control the grain growth and removal of residual porosity.

3.2. Luminescent properties of Yb:RE_2O_3 SHS-powders and ceramics

To study the luminescent properties, powders calcined at 1100°C for one hour and ceramic samples sintered in vacuum at 1750 °C and further calcined in air at 1300°C were chosen. Figure 3 presents luminescent spectra of the 5% Yb:Sc_2O_3, 5% Yb:Y_2O_3 and 5% Yb:Lu_2O_3 SHS-powders and ceramics .

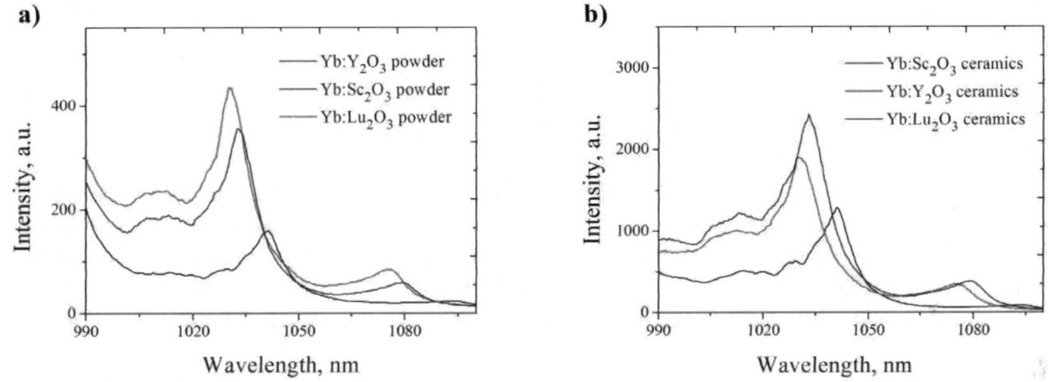

Figure 3. Luminescence spectra of 5%Yb:Sc_2O_3, 5%Yb:Y_2O_3 and 5%Yb:Lu_2O_3 (a) powders and (b) ceramics

The efficiency of the Yb^{3+} ions emission for ceramic samples is much higher compared to respective powders. Nevertheless peak positions and their intensity relations are the same. The calculated characteristics for the most intensive luminescence band (at 1030-1050 nm region) are given in the table 1. The position of the peaks maxima corresponds to the literature values [21], the width of the bands for variety oxides differs insignificantly.

Table 1. The values of the luminescence most intensive peak maxima, full width at half-maximum (FWHM) and the luminescence decay time of the 5%Yb:Sc_2O_3, 5%Yb:Y_2O_3, 5%Yb:Lu_2O_3 powders

Yb^{3+}:RE_2O_3	FWHM, nm	Luminescence peak maximum, nm	τ_1, μs	τ_2 (ceramics), μs
Y_2O_3	17.0	1031.0	58	677 (950)
Lu_2O_3	18.0	1033.6	26	735 (1150)
Sc_2O_3	16.4	1041.0	6	54 (1000)

The kinetics of decay of the ytterbium ions luminescence in Y_2O_3, Lu_2O_3 and Sc_2O_3 powders and ceramics is shown in figure 4. The luminescence decay in powder samples can be divided into two stages: a rapid decay in the beginning with a characteristic time τ_1, corresponding to the excitation energy transfer to closely located acceptors and a much slower decay (τ_2), forming a long tail on the oscillogram and corresponding to energy transfer between distant ions [18]. The luminescence decay time of synthesized powders (see figure 4a) is somewhat lower than the literature values [1, 19], which can be attributed to the imperfection of the crystalline structure and the presence of intermediate products of the starting components interaction, as was previously noted by other authors [20]. This influence is mostly typical for 5%Yb:Sc_2O_3 powders. Apparently, the residual carbon- and hydrogen-containing groups (hydroxyl OH-, carboxylic COO-) do not only remain in the powders after the precursor combustion, but also are not being removed during high-temperature annealing.

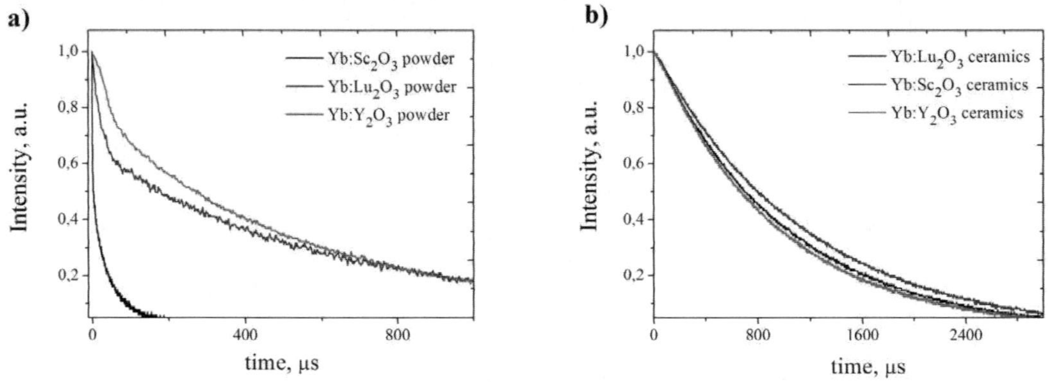

Figure 4. Kinetics of luminescence decay in 5% Yb:Sc_2O_3, 5% Yb:Y_2O_3 and 5% Yb:Lu_2O_3 (a) powders annealed at 1100°C and (b) ceramics sintered at 1750°C

Quite different picture is observed for ceramic samples (figure 4b). The luminescence decay is approximated by one exponent, and the obtained luminescence lifetime values are higher than the literature values for single crystals [21]. Such effect can be explained by two factors. On the one hand, the vacuum sintering conditions at high temperatures facilitate the removal of impurity groups that introduce defects into the crystal lattice of rare-earth element oxides and increasing the crystalline structure ordering.

When the samples are illuminated by laser radiation at a wavelength of 975 nm, green glow is observed in all the studied samples, and the relative intensity of this glow for ceramics is noticeably higher than for powders. Analysis of the anti-Stokes luminescence spectra of the studied samples in the optical range shows the presence of luminescence bands typical for Er^{3+} ions. The luminescence spectra of 5% Yb:Lu_2O_3 ceramic and powder samples are presented in figure 5, which clearly shows the increase in the intensity of green luminescence in ceramics compared to the powder. The presence of Er^{3+} impurity ions can lead not only to the appearance of anti-Stokes luminescence bands, but also influence the kinetics of the Yb^{3+} ions IR-luminescence. In this case the Yb^{3+} ions act as sensitizers, and Er^{3+} act as activators in the process of inter-ion excitation transfer resulting in green luminescence. Since $^4I_{11/2}$ in Er^{3+} and $^2F_{5/2}$ in Yb^{3+} levels are very close, it will not be mistakenly to assume that the energy transfer coefficients from the sensitizer to the activator and back for the singly excited states are the same. Then, according to [22], the IR-luminescence of both ion types will decay with the same time constant:

$$\frac{1}{\bar{\tau}} = \frac{\tau_S^{-1} C_S + \tau_1^{-1} C_A}{C_S + C_A},$$

where C_S and C_A are the total ion concentrations of the sensitizer and activator, τ_S^{-1} and τ_1^{-1} are the probabilities of radiative transitions of the sensitizer and activator ions from singly excited states. For Er^{3+} the radiative lifetime of $^4I_{11/2}$ level is about 8-9 ms and for $^2F_{5/2}$ level of Yb^{3+} is within 0.5-0.9 ms. Therefore, the presence of Er^{3+} ions in powders and ceramics would increase the observed luminescence decay time according to the formula $\tau_{lum.dec.}^{-1} = \frac{1}{\bar{\tau}} + W_{n/r}$, ($W_{n/r}$ - non-radiative transition probability) which we observe in the experiment.

The difference in the intensities of anti-Stokes luminescence in powders and ceramics can be explained by the fact that on the one hand the crystalline structure of the powders is strongly disordered, and on the other hand, the average interionic distance in ceramics is somewhat smaller, and since the probability of the interionic interaction is proportional to $1/R^6$, the effect of increasing anti-Stokes luminescence in ceramics can be significant. A detailed study of the proposed mechanisms of the influence of rare-earth element impurities on the IR-luminescence kinetics will be further considered.

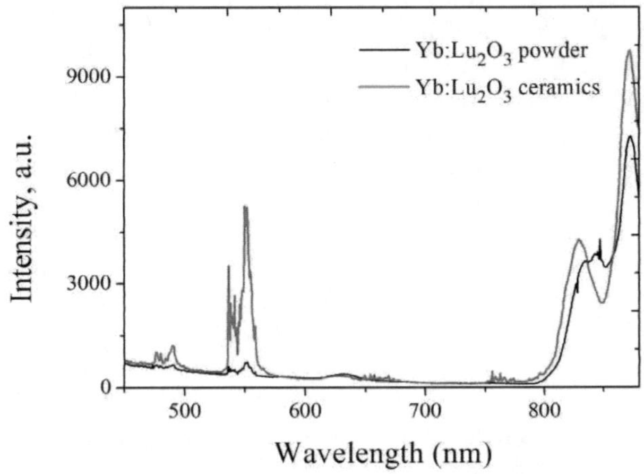

Figure 5. Anti-Stokes luminescence spectra of 5% Yb: Lu_2O_3 ceramic and powder samples

4. Conclusion

In this work, the self-propagating high temperature synthesis of ytterbium doped rare earth (Y, Sc, Lu) oxides powders and comparative study of their structural and luminescent properties are reported. The results show that particles of the SHS-derived powders exhibit cubic crystalline structure and combined in loose agglomerates which can be crushed down to submicron size via attrition milling.

Crystal structure defects in the Yb^{3+}:RE_2O_3 powders are highly affected on the emission efficiency and luminescence lifetime. This influence disappears during high-temperature sintering of the obtained powders into dense ceramics. In this connection it can be concluded, that the developed SHS techniques for the synthesis of rare earth oxide powders are promising for the producing of new optical ceramic materials of visible and near-IR ranges.

Acknowledgements

This study was financially supported by the Russian Foundation for Basic Research within the framework of the scientific project (№ 16-33-60153mol_a_dk)

References

[1] Lupei V, Lupei A and Ikesue A 2005 Transparent Nd and (Nd, Yb)-doped Sc_2O_3 ceramics as potential new laser materials *Appl. Phys. Lett.* **86** 111 – 18
[2] Sanghera J *et al* 2011 10% Yb^{3+}-Lu_2O_3 Ceramic Laser with 74% Efficiency *Opt. Lett.* **36 (4)** 576 – 78
[3] Snetkov I L, Mukhin I B, Balabanov S S, Permin D A and Palashov O V 2015 Efficient lasing in Yb:$(YLa)_2O_3$ceramics *Quant. Electronics* **45 (2)** 95
[4] Tokurakawa M, Shirakawa A, Ueda K, Yagi H, Yanagitani T, Kaminskii A, Beil K, Kränkel C

and Huber G 2012 Continuous wave and mode-locked $Yb^{3+}:Y_2O_3$ ceramic thin disk laser *Opt. Express* **20 (9)** 10847

[5] Li J G, Ikegami T and Mori T 2005 Fabrication of transparent, sintered Sc_2O_3 ceramics *J. Am. Ceram. Soc.* **88** 817–21

[6] Li J G, Ikegami T, Mori T and Yajima Y 2003 Monodispersed Sc_2O_3 precursor particles via homogeneous precipitation: synthesis, thermal decomposition, and the effects of supporting anions on powder properties *J. of Materials Research* **18** 1149–56

[7] Liu W, Kou H, Li J G, Jiang B and Pan Y 2015 Transparent Yb: $(Lu_xSc_{1-x})_2O_3$ ceramics sintered from carbonate co-precipitated powders *Cer. International* **41** 6335–39

[8] Nengli W 2010 Synthesis of La^{3+} and Nd^{3+} co-doped yttria nanopowder for transparent ceramics by oxalate precipitation method *J. of Rare Earths* **28 (2)** 232

[9] Khasanov O, Osipov V, Dvilis E, Kachaev A, Khasanov A and Shitov V 2011 Nanoscaled grain boundaries and pores, microstructure and mechanical properties of translucent Yb:$[Lu_xY_{(1-x)}O_3]$ ceramics *J. of Alloys and Compounds* **509 (1)** S338 – 42

[10] Ivanov M, Kalinina E, Kopylov Yu, Kravchenko V, Krutikova I, Kynast U, Li J G, Leznina M and Medvedev A 2016 Highly transparent Yb-doped (LaxY1 x)2O3 ceramics prepared through colloidal methods of nanoparticles compaction *J. of the European Ceram. Soc.* **36** 4251–59

[11] Baker C, Kim W, Sanghera J, Goswami R, Villalobos G, Sadowski B and Aggarwal I 2012 Flame spray synthesis of Lu_2O_3 nanoparticles *Materials Lett.* **66 (1)** 132–34

[12] Balabanov S S, Yavetskiy R P, Belyaev A V, Gavrishchuk E M, Drobotenko V V, Evdokimov I I, Novikova A V, Palashov O V, Permin D A and Pimenov V G 2015 Fabrication of transparent $MgAl_2O_4$ ceramics by hot-pressing of sol-gel-derived nanopowders *Ceram. International* **41 (10)** 13366–71

[13] Linsheng W, Suqin L, Kelong H, Xinyu Y, Youming Y and Zhi Z 2011 Characterization and photoluminescence of Lu_2O_3-Eu^{3+} nano-phosphor prepared by modified solution combustion method *J. of Rare Earths* **29 (11)** 1049 – 52

[14] Zou X, Yi H, Zhou G, Chen S, Yang Y and Wang S 2011 Highly Transmitting ZrO_2-Doped Lu_2O_3 Ceramics from Combustion Synthesized Powders *J. Am. Ceram. Soc.* **94 (9)** 2772 – 74

[15] Balabanov S S, Gavrishchuk E M, Kut'in A M and Permin D A 2011 Self-propagating high-temperature synthesis of Y_2O_3 powders from $Y(NO_3)_{3x}(CH_3COO)_{3(1-x)}$•$nH_2O$ *Inorg. Mat.* **47 (5)** 484 – 88

[16] Chen W *et al* 2014 Luminescence of $La_{0.2}Y_{1.8}O_3$ nanostructured scintillators *Opt. Lett.* **39 (19)** 5705 – 08

[17] Permin D A, Novikova A V, Gavrishchuk E M, Balabanov S S and Sorokin A A 2017 Self-propagating high-temperature synthesis of Lu_2O_3 powders for optical ceramics *Inorg. Mat.* **12** 1330 – 35.

[18] Zhang H, Yang Q, Lu S and Shi Z 2012 Structural and spectroscopic characterization of Yb^{3+} doped Lu_2O_3 transparent ceramics *Opt. Materials* 969 – 72

[19] Osipov V V, Solomonov V I, Spirina A V, Vovkotrub E G and Strekalovskii V N 2014 Raman scattering and luminescence of yttria nanopowders and ceramics *Optics and Spectroscopy* **116 (6)** 946 – 55

[20] Zych E 2001 On the reasons for low luminescence efficiency in combustion-made Lu_2O_3:Tb *Opt. Materials* **16 (4)** 445 – 52

[21] Peters V 2001 *Growth and spectroscopy of Ytterbium-doped sesquioxides* (Hamburg: University of Hamburg, Shaker)

[22] Kazaryan K A, Timofeev Yu P and Fok M V 1986 *Antistoksovo preobrazovanie izlucheniya v luminoforah s redkozemelnymi ionami* (Tr. FIAN) **175** *(in russian)*

IOP Conf. Series: Materials Science and Engineering **347** (2018) 012009 doi:10.1088/1757-899X/347/1/012009

3D printing biodegradable scaffolds with chitosan materials for tissue engineering

K N Bardakova[1, 2, 4], T S Demina[3], E A Grebenik[2], N V Minaev[1], T A Akopova[3], V N Bagratashvili[1] and P S Timashev[1, 2]

[1] Institute of Photonic Technologies, Federal Research Centre "Crystallography and Photonics," Russian Academy of Sciences, 2 Pionerskaya Str., Moscow 142190, Troitsk, Russia
[2] Institute for Regenerative Medicine, Sechenov First Moscow State Medical University, 8-2 Trubetskaya st., Moscow, 119991, Russia
[3] Enikolopov Institute of Synthetic Polymer Materials, Russian Academy of Sciences, 70 Profsoyuznaya Str., Moscow 117393, Russia

E-mail: arie5@yandex.ru

Abstract. Chitosan-g-oligo (L,L-lactide) copolymer was synthesized through a solvent-free reaction in an extruder. Three-dimensional scaffolds based on photosensitive composition contained the synthetized copolymer were formed by two-photon polymerization. The optimum ratio of components, methods of preparation of photopolymerizable mixtures, parameters of the laser structuring and procedure of washing from unbound crosslinkers have been optimized. Chitosan scaffolds were non-cytotoxic and might therefore be a suitable candidate for treating spinal cord injuries and other neuronal degenerative diseases.

1. Introduction

One of the principal socially important tasks of regenerative medicine is to develop methods for correcting violations of the integrity of the pathways of the spinal cord caused by exchange degenerative, traumatic, inflammatory and other processes, followed by reduction of its functional activity [1]. In recent decades, special carriers of cells or scaffolds are actively being developed to solve the problem of stimulation the reparation processes in the body.

The choice of material for scaffold fabrication is of primary importance. Its main property is a compatibility with the spinal cord environment. The classification of materials for spinal cord is enormous - it can be various natural and synthetic materials, hydrogels, various functionalized materials (surface charged, drug-delivery) [2]. Chitosan is widely used for preparing scaffolds for regenerative medicine due to its ability to bioresorption and biocompatibility. There are known works in which lyophilized chitosan scaffolds were successfully tested as candidates for spinal cord regeneration [3] It should be noted that such a technology scaffolds fabrication - freeze drying - has one significant drawback: an insufficient rigidity of the scaffold obtained and the impossibility to tailor mechanical characteristics. As is known, depending on the hardness of the material, the phenotype of mesenchymal stem cells changes toward a certain lineage [4]. Also, an additional fixation of the spine is required in a surgical operation using such a scaffold [5].

In the present work, we examine chemically modified photosensitive chitosan-based materials for their suitability for microstructuring via two-photon polymerization (2PP) and fabrication of scaffolds for spinal cord regeneration. The natural polymer was used in combination with a biodegradable synthetic polymer - polyethylene glycol diacrylate. This approach allows to adjust the physico-chemical properties of the scaffold (e.g., mechanical properties, biodegradation rate) and could enhance the 2PP processing time. Also, 2PP has numerous advantages including the ability to produce structures with well-defined, user-controlled geometries, structure with high—yet scalable—resolution, and application of a variety of biocompatible materials [6].

2. Materials and methods

2.1. Materials

Chitosan-g-oligo (L,L-lactide) copolymer was prepared by solid-state synthesis in a semi-industrial corotating twin-screw extruder (Berstorff ZE40, Germany) by reactive blending of solid powders of chitosan and oligo (L, L-lactide) at 55 °C according to the protocol of Demina et al. [7].

The resulting graft copolymer of chitosan and oligo (L, L-lactide) was used as the main component of a photosensitive composition for forming hydrogel scaffolds. The synthesized chitosan-co-oligo (L, L-lactide) was dissolved in 3 vol.% acetic acid for 12 hours with constant stirring at room temperature to obtain 4.9 wt.% copolymer solution. The insoluble fraction was separated by centrifugation (40 minutes, 9000 rpm), after which the solution was filtered through a membrane with a pore size of 0.45 μm. Polyethylene glycol diacrylate (PEG-DA, M2000, Sigma-Aldrich, Germany) was further added to the copolymer solution as an additional cross-linking agent and a biocompatible Irgacure 2959 (BASF Kaisten AG, Germany) as a photoinitiator (see Figure 1). Stirring of the photosensitive system was carried out for 20 hours at 35°C until a homogeneous solution was obtained. The ratio of components in the final photosensitive composition was 5/5/1 for chitosan- g -oligo (L,L-lactide), PEG-DA and Irgacure 2959, respectively. The storage time of the photosensitive composition is 3 days.

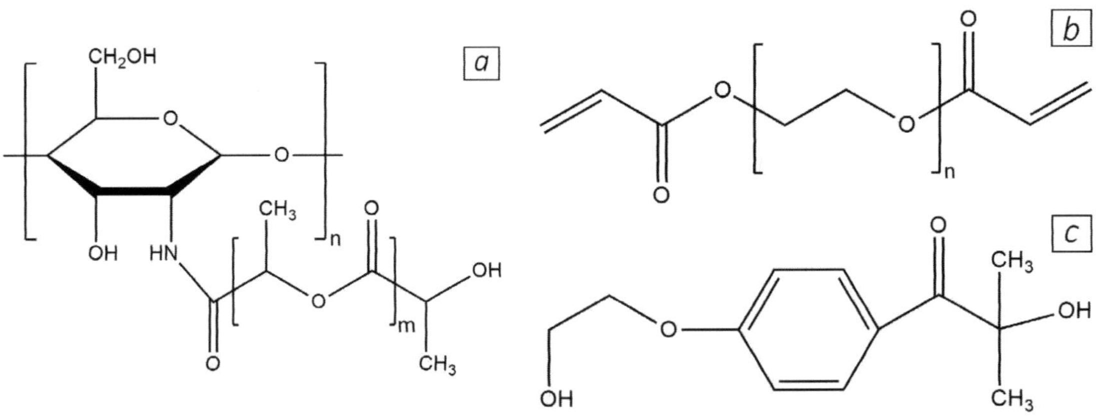

Figure 1. Chemical structure of chitosan-g-oligo (L,L -lactide) copolymer unit (a), polyethylene glycol diacrylate (b) and Irgacure 2959 (c).

2.2. UV-Vis spectroscopy

UV-Vis spectroscopy of 5 % solution of chitosan-g-oligo (L,L-lactide) copolymer in distilled water and this solution with the addition of 5 % PEG-DA were recorded in the 200-500 nm region on a Cary 50 spectrophotometer. The measurements were performed in quartz cuvettes with an optical path length of 1 mm.

2.3. Scaffold fabrication and characterization

Two-photon-induced microstereolithography method was used to form three-dimensional scaffolds from a photosensitive composition based on chitosan-g-oligo(L,L-lactide) copolymer [8,9]. This approach allows the creation of a three-dimensional model with micron resolution using a targeted amount of photosensitive composition without the need to apply separate layers. The photosensitive composition was placed between two glasses and was limited to a silicone spacer to form three-dimensional scaffolds according to a predetermined three-dimensional model. The process of photopolymerization was carried out under femtosecond laser radiation TEMA-100 (Avesta-project, Russia). Three-dimensional scaffold was formed directly in the bulk of the material. Irradiation was carried out with the use of microscopic planar lens 4X with a small number aperture (n.a. 0.1), which allowed to carry out photocuring process in workspace diameter of 4mm. The volume of the sample was layer-by-layer filled with an array of polymerized "lines" with a small overlapping. The average laser radiation power was 100 mW, the velocity of the laser beam moving along the sample volume was 1 m/s. The scaffold was washed from the non-crosslinked photosensitive composition by cyclic washing with deionized water for 2 hours.

For cell viability experiments the composition based on chitosan-*g*-oligo(L,L-lactide) copolymer was photocured to form films. The photosensitive composition (100 μl) was placed in a well of an aluminum spacer, and the top and bottom were limited with a release film based on polytetrafluoroethylene. As a source of ultraviolet radiation, a force matrix assembly of ultraviolet diodes with a wavelength of 365 nm (Epileds) was used. It consists of 50 separate diodes, which are placed on a single substrate on a powerful radiator. The total electrical power of the diode assembly was 50 W. Irradiation was performed on both sides for 15 minutes.

2.4. Biocompatibility evaluation

Dissociated rat cortical neurons were cultured on the film surface in the Neurobasal medium (Gibco) supplemented with 2% B27 (v/v) and 0.25% GlutaMAX (v/v) in a humidified incubator in an atmosphere of 5% carbon dioxide at 37°C. After 10 days in culture, the cells were loaded with 2 μM Calcein AM fluorescent probe and then rinsed twice with PBS. The fluorescence signal was accessed using a fluorescence microscope (Olympus, Japan) in order to visualize living cells.

3. Results

3.1. 3D microfabrication

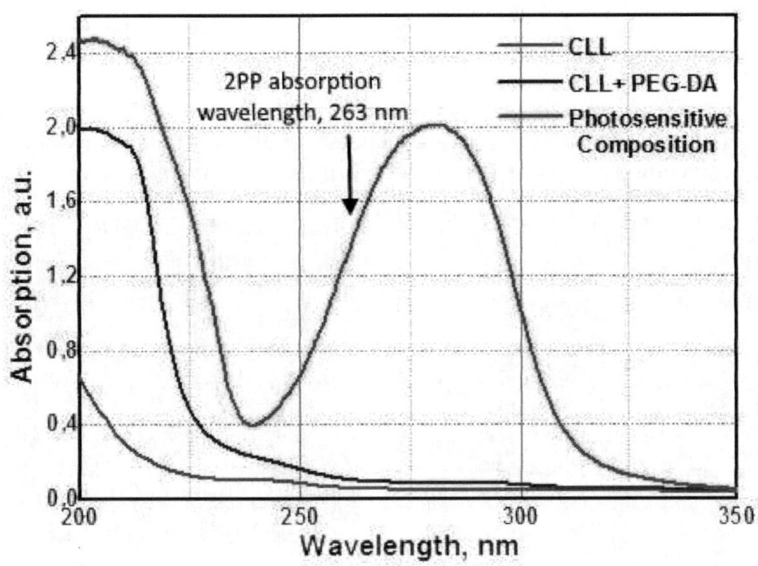

Figure 2. Electronic absorption spectra of 5 wt % solution of chitosan-*g*-oligo(L,L-lactide) copolymer (CLL) in distilled water , CLL with PEG-DA (5 wt %) and the final photosensitive composition.

In figure 2 shows the absorption spectra of chitosan-g-oligo (L, L-lactide) copolymer, a mixture of a copolymer with PEG-DA and a photosensitive composition with an introduced photoinitiator. Chitosan-g-oligo (L, L-lactide) copolymer and its mixture with PEG-DA show no absorption in the UV region at 263 nm (2PP absorption wavelength). This allows us to use them for fabrication copolymer-based hydrogels by two-photon-induced microstereolithography method.

We developed a three-dimensional model for treating spinal cord injuries. It is a truncated cylinder with a length of 1000 μm and a height of 600 μm. There is an array of through holes 30 μm in diameter on the convex side of the scaffold. Each structure required 40 min for fabrication. Then, the scaffolds were washed of unreacted components by cyclic immersion in aqueous medium at various pH. The final scaffolds possessed dimensions increased in 1.8-2 times compared to the pre-set ones due to swelling during the washing process. It should also be noted that there was a slight bending of the obtained scaffold in comparison with the given model that could be caused by the heterogeneity of the cross-linking of the material and the geometric dimensions of the scaffold, as well as the resulting internal stresses associated with the geometry. The swelling behaviour of the hydrogels was preliminary adjusted using the amphiphilic graft copolymers as well as addition of proper amount of synthetic component, i.e. PEG-DA. As a result, we got a well-structured 3D hydrogel with good handling properties, which was of great importance for future in vitro and in vivo experiments.

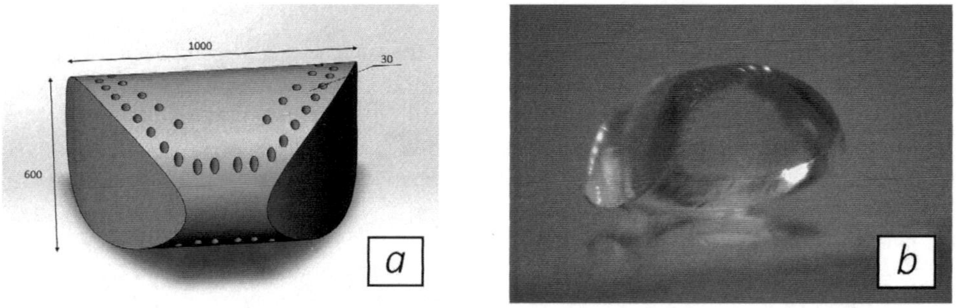

Figure 3. Three-dimensional model (a), and a scaffold fabricated by two-photon-induced microstereolithography (b).

3.2. Biocompatibility evaluation
It is shown that the films provide a high survival rate of cortical neurons and the formation of neural networks, and thus could be considered biocompatible and suitable for neuroregeneration. The image of Calcein AM-stained cells cultured on the surface of the films is shown in figure 4.

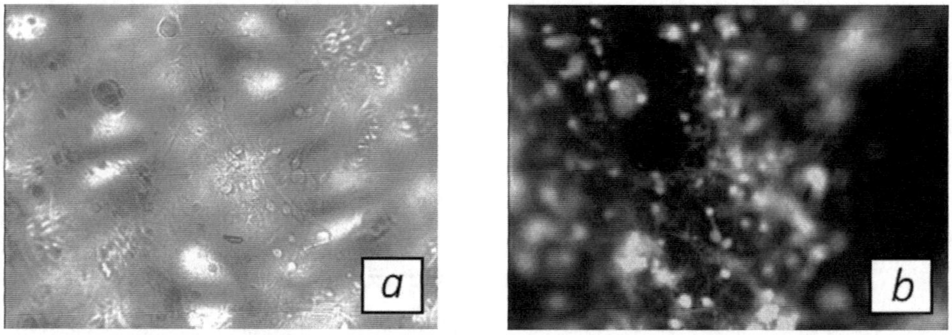

Figure 4. Rat cortical neurons on the polymeric film surface after 10 days in culture: Brigthfield (a) and Calcein AM-assisted fluorescence microscopy (b). Magnification, 40×.

4. Conclusion
In this work, the two-photon-induced microstereolithography method has been used to prepare three-dimensional chitosan scaffolds. Scaffolds produced were shown to be neurocompatible and might

therefore be a suitable candidate for treating spinal cord injuries and other neuronal degenerative diseases.

Acknowledgments
This work was partially supported by Russian Scientific Foundation grant 17-15-01487 (in a part of biocompatibility evaluation of the scaffolds) and by the Russian Foundation for Basic Research grant 16-02-00248-a (in a part of the polymer synthesis and scaffold design).

References
[1] Bickenbach J, Officer A, Shakespeare T and Groote P 2013 *International perspectives on spinal cord injury* (Geneva: World Health Organization)
[2] Kohane D S and Langer R 2008 *Pediatric Res.* **63** 487-91
[3] Zhang J et al. 2016 *Cell Tissue Res.* **366** 129-42
[4] Engler A J et al. 2006 *Cell* **126** 677-89
[5] Rooney G E et al. 2008 *Spine* **33** E914
[6] Koroleva A et al. 2012 *Biofabrication* **4** 025005
[7] Demina T S et al. 2017 *Polymers* ***9*** 302
[8] Timashev P S et al. 2016 *Appl. Biochem. Microbiol.* **52** 508-14
[9] Kufelt O et al. 2014 *Biomacromolecules* **15** 650-9

IOP Conf. Series: Materials Science and Engineering **347** (2018) 012010 doi:10.1088/1757-899X/347/1/012010

Synthesis of copper oxides films via anodic oxidation of copper foil followed by thermal reduction

D S Zimbovskiy, A I Gavrilov and B R Churagulov

Lomonosov Moscow State University, Moscow, 119991, Russia

E-mail: ZimbovskiyDS@gmail.com

Abstract. Copper oxides films were successfully synthesized via anodic oxidation of copper foil at various conditions followed by thermal reduction of oxidized samples in inert nitrogen atmosphere. The morphology and phase composition of final and intermediate products were determined by scanning electron microscopy, X-ray diffraction and Raman spectroscopy techniques. Optimal synthetic conditions of Cu_2O films on copper foil surface were established.

1. Introduction

Quick growth of industrial activity in last decades compels manufactures to pay great attention to make a new way of energy supply with the decreasing of environment pollutants [1, 2]. Hydrogen generation via water splitting under sunlight on the surface of semiconductor photoelectrodes is one of the most promising way because it lets to produce fuel with the minimal environmental impact [3, 4]. Photo-electrochemical water splitting process often carries out in photo-electrochemical cells (PEC) which in most cases consist of three electrodes: working (semiconductor), counter (more often is platinum), and reference (silver chloride, saturated calomel etc) electrodes. During water decomposition process, semiconductor surface is illuminated and as a result electrons and holes are generated. Charge carriers induce oxidation/reduction reactions to produce oxygen and hydrogen (equations 1 – 3) [5, 6]:

$$\text{Oxidation: } H_2O + 2h^+ \rightarrow 2H^+ + {}^1\!/_2\, O_2 \quad (1)$$

$$\text{Reduction: } 2H^+ + 2\bar{e} \rightarrow H_2 \quad (2)$$

$$\text{The overall reaction is: } H_2O \xrightarrow{h\nu} H_2 + {}^1\!/_2\, O_2 \quad (3)$$

Among various semiconductor materials for PEC photoelectrodes, cuprous oxide (Cu_2O) is one of the most attractive materials because of its cheapness, environment safety, good value of band gap (2,0 – 2,2 eV) which provide absorbance of considerable part of solar spectrum [7-9]. In order to provide all necessary requirements imposed for semiconductor materials for PEC applications, a variety of Cu_2O structures (wires, cubes, polyhedral, flower-like structure, etc.) [10-13] and copper oxides composites have been obtained by different techniques, such as hydrothermal, microwave-assisted hydrothermal, electrodeposition, chemical oxidation with thermal reduction, sputtering, etc. [9, 14-17].

Anodic oxidation is a surface treatment low-cost method, which can provide synthesis of different nanostructured materials on metals' surface. By this method shape, morphology and size of structures can be modulated by change of concentration, temperature, pH, etc. [18, 19]

In this work Cu_2O films on copper foil surface (ready effective photoelectrode for water splitting process) have been successfully synthesized via simple and cheap two step method. The first step was a galvanostatic anodic oxidation of copper foil in alkaline solution over a range of current densities in order to obtain $Cu(OH)_2$ wires. The second step was the thermal reduction of wires under inert atmosphere (N_2), as a result Cu_2O film covers up surface of copper foil.

2. Experimental

2.1. Materials

All chemical reagents used are analytical grade (Copper foil (Cu, 99.99% purity, 20x20x0.5 mm), sodium hydroxide (NaOH, Pure)). All solutions used in this research were prepared with deionized water.

2.2. Anodic oxidation followed by thermal reduction

Cuprous oxide films were fabricated via anodic oxidation of copper foil followed by thermal reduction of oxidized sample as shown in figure 1.

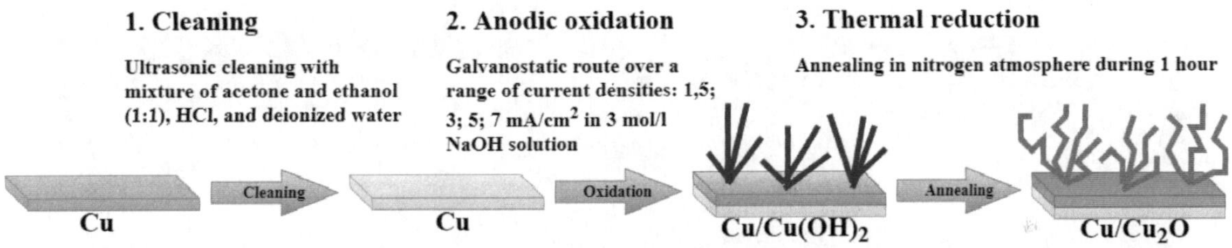

Figure 1. The synthesis scheme of Cu_2O films.

$Cu(OH)_2$ wires have been obtained according to methods, described in [9, 19]. Before the anodization process, a square copper plates had been ultrasonically cleaned in mixture of acetone and ethanol (1:1), HCl (10%) and deionized water. The anodic oxidation was performed with two-electrode cell (anode – copper plate, cathode – stainless steel net) in 3 mol/l NaOH solution over a range of current densities: 1,5; 3; 5; 7 mA/cm^2 with and without stirring for 30 or 10 minutes (two last samples). The obtained samples were extracted from the solution, rinsed with deionized water and ethanol, and dried in air. Thermal reduction was carried out for samples, obtained by anodization at 5 mA/cm^2 for 10 minutes without any stirring. Annealing was performed in tube furnace (Naberthem R 50/250/12) at various temperatures (200 – 600°C) in inert atmosphere (N_2) for 1 hour.

2.3. Characterization

Morphology of fabricated films was examined by scanning electron microscopy (SEM) with scanning electron microscope Supra 50 VP (LEO) operated at 5 – 20 kV. X-ray diffraction (XRD) patterns were obtained using X-ray diffractometer Rigaku D/MAX 2500, with $Cu_{K\alpha}$ radiation (λ = 1,54178Å) in the range of 2θ from 10° to 80°. The Raman spectra were recorded by Renishaw InVia system equipped with an Ar laser (514 nm, 20 mW, 10% of max power).

3. Results and discussion

3.1. SEM characterization

SEM images of samples, which were obtained by anodic oxidation at different conditions, are presented on figure 2.

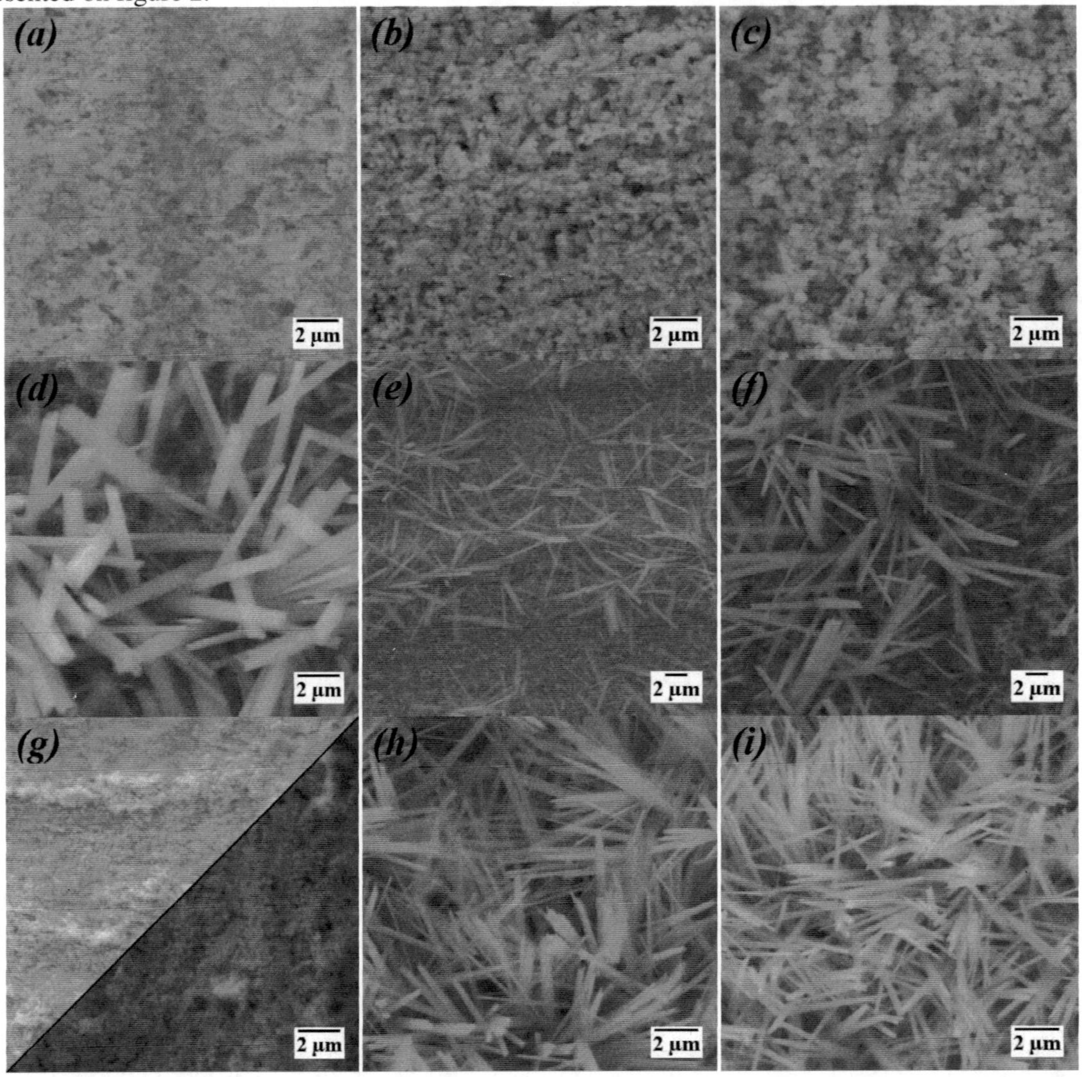

Figure 2. SEM images of copper plates after anodization at: (*a*) 1,5, (*b*) 3, (*c*) 5, (*d*) 7 mA/cm^2 with solution stirring during 30 minutes; (*e*) 1,5, (*f*) 3, (*g*) 5 (top) and 7 (bottom) mA/cm^2 without stirring during 30 minutes; (*h*) 5 mA/cm^2 during 10 minutes; (*i*) 7 mA/cm^2 during 5 minutes.

Anodic oxidation of copper foil leads to the formation of films on the substrates surfaces with different morphology. Anodization at 1,5 – 5 mA/cm^2 during 30 minutes with solution stirring (figure 2 *a* – *c*) follows the formation of dense films, while increasing of current density up to 7 mA/cm^2 (figure 2 *d*) leads to the growth of wires on the copper foil surface with a length (L) of 5 – 7 μm and thick (D) of about 600 – 900 nm.

Without stirring wires on the surface of copper foil start to grow even at 1,5 mA/cm^2 (figure 2 *e*) and the larger a value of current density is, the more wires form on the surface. Therefore, convection currents, which take place during solution stirring, allow to prevent wires growth on copper foil surface. When the current density reach the values 5 and 7 mA/cm^2 (30 minutes) the film tears off from substrate and a thin layer of anodization products stays on surface (figure 2 *g*). To protect the

IOP Conf. Series: Materials Science and Engineering **347** (2018) 012010 doi:10.1088/1757-899X/347/1/012010

films from separation at higher current densities the time of process was decreases to 10 and 5 minutes respectively. By applying these synthesis parameters, films with the largest density of wires on the surface have been obtained (figure 2 *h*, *i*). Wires with L = 3 – 5 μm and D = 200 – 400 nm have been formatted at current densities 1.5, 3 mA/cm^2 and 5 mA/cm^2 during 30 minutes anodization process. Increase of current density to 7 mA/cm^2 led to formation of wires reaching the length L ~ 9 μm and the same values of thickness.

SEM images of samples, which have been fabricated by thermal treatment of oxidized copper foil (5 mA/cm^2 10 minutes) at various temperatures of 200 – 600°C in nitrogen atmosphere during 1 hour, are displayed in figure 3.

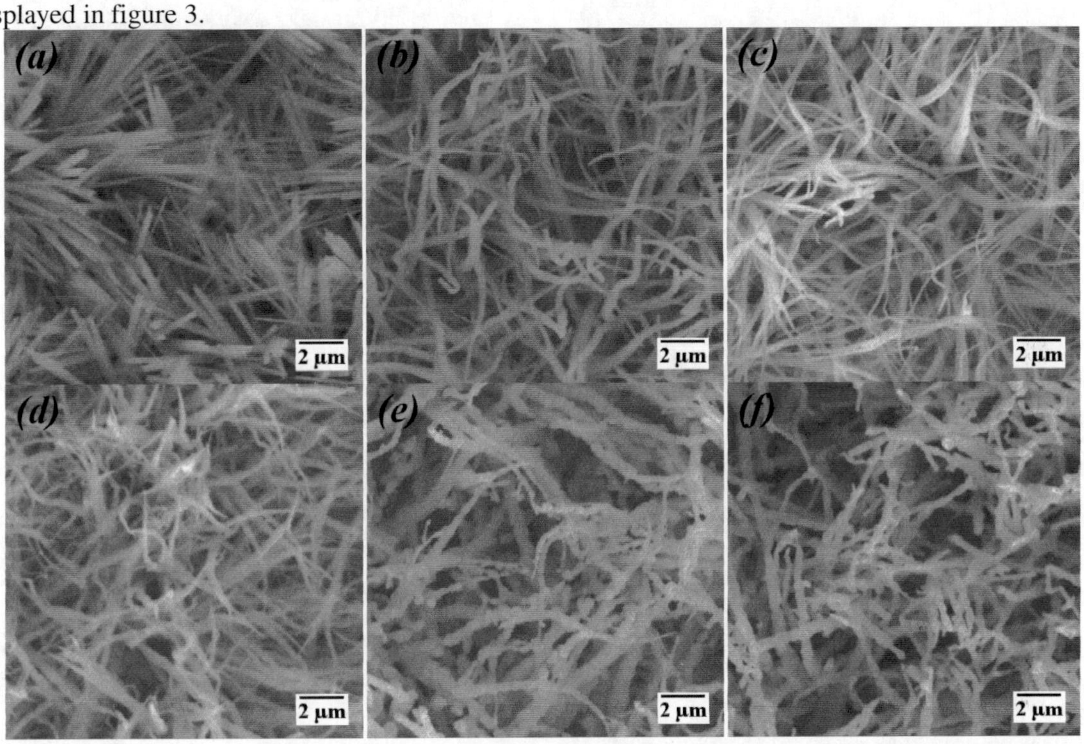

Figure 3. SEM images of samples (*a*) after anodization 5 mA/cm^2 during 10 minutes (source sample); after annealing of source sample during 1 hour at (*b*) 200°C, (*c*) 300°C, (*d*) 400°C, (*e*) 500°C, (*f*) 600°C.

All wires have been curved after thermal treatment at temperatures 200 – 600°C in nitrogen atmosphere during 1 hour. Annealing at temperatures from 200 to 400°C leads to bending of source wires, but the dimensions of wires still remain the same (figure 3 *b* – *d*). At the temperatures 500 and 600°C apparent change in the shape of wires to coral-like structures can be observed (figure 3 *e*, *f*), moreover length and thick are changed (L = 5 – 7 μm, D = 300 – 500 nm).

3.2. XRD characterization

The phase composition of the samples were determined by using of XRD method, as shown in Figure 4. The XRD pattern of copper plate is also displayed for comparison.

IOP Conf. Series: Materials Science and Engineering **347** (2018) 012010 doi:10.1088/1757-899X/347/1/012010

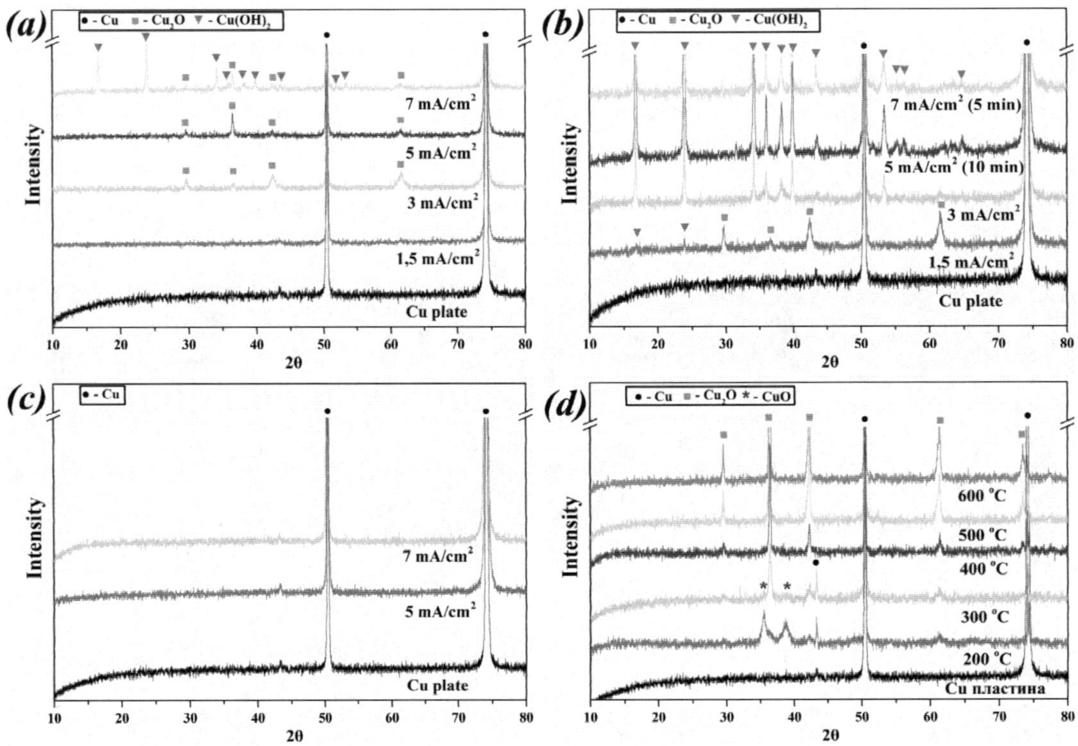

Figure 4. X-ray diffraction patterns of oxidized samples (*a* – *c*) and after thermal reduction (*d*).

The XRD pattern of the samples, obtained after anodic oxidation at 3 and 5 mA/cm^2 under solution stirring (figure 4 *a*), shows that all of the peaks (except peaks, which belong to copper substrate) can be readily indexed to cubic-phase Cu_2O (JCPDS card No. 78-2076). If current density during anodic oxidation process achieves value 7 mA/cm^2, then the diffraction peaks indexed to orthorhombic-phase $Cu(OH)_2$ (JCPDS card No. 72-140) appear along with the same peaks, which belong to Cu_2O and Cu substrate.

Anodization without stirring (figure 4 *b*) leads to formation of $Cu(OH)_2$ phase from 1,5 mA/cm^2, this reveals that convection currents, during solution stirring, prevent from appearing $Cu(OH)_2$ phase on substrate's surface. XRD pattern of the sample, which have been fabricated at 1,5 mA/cm^2 contains peaks, which can be attributed to Cu_2O, Cu and $Cu(OH)_2$ phases, and the larger a value of current density is, the higher intensity of peaks is which belong to $Cu(OH)_2$, in addition with growth of current density diffraction peaks of Cu_2O are disappear. But process of anodic oxidation are limited by tear off of film from substrate surface: anodization at 5 and 7 mA/cm^2 during 30 minute lead to tear off of $Cu(OH)_2$ film from surface and layer of roentgenamorphous product are fabricated on this place (figure 4 *c*). To prevent $Cu(OH)_2$-foil from destructing during oxidation at 5 and 7 mA/cm^2, the time of process was reduce to 10 and 5 minutes respectively and in this case XRD patterns include diffraction peaks of $Cu(OH)_2$ phase only, moreover peaks of $Cu(OH)_2$ phase for sample, which have been obtained at 5 mA/cm^2 during 10 minutes, have higher intensity compared to others samples, which have been obtained in another conditions. Decrease of intensity of peaks for sample which have been obtained by anodic oxidation at 7 mA/cm^2 during 5 minutes may be due to processes of tearing off foil from substrate's surface.

Transformation from Cu_2O to $Cu(OH)_2$ during anodization of copper foil, according to [20], is caused by formation of Cu_2O phase at the initial time. Increase of current density leads to formation $Cu(OH)_2$ phase on the surface preferably.

IOP Conf. Series: Materials Science and Engineering **347** (2018) 012010 doi:10.1088/1757-899X/347/1/012010

Thermal reduction of oxidized samples leads to disappearing of diffraction peaks of $Cu(OH)_2$ phase and appearing of peaks, which belong to Cu_2O and CuO phases that indicate the transformation of $Cu(OH)_2$ through thermal reduction in nitrogen atmosphere was completed (figure 4 *d*). This transformation according to [9] is caused by processes of dehydrate and deoxidization of source sample. XRD patterns for samples, after annealing at 200 and 300°C, have diffraction peaks from CuO phase at 35,56° ($\bar{1}11$) and 38,74° (111) (JCPDS card No. 89-5895) and with the increasing of the temperature of thermal treatment, this diffraction peaks reduce and disappear when temperature of annealing reach a value 400°C. At the same time the intensities of peaks, which can be attributed to Cu_2O phase, start to grow with increasing of temperature from 300°C and reach the high value at 600°C.

All this variations in structures correlate with the finding of SEM, and also confirm the formation of curl-like and coral-like Cu_2O wires.

3.3. Raman characterization

The transformation of the films and the effect of thermal treatment were also studied by Raman spectroscopy (figure 5).

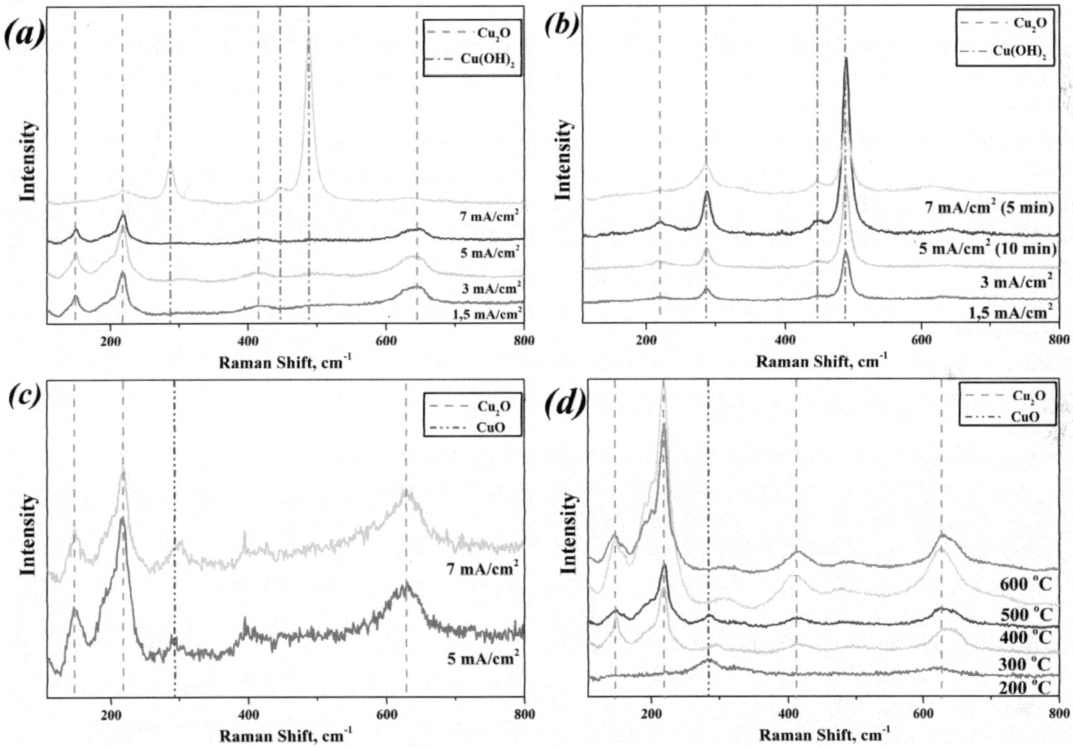

Figure 5. Raman spectra of oxidized samples (*a* – *c*) and after thermal reduction (*d*).

Spectra of samples, which have been obtained via anodic oxidation with solution stirring, include phonon modes of $Cu(OH)_2$ and Cu_2O (figure 5 *a*). Two peaks at 150 cm^{-1} can be attributed to IR active modes $\Gamma_{15}^{-(1)}$ and $\Gamma_{15}^{-(2)}$ accordingly. The peaks at 218 cm^{-1} originated from the second-order Raman-allowed mode of the Cu_2O crystals Γ_{25}^{+}. The peak at 416 cm^{-1} is assigned to four-phonon mode $3\Gamma^{-12} + \Gamma^{-25}$ [9, 21-23]. With increasing of current density, the strongest peak of 218 cm^{-1} from Cu_2O became smaller. For sample, which have been fabricated after anodic oxidation at 7 mA/cm^2, peaks 286, 445 and 488 cm^{-1}, which can be attributed to $Cu(OH)_2$, appear [23, 24].

In spectra from sample, which have been fabricated by anodic oxidation without solution stirring (figure 5 *b*), peaks of $Cu(OH)_2$ phonon modes appear at current density from 1,5 mA/cm^2 and achieves the maximum value of the intensity of peak at 488 cm^{-1} for sample at 5 mA/cm^2. The results correlate with data obtained from XRD, and confirmed that the best crystallization of $Cu(OH)_2$ can be achieved by oxidation at 5 mA/cm^2 without solution stirring. Moreover in Raman spectra for samples, from surfaces of which the films torn off (figure 5 *c*), peaks at 150, 218, 630 cm^{-1}, which can be attributed to phonon modes of Cu_2O, and 298 cm^{-1}, which can be attributed to phonon modes of CuO (phonon mode A_g) [25], are present. This indicate that after the tearing off of $Cu(OH)_2$ films on the surfaces of the samples the layer of roentgenamorphous product which consist from mixture of cuprous and cupric oxides stays.

Raman spectra of annealed sample contain one wide peak at 298 cm^{-1} which attribute to phonon mode of CuO and with the growing of annealing temperature intensity the peak reduce and disappear at 500°C. In the same time with the increase of temperature of treatment peaks at 150, 218, 413 and 630 cm^{-1}, which belong to phonon modes of Cu_2O, they appear (from 300°C) and increase, the high value is reached at 500°C. According to [9], the presence of CuO phase is associated with the structural transformation through dehydrate of $Cu(OH)_2$ and subsequent deoxidization of CuO to Cu_2O.

4. Conclusions

Copper oxides films were successfully synthesized via anodic oxidation of copper foil at various conditions followed by thermal reduction of oxidized samples in inert nitrogen atmosphere. The morphology and phase composition of final and intermediate products were determined by scanning electron microscopy, X-ray diffraction and Raman spectroscopy techniques. Optimal conditions, for obtaining of $Cu(OH)_2$ foil on copper foil surface, is 5 mA/cm^2 during 10 minutes without solution stirring. Thermal reduction of the $Cu(OH)_2$ foil lead to transfer to Cu_2O foil through dehydrate and deoxidization processes under nitrogen atmosphere and from XRD data the optimal temperature of annealing is 500°C.

Acknowledgments

This work is supported by Russian Foundation for Basic Research (16-03-00488).

References

[1] Borges M E, Sierra M, Cuevas E, García R D and Esparza P 2016 *Sol. Energy* **135** 527
[2] Chaudhary D, Vankar V D and Khare N 2017 *Sol. Energy* **158** 132
[3] Yang Y, Xu D, Wu Q and Diao P 2016 *Sci. Rep.* **6** 1
[4] Fujishima, A. and Honda K 1972 *Nature* **238** 37
[5] Navarro R M, del Valle F, Villoria de la Mano J A, Álvarez-Galván M C and Fierro J L G 2009 *Adv. Chem. Eng.* **36** 111
[6] Lianos P 2017 *Appl. Catal. B Environ.* **210** 235
[7] Elmezayyen A S, Guan S, Reicha F M, El-Sherbiny I M, Zheng J and Xu C 2015 *J. Phys. D. Appl. Phys.* **48** 175502
[8] Luo J, Steier L, Son M K, Schreier M, Mayer M T and Grätzel M 2016 *Nano Lett.* **16** 1848
[9] Hsu Y, Yu C, Chen Y and Lin Y 2013 *J. Power Sources* **242** 541
[10] Tan Y W, Xue X Y, Peng Q, Zhao H, Wang T H and Li Y D 2007 *Nano Lett.* **7** 3723
[11] Zhao X, Bao Z, Sun C and Xue D 2009 *J. Cryst. Growth* **311** 711
[12] Zhang Y, Deng B, Zhang T, Gao D and Xu A 2010 *J. Phys. Chem. C* **114** 5073
[13] Li S-K, Guo X, Wang Y, Huang F-Z, Shen Y-H, Wang X-M and Xie A-J 2011 *Dalton Trans.* **40** 6745
[14] Lan X, Zhang J, Gao H and Wang T 2011 *CrystEngComm* **13** 633
[15] Wang B, He J, Liu F and Ding L 2017 *J. Alloys Compd.* **693** 902
[16] Laidoudi S, Bioud A Y, Azizi A, Schmerber G, Bartringer J, Barre S and Dinia A 2013 *Semicond. Sci. Technol.* **28** 115005
[17] Holzschuh H and Suhr H 1990 *Appl. Phys. A Solids Surfaces* **51** 486

[18] Allam N K and Grimes C A 2011 *Mater. Lett.* **65** 1949–1955
[19] La D D, Park S Y, Choi Y W and Kim Y S 2010 *Bull. Korean Chem. Soc.* **31** 2283–2288
[20] Gomez Becerra J, Salvarezza R C and Arvia A J 1988 *Electrochim. Acta* **33** 613
[21] Deng Y, Handoko A D, Du Y, Xi S and Yeo B S 2016 *ACS Catal.* **6** 2473
[22] Singhal A, Pai M R, Rao R, Pillai K T, Lieberwirth I and Tyagi A K 2013 *Eur. J. Inorg. Chem.* 2640
[23] Balkanski M, Nusimovici M A and Reydellet J 1969 *Solid State Commun.* **7** 815
[24] Wang W, Zhuang Y, Wang Z, Zhang Y, Zhong X, An N and Zhang G 2009 *Mater. Lett.* **63** 2432
[25] Debbichi L, Marco de Lucas M C, Pierson J F and Krüger P 2012 *J. Phys. Chem. C* **116** 10232

Morphology and dispersion of FeCo alloy nanoparticles dispersed in a matrix of IR pyrolized polyvinyl alcohol

A A Vasilev[1,2], E L Dzidziguri[1], D G Muratov[2], N A Zhilyaeva[2], M N Efimov[2] and G P Karpacheva[2]

[1] Department of Functional Nanosystems and High-Temperature Materials, National University of Science and Technology "MISiS", Moscow 119049, Russia
[2] Chemistry of Polyconjugated Systems Laboratory, A.V. Topchiev Institute of Petrochemical Synthesis RAS, Moscow 119991, Russia

E-mail: raver.vasiljev@mail.ru

Abstract. Metal-carbon nanocomposites consisting of FeCo alloy nanoparticles dispersed in a carbon matrix were synthesized by the thermal decomposition method of a precursor based on polyvinyl alcohol and metals salts. The synthesized powders were investigated by X-ray diffraction (XRD), X-ray fluorescent spectrometry (XRFS), transmission electron microscopy (TEM) and scanning electron microscopy (SEM). Surface characteristics of materials were measured by BET-method. The morphology and dispersity of metal nanoparticles were studied depending on the metals ratio in the composite.

1. Introduction

At present days the development of new materials with predefined properties is relevant task in both scientific and applied aspects. Metal-carbon nanocomposites based on pyrolyzed polymers are attractive due to the diversity and uniqueness of their physico-chemical properties providing their high potential for commercial application [1 – 3].

It is well known that Fe and Co have a great importance in different scientific fields and engineering due to their catalytic and magnetic properties [4 – 8]. Fe-Co bimetallic system allows to improve various physico-chemical properties significantly. For example, Fe-Co bimetallic system is characterized by high Curie temperature, large saturation magnetization, great magnetic permeability [2, 5 – 9] and gives rise to a synergistic effect in catalysis [5, 10].

Using of carbon based materials is one of the best way to stabilize nanoparticles. The carbon encapsulation increases electrical conductivity, mechanical performance, capability of nanoparticles to absorb the electromagnetic waves, biocompatibility of nanoparticles and decreases interparticle interaction [3, 7, 11]. In addition, carbon is a good catalyst support material due to its various porous structure, resistance to acid and basic environments, low cost and easy accessibility. Furthermore, metal nanoparticles stabilized by carbon materials can be considered as models of the dimensional effect influence on catalytic activity [12]. Thus, metal-carbon nanocomposites based on Fe and Co are promising as catalysts, absorber of electromagnetic waves, carriers for magnetically-driven drug delivery and in development new magnetic materials based on these nanocomposites [13 – 17].

The goals of this work were the preparation of metal-carbon nanocomposites based on FeCo nanoparticles immobilized in IR pyrolized polyvinyl alcohol support and the investigation of the phase formation, morphology and dispersity of FeCo bimetallic nanoparticles depending on the intensity of IR-annealing and metals ratio in the composite.

2. Materials and methods

Metal-carbon nanocomposites were prepared under the conditions of infrared (IR) pyrolysis of a precursor based on solution of polyvinyl alcohol (PVA) and iron and cobalt hydronitrates ($Fe(NO_3)_3 \cdot 9H_2O$ and $Co(NO_3)_2 \cdot 6H_2O$) with the follow ratio of metal components 9 : 1; 3 : 1; 1 : 1; 1 : 3; 1 : 9 (mole) at different temperatures (300 – 700 °C). The concentration of PVA in water was 5 % wt. and the summary metal concentrations in polymer was 20 % wt.

The pyrolysis was carried out in the quartz reactor of IR heating device. IR annealing was accomplished at two-step mode: preliminary annealing at presence of the air at 150 and 200 °C for 15 minutes in order to remove the solvent and start the initial structuring of PVA followed by the main annealing in an argon atmosphere at a necessary temperature. The main annealing step takes 2 minutes at desired temperature.

The size and morphology of the metal nanoparticles were evaluated by TEM (LEO 912 AB) and SEM (JSM 6700F) analysis. Histograms of particle size distribution were obtained via estimation of particle size from TEM micrographs.

XRD analysis was carried out at room temperature on a DIFRAY-401 diffractometer using Cr-Kα radiation source with the Bragg-Brentano focusing geometry. Spherical crystallite diameter distribution was estimated from XRD data.

The elemental composition of the samples was determined by X-ray analytical microscope RAM-30μ.

The N_2 adsorption/desorption isotherms of metal-carbon nanocomposites were measured with an accelerated surface area and porosimetry system (ASAP-2020 «Micromeritics») for determining the surface areas.

3. Results and discussion

The elemental composition of the composites determined by X-ray fluorescence analysis is given in table 1. It was found that initial and measured metals content were similar.

Table 1. Metal content of the IR-PVA/Fe-Co composites

Ratio of metals Fe : Co	Annealing temperature, °C	Fe content, at. %	Co content, at. %
9 : 1	700	90	10
3 : 1	700	75	25
1 : 1	300	48	52
1 : 1	400	48	52
1 : 1	500	49	51
1 : 1	600	49	51
1 : 1	700	49	51
1 : 3	700	24	76
1 : 9	700	7	93

Figure 1 shows the TEM images of samples obtained at different temperatures with the equal ratio of components. FeCo spherical nanoparticles dispersed in translucent carbon matrix are observed and have a uniform distribution. Figure 2 demonstrates the TEM images of composites synthesized at 700 °C with different ratio of components. Morphology of metal nanoparticles are the same at different ratio of components, however the carbon matrix uniformity is changed in the sample with the ratio of metals Fe : Co = 1 : 9.

Figure 3 shows the SEM images at high magnification on which nanoparticles also have spherical forms localized on the carbon matrix surface. The particle size distribution histograms were obtained

IOP Conf. Series: Materials Science and Engineering **347** (2018) 012011 doi:10.1088/1757-899X/347/1/012011

via estimation of particle size from TEM micrographs (figure 4) and described by the log-normal distribution function. Table 2 represents the average particle sizes D_{el}.

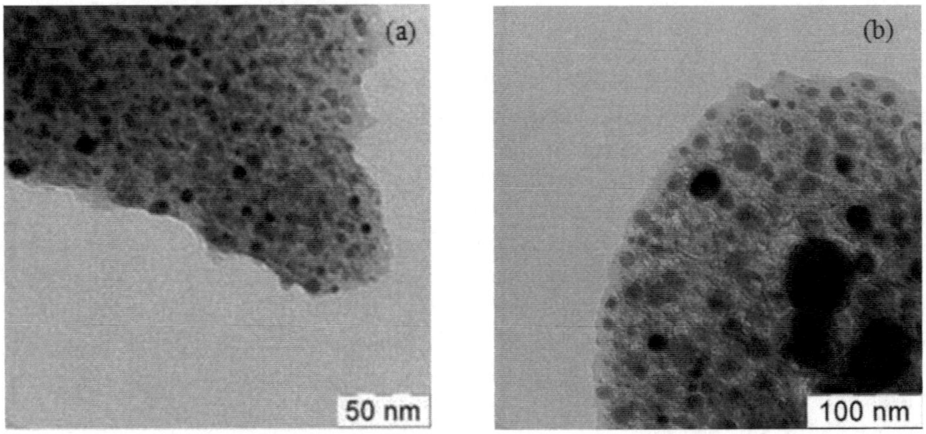

Figure 1. TEM micrograph of IR-PVA/Fe-Co samples with the equal ratio of components obtained at: **(a)** 600 °C; **(b)** 700 °C

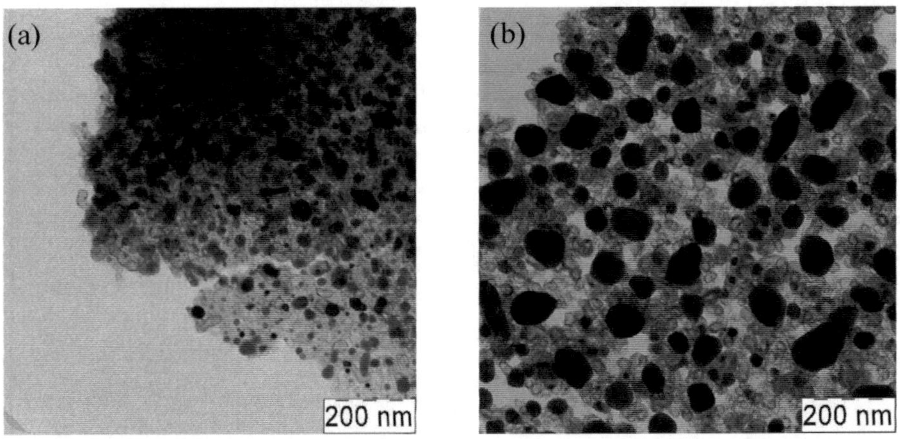

Figure 2. TEM micrograph of IR-PVA/Fe-Co samples obtained at 700 °C: **(a)** Fe : Co = 9 : 1; **(b)** Fe : Co = 1 : 9

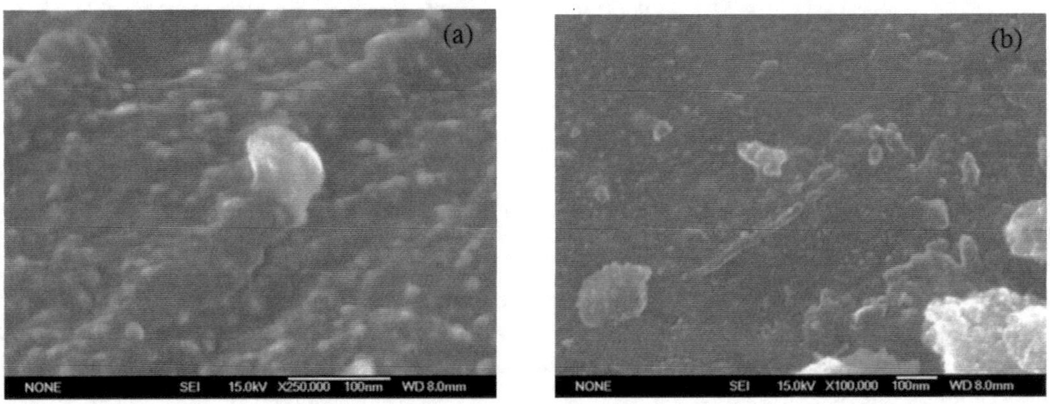

Figure 3. SEM micrographs of IR-PVA/Fe-Co samples with the equal ratio of components obtained at: **(a)** 600 °C; **(b)** 700 °C

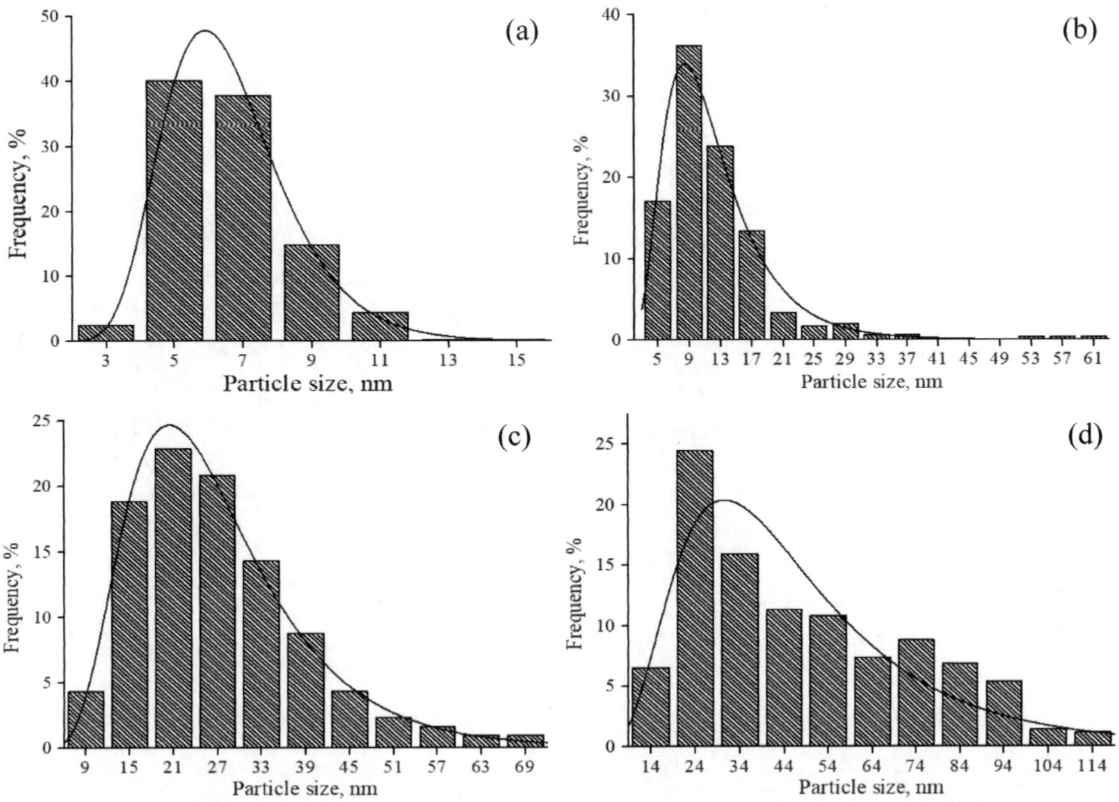

Figure 4. Particle size distribution histograms of FeCo nanoparticles obtained at: **(a)** 600 °C (Fe : Co = 1 : 1); **(b)** 700 °C (Fe : Co = 1 : 1); **(c)** 700 °C (Fe : Co = 9 : 1); **(d)** 700 °C (Fe : Co = 1 : 9)

XRD analysis of samples confirms the formation of FeCo solid solution. For a better visual perception the XRD patterns are shifted relative to each other by the same scale along the intensity axis on figure 5. FeCo nanoparticles with body-centered cubic (bcc) structure start forming at 400 °C. In the temperature range of the IR-annealing from 300 to 500 °C the XRD patterns peaks corresponding to iron and cobalt oxides are observed, which are superimposed on one another.

Figure 6 represents the reflection lines corresponding to the maximum intensity peak of FeCo solid solution, the crystal plane of (110). It is observed that the FeCo reflection peaks at the range from 600 to 700 °C have a symmetrical shape and are aligned relative to each other, whereas at 500 °C the diffraction maximum is shifted towards larger angles. Therefore, FeCo solid solution formation with unequal ratio and with a smaller lattice constant occurs at 500 °C. The observed iron oxide peaks at 500 °C indicate that FeCo solid solution formation is incomplete. The diffraction maximum corresponding to the graphite appears on the diffraction pattern from 600 °C. The crystalline phase amount of carbon is increased with rising temperature.

Figure 7 shows XRD patterns of composites synthesized at 700 °C with different ratio of metals. FeCo solid solution with bcc structure is formed in all samples except the composite with the ratio of component Fe : Co = 1 : 9. FeCo phase based on faced-centered cubic structure (β-Co) is becoming visible for the samples containing more than 70 at. % of cobalt in the solid solution. The XRD pattern of composite containing 90 at. % of cobalt characterized by the presence of FeCo nanoparticles based on β-Co lattice only.

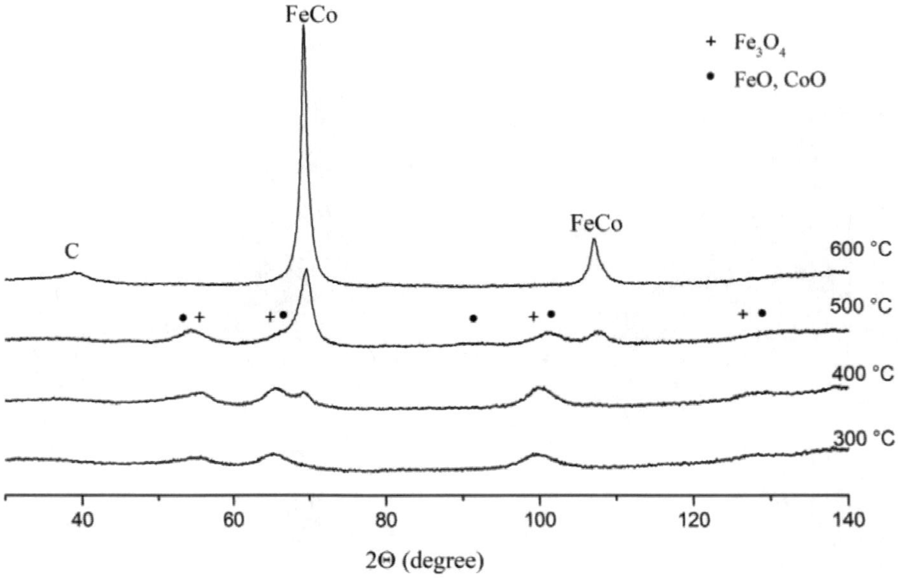

Figure 5. X-ray diffraction patterns of IR-PVA/Fe-Co samples

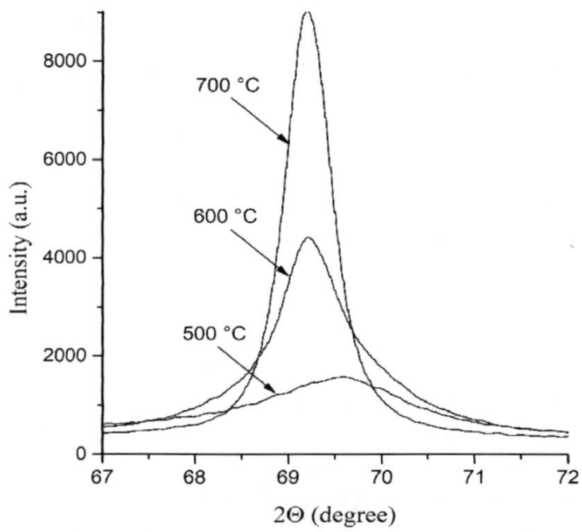

Figure 6. X-ray diffraction patterns of IR-PVA/Fe-Co samples with the equal ratio of components obtained at 500 – 700 °C, the crystal plane of (110)

All picks have a symmetrical shape and are shifted toward larger angles with increasing cobalt content in the samples (figure 8). It indicates that the lattice constant is decreased and the chemical composition of solid solution is changed.

The coherent scattering region (CSR) distribution was calculated according to the method of Selivanov and Smislov [18]. The sample obtained at 600 °C has the most narrow distribution and smaller size of the CSR. The average crystallite size for all composites obtained at 700 °C is approximately the same (20 – 24 nm) and doesn't depend on the ratio of components.

According to the table 2 nanoparticles consist of a single crystallite in all samples. In the composite annealed at 700 °C, the average crystallite size larger than the average particles size. This can be explained by the locality of the TEM method, while X-ray analysis is an integral.

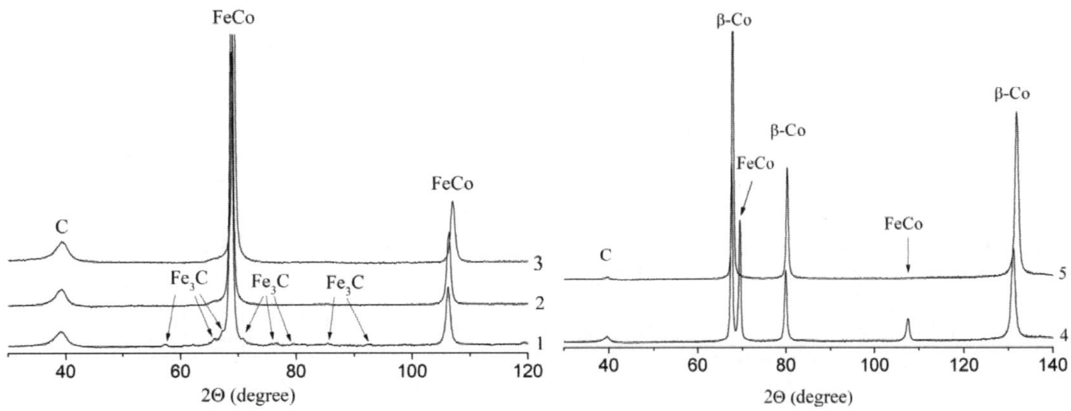

Figure 7. X-ray diffraction patterns of IR-PVA/Fe-Co samples synthesized at 700 °C: **(1)** Fe : Co = 9 : 1; **(2)** Fe : Co = 3 : 1; **(3)** Fe : Co = 1 : 1; **(4)** Fe : Co = 1 : 3; **(5)** Fe : Co = 1 : 9

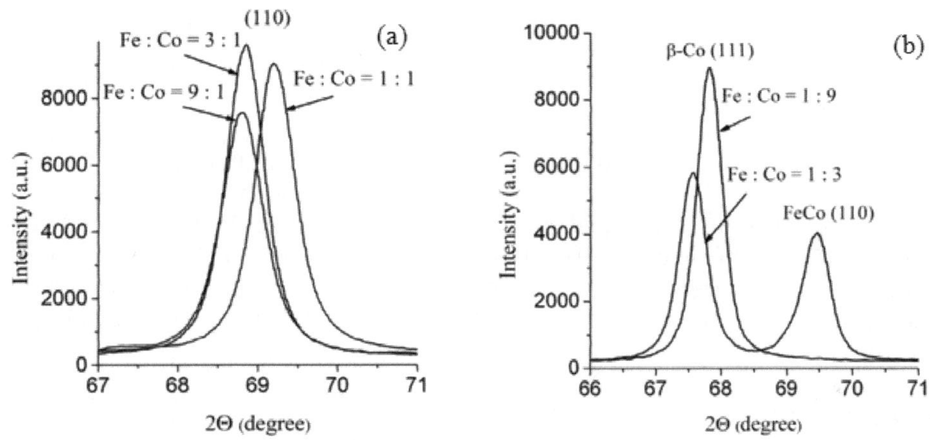

Figure 8. X-ray diffraction patterns of IR-PVA/Fe-Co samples with different ratio of components obtained at 700 °C, the crystal planes of (110) and (111)

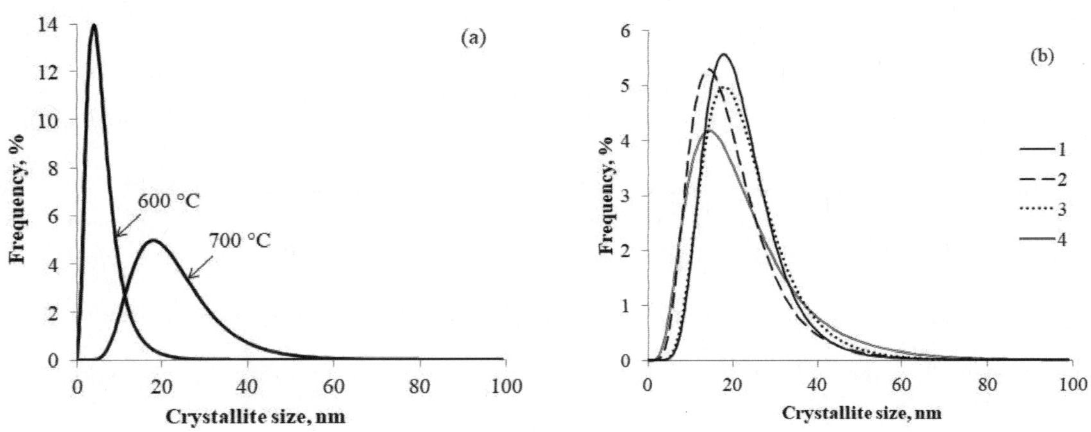

Figure 9. Crystallite size distribution of the FeCo phase (BCC): **(a)** with the equal ratio of metals; **(b)** with different ratio of metals: **(1)** Fe : Co = 9 : 1; **(2)** Fe : Co = 3 : 1; **(3)** Fe : Co = 1 : 1; **(4)** Fe : Co = 1 : 3

The specific surface area was determined from nitrogen adsorption-desorption isotherm data and collected in table 2. The specific surface area is higher for samples with a predominant Fe content. This fact corresponds to the results obtained in [19, 20] where iron-containing carbon composite characterized by higher surface area in comparison to cobalt-containing material. Hence, the metals affect on porosity and structure of the metal-carbon composites.

Table 2. Dimensional characteristics of FeCo nanoparticles and specific surface area of nanocomposites

Ratio of metals Fe : Co	Annealing temperature, °C	Metal phase	Specific surface area, m^2/g	D_{el}, nm	D_{CSR}, nm
9 : 1	700	BCC	195	27	23
3 : 1	700	BCC	187	25	21
1 : 1	600	BCC	-	7	7
1 : 1	700	BCC	21	12	24
1 : 3	700	BCC FCC	58	17	24 48
1 : 9	700	FCC	17	48	-

4. Conclusions.

The obtained research results show that FeCo nanoparticles with bcc structure start forming at 400 °C. The metallic phase of samples synthesized at 600 and 700 °C consists of FeCo solid solution only. Nanoparticles have spherical form and uniform distribution in a carbon matrix. Equal ratio of components and iron predominance in the samples lead to formation of FeCo nanoparticles with the structure based on α-Fe lattice whereas an excess of cobalt leads to form a FCC-lattice based on β-Co. The sample at the ratio of Fe : Co = 1 : 9 is characterized by larger particle sizes and destruction of carbon matrix uniformity.

Acknowledgements

The equipment of the collective exploitation center "New petrochemical processes, polymer composites and adhesives" was used.

This work was done as part of TIPS RAS (№ 79) State Plan.

References

[1] Kulikova M V, Zemtsov L M, Sagitov S A, Efimov M N, Krylova A Y, Karpacheva G P, Khadzhiev S N 2014 *Solid Fuel Chemistry* **48** 32

[2] Zaharov Yu A, Pugachev V M, Popova A N 2013 *Bulletin of the Russian Academy of Sciences: Physics* **77** 142

[3] Afghahi S, Shokuhfar A 2014 *Journal of Magnetism and Magnetic Materials* **370** 37

[4] Sourmail T 2005 *Progress in Materials Science* **50** 816

[5] Sedighi B, Feyzi M, Joshaghani M 2015 *Journal of the Taiwan Institute of Chemical Engineers* **50** 108

[6] [Chokprasombat K, Harding P, Pinitsoontorn S, Maensiri S 2014 *Journal of Magnetism and Magnetic Materials* **369** 228

[7] Nautiyal P, Seikh M M, Lebedev O I, Kundu A K 2015 *Journal of Magnetism and Magnetic Materials* **377** 402

[8] Karipoth P, Thirumurugan A, Joseyphus R J 2013 *Journal of Colloid and Interface Science* **404** 49

[9] Chaubey G S, Barcena C, Poudyal N, Rong C, Gao J, Sun S, Liu J P, Am J. 2007 *Chemical Society* **129** 7214

[10] Díaz J A, Romero A. García-Minguillán A M 2015 *Fuel Processing Technology* **138** 455

IOP Conf. Series: Materials Science and Engineering **347** (2018) 012011 doi:10.1088/1757-899X/347/1/012011

[11] Castrillon M, Mayoral A, Urtizberea A, Marquina C, Irusta S, Meier J G, Santamaria J 2013 *Nanotechnology* **24** 505702
[12] Pomogailo A D, Rozenberg A S, Uflyand I E 2000 *Metal nanoparticles in polymers* (Moscow: Khimiya) p 672
[13] Galaburda M V, Bogatyrov V M, Tomaszewski W, Oranska O I, Borysenko M, Skubiszewska-Zieba J, Gun'ko V 2017 *Colloids and Surfaces A* **529** 950
[14] Xing W, Zhue S, Gao X 2009 *Materials Letters* **63** 1177
[15] Liu L, Concepcion P, Corma A 2016 *Journal of Catalysis* **340** 1
[16] Tsurin V A, Ermakov A E, Uimin M A, Mysik A A, Schegoleva A A, Gaviko V S, Maikov V V 2014 *Physics of the Solid State* **56** 287
[17] Kozhitov L V, Kozlov V V, Kostikova A V, Popkova A V 2013 *Russian Microelectronics* **42** 498
[18] Selivanov V N, Smislov E F 1993 *Crystallography Reports* **38** 174
[19] Wu M, Zhang E, Guo Q, Wang Y, Qiao J, Li K, Pei P 2016 *Applied Energy* **175** 468
[20] Han C, Bo X, Liu J, Li M, Zhou M, Guo L 2017 *Journal of Alloys and Compounds* **710** 57

IOP Conf. Series: Materials Science and Engineering **347** (2018) 012012 doi:10.1088/1757-899X/347/1/012012

Suspensions on the basis of stabilised zirconium oxide for three-dimensional printing

P S Sokolov[1], D A Komissarenko[1], I A Shmeleva[1], I V Slyusar[1], G A Dosovitskiy[1], P V Evdokimov[2], V I Putlyaev[2], and A E Dosovitskiy[3]

[1] NRC «Kurchatov Institute» – IREA, Bogorodskiy val str. 3, Moscow, 107076, Russia
[2] Moscow State University, Vorobievi Gory 1, Moscow, 119991, Russia
[3] NeoChem, Profsoyuznaya str. 115-2-331, Moscow, 117647, Russia

E-mail: sokolov_ps@irea.org.ru

Abstract. Present work considers the first results on rheological and photo-curing behaviour of suspension consisting of nanocrystalline stabilised zirconium dioxide powders (19 - 27 vol. %) and a liquid UV-photosensitive organic monomer. At ambient temperature compositions showed a viscosity of 2.5 and 0.8 Pa×s at 10 and 100 s^{-1} shear rates, respectively. Printability of these compositions was subsequently investigated by using an stereolithography machine Ember (Autodesk). 3D objects were later sintered in a separate furnace into dense translucent ZrO_2 ceramics.

1. Introduction

Stereolithography (SLA) is a technology that opens the possibility to fabricate complex 3D construction and functional ceramic components without tooling and with a good dimensional resolution. There have been some reports [1 - 6] on successful using the SLA method for production of complex-shape ceramic, such as Al_2O_3 [1], ZrO_2 [2, 3], yttrium aluminium garnet [4] and calcium phosphates [1, 5, 6]. This has been done via ultraviolet curing of a highly concentrated suspension of ceramic particles in a photopolymerisable liquid. The suspension must contain as much as possible (at least 30 vol %) ceramic powders to provide subsequent sintering with a viscosity less than 3 Pa*s at low shear rates to promote self-leveling and recoating. In this paper, the rheological and photo-curing behavior suspensions and the influence of the ZrO_2 particle size on viscosity, have been investigated.

2. Experimental

2.1. Starting materials

High-purity, homogeneous stabilised zirconia powders were synthesised by co-precipitation method and then annealed at different temperatures [7]. Their main properties are presented in Table 1. Four grades of 8 mol% Y_2O_3 stabilised ZrO_2 powders with different particle size and specific surface area (*SSA*) were prepared. Average crystallite size (*D*) was calculated from the broadening of X-ray diffraction peaks. The *SSA* and pore volume (V_{pore}) were measured using the Brunauer-Emmett-Teller (BET) and Barret-Joyner-Halenda (BJH) method, respectively. The medium aggregate size (d_{50}) was measured by the laser diffraction. Tap density (ρ_{tap}) of powders was estimated according to GOST

25279-93 (ISO 3953). The moisture (*MC*) of the zirconia powders was determined from weight loss on annealing at 378-383 K for 8 h, an average from three experiments was used for each sample.

A photopolymerisable suspension (slurry) is usually consists from acrylic monomer, ceramic powder, dispersant and photoinitiator (PI) [1-6]. In the present work 1,6-hexanediol diacrylate (HDDA, technical grade, Aldrich) was chosen as a monomer. HDDA is a fast curing monomer with a low-viscosity (below 9 mPa*s at 293 K) and relatively high refractive index (n = 1.46). It has low volatility and low odour. The commercially available substances Triton X-100 (Aldrich) and w969 (BYK) were selected as a dispersant due to their good wetting and dispersing properties. Triton X-100 is a nonionic surfactant that has a hydrophilic polyethylene oxide chain and an aromatic hydrocarbon hydrophobic group. BYK w969 is a mixture of 2-Phenoxyethanol and alkanolammonium salt of an acidic polyester. Dispersant with 0.5-12.0 wt% concentration was added into the mixture, on a dry weight of stabilized ZrO_2 powders. The PI was Irgacure 2100 (BASF). Irgacure 2100 is a liquid blend of acylphosphine oxides for radical polymerisation. We used 0.5 or 1.0 wt% of PI with respect to the mass of monomer. No dye or UV light absorber or additional inhibitor of polymerisation was used.

Table 1. Main physical properties of stabilised zirconium dioxide powders

	SSA (m^2/g)	V_{pore} (cm^3/g)	D(nm)	d_{50} (μm)	ρ_{tap} (g/cm^3)	*MC* (%)
8YSZ1	48.5	0.205	14	1.09	1.04	1.57
8YSZ2	21.7	0.120	25	0.91	1.19	0.61
8YSZ3	12.4	0.092	36	0.83	1.29	0.44
8YSZ4	5.9	0.016	124	0.81	1.49	0.18

2.2. *Preparation of UV-curable suspensions*

The preparation method of the UV-curable suspension used in this work contains a few steps. First, photoinitiator and calculated amount of dispersant were added to the HDDA. The mixture was kept under stirring at room temperature and in the dark for 1 h. Second, the stabilised ZrO_2 powder, after sieving through a 70 μm mesh and dried for 8 h at 373 K, was added in small increments to mixture using a laboratory magnetic stirrer. Then the suspensions were mixed additional 1 h to produce a well-blended system. The stir bar was of 25 mm in length and of 5 mm in diameter, and the mixing speed was 100 rpm. The mixing process was carried out in a 25 ml glass. Finally, bubbles were removed from the completed suspension by vacuum deairing under 10-30 mbar pressure about 10 min under continued stirring.

2.3. *3D printer*

A commercially available open-source desktop 3D printer [8], the Ember (Autodesk, USA), was used for this study. The Ember uses a 3D printing method in vat photopolymerisation known as digital light processing. The main difference with classical laser stereolithography is a digital light projector which is used as a UV light source. The projector shines light below the vat through an optical window. The optical window in the bottom of the vat has layer of polydimethylsiloxane. The projector has a wavelength of 405 nm with a total optical powder of 5 W. The Ember's UV light source was monitored using special UV light meter, the model 222 (G&R Labs, USA). Ember has a resolution of 50 μm [8] in the plane parallel to printing surface (XY resolution). The photo-curing behaviour of prepared ceramic suspensions was examined on the same printer.

2.4. *Washing, binder burnout and sintering*

After printing, the green body was rinsed with inert solvent in a heated ultrasonic bath to remove any uncured residual monomers, dried at room condition and then post-cured using a 36 W UV nail exposure chamber with Hg bulb (365 nm) for a minimum of 2 min. The debinding of green body was performed with a heating rate of 1 K/min up to 1073 K in air. Finally green body was sintered at 1873 K for 2 h with a heating rate of 6 K/min. The microstructure of sintered samples was observed by scanning electron microscopy (SEM) using JSM 7100 F (Jeol, Japan) instrument.

2.5. Rheometric measurement
Rheological characterisation of the suspension was performed at room temperature with rheometer Physica MCR 52 (Anton Paar, Austria). A plate-and-plate geometry with a measuring disk with diameter of 25 mm and a gap of 0.200 mm was used.

2.6. TGA-DSC measurement
Thermal gravimetric analysis (TGA) and differential scanning calorimetry (DSC) of cured suspensions of 5-10 mg in alumina crucibles was carried out on a simultaneous analyser SDT Q600 (TA Instruments, USA) under dynamic air atmosphere, at a constant heating rate of 2, 3, 5, 10 and 20 K/min from room temperature up to 1073 K. The DSC-TGA samples are flat disk with dia. of 4 mm and thickness of 100 μm.

3. Results and discussion

3.1. Rheological behaviour of the ceramic suspensions
Suspensions containing cubic ZrO_2 with various amount of dispersant showed different viscosities (Figure 1). It was observed that the optimal concentration of dispersant was about 3 wt%. The relation between shear stress and shear rate is not linear. Viscosity strongly depends on shear rate. All HDDA suspensions with 8YSZ as a filler showed the non-Newtonian (pseudoplastic or shear-thinning) behavior). Thus, 8YSZ3 powder containing suspension with a lowest viscosity was chosen for further experiments.

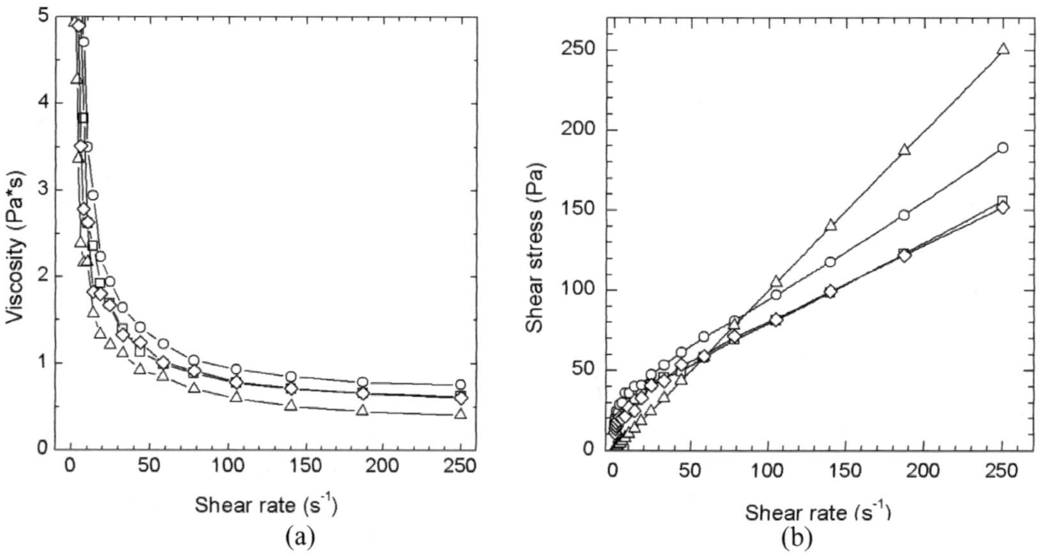

Figure 1. Representative rheological curves. (*a*) Viscosity curves and (*b*) flow curves of HDDA/Triton X-100 with stabilised zirconia powders: squares – 8YSZ1 (21 vol%), circles – 8YSZ2 (23 vol%), triangles – 8YSZ3 (27 vol%) and diamonds – 8YSZ4 (26 vol%).

No use degassing under vacuum leads to an increase of suspensions viscosity at a low shear rate up to 50 Pa*s. The dispersants Triton X-100 and BYK w969, in spite of their different nature, showed a similar reducing effect on the viscosity of suspensions. Without a dispersant the highest loading of ceramic fillers did not exceed 10-12 vol%. Addition of PI did not change the viscosity of the suspensions but slightly decrease loading. Addition of the inert diluents (ethylene glycol, diethylene glycol, polyethylene glycol 400 and polypropylene glycol 1025 in the amount of 1-10 wt%) did not effect the viscosity of suspensions but could reduce the filler loading.

3.2. Curable of the ceramic suspensions

The turbidity of highly concentrated ceramic suspensions induces light scattering and limits light penetration into the suspension. One aim of this study was to verify that the cured depth in the ceramic suspension is sufficient to SLA process. Then, the curing ability will be compared to that obtained with the pure monomer.

The polymerisation depth (C_p) is expressed by the Jacob equation (Beer-Lamber law) [6]:

$$C_p = D_p \ln(E/E_c) \quad (1)$$

where D_p is the depth of penetration, E – the exposure energy and E_c - the critical (or minimal) exposure energy to provide polymerisation of the monomer. For a load monomer D_p is a function of the particle size, concentration and refractive index difference between monomer and ceramic powder [6]. E_c is only depends on the PI/monomer system. So for a high refractive index difference (cubic ZrO_2 n = 2.15) of the cured depth is more attenuated and can becomes insufficient for SLA process [6].

Two suspensions with 0.5 or 1.0 wt% of photoinitiator with respect to the HDDA were characterized for curable. Experimental cure depths curves are plotted in Figure 2 on a semi-logarithmic scale. For comparison pure HDDA without powder loading have the depth of cure more 1 mm in the same condition. The D_p values calculated from the slopes are found to be 59 ± 2 and 69 ± 6 μm for 0.5 and 1.0 wt% of PI, respectively. The average E_c values are to be 49 ± 4 and 46 ± 4 mJ/cm², for 0.5 and 1.0 wt% of PI, respectively. These curves allow us to choose the optimal print settings (time and dose of irradiation, Z-layer thickness – 100 μm, *etc*).

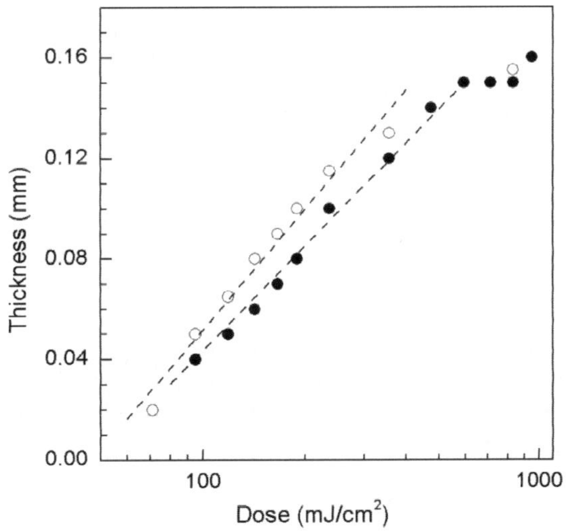

Figure 2. Dependence of polymerisation depths *versus* energy dose for 26 vol% 8YSZ3 suspensions with 3.6 wt% of BYK w969. Open symbols are suspension with 1 wt% of PI, solid — 0.5 wt% of PI. Cure depth measurement was performed using 405 nm source with 23 ± 0.5 mW/cm² UV light intensity.

3.3. Thermal decomposition of cured 8YSZ/HDDA suspensions

TGA was applied to identify temperature regions of rapid mass loss of the green body of ceramic mold. Figure 3 represents the weight loss and DSC curve as a function of temperature for constant heating rate of 2 K/min. Noticeable weight loss was observed in the range of 400-700 K. All the organic binders are removed below 800 K in air (Table 2). The exothermic events demonstrate that the degradation involves oxidation reaction. The samples after DSC-TGA measurement with heating rate above 3 K/min have cracks.

IOP Conf. Series: Materials Science and Engineering **347** (2018) 012012 doi:10.1088/1757-899X/347/1/012012

Table 2. Thermal events of decomposition of cured 8YSZ/HDDA suspensions (0.5 wt% PI).

Rate	First event			Second event		
	T_{b1} (K)	T_{p1} (K)	ΔQ_1 (kJ/g)	T_{b2} (K)	T_{p2} (K)	ΔQ_2 (kJ/g)
2	602	634	1.349	668	683	0.295
3	713	650	1.172	681	694	0.266
5	626	668	0.779	693	706	0.195
10	601	696	2.290	–	728	0.572
20	598	715	2.23	–	753	0.567

Note: T_b - temperate of the beginning of the thermal event, T_p – peak temperate; ΔQ – heat effect.

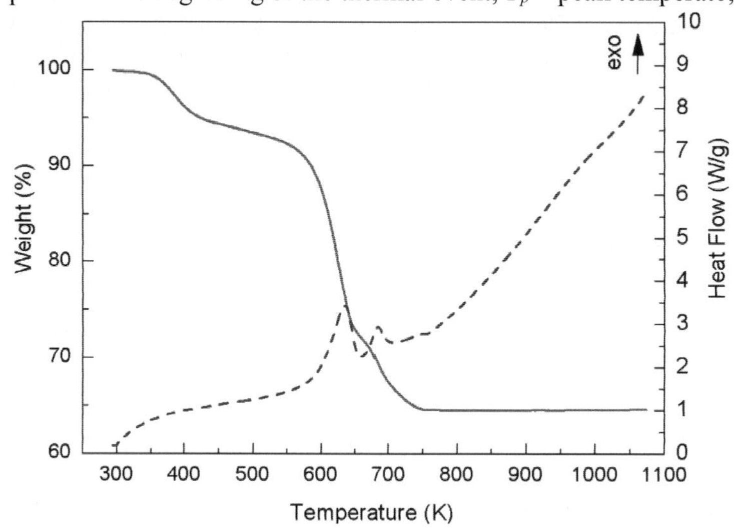

Figure 3. TGA and DSC curve at 2 K/min of pyrolysis of cured suspension 8YSZ3/BYK/HDDA with 0.5 wt% of PI in 25 vol% stabilised ZrO_2 printed by stereolithography showing mass loss accompanied by two exothermic reactions.

The kinetic investigation of thermal degradation was achieved from DSC data by the application of Kissinger [9] and Takhor [10] methods. The activation energy (E_a) is an important parameter associated with nonisothermal decomposition. The E_a can be derived from the peak temperature T_p with heating rates by the Kissinger approch [9] as follows:

$$\frac{d[\ln(\phi / T_p^2)]}{d(1/T_p)} = -\frac{E_a}{R} \quad (2)$$

and by the Takhor model [10] as:

$$\frac{d[\ln(\phi)]}{d(1/T_p)} = -\frac{E_a}{R} \quad (3)$$

where φ is the heating rate, T_p is the thermal decomposition peak temperature, and R is the universal gas constant. Plots of $\ln(\varphi)$ *versus* $1/T_p$ are shown in Figure 4a (Takhor model) and plots of $\ln(\varphi/T_p^2)$ *versus* $1/T_p$ are shown in Figure 4b (Kissinger model), and they are fitted to straight lines. The E_a values calculated from the slopes are found to be 94 ± 6 and 130 ± 6 kJ/mole for 0.5 wt% of photoinitiator (for first and second thermal event respectively) for Kissinger model. The comparable E_a of 0.5 and 1.0 wt % of PI suggested that polimer structure did not significantly affect the energy required to decomposition.

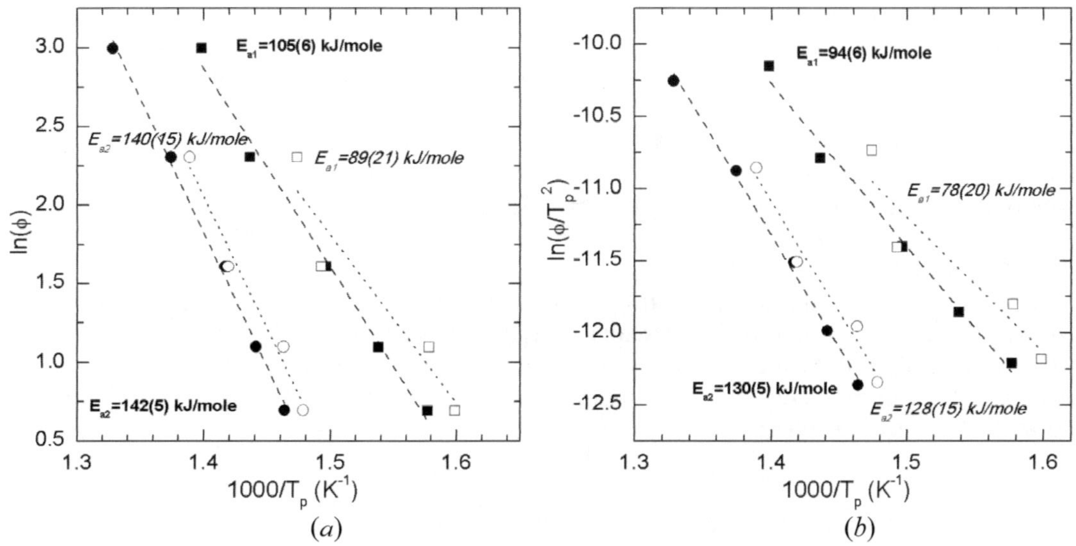

Figure 4. Kinetic of thermal decomposition of cured 8YSZ3/BYK/HDDA suspensions. Open symbols are suspension with 1 wt% of PI, solid — 0.5 wt% of PI. Squares are first thermal events, circle — second thermal events. (*a*) Takhor plot, (*b*) Kissinger plot with various heating rate. $\mathbf{E_a}$ activation energy for 0.5 wt% of PI, E_a – for 1.0 wt% of PI.

In order to verify the printability of the best suspension (8YSZ3 powder with load 26 vol%) the reticulated variant of the Kelvin structure [6] was prepared (Figure 5). As disadvantages of suspensions based on HDDA, it is important to note the interaction with polydimethylsiloxane and the some brittleness of the green body after polymerisation, it may be necessary to modify the suspension composition or add plasticizer. One can see some cracks in the green body which also present in the sintering ceramic. According to SEM study the 3D printed ceramic material shows the typical fine grained microstructure of Y_2O_3-stabilised zirconia with grain a few micrometers.

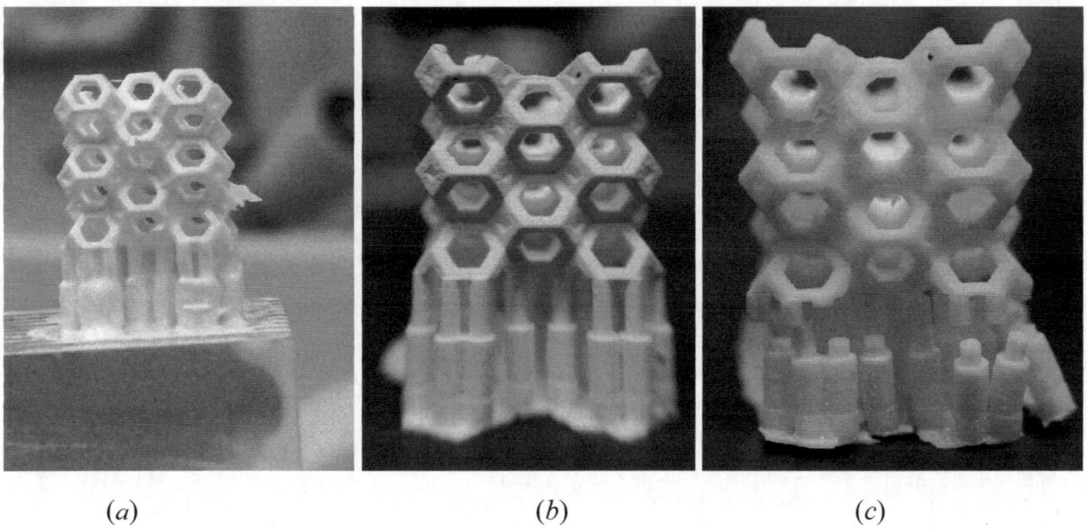

Figure 5. (*a*) The green body immediately after printing; (b) after post-curing and (c) sintering ceramic. The XY dimension of 9.8×9.8 mm (*a*,*b*), linear shrinkage is about 33% (*c*).

4. Conclusions

Suspensions of HDDA and fine zirconia powders with dispersant show good loading (of around 25 vol.% stabilised zirconia), good viscosity and stability 24-48 h at least, that allows to use them for stereolithography 3d printing.

The dispersants Triton X-100 and BYK w969 show the good liquefier properties. The optimal concentration of dispersant is about 3 wt% according to dry weight of zirconia powder. The better concentration of the photoinitiator Igracure 2100 is 0.5 wt% by weight of HDDA.

Acknowledgments

The authors would like to thank E.I. Panarina, S.K. Belus and M.A. Timokhin for technical assistance and A.M. Nemeryuk for valuable discussions. Measurements were performed using equipment of IREA shared analytical facilities center. This work was supported by Ministry of Science and Education of the Russian Federation, subsidy agreement N° 14.625.21.0040 dated 29.09.2016, project identifier RFMEFI62516X0040.

References

[1] Pfaffinger M, Mitteramskogler G, Gmeiner R and Stampfl J 2015 Thermal debinger of ceramic-filled photopolymers *Mater Sci Forum* **825-826** 75

[2] Mitteramskogler G, Gmeiner R, Felzmann R, Gruber S, Hofstetter C, Stampfl J, Ebert J, Wachter W and Laubersheimer J 2014 Light curing strategies for lithography-based additive manufacturing ofcustomized ceramics *Additive Manufacturing* **1-4** 110

[3] Harrer W, Schwentenwein M, Lube T and Danzer R 2017 Fractography of zirconia-specimens made using additive manufacturing (LCM) technology *J Eur Ceram Soc* **37** 4331

[4] Dosovitskiy G A, Karpyuk P V, Evdokimov P V, Kuznetsova D E, Mechinsky V A, Borisevich A E, Fedorov A A, Putlyaev V I, Dosovitskiy A E and Korjik M V 2017 First 3D-printed complex inorganic polycrystalline scintillator *CrystEngComm* **19** 4260

[5] Putlyaev V I, Evdokimov P V, Safronova T V, Klimashina E S and Orlov N K 2017 Fabrication of osteoconductive $Ca_{3-x}M_{2x}(PO_4)_2$ (M= Na, K) calcium phospahate bioceramics by stereolitographic 3D printing *Inorg Mater* **53** 529

[6] Ievlev V M, Putlyaev V I, Safronova T V and Evdokimov P V 2015 Additive technologies for making highly permeable inorganic materials with tailored morphological architectonics for medicine *Inorg Mater* **51** 1295

[7] Sokolov P S, Karpyuk P V, Dosovitskiy G A, Volkov P A, Lyskov N V, Slyusar I V and Dosovitskiy A E 2018 Nanostructured stabilized zircomia powders for solid oxide fuel cells *Russian Journal of Electrochemistry* Accepted for publication

[8] Adzima B 2015 The Ember Printer: An Open Platform for Software, Hardware, and Materials Development. uv.ev WEST Conference; Redondo Beach, CA.

[9] Kissinger H E 1956 Variation of peak temperature with heating rate in differential thermal analysis *J Res Natl Inst Stand Technol* **57** 2712

[10] Takhor R L 1971 Advance in nucleation and crystallization of glasses; American Ceramics society, Columbus

IOP Conf. Series: Materials Science and Engineering **347** (2018) 012013 doi:10.1088/1757-899X/347/1/012013

Ce:YAG ceramics: the influence of the synthesis technology features on the luminescent and the optical properties

K E Lukyashin[1], V A Shitov[1], A I Medvedev[1,2], A V Ishchenko[2], V S Shevelev[2], B V Shulgin[2], L R Basyrova[2]

[1] Institute of Electrophysics UrB RAS, Ekaterinburg 620016, Russia

[2] Ural Federal University, Ekaterinburg 620002, Russia

E-mail: kostya@iep.uran.ru

Abstract. In this paper, we report on the dependence of the luminescent and the optical properties on the synthesis conditions of the transparent 0.1 at.% Ce:YAG and 1 at.% Ce:YAG ceramics. The ceramics were produced from the nanopowders with a diameter of about 10–15 nm by means of the laser method. The fundamental difference between the two described methods is in the synthesis of the main phase YAG: directly during the vacuum sintering (1 – the first method) and before the vacuum sintering (2 – the second method). For this purpose, the transparent samples (Ø10×2 mm) with the optical transmittance ranging from 58 to 82% at the wavelength of 600 nm were obtained. The first method was proven to be the most preferable in terms of the exact dosage of the dopant which gives the samples the best scintillation characteristics. In a point of fact atom of cerium can potentially leave the material at any or at a certain stage of the ceramics synthesis, reducing the total concentration of Ce^{3+} in YAG.

1. Introduction

The most of the scintillation materials used in the detectors of the ionizing radiation are used for the single-crystal form. It is well known that the growth of the scintillation crystals with the high melting temperature is a labor-intensive and expensive process. One of the perspective approaches to it is the replacement of the monocrystals with the ceramic materials [1].

The important factors influencing the scintillation properties are the high optical quality of material, the high effective atomic number, the low scintillation decay time, the uniformity of the activator ions' distribution in the crystal lattice. The ceramic technology holds out the high-transparent and the large-size garnet materials production opportunity at the lower temperatures and the homogeneous distribution or the gradient-controlled activator ions distribution [2, 3]. In this paper, we are mainly presenting the dependence between the research results on the optical and the luminescent properties of one of the known scintillation ceramic materials, namely, the aluminum-yttrium garnet $Y_3Al_5O_{12}$ activated by Ce^{3+} (Ce^{3+}:YAG) with the Ce^{3+} dopant concentration of 0.1 and 1 at.% on the synthesis technology features. We assume that the ceramics synthesis is the main link to a simultaneous achievement of the high-doping level and the uniformity of the activator ions distribution in the crystal lattice.

2. Experimental

We used the submicron commercial powders with >99,99% purity. The ceramic general flow chart is presented in table 1. The samples of the transparent Ce^{3+}:YAG ceramics were fabricated from our own production-received nanopowders Al_2O_3, Y_2O_3, 1 at.% Ce^{3+}:Y_2O_3 with the diameter of about 10–15 nm by means of the laser method [4]. The nanopowders were produced in the metastable phases. For their transformation in the main phases, the nanopowders Al_2O_3 and Y_2O_3, 1 at. % Ce^{3+}:Y_2O_3 were annealed at 1200 and 900 °C for 3 hours in the atmospheric muffle electric furnaces with the silicon-carbide heaters, respectively. The nanopowders Al_2O_3, Y_2O_3, 1 at.% Ce^{3+}:Y_2O_3 were used in the required stoichiometric proportion ((Ce + Y):Al = 3:5) for the synthesis of Ce^{3+}: YAG with the necessary concentration (0.1 and 1 at. %) of cerium. The mix of the nanopowders was mixed on a bead mill with an inclined axis in a plastic bottle for 48 hours using the YSZ balls. The mass ratio of the nanopowder: balls: ethanol was 1:4:8. Tetraethoxysilane (TEOS, $(C_2H_5O)_4Si$) of 0.5 wt.% (0.14 wt.% SiO_2) was used as a sintering aid in all samples. After that, the mix was evaporated in the rotational vacuum evaporator and was annealed at 600 °C for 3 hours to remove the alcohol fragments.

The received powders were molded into compacts (disks Ø14×4 mm) at 200 MPa in a metal mold to a density of 40–50% from the theoretical density of YAG (4.55 g/cm^3) by means of the cold uniaxially press (CUP) method. The compacts were annealed in the atmosphere at 800 °C for 3 hours to remove the organic impurity. The compacts were sintered in a high-temperature vacuum furnace with graphite heaters under 5×10^{-5} mbar vacuum and at 1700–1780 °C for 20 hours. After the vacuum sintering, the samples obtained the black color because of the oxygen deficiency in YAG structure. The samples were annealed in the atmosphere at 1300 °C for 5 hours to complete the oxygen deficiency.

Table 1. Ceramic general flow chart.

First method		**Second method**
Raw materials of yttrium oxide, cerium oxide and aluminum oxide with purity >99,99%		
Produce of nanopowder Al_2O_3 and Y_2O_3, 1 at.% Ce^{3+}:Y_2O_3 (10–15 nm) [4]		
Annealing of nanopowder (1200 and 900 °C/3 h)		
Mixing of nanopowder (Y+ Ce):(Al) = 3:5 (48 h) in ethyl alcohol (C_2H_5OH)		
Addition 0.5 wt.% TEOS, $(C_2H_5O)_4Si$ (0.14 wt.% SiO_2)		
Mix evaporation		
Annealing (600 °C/3 h)		
-		Briquetting (20%)
Passing (200 mesh)		Powders' mix transformation into YAG (1200 °C/3 h)
-		Milling of bricket (C_2H_5OH)
-		Mix evaporation
-	Annealing (600 °C/3 h)	
Cold uniaxially press (200 MPa)		
Annealing (800 °C/3 h)		
Vacuum sintering (1700 °C ~ 20 h)		(1780 °C ~ 20 h)
Annealing 1300 °C/5 h		
Optical polishing		
samples No.**2420 – 0.1 at.% Ce** No.**2405 – 1 at.% Ce**	**samples** No.**2421 – 0.1 at.% Ce,** No.**2406 – 1 at.% Ce**	**samples** No.**1854 – 0.1 at.% Ce,** No.**1851 – 1 at.% Ce**

During this procedure, the samples were synthesized in two different methods. The fundamental difference between these methods is in the synthesis of the main phase of the aluminum-yttrium garnet: directly during the vacuum sintering (1 – the first method) and before the vacuum sintering (2 – the second method). In the first method, the powders' mix after annealing have passed through a

200 mesh screen additionally and then were annealed (samples No.2406 – 1 at.% Ce, No.2421 – 0.1 at.% Ce), or were not annealed (samples No.2405– 1 at.% Ce, No.2420 – 0.1 at.% Ce) in the air atmosphere at 600 °C for 3 hours to remove the organic impurity. At the same time, the temperature of the vacuum sintering was 1700 °C for 20 hours. In the second method, even before the vacuum sintering (samples No.1851 – 1 at.% Ce, No.1854 – 0.1 at.% Ce), and after the annealing the powders' mix was transformed into YAG in briquettes with the density of 20 % from the theoretical density at 1200 °C for 3 hours in the atmospheric furnaces. Further, the briquettes were crushed and ball milled similarly to the procedure of mixing powders. At the same time, the temperature of the vacuum sintering was 1780 °C for 20 hours. After that, the ceramics were polished. The methods, in which we received optical ceramics earlier [5] underlie the synthesis of these ceramics.

The specific surface area of powders (BET analysis) was determined by the automated gas adsorption analyzer (TriStar 3000, Micromeritics Instrument Corporation, USA). The structures of the sintered ceramics were determined by an X-ray diffraction (XRD) on the copper radiation (Cu $K_{\alpha 1,2}$ λ=1,542Å) with the graphite monochromator on a secondary beam (D8 Discover GADDS, Bruker AXS, Germany). The processing of the XRD results was carried out via the full-height analysis program named TOPAS 3. The surface morphology was studied with the optical microscope (BX51TRF-5, Olympus Corp., Japan). The optical transmittance and the absorption spectra of the obtained Ce^{3+}:YAG ceramics samples were measured using a UV-2450 (Shimadzu, Japan) optical dual-beam spectrophotometer (200–900 nm). The X-ray luminescence (XRL) spectra were measured using a monochromator MDR-23 equipped by an X-ray source of URS-1,0 (40 kV, 10 mA, W-anode) and Hamamatsu R928 and FEU-106 photomultipliers (experimental physics department, UrFU, Russia).

3. Results and discussion

The results of the implementation of the above-stated procedures by means of the solid-state reaction method, 6 samples (Ø10×2 mm) with the Ce^{3+} dopant concentration of 0.1 and 1 at.% are fabricated (figure 1).

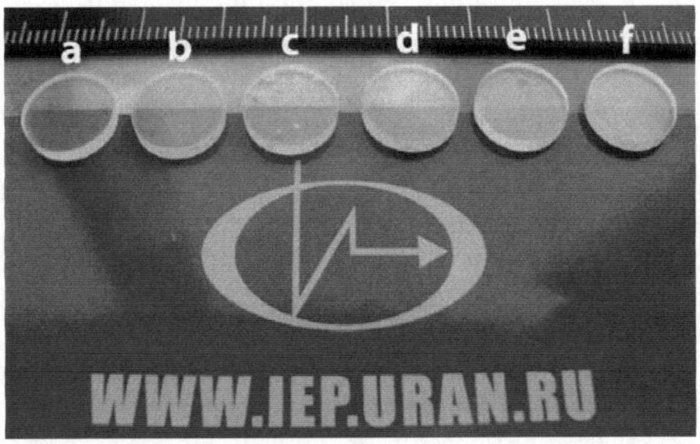

Figure 1. Photograph of Ce^{3+}:YAG ceramics samples: the first method: No.2420 – 0.1 at.% Ce (a), No.2405 – 1 at.% Ce (b); annealing after sifting samples No.2421 – 0.1 at.% Ce (c), No.2406 – 1 at.% Ce (d); the second method: No.1854 – 0.1 at.% Ce (e), No.1851 – 1 at.% Ce (f).

As it was already mentioned in this paper, the fundamental difference between the two methods is in the synthesis of the main phase of the aluminum-yttrium garnet from the yttrium and aluminum oxides powders: directly during the vacuum sintering (1 – the first method) and before the vacuum sintering (2 – the second method). This difference is visible on an X-ray phase pattern (XRD) of the compacts: the first method – figure 2a, the second method – figure 2b.

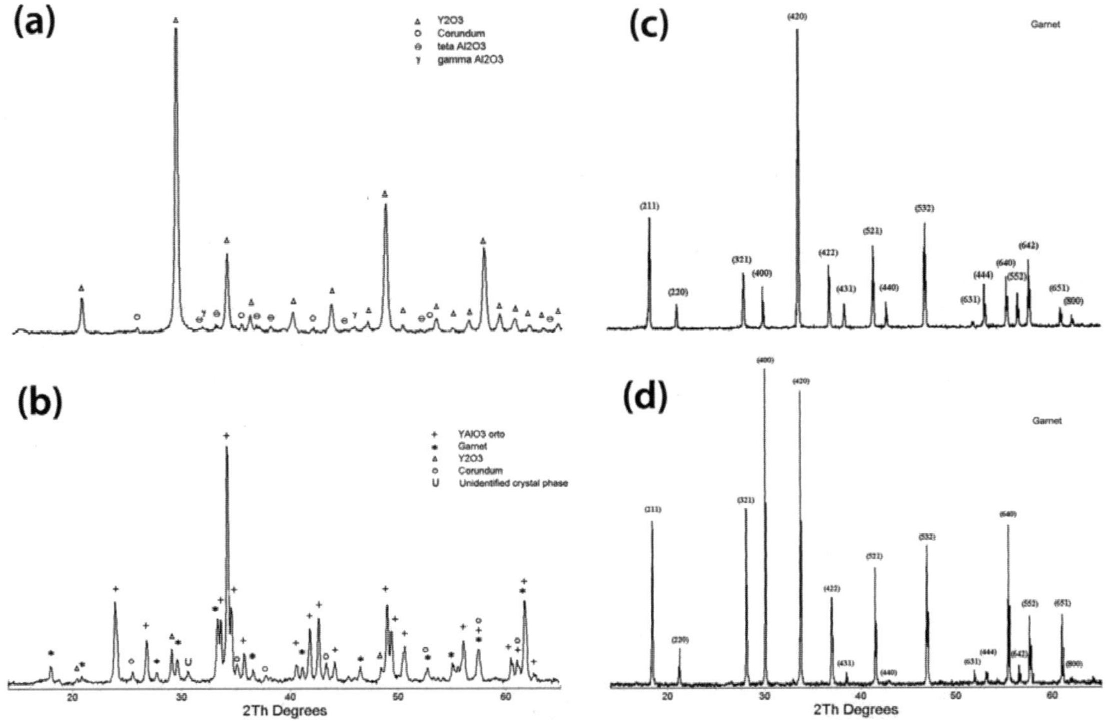

Figure 2. XRD pattern of compacts: the first method (a), the second method (b); and ceramics: the first method (c), the second method (d).

The compact (the first method) was annealed at 800 °C for 10 hours. According to the technology, the removal of the organic chemistry annealing for 3 hours has been carried out, but for the precise fixation of the reactions lack between the yttrium and aluminum oxides the time has been increased to 10 hours. The research has shown that there are no reactions between the oxides (figure 2a). Herewith, XRD shows the existence of the cubic phase Y_2O_3 (60 wt.%) and the crystallites average sizes (OKR) = 43(5) nm, the existence of aluminum oxide various phases: rhombohedral corundum (10 wt.%) and OKR = 48(10) nm, monoclinic θ-phase Al_2O_3 (20 wt.%) and OKR = 22(2) nm, cubic γ-phase Al_2O_3 (10 wt.%) and OKR = 29(3) nm. The existence of a large number of various aluminum oxide phases is caused by the pre-annealing of the nanopowder and the compact.

The compact (the second method) was also annealed at 800 °C for 10 hours. In XRD evidence (figure 2b) it contains the next phases: the orthorhombic perovskite phase $YAlO_3$ (65 wt.%) and OKR = 90(20) nm, the rhombohedral corundum (17 wt.%) and OKR = 70(10) nm, the cubic garnet $Y_3Al_5O_{12}$ (14 wt.%) and OKR = 78(6) nm, the cubic Y_2O_3 (4 wt.%) and OKR = 54(6) nm, the unidentified crystal phase (<1 wt.%). Most probably, the unidentified crystal phase is one of aluminum oxide modifications. As we can see from XRD (figure 2b), the parent compact has been received from the reacted oxide powders which have formed a phase cubic garnet $Y_3Al_5O_{12}$ required for orthorhombic perovskite phase $YAlO_3$ and the non-reacted parent material: the yttrium and the aluminum oxides. Besides, we also observe crystallites size (OKR) increase and phase transformations natural to heat treatment.

As a result of both ways, after vacuum sintering, 100% YAG phase ceramics are fabricated that is confirmed by XRD (figure 2c, 2d).

One of the important ceramics synthesis stages is the powder molding stage, therefore, the nanopowders compaction considerably differing from the same pressure molding (in this case is 200 MPa) is also notable. So, for example, in the first method, the density of a compact was about

2.02 g/cm^3 (44.4% of YAG density), for the second method – 2.18 g/cm^3 (47.9% o of YAG density). The powders in the second case (second method) were coarser after the calcination and the milling compared to the initial nanopowders in the first method. The nanopowders compaction problem still exists and is under study [6, 7]: there is less size particle so much the worse they consolidate. The density of the initial powder particles affects the final characteristics of the optical ceramics.

Besides, the final dispersion of particles on the choice of the corresponding YAG synthesis temperature was also considered. In the first method, the temperature of 1700 °C has been chosen, as the smaller powder was used, and respectively, the more active in terms of sintering because of the more developed particles surfaces. For the second method, the temperature of 1780 °C has been chosen, as the coarser powder was used. The powder was coarsened even after the milling because of the powder agglomeration and the phase transformations. On BET evidence powders analysis for the first method (before the CUP stage), the surface area yttrium oxide is 17.92 m^2/g, aluminum oxide is 57.5 m^2/g. On BET evidence powders analysis for the second method (before a CUP stage) surface area is 11.7 m^2/g. Data of the powders BET analysis is agreed with XRD crystallites average sizes (OKR) fore-quoted (figure 2a, 2b) where OKR of materials phases in a compact much greater than the parent nanopowders OKR sizes.

For a microstructure research, the ceramic's side plane was polished and then was subjected to a thermos-etching in the air at a temperature of 1350 °C for 5 hours. The research has shown that the ceramics average grain sizes do not depend on the method of fabricating. There is a dependence between the samples and the various concentration of ions Ce^{3+} (figure 3). The average grain size decreases from 24–27 μm (0.1 at.% Ce) up to 17 μm (1 at.% Ce). The grain size decrease, most likely, is related to a higher dopant Ce^{3+} concentration, which inhibits the grain growth because of the various cerium Ce^{3+} ionic radiuses in comparison with the yttrium Y^{3+} ionic radius.

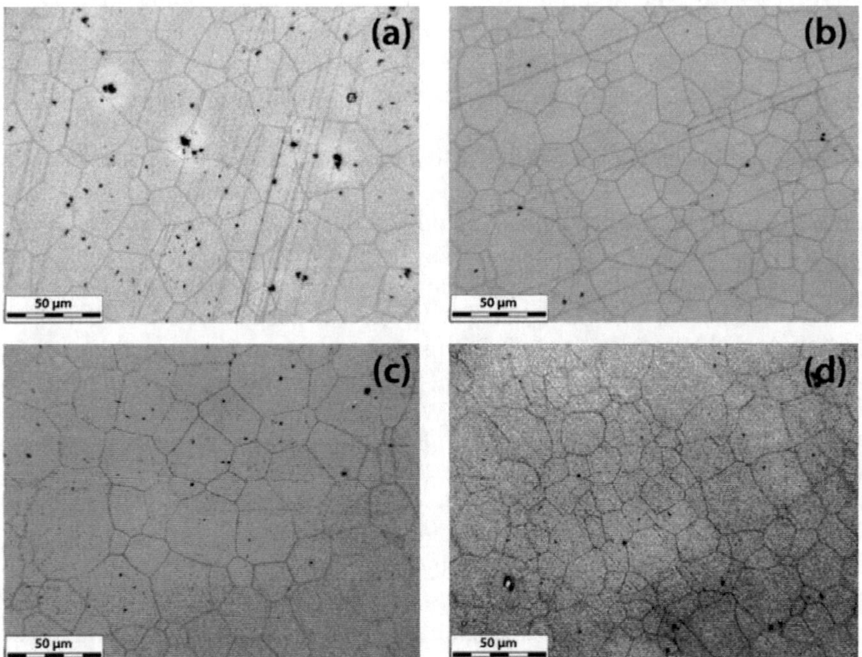

Figure 3. The photographs of the polished Ce^{3+}: YAG ceramics samples surfaces:
the first method: 0.1 at. % Ce (a), 1 at. % Ce (b);
the second method: 0.1 at. % Ce (c), 1 at. % Ce (d).

The optical absorption and transmission spectra of the investigated samples are shown in figure 4a and 4b, respectively. The spectra demonstrate the noticeable absorption bands with the maxima at 228, 260, 304, 338, and 457 nm, associated with 4f→5d transitions in Ce^{3+} ions.

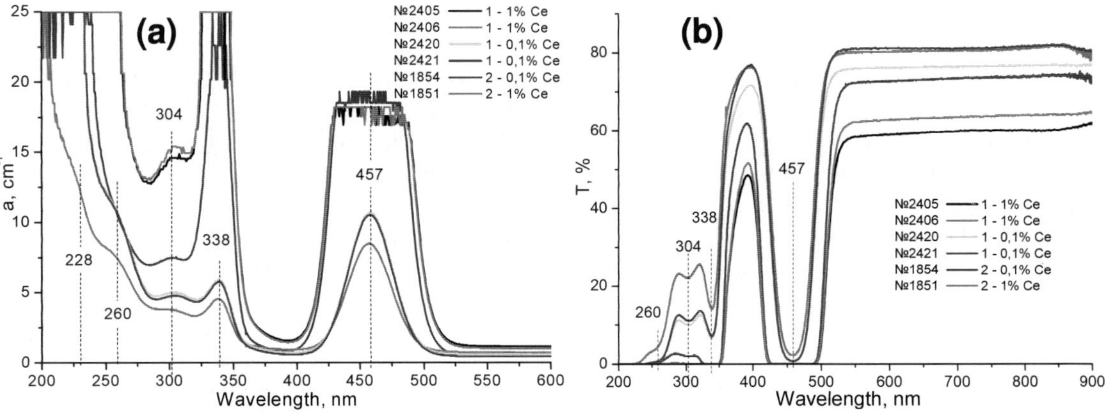

Figure 4. The optical absorption (a) and transmittance (b) spectra of ceramic Ce:YAG samples.

As can be seen from the optical absorption spectra (figure 4a), the intensity of the above absorption bands of the samples synthesized in the framework of one method essentially depends on the concentration of the dopant. However, it is also seen that the intensity of the absorption bands in the samples with the same concentration of charged dopant and synthesized by different methods, is different (for example, spectra for samples No.2420, 2421 (first method) and No.1854 (second method), figure 4a). As a result of this analysis of the optical absorption spectra, it can be concluded that the samples synthesized by means of the second method contain a smaller amount of Ce^{3+} ions. Based on this, the first method can be considered as the most preferable in terms of the exact dosage of the dopant.

The optical transmission spectra (figure 4b) reveal that the transmittance of the ceramics samples at a 600 nm wavelength was 58 to 82%. We also noted that the transmission coefficient in the 500–900 nm region (figure 4b), which determines the optical quality of the samples, directly correlates to the concentration of the Ce^{3+} ions in YAG lattice. The low value of the cerium ions concentration corresponds to the high optical quality of the samples. We assume that one of the reasons for this effect can be the influence of the grain size on the optical properties of the ceramic samples. As mentioned above, with increasing Ce^{3+} concentration the average grain size in the sample is decreased. With a smaller grain size, the additional light-scattering centers have appeared leading to a degradation of the optical quality of the samples. We believe that an increase in the vacuum sintering time should lead to an increase in the grain size, which causes an increase in the ceramics transparency with a higher concentration of the Ce^{3+} dopant.

The following effect should also be noticed. The samples with the identical dopant concentrations synthesized with the first method (samples No.2420, No.2421 (0.1% Ce) and No.2405, No.2406 (1% Ce)) have a different transmission coefficient at 500–900 nm wavelength region (see figure 4b). The grain sizes in the pairs of samples mentioned above are the same. The difference in pairs of the samples lies in the presence or absence of the additional annealing having passed through a 200 mesh screen during the synthesis (see table 1). We believe that the samples that have not undergone the additional annealing contain more pores and defects caused by the contamination of the organic impurity in the initial powder.

The XRL spectra of the samples were measured in the range of 200–800 nm (figure 5). It is evident from the figure that all the samples exhibit an intense luminescence with a band maximum at the 525–538 nm region corresponding to the radiative *d-f* transitions ($^2D \rightarrow ^2F_{5/2}$ and $^2D \rightarrow ^2F_{7/2}$) in the Ce^{3+} ions. On the normalized XRL spectra (figure 5b), it is demonstrated that the increase in the concentration of the dopant in the samples leads to the maximum shift of the luminescence spectrum from 525 to 538 nm. In the 240–460 nm region, a luminescence band associated with the decay of the self-trapped excitons (STE) in the YAG host [8]. The complex shape of the STE band is due to the presence of the absorption bands from Ce^{3+} in this region. In addition, the luminescence bands of the uncontrolled rare earth ions Gd^{3+} and Tb^{3+} (figure 5) are observed in the XRL spectra. The presence of such contamination in negligible quantities does not affect the main luminescence band of Ce^{3+} ions.

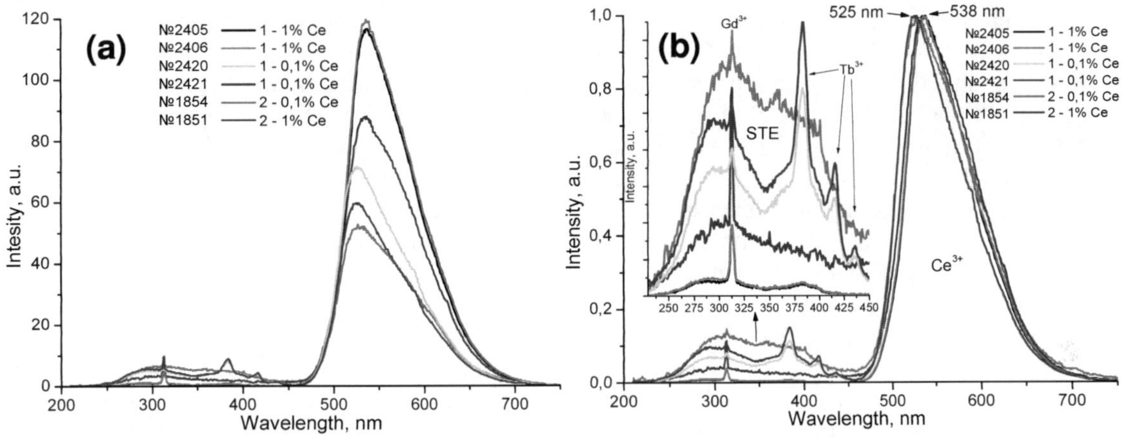

Figure 5. The un-normalized (a) and normalized (b) XRL spectra of ceramic Ce^{3+}:YAG samples.

The investigated samples with the highest luminescence intensity (figure 5a) are samples synthesized in the framework in the first method with 1% of Ce (No.2405 and No.2406). Among the samples with a cerium concentration of 0.1%, the sample with highest intensity is No.2420. It is interesting to note that this sample has a worse optical quality than similar samples with 0.1% of Ce. However, in this case, a decrease in the optical quality of the sample compensated by an increase in the concentration of the dopant. For a more accurate estimation of the relative content of Ce^{3+} in YAG ceramics, the attention should be drawn to the luminescence band of the STE. The intensity of the STE band in this case indirectly indicates the relative concentration of the cerium ions in YAG (inset in figure 5b). The higher is the concentration of cerium ions in YAG, the lower is the intensity of the emission band of the STE. The intensity of the emission band of the STE for sample No.2420 is lower than that of samples No.2421 and 1854, which means that the concentration of Ce^{3+} ions in it is greater.

As a result of the similar STE band intensity analysis in the XRL spectra, considering at the same time the results of the optical absorption and the spectra transmission analysis (as mentioned above), we may conclude that the content of the Ce^{3+}ions in YAG lattice in the second method of the ceramics synthesis is less.

4. Conclusions

We have fabricated the samples of the 0.1 at.% Ce:YAG and 1 at.% Ce:YAG sized Ø10×2 mm ceramics generated from the nanopowders with the diameter of about 10–15 nm by means of two different methods. The fundamental difference between the two methods is in the synthesis of the main phase of the aluminum-yttrium garnet: directly during the vacuum sintering process (1 – the first method), and before the vacuum sintering (2 – the second method). The result of the research carried by means of the XRD method has revealed that during the synthesis of the samples at the temperature of 800 °C no reactions between the initial oxides have been detected. Upon the increase in the temperature of the synthesis to 1200 °C, the samples contained 14% YAG and 66% YAP. The temperatures of 1700 °C (in the first method) and 1780 °C (in the second method) caused the formation of the single-phase compositions of YAG.

It has been demonstrated that the increase in the concentration of the dopant from 0.1 at.% Ce^{3+} to 1 at.% Ce^{3+} in the samples synthesized by means of these two methods has caused the average grain size to reduce from 24–27 μm to 17 μm, which in turn, is one of the reasons for the decrease in the transparency of the ceramics' samples in the wavelength ranged 500–600 nm from 82 to 58%. During the synthesis of the main phase of the aluminum-yttrium garnet in ceramics directly during the sintering of the samples, the absence of the intermediate annealing phase after the passing (200 mesh) of the powder before pressing (CUP) leads to a decrease in the transparency of the finished ceramics.

The luminescent-optical measurements have shown that the brightest glow among the synthesized ceramics was obtained by the samples synthesized by means of the first method at a 1% Ce. The shift of the maximum XRL spectrum from 525 to 538 nm with the increase of the dopant concentration from 0.1 at.% Ce^{3+} to 1 at.% Ce^{3+} has been detected. In the XRL spectra synthesized samples the dependence of the variation intensity effect of the STE luminescence band on the concentration of the Ce^{3+} ions in YAG has been discovered, which allows to estimate the relative concentration of the Ce^{3+} ions in YAG with more certainty than by means of the spectra of the absorption and transmission. It has been demonstrated that the samples synthesized in the second method contain fewer amounts of Ce^{3+} eons than those synthesized in the first method, whereas the initial dopant concentration in the ready-made samples is the same.

The above allows us to conclude that:

Firstly, the given comparison of the two methods to the synthesis of the Ce:YAG ceramics, the atom of cerium Ce can be not only in the condition of Ce^{3+} but can potentially leave the material at any or a specific stage of the ceramics synthesis reducing the final concentration of Ce^{3+} in YAG;

Secondly, the first method of the synthesis (the synthesis of the main phase of the aluminum-yttrium garnet directly during the vacuum sintering) is the most preferable in terms of the exact dosage of the dopant and from the point of view of simplicity, time and energy consumption and, consequently, the cost-efficiency the of optical scintillation ceramics production technology based on YAG.

Acknowledgments
The work was fulfilled in the frame of state task project №0389-2016-0002, 2018-2020 year. The authors gratefully acknowledge Dr V.V. Platonov and R.N. Maksimov from Institute of Electrophysics UrB RAS for laser synthesis of nanopowders and optical polishing, respectively.

References

[1] Greskovich C, Duclos S 1997 *Annu. Rev. Mater. Sci.* **27** (1) 69–88
[2] Osipov V, Maksimov R, Shitov V et al 2017 *Opt. Mater.* **71** 45 – 49
[3] Lee S, Kochawattana S and Messing G 2006 *J. Am. Ceram. Soc.* **89** (60) 1945 – 1950
[4] Osipov V, Kotov Yu, Ivanov M et al 2006 *Las. Phys.* **16** 116 – 125
[5] Bagayev S, Osipov V, Solomonov V et al 2012 *Opt. Mater.* **34** (8) 1482 – 1487
[6] Boltachev G, Lukyashin K, Shitov V and Volkov N 2013 *Physical review E* **88** 012209
[7] Boltachev G, Volkov N, Lukyashin K et al 2017 *IOP Conference Series: Materials Science and Engineering* **208** (1) 012008
[8] Osipov V, Ishchenko A, Shitov V et al 2017 *Opt. Mater.* **71** 98 – 102

IOP Conf. Series: Materials Science and Engineering **347** (2018) 012014 doi:10.1088/1757-899X/347/1/012014

Study of strength kinetics of sand concrete system of accelerated hardening

A V Sharanova, D A Lenkova and A D Panfilova

Institute of Environmental Management, Urban Development and Spatial Planning, Immanuel Kant Baltic Federal University, 14 Nevskogo street, Kaliningrad 236016, Russia

E-mail: a.sharanova@gmail.com

Abstract. Methods of calorimetric analysis are used to study the dynamics of the hydration processes of concretes with different accelerator contents. The efficiency of the isothermal calorimetry method is shown for study of strength kinetics of concrete mixtures of accelerated hardening, promising for additive technologies in civil engineering.

1. Introduction

Nowadays the use of additive technologies in construction industry is one of the most actual subjects of research. Growth of interest in such technologies in construction is conditioned by a number of factors: high level of production automation, improvement of production quality, acceleration of creation processes, possibility of optimization of CAD models, reduction of production wastes [1]. These factors represent a great interest for the application of additive technologies in civil engineering and use of 3D prototyping in place of the traditional methods of construction.

The main problem for the use of additive technologies in construction is the materials used, since the ordinary concrete does not meet the requirements of 3D printing productivity [2]. Special composition for building printing should harden quickly enough to withstand the weight of subsequently deposited layers without deformation. Composites that harden for a long time slow down the process of erecting the building and cannot meet the requirements of 3D printing productivity [1]. There is a need to select and determine the optimal composition of accelerating additives to reduce the setting time of the concrete mixture.

A large number of recent works on the control of setting time show the relevance of this subject of research. Shakor P [3] uses a printed composition based on a mixture of ordinary Portland cement (32.2%) and high aluminate cement (67.8%) with the addition of a lithium carbonate accelerator (4.5%). Lithium carbonate provides hardening acceleration, high early strength and good adhesion. Kazemian A et al [4] examine the effect of the accelerating additive based on calcium chloride on the workability of the concrete mixture for the building 3D printer. It is noted that increasing the dose of accelerator from 2% to 3% has no significant effect on the loss of workability of concrete in the first 90 minutes after the preparation of the mixture. In addition to the accelerators, retarders of setting time are also widely used to maintain the concrete mixture in the liquid state for the time necessary for the extrusion of the material. Retarders and accelerators of the setting time are added at different stages of production [5]. Sulphoaluminate cement could be used as an accelerator. In the work [6] a mixture of ordinary Portland cement (93%) with a sulphoaluminate cement additive (7%) is studied as a working

IOP Conf. Series: Materials Science and Engineering **347** (2018) 012014 doi:10.1088/1757-899X/347/1/012014

material for a building 3D printer. The use of sulphoaluminate cement reduces the start and the end of setting. In the same study, a calorimetric analysis of cement paste was performed to determine the effect of the addition of sulfoaluminate cement on the hydration process. The total heat of the mixture is close to the heat release of pure Portland cement, with the exception of reduction of the induction period, with leads to an acceleration of hardening.

One of the effective methods for estimating the kinetics of cement hydration is the isothermal calorimetry method [8-10]. In this study, the efficiency of the isothermal calorimetry is evaluated to determine the optimum concentration of the accelerating additive.

2. Materials and Methods

To evaluate the strength properties of sans concrete, mechanical tests were carried out on beam samples made from compositions with different contents of the accelerating additive. The flexural tensile strength and compressive strength were measured in the first 18 and 24 hours after the samples were made. At the same time, the calorimetric analysis of the studied compositions was performed.

2.1. Materials

Four compositions of sand concrete with different accelerator contents were used. For the preparation of sand concrete the following materials were used: 1) sand for construction: fraction 0-3 мм, size modulus M_S=3.15, mining: Kaliningrad region; 2) Portland cement Eurocem 500 super, CEM I 42.5 H, LLC "Petersburg cement"; 3) hardening accelerator of concrete and mortars "SPEED Vincent polyline", component: calcium chloride 25-35%.

Calculation of the concrete composition was made in accordance with GOST 26633-2015 using the "Recommendation on the selection of heavy and sand concretes" [7]. The accelerator was used in dosages of 1L, 2L and 3L for 50 kg of cement, respectively, which for water/binder ratio W/B=0.5 corresponds to 4%, 8% and 12% of water volume. The consumption of materials per 1m^3 of concrete mixture is given in table 1. The samples were hardened in a chamber of normal hardening at a temperature of 20±1 °C and an air humidity not less than 90%.

Table 1. Consumption of materials.

No. of composition	Cement (kg/m^3)	Sand(kg/m^3)	Water (kg/m^3)	Accelerator (L)
1	459.6	1470.7	229.8	0
2	459.6	1470.7	220.6	9.2
3	459.6	1470.7	211.42	18.38
4	459.6	1470.7	202.23	27.57

2.2. Mechanical testing

To carry out mechanical tests for flexural tensile strength and compressive strength, control beam samples measuring 40×40×160 mm were made (GOST 310.4-81). Strength tests of samples aged 18 and 24 hours were conducted using servo-hydraulic testing machine ToniPrax by ToniTechnik Company.

2.3. Calorimetric analysis

Calorimetric analysis of the four test compositions was carried out in accordance with GOST 310.5-88 using an 8-channel isothermal calorimeter TAM Air at a constant temperature 20°C inside the measuring channel. Standard ampoules of 20 ml were used for the analysis. Dry cement weighing 6 g was placed in an ampoule, then, using a special device, 3g of liquid (water + accelerator of the necessary concentration) were added to the ampoule and thoroughly mixed with a hand-bladed stirrer. Thus, water/binder ration of 0.5 was maintained, which was used for the preparation of samples for mechanical testing. Immediately after adding liquid and stirring, the ampoules with the mortars were sealed and placed in a calorimeter for analysis.

3. Results and Discussion

3.1. Mechanical testing

The test results are shown in table 2.

Table 2. Results of mechanical tests.

No. of composition	Age (hours)	$R_{tens.}$(MPa)	$R_{comp.}$(MPa)	Age (hours)	$R_{tens.}$(MPa)	$R_{comp.}$(MPa)
1	18	0.63	4.09	24	1.41	8.57
2	18	0.63	3.88	24	0.94	6.48
3	18	0.94	6.66	24	1.56	10.95
4	18	0.86	5.53	24	1.02	7.56

Dynamics of the tensile strength $R_{tens.}$ and compressive strength $R_{comp.}$ of the test samples are shown in figures 1 and 2 by the dependencies of the strength limits on the setting time. Figures 3 and 4 show the tensile strength and compressive strength dependencies on the concentration of hardening accelerator.

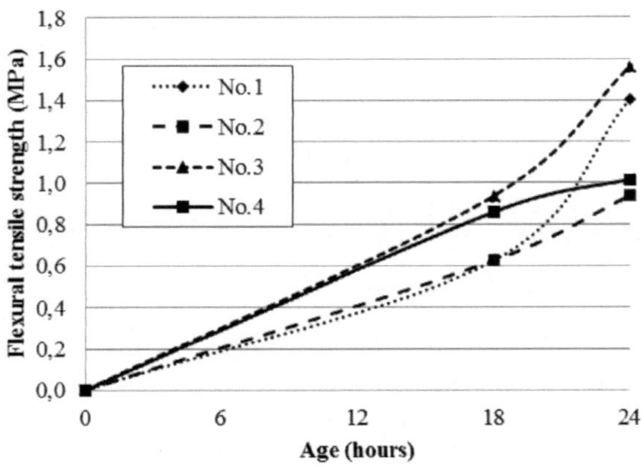

Figure 1. Graph of tensile strength growth.

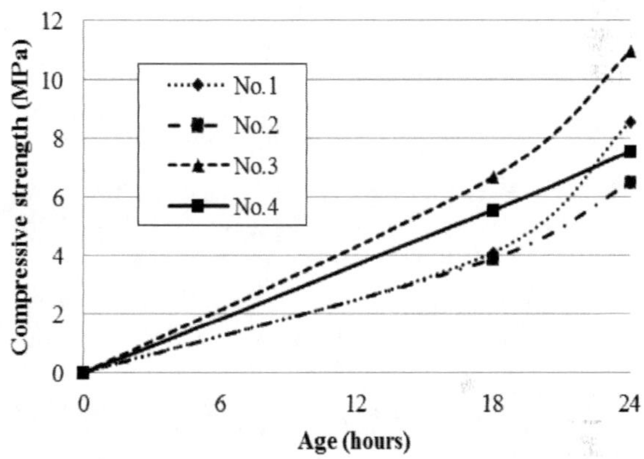

Figure 2. Graph of compressive strength growth.

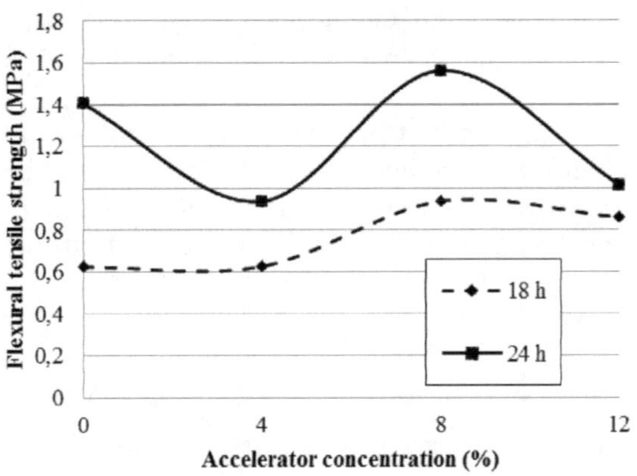

Figure 3. Dependence of $R_{tens.}$ on the accelerator concentration.

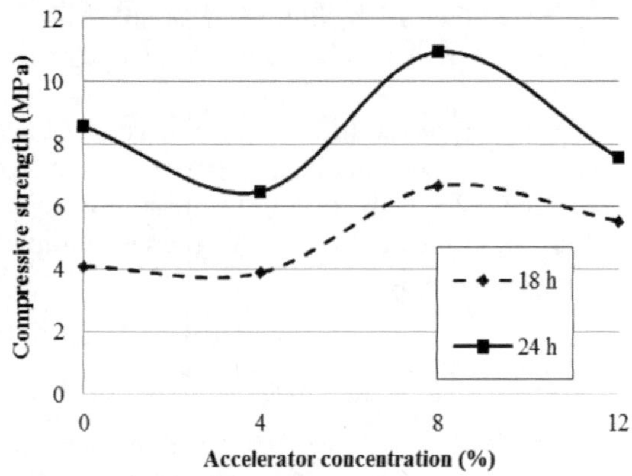

Figure 4. Dependence of $R_{comp.}$ on the accelerator concentration.

Analyzing the graphs in figures 3 and 4, we conclude that there is a nonlinear dependence of the strength on the accelerator concentration. There is an optimum value of the accelerator concentration, which ensures a rapid increase in the strength of concrete without loss of final strength characteristics. In this study, composition No.3 can be considered as optimal.

3.2. Calorimetric analysis

Isothermal calorimetry allows to accurately measure the heat flow change in time. Due to isothermal conditions during the measurement, it is possible to detect all the exothermic reactions associated with the hydration of fresh cementitious materials [8]. The heat flow graph and heat release graph of the test samples are shown in figures 5 and 6 respectively.

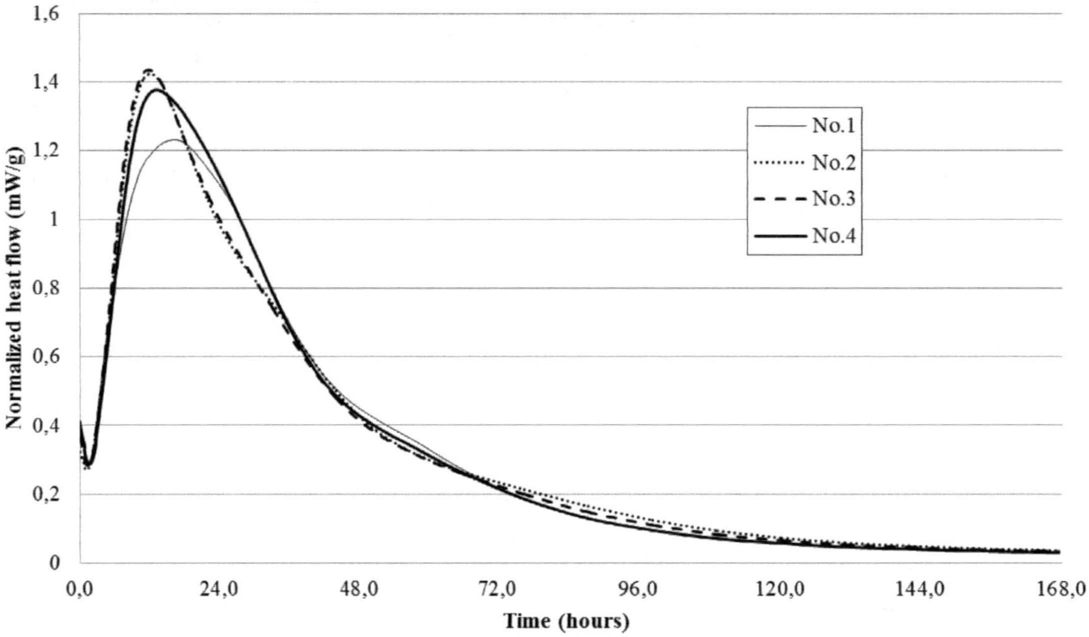

Figure 5. Heat flow graph.

Peaks on the graph of heat flow indicate the passage of the main hydration reaction of cement, or so-called period of acceleration of reactions [9]. The higher is the peak, the more intense is the reaction. The analysis of the heat flow curves in figure 5 shows the increased heat dissipation rate at the early stage of hydration in samples No.2 and No.3 in relation to samples No.1 and No.4. This indicates a real acceleration of the hydration process of cement when the accelerating additive is injected.

Analysis of the heat release curves, obtained using the isothermal calorimetry method, allows to evaluate the dynamic characteristics of the hydration processes of compositions with and without an accelerator [10]. According to the summary heat release shown in figure 6, the accelerating effects of additives in the control ages of the test compounds were calculated. The results of the calculations are presented in table 3.

Table 3. Determination of the accelerating effect.

Total heat energy (J/g)	Composition No.1	Composition No.2		Composition No.3		Composition No.4	
	Age (hours)	Age (hours)	Acceleration (%)	Age (hours)	Acceleration (%)	Age (hours)	Acceleration (%)
56.8	18	17	5.6	16.3	9.4	16.5	8.4
82.2	24	22.6	5.8	22.2	7.5	22.4	6.7

IOP Conf. Series: Materials Science and Engineering **347** (2018) 012014 doi:10.1088/1757-899X/347/1/012014

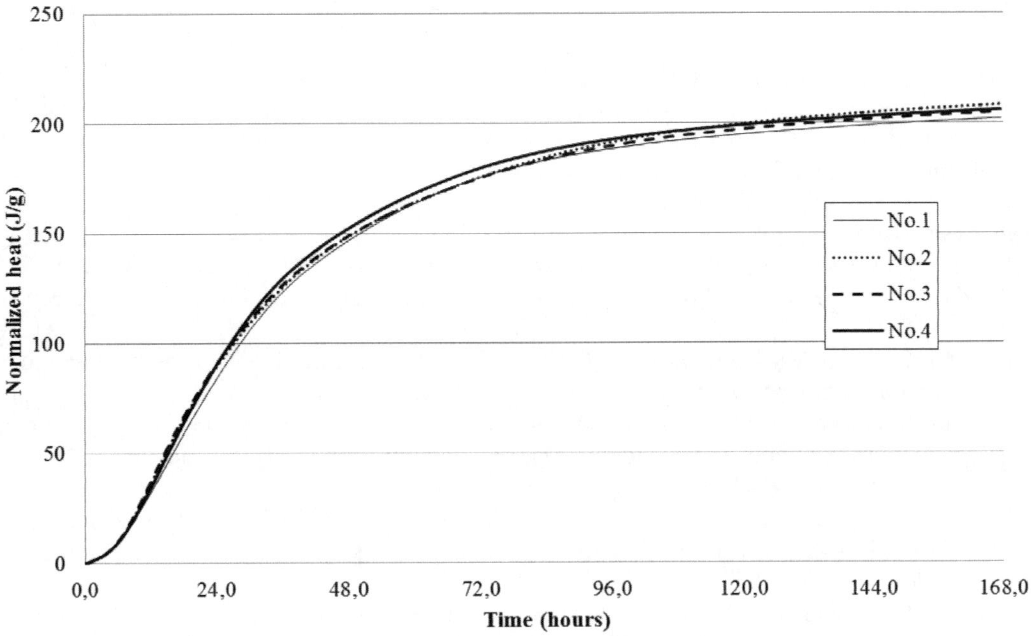

Figure 6. Heat graph.

According to the data in table 3, we conclude that the greatest accelerating effect from the introduction of the additive is observed in the composition No.3.

The graphs in figures 1, 2 and 5 show that one of the composition is the most active in the strength set at the early stages of hardening (18 and 24 hours) compared to others, and the data in table 3 confirm this.

The hardening of a concrete mixture is a highly exothermic physicochemical process. Calorimetric tests allow us to investigate the dynamics of thermal processes.

The results of the evaluation of the strength characteristics and the thermodynamics of the hardening process correlate well with each other. This allows us to conclude that the method of isothermal calorimetry is applicable for studying the strength kinetics of sand concrete systems of accelerated hardening.

4. Conclusion

This paper presents the results of the study of the strength kinetics of sand concrete systems of accelerated hardening. The samples of sand concrete with different contents of the accelerating additive were subjected to mechanical strength tests at the age of 18 and 24 hours after manufacturing. Also, the calorimetric analysis of concrete mixtures with different accelerator contents was carried out. According to the results of mechanical tests, the composition with the best strength parameters at the early stages of hardening was determined. This allows us to state that there is an optimum value of the accelerator concentration, which ensures a rapid increase in the strength of concrete without loss of final strength characteristics. Calorimetric analysis confirmed the accelerating effect of the additive at the stage of cement hydration. According to the heat release graph, the accelerating effect of the additives was calculated. The composition with the optimum accelerator concentration was determined, this composition corresponds to the results of mechanical tests.

Analysis of the results of mechanical and calorimetric tests showed that the method of isothermal calorimetry is applicable for studying the strength kinetics of sand concrete systems of accelerated hardening. It becomes possible to specify the direction of searching for methods of nondestructive testing of the degree of materials strength setting in the printing process by the methods of additive technologies in construction.

Acknowledgments

The reported study was carried out within the framework of the project "Materials for additive technologies" under the scientific supervision of Dr. Dmitrieva Maria. The project was implemented within the framework of support of scientific groups of students, postgraduates, young scientific and pedagogical workers, at the expense of the subsidy provided under the Program of Competitive Growth of Immanuel Kant Baltic Federal University.

References

[1] Vatin N, Chumadova L, Goncharov I, Zykova V, Karpenya A, Kim A and Finashenkov E 2017 3D printing in construction *Construction of Unique Buildings and Structures* **1(52)** 27–46 (rus)

[2] Hager I, Golonka A and Putanowicz R 2016 3D printing of buildings and building components as the future of sustainable construction? *Procedia Engineering* **151** 292–99

[3] Shakor P, Sanjayan J, Nazari A and Nejadi S 2017 Modified 3D printed powder to cement-based material and mechanical properties of cement scaffold used in 3D printing *Construction and Building Materials* **138** 398–409

[4] Kazemian A, Yuan X, Cochran E and Khoshnevis B 2017 Cementitious materials for construction-scale 3D printing: laboratory testing of fresh printing mixture *Construction and Building Materials* **145** 639–47

[5] Malaeb Z, Hachem H, Tourbah A, Maalouf T, El Zarwi N and Hamzeh F 2015 3D concrete printing: machine and mix design *Int. Journal of Civil Engineering and Technology* **6** 14–22

[6] Khalil N, Aouad G, El Cheikh K and Remond S 2017 Use of calcium sulfoaluminate cements for setting control of 3D-printing mortars *Construction and Building Materials* **157** 382–91

[7] Brusser M I, Kaprielov S S, Podmazova S A, Titova L A and Sheinfeld A V 2016 *Rekomendacii po podbory sostavov betonnih smesei dlia tiajelikh i melkozernistih betonov* (Moscow: NIC Stroitelstvo) p 100 (rus)

[8] Boháč M, Kubátová D, Nečas R, Zezulová A, Masárová A and Novotný R 2016 Properties of cement pastes with zeolite during early stage of hydration *Procedia Engineering* **151** 2–9

[9] Adamtsevich A and Pustovgar A 2015 *Sukhie stroitelnye smesi* **4** 26–29 (rus)

[10] Adamtsevich A, Pashkevich S and Pustovgar A 2013 *Ingenerno-stroitelnyi journal* **3** 36–42 (rus)

IOP Conf. Series: Materials Science and Engineering **347** (2018) 012015 doi:10.1088/1757-899X/347/1/012015

Biodegradable materials containing recycled polymers

M V Podzorova[1], Yu V Tertyshnaya[2] and A A Popov[1,2]

[1] Plekhanov Russian University of Economics, 36 Stremyanny per., Moscow, 117997, Russia

[2] Emanuel Institute of Biochemical Physics, Russian Academy of Sciences, 4 Kosygina str., Moscow, 119991, Russia

E-mail: mariapdz@mail.ru

Abstract. The work is devoted to study the effects of different environmental factors such as water, oxygen and, light composition based on polylactide and polyethylene of low density with the addition of oxidized polyethylene, as an analog of recycled materials. Established that in the composition polylactide – polyethylene at the first stage the significant impact of moisture and UV light. The influence of UV radiation on polylactide destruction was proved by differential scanning calorimetry (DSC). It is found that polylactic acid is oxidized slower than polyethylene.

1. Introduction

At the present time, the effect of temperature, water, oxygen, light, microorganisms on biodegradable polymers such as polylactide [1-4], polyhydroxybutyrate [5-7], polycaprolactone and composites on their basis are studied intensively. Polylactide or polylactic acid (PLA) is the polyester with a good range of physical and mechanical properties, biocompatibility with the human body, as well as the ability to biodegrade in the environment [4]. At low temperature, microbiological degradation of PLA is strongly retarded [8].

The main problem for those blends is the low compatibility between systems. The idea of using highly degraded polyethylene as a compatibilizer has good prospects [9]. In this study highly oxidated low density polyethylene (LDPEox) was used as a model of recycled polyethylene. The idea was to improve adhesion between PLA and neat low density polyethylene (LDPE) by the addition of LDPEox. Furthermore, the obtained triple blends, made of PLA, LDPE, LDPEox should be more prone to biodegradation than the double ones, made of PLA and LDPE. The influence of the environmental factors on the polymeric materials is extremely important [10-12]. That is why it is important to determine the influence of ultraviolet radiation, water absorption and thermal oxidation on the polymer degradation.

2. Experimental

In this study the authors evaluated the influence of ultraviolet radiation, incubation in the soil, and water absorption of double blends made of PLA and LDPE, and triple blends made of PLA, LDPE and oxidized LDPE. Different methods were used in this study: differential scanning calorimetry (DSC), thermal oxidation, influence of ultraviolet and water absorption.

Blends of LDPE (TM 15803-020 from Neftekhimsevilen, OJSC, Russia) with PLA (4032D from Nature works, USA) in a wide range of compositions were studied. The PLA content in the polymer matrix was 0, 20, 30, 50, 70, 80, 100 wt%. There were also made three triple blends with 10, 20, 30 wt% of LDPEox (LDPE held in air, in an open vessel, at a temperature of 90°C for 250 hours, TM 15803-020 from Neftekhimsevilen OJSC, Russia) added to 60, 50, 40 wt% of neat LDPE respectively and 30 wt% of PLA. The blends were prepared by the Plasticorder PLD-651 (Brabender) at a temperature of 180°C for 6 minutes. After cooling, the samples of the obtained material were milled using a knife mill RM-120 and then compressed in a manual hydraulic press PRG-10 at 180°C under a load of 7 kN followed by rapid cooling. As a result, film round samples with a diameter of 8 cm and thickness of 100±10 μm were obtained.

The degree of crystallinity of polymers in the blends was studied using a differential scanning calorimeter Netzsch DSC 214 Polyma at a heating rate of 8 K/min and sample weight of 10±0.1 mg. The temperature scale was calibrated against the indium standard (Tm = 156.6°C, ΔH = 28.44 J/g). The water absorption degree was determined in distilled water at 25°C according to DIN EN ISO 62:2008-05.

The influence of ultraviolet on the obtained samples was studied by a lamp Vilber Lourmat VL-6.LC at λ=254 nm for 100 hours. The experiment on thermal oxidation was performed in a static manometric installation at 110°C and oxygen pressure of 300 mm Hg with absorption of gaseous reaction products by solid KOH.

The mechanical properties were examined by a tensile compression testing machine Devotrans DVT GP UG 5 (Turkey) according to ISO 527-1:2012. The tensile strength and elongation at break were determined. The data were averaged by six samples.

The work was carried out using scientific equipment of the Center of Shared Usage «New Materials and Technologies of Emanuel Institute of Biochemical Physics» and Joint Research Center of Plekhanov Russian University of Economics. This study was financed by a grant from the Plekhanov Russian University of Economics.).

3. Results and discussion

It was found that the addition of a small amount of LDPEox to the blend of PLA and LDPE does not influence their melting temperatures (method DSC) but decreases the crystallinity degree of LDPE and slightly increases the crystallinity of PLA (Table 1).

Table 1. Thermophysical characteristics of PLA/LDPE and PLA/LDPE/LDPEox blends.

Blends PLA/LDPE, wt.%	T_m, °C	ΔH_m, J/g	$\chi_{кр}$, %
0/100	103	61	21
30/70	164/105	49/46	52/16
100/0	165	42	45
Blends PLA/LDPE/LDPEox, wt.%	T_m, °C	ΔH_m, J/g	$\chi_{кр}$, %
30/60/10	165/104	47/40	50/14
30/50/20	165/104	44/41	47/14
30/40/30	165/104	41/47	45/16

It has been established that at the initial stage of LDPE-PLA blends biodegradation two factors have the main influence: water and ultraviolet radiation. Water absorption of LDPE-PLA blends is rather low, but it is enhanced with the addition of oxidized LDPE to the blend (figure. 1). The maximum degree of water absorption is observed for the mixture containing 30% PLA, 40% LDPE and 30% recycled LDPE. But nevertheless the composition can be attributed to the hydrophobic materials, since the degree of water absorption is less than 10%.

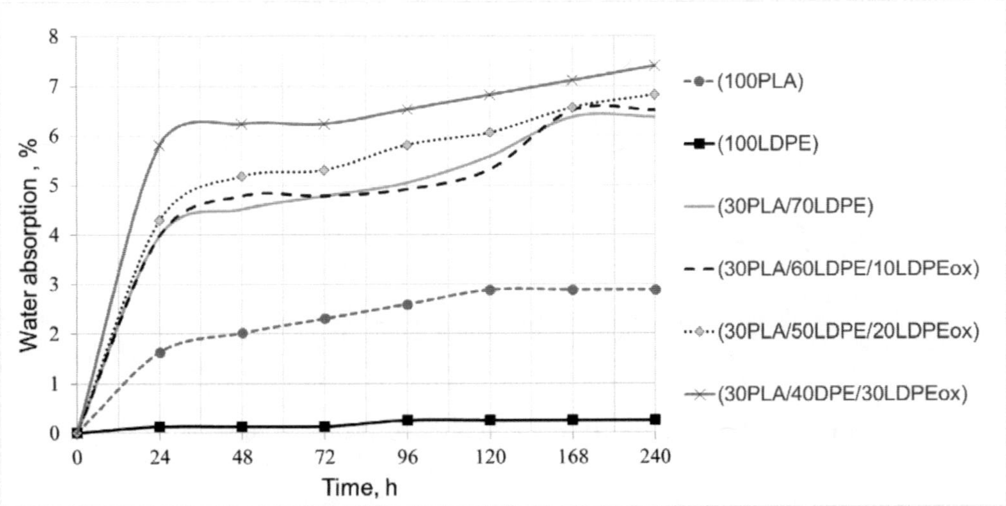

Figure 1. Kinetic curves of water absorption of blends.

During the experiment, ultraviolet radiation does not affect the structure of PE but has a strong influence on the structure of LA that was confirmed visually by samples cracking (figure 2) and by DSC – melting temperature decreases after radiation (figures 3-6). It is observed that compositions without oxidized LDPE are more resistant to UV radiation than compositions with oxidized LDPE.

Figure 2. Film sample of PLA after 50 hours of ultraviolet radiation.

Thermophysical parameters of PLA before and after UV treatment are studied. The change in the degree of crystallinity after exposure to UV radiation has been determined (figure 3). It is shown at the Figure 1 that as the time of exposure to UV radiation increases, the degree of crystallinity decreases. After 100 hours, this value drops by half, from 45 to 24%.

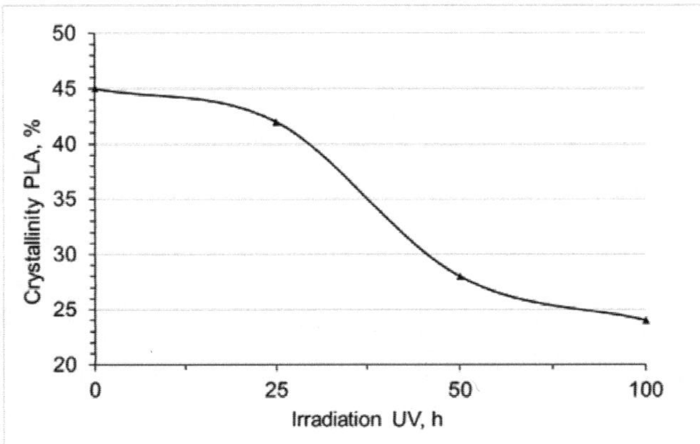

Figure 3. Degree of crystallinity of PLA after UV radiation.

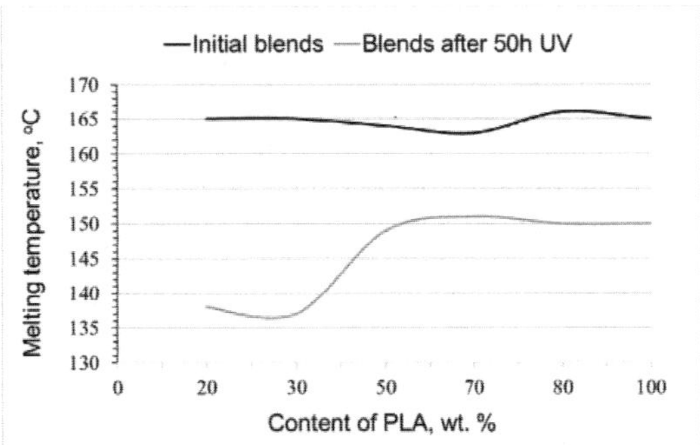

Figure 4. PLA melting temperature before (black) and after (grey) ultraviolet radiation.

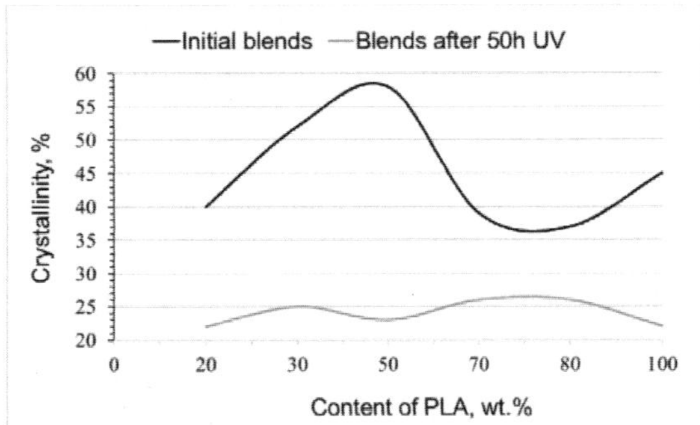

Figure 5. PLA crystallinity degree before (black) and after (grey) ultraviolet radiation.

In the process of UV degradation the PLA melting temperature changes significantly. Under the action of UV radiation (wavelength of 254 nm) Tm decreases from 163 to 152 °C after 50 hours, and to 135 °C after 100 hours (figure 6). The decrease in the melting temperature is explained by the shortening of macrochains due to oxidative degradation.

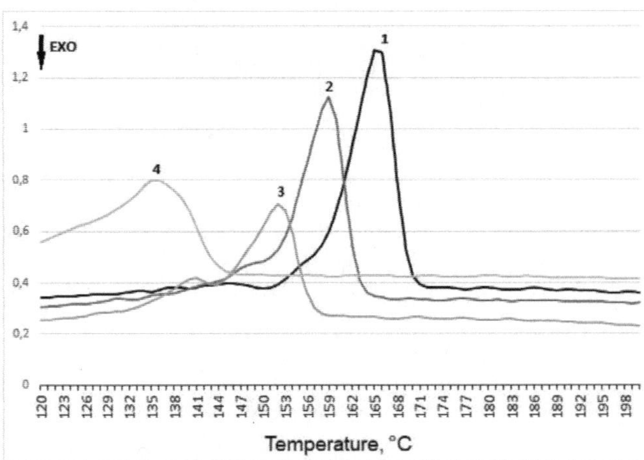

Figure 6. Change in the melting temperature of PLA: 1 – initial sample, 2 – after 25 hours, 3 – after 50 hours, 4 – after 100 hours of UV radiation.

It was also found that PLA had a lower oxidation rate than LDPE. Triple blends (with oxidized LDPE) had a higher rate of oxidation than the double ones (without oxidized LDPE) (figure 7). PLA has a fairly high crystallinity and is probably dense amorphous phase, low permeability for oxygen. If the temperature of the experiment increases, the oxidation process is rapidly moving to destruction.

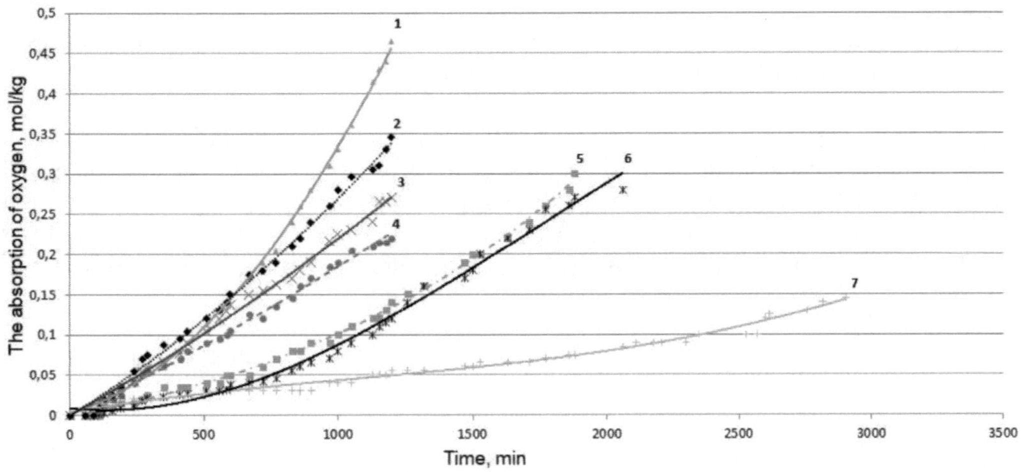

Figure 7. Kinetic curves of thermal oxidation of blends (1 - LDPE, 2 – 30PLA/40LDPE/30LDPE$_{ox}$, 3 - 30PLA/50LDPE/20LDPE$_{ox}$, 4 - 30PLA/60LDPE/10LDPE$_{ox}$, 5 – 30PLA/70LDPE, 6 – 50PLA/50LDPE, 7 – PLA).

The mechanical characteristics of the PLA after UV exposure reduced significantly. It was found that after 10 hours of UV exposure, the mechanical parameters of the PLA fell dramatically, it became fragile, and therefore the study after 25 hours of exposure was impossible to occur (figure 8).

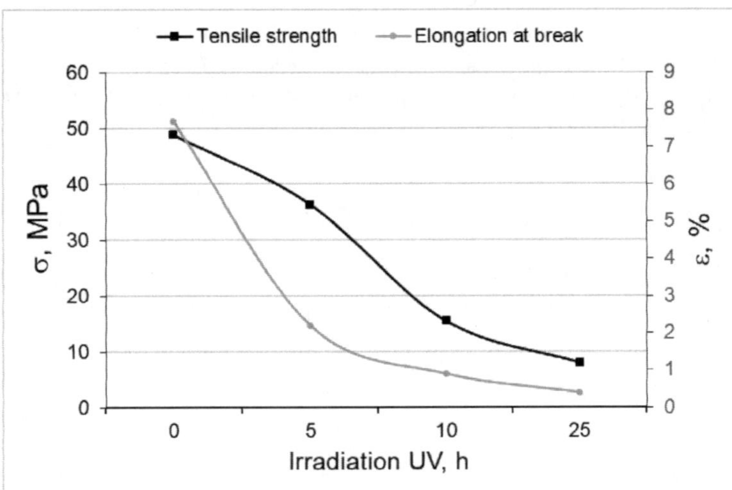

Figure 8. Mechanical characteristics of PLA in dependence to the time of exposure.

4. Conclusion

PLA becomes brittle after 50 hours of exposure to UV radiation. After 100 hours of photo-oxidation, cracking of all samples was noted, as well as the sharp decrease in the melting temperature (from 163 to 135 °C) and the degree of crystallinity (from 45 to 24%).

The tensile strength and the elongation at break of the PLA after 25 hours of exposure to UV radiation was sharply reduced.

At the infrared spectrum before and after 50 hours of UV irradiation, the optical density of the characteristic regions and PLA bands was changed. It indicates the oxidation and destruction of the material.

It is possible to deduce that all the double (LDPE-PLA) and triple (LDPE-PLA-LDPEox) blends satisfy the technological requirements and are suitable for producing goods. The addition of oxidized LDPE into PLA/LDPE blends changes their properties significanly. Blends with oxidized LDPE degrade easier because they are less stable to oxidation, more prone to water absorption and LDPE in such blends has a less degree of crystallinity. In order to accelerate the process of biodegradation in the environment, it is possible to radiate these blends by ultraviolet at the initial stage.

References

[1] Podzorova M V, Tertyshnaya Yu V, Pantyukhov P V, Shibryaeva L S, Popov A A and Nikolaeva S 2016 *AIP Conference Proceedings*. **1783**, 020185.

[2] Podzorova M V, Tertyshnaya Yu V, Monakhova T V and Popov A A 2016 *Rus J Phys Chem B*. **10(5)**, 825-829.

[3] Tertyshnaya Yu V, Karpova S G, Shatalova O V, Krivandin A V and Shibryaeva L S 2016 *Polym Sci Ser A*. **58(1)**, 50-56.

[4] Podzorova M V, Tertyshnaya Yu V and Popov A A 2014 *Rus J Phys Chem B*. **8(5)**, 726-732.

[5] Tertyshnaya Yu V and Shibryaeva L S 2013 *Polym Sci Ser B*. **55(3-4)**, 164-168.

[6] Tertyshnaya Yu V, Shibryaeva L S and Ol'khov A A 2015 *Rus J Phys Chem B*. **9(3)**, 498-503.

[7] Ol'khov A A, Iordanskii A L and Danko T P 2015 *J Polym Eng*. **35(8)**, 765–771.

[8] Rudnik E and Briassoulis D 2011 *J Polym Environ*. **19(1)**, 18-39.

[9] Kazemi-Najafi S and Englund K R 2013 *J Appl Polym Sci*. **129(6)**, 3404–3410.

[10] Tertyshnaya Yu V, Karpova S G and Popov A A 2017 *Rus J Phys Chem B*. **11(3)**, 531-537.

[11] Olkhov A, Pantyukhov P, Zykova A, Karpova S, Kucherenko E and Iordanskii A 2017 *IOP Conference Series: Materials Science and Engineering*. **175**, 012007.

[12] Mastalygina E E, Varyan I A, Kolesnikova N N and Popov A A 2016 *AIP Conference Proceedings*. **1736**, 020097.

IOP Conf. Series: Materials Science and Engineering **347** (2018) 012016 doi:10.1088/1757-899X/347/1/012016

Formation of biodegradated polymers as components of future composite materials on the basis of shape memory alloy of medical appointment

E O Nasakina, A S Baikin, K V Sergiyenko, M A Kaplan, S V Konushkin, A D Yakubov, A V Izvin, M A Sudarchikova, M A Sevost'yanov and A G Kolmakov

Baikov Institute of Metallurgy and Material Sciences, 49 Leninsky Avenue, Moscow, Russia

Email: nacakina@mail.ru

Abstract. The processes of formation of polymer polylactide or polyglycylidactide films for the subsequent creation of a layered composite with a biodegradable layer on the basis of a nickel-free shape memory alloy TiNbTaZr are studied. The structure of the samples was determined using an SEM. The correspondence of morphology of surfaces of and the substrate itself is noted. High adhesion of the polymer to the future basis of the developed composite material is supposed. The formed films is homogeneous and amorphous throughout the polymer volume. By varying the volume of solutions, it is possible to obtain films of a given thickness for any type of polymer, its molecular weight, and the solution concentration of the polymer in chloroform. Poly (glycolide-lactide) should be more plastic than polylactide.

1. Introduction

Medicine always appeals to other sciences in the search for new, more successful tools. Alloys with the shape memory effect (SME) and superelasticity are now one of the most interesting materials in the endoscopic implantology - when creating implants like Stent and Kawa-filter [1-12]. First, they have mechanical properties similar to the behavior of living tissues, which, when implant is installed and work in the body, provides less damage to the implant itself and the tissue damage. Secondly, the SME provides a "self- expansion" of the product, facilitating smaller dimensions of it in the delivery state, eliminating the need for additional means of their deployment and slowing down this process, which ultimately leads to a less traumatic effect of operations. But in addition to mechanical, there are still many characteristics of materials that are important in the production of these implants. And increasingly, multicomponent materials are being created now to improve and combine various functional properties. In particular, the creation of a composite material with a surface layer of a biodegradable biopolymer on the basis of an alloy with an SME will provide medical products with the possibility of local therapeutic effects through controlled release of drugs for a given time [13-15]. By correctly choosing a biodegradable polymer as a drug delivery system, it is possible to control the rate of release and concentration of the drug.

In this paper, we consider the formation of polymer components for layered composites on shape memory alloy substrates with biodegradable surface layers of polylactide and poly(glycolide-lactide).

IOP Conf. Series: Materials Science and Engineering **347** (2018) 012016 doi:10.1088/1757-899X/347/1/012016

2. Materials and methods

Polylactide / polyglycolide are soluble in most solvents, including chlorinated solvents. In the work, chloroform was chosen as the solvent. In comparison with other solvents, it exhibits a higher degree of dissolution of the polymers used, even at room temperature. It is assumed that in the process of obtaining polymer coatings, it will completely evaporate, leaving no negative biological effects in the final product.

By the method of casting from a polymer solution, it is possible to efficiently obtain bioresorbable layers, including with medicinal filler. By varying the polymer concentrations in the solvent, it is possible to control the film thickness, and hence the resorption time. The nature of the polymer and its molecular weight should also influence the film thickness and structure.

Poly-D,L-lactide (PLA) and Poly-glycolide-D,L-lactide (PGLA, molar ratio 30/70) of molecular weights of 45, 90 and 180 kDa were used as starting materials of the biodegradable surface layers of future biocompatible composites.

To create model films, we prepared hitches of polymers weighing 2, 6 and 10 g (± 0.01 g) per 200 ml of chloroform. The solution was heated to 80 °C and stirred until homogeneous for 1 hour. The resulting solution was kept for another 5 minutes at 80 °C and pour into glass pallets. Drying was carried out for 2 days in air at 37 °C in a thermostat. At the end of the drying, the resulting films were removed.

3. Results and discussion

The obtained samples were assigned ciphers intended for independent research in the future. The corresponding characteristics are listed in table 1. Figures 1-5 show SEM images of the surface and cross-sections of model coatings - polylactide and poly(glycolide-lactide) films.

Table 1. Characteristics of samples of model films

Cipher	Polymer	Molecular weight, kDa	Concentration, g per 100 ml chloroform	Thickness, μm
1.1	PLA	45	1	0.042
1.2			3	0.107
1.3			5	0.339
2.1		90	1	0.038
2.2			3	0.125
2.3			5	0.262
3.1		180	1	0.039
3.2			3	0.082
3.3			5	0.152
4.1	PGLA	45	1	0.066
4.2			3	0.155
4.3			5	0.224
5.1		90	1	0.063
5.2			3	0.148
5.3			5	0.258
6.1		180	1	0.032
6.2			3	0.154
6.3			5	0.252

IOP Conf. Series: Materials Science and Engineering **347** (2018) 012016 doi:10.1088/1757-899X/347/1/012016

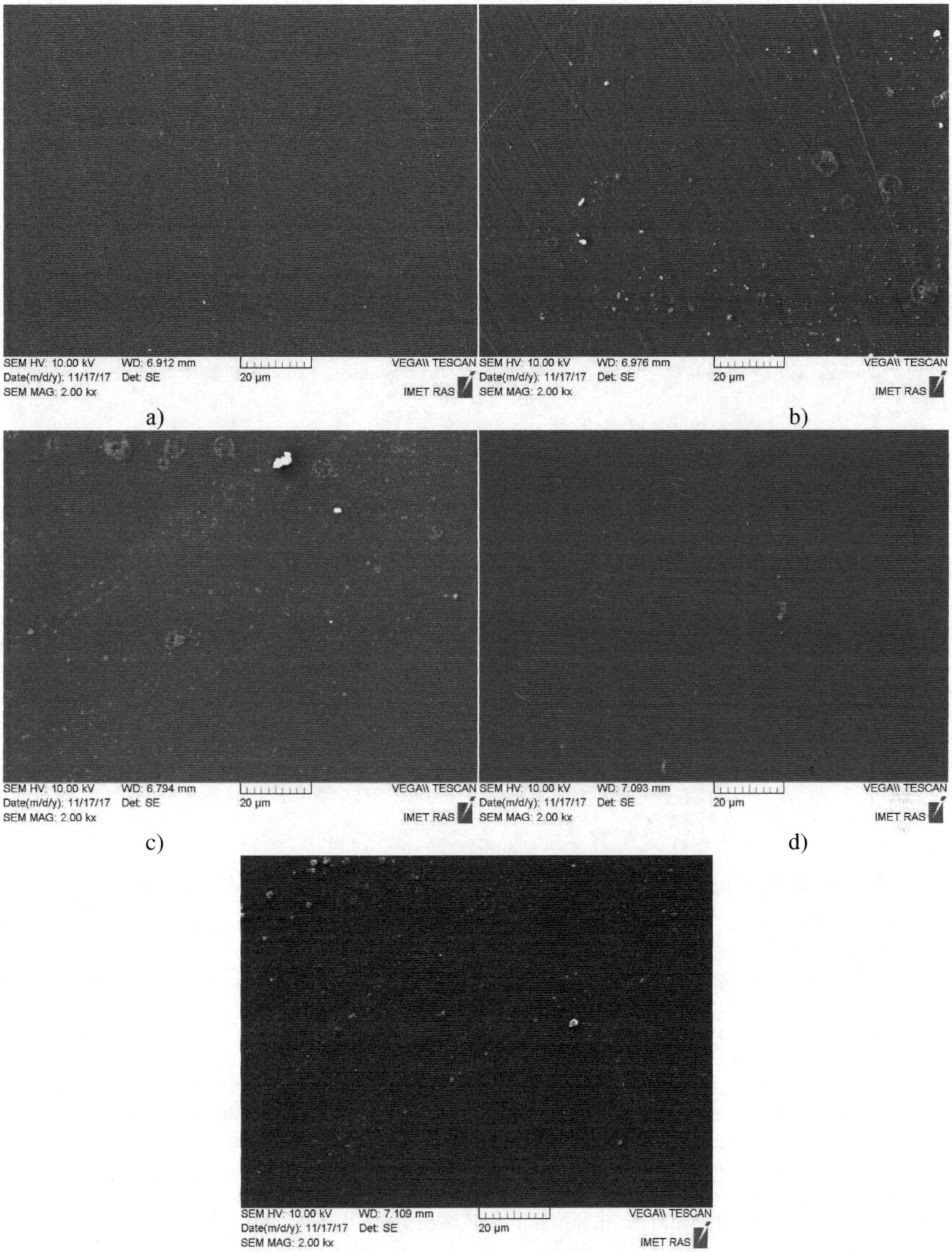

e)

Figure 1. SEM images of the samples surfaces a) 1.1, b) 3.1, c) 5.1, d) 4.1, e) 4.2

IOP Conf. Series: Materials Science and Engineering **347** (2018) 012016 doi:10.1088/1757-899X/347/1/012016

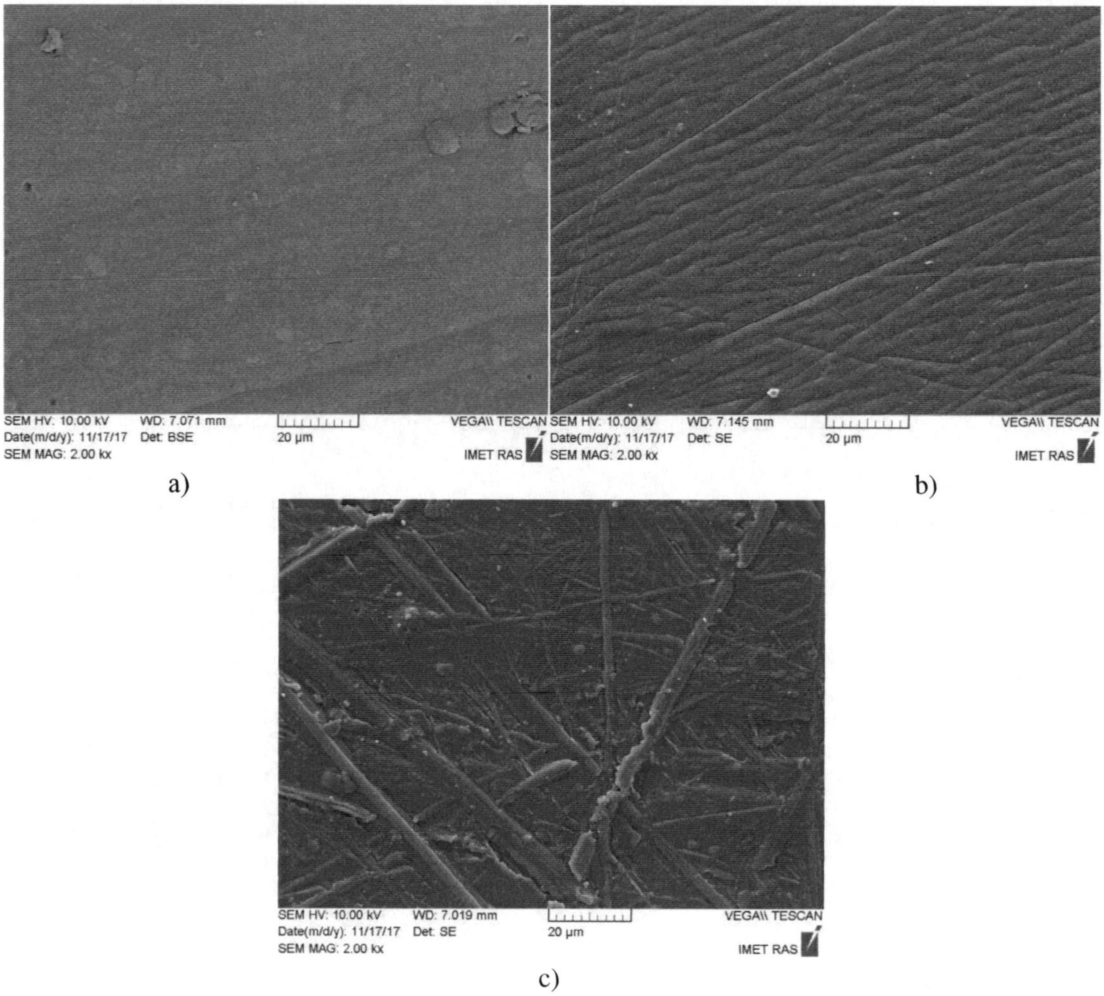

Figure 2. SEM image of the sample surface facing the substrate: a) 1.2, b) 2.1, c) 3.3

Obviously, the surface of polylactide and poly(glycolide-lactide) model films is a characteristic glassy surface without significant defects (figure 1). Neither the nature of the polymer (figure 1 a, d) nor its molecular mass (figure 1 a, b and c, d, respectively) nor the concentration in the solvent (figure 1 d, e) does not exert any significant influence.

An interesting observation is that in the case of analysis of the polymer film surface facing the substrate on which it was grown (figure 2), a clear correspondence of the morphology of these surfaces with a complete repetition of the initial relief is evident. The nature and amount of the polymer also do not affect the degree of repetition of morphology. Thus, an extremely high ability of polymers to fill the cavities of any roughness is observed, which should promote high adhesion of the polymer to the future basis of the developed composite material. This also leads to the assumption of high homogeneity and fluidity of the solution, which contributes to the best characteristics of the structure of the resulting film and the manufacturability of the process.

The thickness of the model films (figures 3-5, table 1) was 32 to 339 μm, depending on the type of polymer used, its molecular weight, and the concentration of the polymer solution in chloroform. There was no linear dependence. At the same time changing the volume of solutions it is possible to obtain films of the same thickness under any of these conditions.

IOP Conf. Series: Materials Science and Engineering **347** (2018) 012016 doi:10.1088/1757-899X/347/1/012016

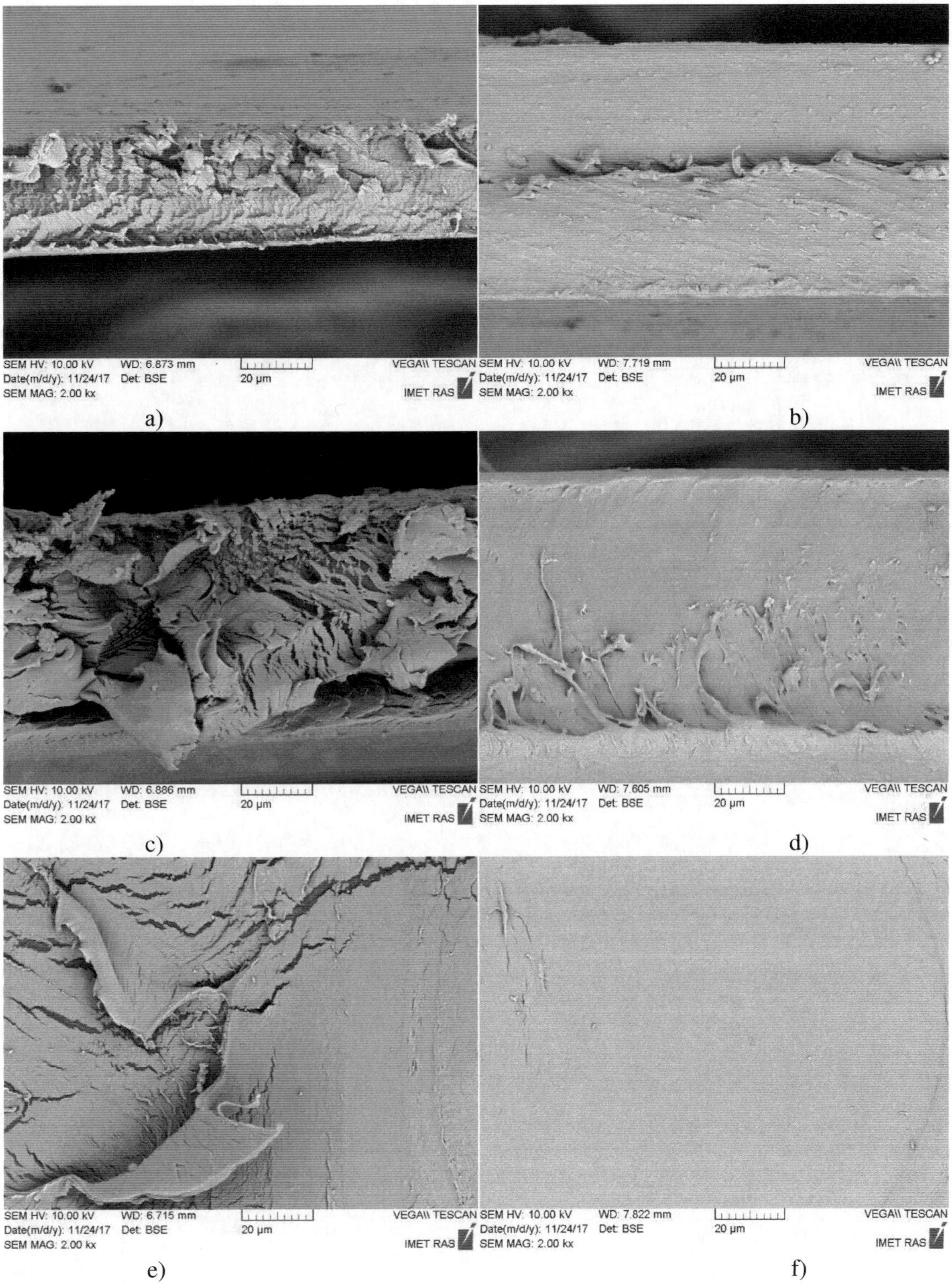

Figure 3. SEM image of the cross section of the sample a) 1.1, b) 4.1, c) 1.2, d) 4.2, e) 1.3, f) 4.3

IOP Conf. Series: Materials Science and Engineering **347** (2018) 012016 doi:10.1088/1757-899X/347/1/012016

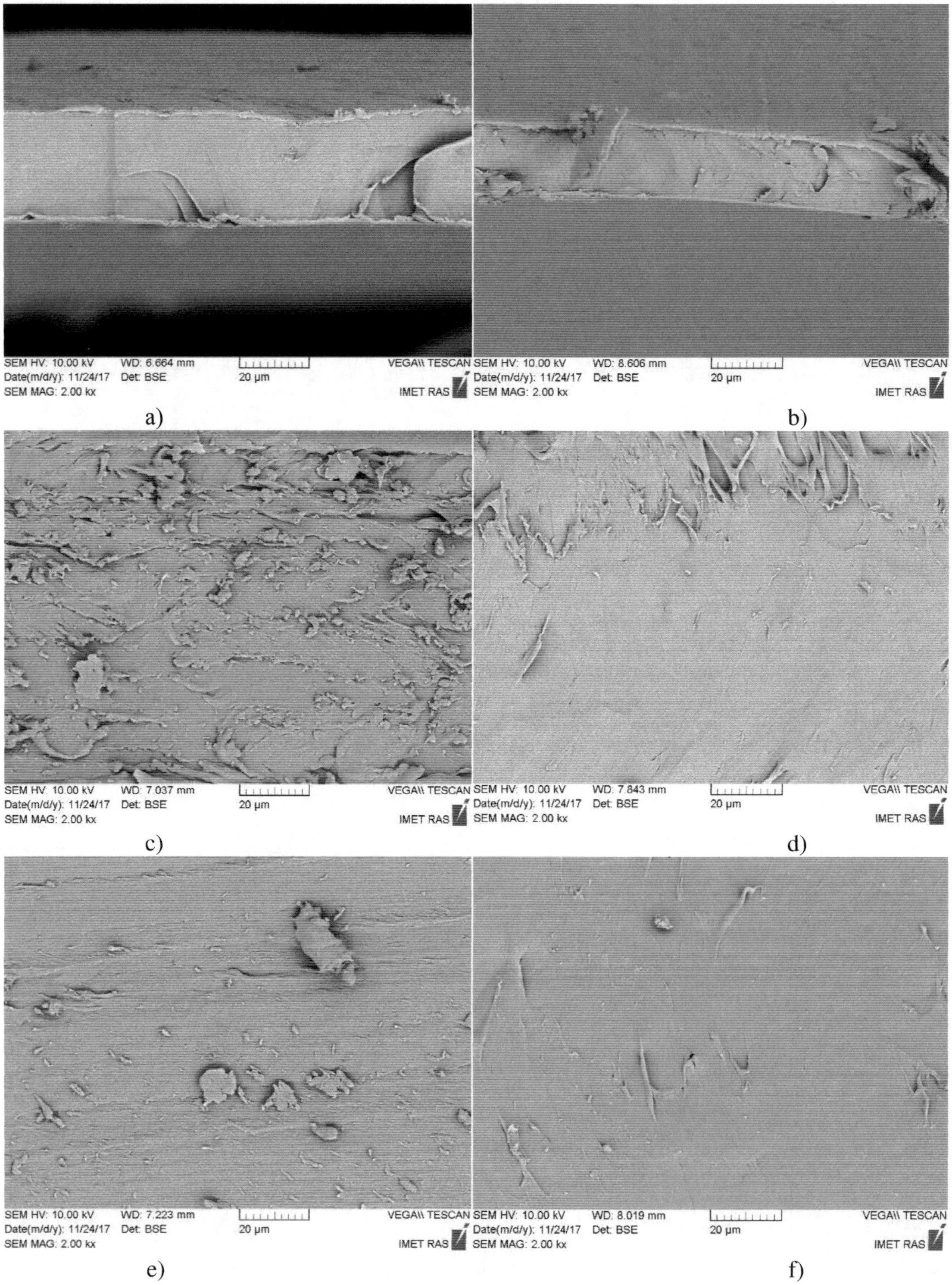

Figure 4. SEM image of the cross section of the sample a) 2.1, b) 5.1, c) 2.2, d) 5.2, e) 2.3, f) 5.3

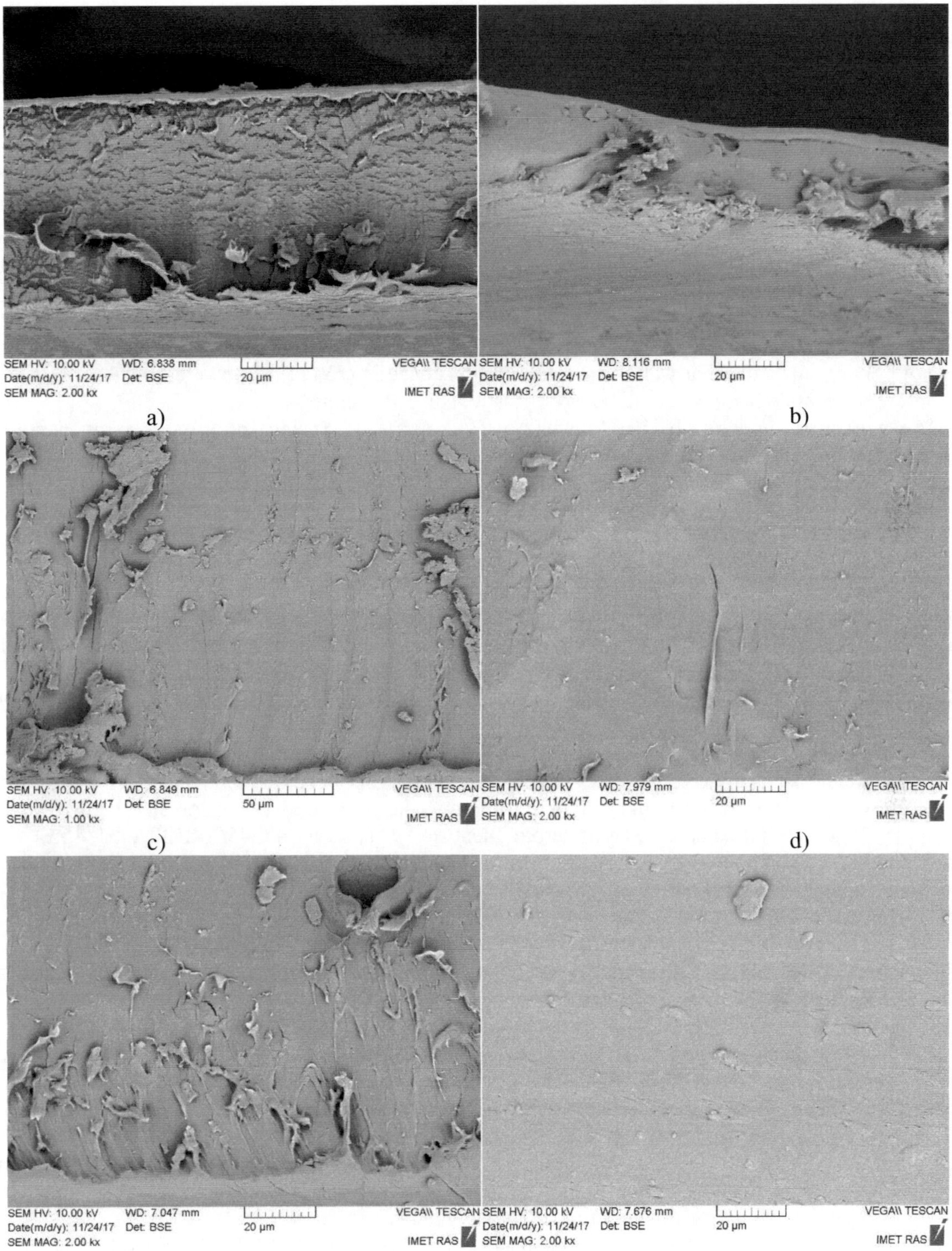

Figure 5. SEM image of the cross section of the sample a) 3.1, b) 6.1, c) 3.2, d) 6.2, e) 3.3, f) 6.3

Uniformity of the composition without appreciable defects is also observed throughout the polymer volume. The formed films are amorphous. It was observed that, judging by the nature of the cross-sections after cutting (figures 3-5), poly(glycolide-lactide) should exhibit greater plasticity than polylactide: the number of sharp marks from mechanical deformation practically reduces to zero.

4. Conclusions

1. Thin polymeric films of polylactide and polyglycylidactide as the components for the subsequent formation of layered composites based on the shape memory alloy with drug filling were obtained and studied.

2. The surface of the polymer film facing the substrate on which it was grown reproduces completely the morphology of the surface of this substrate. Thus, an extremely high ability of polymers to fill the cavities of any roughness is observed, which should promote high adhesion of the polymer to the future basis of the developed composite material.

3. The formed films is homogeneous and amorphous throughout the polymer volume.

4. By varying the volume of solutions, it is possible to obtain films of a given thickness for any type of polymer, its molecular weight, and the solution concentration of the polymer in chloroform.

5. Poly (glycolide-lactide) should be more plastic than polylactide.

The work was supported by the Ministry of Education and Science of Russia (grant identifier RFMEFI60417X0196).

References

[1] Petrini L, Migliavacca F 2011 Biomedical Applications of Shape Memory Alloys *Journal of Metallurgy* **2011** 1-15

[2] Duerig T W, Melton K N, Wayman C M, Stockel D 1990 *Engineering aspects of shape–memory alloys* (Oxford: Butterworth Heinemann Ltd) pp 181 – 194

[3] Shabalovskaya S 1996 On the nature of the biocompatibility and medical applications of NiTi shape memory and superelastic alloys *Bio Med Mater Eng* **6** 267 – 289

[4] Marjan Bahrami Nasab, Mohd Roshdi Hassan 2010 Metallic Biomaterials of Knee and Hip - A Review *Trends in Biomaterials and Artificial Organs* **24(1)** 69 - 82

[5] Surdell D, Shaibani A, Bendok B, Eskandari M K 2007 Fracture of a Nitinol Carotid Artery Stent That Caused Restenosis *J Vasc Interv Radiol* **18(10)** 1297–1299

[6] Bose A, Hartmann M, Henkes H A 2007 Novel, Self–Expanding Nitinol Stent in Medically Refractory Intracranial Atherosclerotic Stenosis: Wingspan Study *Stroke* **38** 1531–1537

[7] Duerig T W, Tolomeo D E, Wholey M 2000 An overview of superelastic stent design *Minim Invasive Ther Allied Tech* **9** 235–246

[8] Horowitz M B, Purdy P D 2005 The use of stents in the management of neurovascular disease: a historical and present status review *Progress in Neurological Surgery* **17** 18

[9] Rösch J, Keller F S, Kaufman J A 2003 The Birth, Early Years, and Future of Interventional Radiology *J Vasc Interv Radiol* **14** 841-853

[10] Balazic M, Kopac J 2007 Improvements of medical implants based on modern materials and new technologies *J. Achiev. Mater. Manuf. Eng* **25(2)** 31-34

[11] Meier B 2006 Use and Abuse of Coronary Stenting *Hospital Chronicles* **Suppl**. 99–103

[12] Stoeckel D, Pelton A, Duerig T 2004 Self-expanding nitinol stents: material and design considerations *Eur Radiol* **14** 292-301

[13] Kim J H, Shin J H, Shin D H, Moon M W, Park K, Kim T H et al. 2011 Comparison of diamondlike carbon-coated nitinol stents with or without polyethylene glycol grafting and uncoated nitinol stents in a canine iliac artery mode *Br J Radiol* **84** 210-215

[14] Sevostyanov M A, Fedotov A Yu, Kolmakov A G, Zabolotnyi V T, Barinov S M, Goncharenko B A, et al. 2014 Mechanical Properties of Nanostructured Nitinol/Chitosan Composite Material *Inorg. Mater. Appl. Res.* **5(4)** 344–346

[15] Sevost'yanov M A, Fedotov A Yu, Nasakina E O, Teterina A Yu, Baikin A S, Sergienko K V, et al. 2015 Kinetics of the release of antibiotics from chitosan-based biodegradable biopolymer membranes *Doklady Chemistry* **465(1)** 278-280

IOP Conf. Series: Materials Science and Engineering **347** (2018) 012017 doi:10.1088/1757-899X/347/1/012017

Spin splitting in band structures of BiTeX (X=Cl, Br, I) monolayers

D C Hvazdouski, M S Baranava and V R Stempitsky

Lab "Computer-aided design of micro- and nanoelectronic systems", Micro- & Nanoelectronics Department, Belarusian State University of Informatics and Radioelectronics, Minsk, P. Brovki str. 6, 220013, Republic of Belarus

E-mail: gvozdovsky@bsuir.by, baranova@bsuir.by, vstem@bsuir.by

Abstract. In systems with breaking of inversion symmetry a perpendicular electric field arises that interacts with the conduction electrons. It may give rise to electron state splitting even without influence of external magnetic field due to the spin-orbital interaction (SOI). Such a removal of the spin degeneracy is called the Rashba effect. Nanostructure with the Rashba effect can be part of a spin transistor. Spin degeneracy can be realized in a channel from a material of this type without additive of magnetic ions. Lack of additive increases the charge carrier mobility and reliability of the device. *Ab initio* simulations of BiTeX (X=Cl, Br, I) monolayers have been carried out using VASP wherein implemented DFT method. The study of this structures is of interest because such sort of structures can be used their as spin-orbitronics materials. The crystal parameters of BiTeCl, BiTeBr, BiTeI have been determined by the ionic relaxation and static calculations. It is necessary to note that splitting of energy bands occurs in case of SOI included. The values of the Rashba coefficient α_R (in the range from 6.25 to 10.00 eV·Å) have high magnitudes for spintronics materials. Band structure of monolayers structures have ideal Rashba electron gas, i.e. there no other energy states near to Fermi level except Rashba states.

1. Introduction

In the last decade, spintronics has become a priority direction for micro- and nanoelectronics element base development. Appropriate choice of materials and structures based on these materials is one of the main scientific issues faced by the researchers.

Consequently, particular interest arises in relation to materials, which exhibit the Rashba effect [1]. This effect appears in systems with inversion symmetry breaking. It may give rise to electron state splitting even without influence of external magnetic field due to the spin-orbital interaction (SOI).

In the presence inversion symmetry breaking in structure the spin-degenerate parabolic band splits into two spin-polarized bands. This formulation yields new extreme at a momentum offset (k_0) and energy splitting (E_R) that are related to each other via $\alpha_R = 2 \cdot E_R / k_0$ (figure 1). The two Rashba split branches have opposite spins, which can influence the photoexcitation's optical and magnetic properties, a situation that may benefit spintronics because it enhances the spin-to-charge conversion efficiency.

Rashba effect is central physical phenomenon of spin-orbitronics. The operation of transistors can be realized on the basic principles of the effect described above. The spin-transistor construction, proposed by Datta and Das [2], is very similar to the conventional field-effect transistor, where the

SOI in channel is tuned by gate voltage, which is the Rashba effect [3–5]. Furthermore, a typical spin transistor still has not been implemented experimentally.

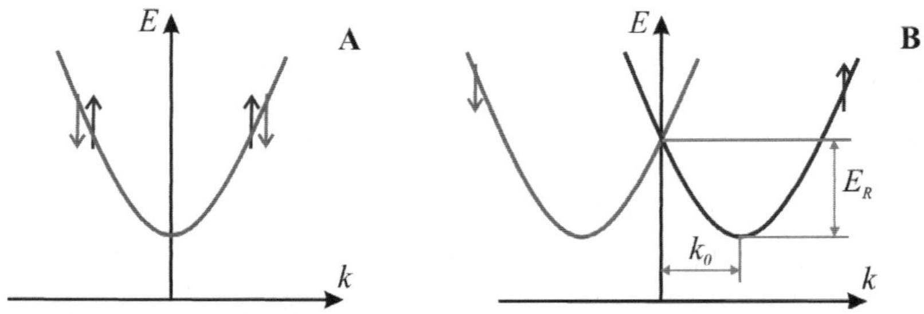

Figure 1. Wave function (a) without and (b) with Rashba spin splitting.

Until now, spin-transistor samples that have been manufactured in laboratory conditions and are based on the Rashba effect have a number of disadvantages, which are primarily associated with low efficiency of used materials.

It is known that the inversion symmetry breaking condition is observed in materials with wurtzite, zins-blende and 2D structures, and in structures, that are grown along crystallographic directions, that ensure the quantum well formation [6–10]. That is one of the possible reasons for the emergence of the Rashba effect.

Unique properties arise in materials with the transition to a two-dimensional structure. A special role is played by relativistic effects. In particular, spin-orbit interactions become important. BiTeX (X=Cl, Br, I) monolayer has a hexagonal crystal structure with breaking the inversion symmetry. Also heavy metals in the chemical compounds provide a large value of α_R. Therefore, the study of such structures is of interest from the point of view of the spin-orbitronics materials.

Right choosing of materials and structures based on this materials is one of the main scientific problem for investigators. Important conditions, which should be satisfied for devices based on these materials, are high Rashba parameter value; ideal Rashba electron gas; compatibility with advanced technological processes of micro- and nanoelectronic device manufacturing.

2. Computational details

Our calculations were performed based on the density functional theory (DFT). The projector-augmented wave (PAW) [11] potentials and Perdew–Burke–Ernzerhof (PBE) [12] functional have been used. A cutoff energy of 500 eV and 11×11×1 k-points grids determined by a fine grid of gamma-centered method [13] in the Brillouin zone have been used. The valence electron configurations for Bi, Te, Cl, Br and I were $[Xe]4f^{14}5d^{10}6s^26p^3$, $[Kr]4d^{10}5s^25p^4$, $[Ne]3s^23p^5$, $[Ar]3d^{10}4s^24p^5$ and $[Kr]4d^{10}5s^25p^5$, respectively. A vacuum layer of 20 Å along z direction was constructed to eliminate the interaction with spurious replica images. The atomic structures were relaxed until the forces on all unconstrained atoms were smaller than 0.01 eV/Å. The DFT-D3 method of Grimme demonstrated demonstrated the smallest difference between the calculated and experimental magnitude of lattice constants and have been used for further calculation. All of calculations have been carried out using VASP (Vienna Ab-initio Simulation Package) [14, 15] wherein implemented methods described above. Structural figures and charge density drawings were produced by VESTA package [16].

3. Crystal structure

The crystal structures of BiTeX (X=Cl, Br, I) have three atoms in unit cell and a layered structure along crystallographic c axis (figure 2). Within unit cell a Bi atom is located between one Te atom and one X atom thus forming a monolayer with three kinds of atoms. This monolayres weakly coupled each other via the van der Waals interactions in bulk structure. The distance between atoms in

IOP Conf. Series: Materials Science and Engineering **347** (2018) 012017 doi:10.1088/1757-899X/347/1/012017

monolayer is less than interlayer distances within bulk structure along c crystallographic direction. Such nature of the bonds provides the possibility of forming a monolayer consisting of three kind of atoms.

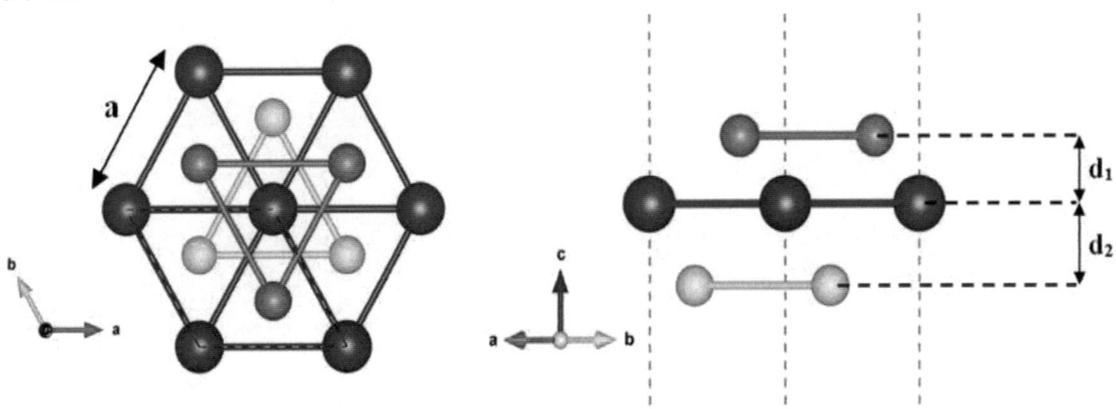

Figure 2. Unit cell of BiTeX (X=Cl, Br, I) monolayer.

Ground states of BiTeX (X=Cl, Br, I) monolayers have been calculated via *ab initio* simulation. The distances of Bi–Te (d_1) and Bi–X (d_2), as well as the calculated lattice constant *a* of the hexagonal unit cell are collected in table 1.

Table 1. Structure parameters of monolayer BiTeX (X=Cl, Br, I).

Structure	Lattice constant *a*, Å	d_1, Å	d_2, Å	Exp. lattice constant *a*, Å
BiTeCl	4.26	1.79	1.62	4.241[17]
BiTeBr	4.28	1.76	1.84	4.266[18]
BiTeI	4.35	1.73	2.08	4.339[19]

The obtained values of the lattice constant are in good agreement with the data from the experimental works [17–19].

BiTeX (X=Cl, Br, I) monolayers have a hexagonal crystal structure and is built up of alternating layers of Te, Bi, and X atoms stacked along the hexagonal axis what breaks the inversion symmetry. The broken inversion symmetry leads to the charge in the crystal is expected to distribute unevenly along the Te–Bi–X layers, giving rise to intrinsic internal potential gradient. Thus BiTeX monolayers have electrical gradient along the normal direction to the monolayer plane.

4. Band structures with spin splitting

Calculations of the band structures for BiTeX (X=Cl, Br, I) monolayers are performed taking into account the spin-orbit interaction (SOI). The electronic band structures of the BiTeI monolayer is depicted in figure 3. Conduction band minimum is locates at the near Γ point. Band structure has an indirect character and has gap of 0.74 eV. Band structures of the BiTeI monolayer consist of six conduction and twelve valence bands. The Fermi level is in the middle of the band gap. Conduction bands mostly compose Bi-6p electronic states. Twelve valence bands form valence bands of which the six bottom ones are the tellurium levels, and the six top ones are iodine levels.

The band structures have significant changes in the case of SOI included. The splitting of the energy bands occurs, i.e spin degeneracy disappears. We note that splitting is significant because the

IOP Conf. Series: Materials Science and Engineering **347** (2018) 012017 doi:10.1088/1757-899X/347/1/012017

structures under study consist of heavy metals. Strongly band splitting are effect from broken inversion symmetry. Because of conduction band minimum locates near the Γ point, thus forming a significant Rashba spin splitting.

The character of band structures of BiTeCl and BiTeBr is similar to character of band structure BiTeI. The band gaps of BiTeCl and BiTeBr are 0.94 and 0.89 eV, respectively (figure 4).

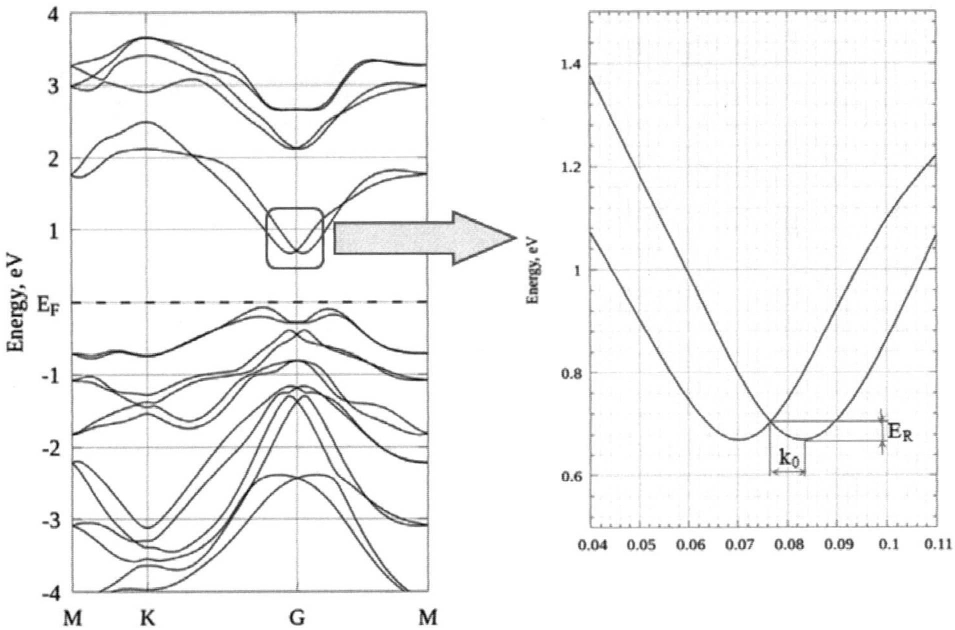

Figure 3. Band structure of BiTeI monolayer with SOI.

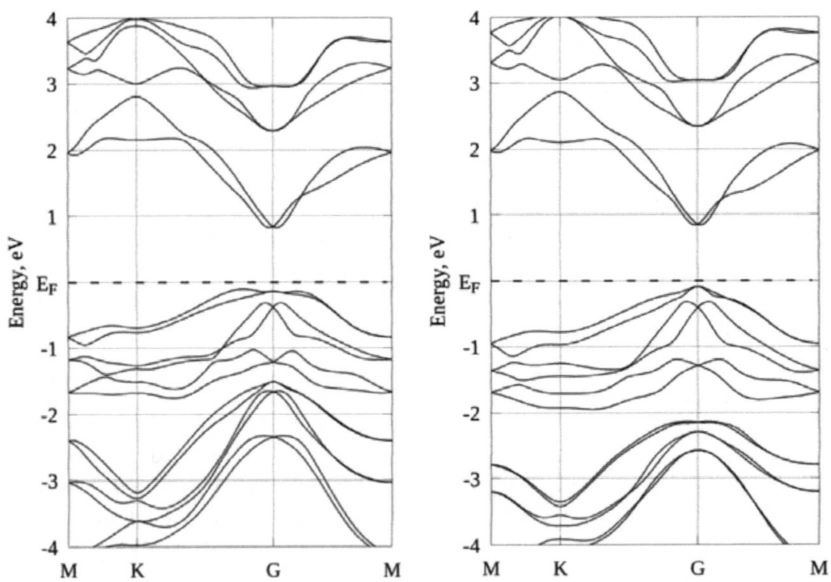

BiTeBr BiTeCl

Figure 4. Band structure of BiTeBr and BiTeCl monolayers with SOI.

Parameters of the Rashba spin splitting including the momentum offset k_0, the Rashba energy E_R and the coupling constant α_R (Rashba parameter) have been calculated (table 2) as

$$E_{\mathrm{R}} = \frac{\hbar^2 k_0^2}{2m^*}, \tag{1}$$

$$k_0 = \frac{\mathrm{m}^* \alpha_{\mathrm{R}}}{\hbar^2}. \tag{2}$$

The characteristic of spin splitting of the conduction band in the BiTeCl monolayer k_0 and E_R are evaluated to be 0.0032 Å^{-1} and 0.01 eV, respectively. Based on k_0 and E_R, the Rashba parameter α_R can be estimated as 6.25 eV·Å (table 2). For the BiTeBr and BiTeI monolayers the evaluated values of k_0, E_R and α_R are 0.048 Å^{-1}, 0.017 meV, 7.08 eV·Å, and 0.068 Å^{-1}, 0.034 meV, 10.0 eV·Å respectively.

Table 2. Parameters of the Rashba spin splitting.

Structure	Calculated			Referencies [20]		
	E_R, eV	k_0, Å^{-1}	α_R, eV·Å	E_R, eV	k_0, Å^{-1}	α_R, eV·Å
BiTeCl	0.010	0.032	6.250	–	–	–
BiTeBr	0.017	0.048	7.080	0.018	0.027	1.310
BiTeI	0.034	0.068	10.000	0.040	0.043	1.860

The values obtained for the Rashba coefficient are in the range from 6.25 to 10.00 eV·Å. Our results are in good agreement with previous theoretical calculations. In standard materials of micro- and nanoelectronics, the value of α_R is of the order of 0.1 eV·Å, which is insufficient for effective application in spintronics.

5. Conclusion

In the present work, the BiTeX (X=Cl, Br, I) compounds were studied by first-principles simulation using density functional theory. The crystal structures of BiTeCl, BiTeBr, BiTeI have been established from the ionic relaxation and static calculations. The band structures of BiTeCl, BiTeBr, BiTeI monolayers have been observed via first principles simulation with spin-orbit interaction included and we note that ones have significant changes in this case. The splitting of the energy bands occurs, i.e spin degeneracy disappears. The values of the Rashba coefficient α_R (in the range from 6.25 to 10.00 eV·Å) have high magnitudes for spintronics materials. Band structures of the investigated compounds have ideal Rashba electron gas, i.e. there no other energy states near to Fermi level except Rashba states. Thus, the BiTeCl, BiTeBr, BiTeI monolayers can theoretically be considered as the spin-orbitronics material for manufacturing gate-controlled devices, such as FETs.

Acknowledgments

This work was supported by the grants 2.53 of Belarusian National Scientific Research Program "Physical Materials Science, Novel Materials and Technologies".

References

[1] Rashba E I 1960 Properties of semiconductors with an extremum loop 1. Cyclotron and combinational resonance in a magnetic field perpendicular to the plane of the loop *Sov. Phys. Sol. State* **2** 1109–22

[2] Das B, Miller D C, Datta S, Reifenberger R, Hong W P, Bhattacharya P K, Singh J and Jaffe M 1989 Evidence for spin splitting in in x $Ga_{1-x}As/In_{0.52}Al_{0.48}As$ heterostructures *Phys. Rev. B* **39** 1411–4

[3] Bychkov Y A and Rashba E I 1984 Properties of a 2D electron-gas with lifted spectral

degeneracy *Jept lett* **39**, 78–81

[4] Manchon A, Koo H C, Nitta J, Frolov S M and Duine R A 2015 New perspectives for rashba spin–orbit coupling *Nat. Mater.* **14** 871–82

[5] Go D *et al.* 2017 Toward surface orbitronics: giant orbital magnetism from the orbital Rashba effect at the surface of *sp*-metals *Sci. Rep.* **7** 46742 doi: 10.1038/srep46742

[6] Luo J, Munekata H, Fang F F and Stiles P J 1990 Effects of inversion asymmetry on electron-energy band structures in GaSb/InAs/GaSb quantum-wells *Phys. Rev. B* **41** 7685–93

[7] Dil J H *et al.* 2008 Rashba-type spin-orbit splitting of quantum well states in ultrathin pb films *Phys. Rev. Lett.* **101** 266802

[8] Yaji K *et al.* 2010 Large Rashba spin splitting of a metallic surface-state band on a semiconductor surface *Nat. Commun.* **1** 17

[9] Wang T H 2017 Wide-range ideal 2D Rashba electron gas with large spin splitting in $Bi2Se_3/MoTe_2$ heterostructure *Npj computational materials* **3** 5

[10] Zhong S, Moore J E and Souza I 2016 Gyrotropic magnetic effect and the magnetic moment on the fermi surface *Phys. Rev. Lett.* **116** 077201

[11] Kresse G and Joubert D 1999 *Phys. Rev. B: Condens. Matter Mater. Phys.* **59** 1758–75

[12] Perdew J P, Burke K and Ernzerhof M 1996, *Phys. Rev. Lett.* **77**, 3865

[13] Monkhorst H J and Pack J D 1976 *Phys. Rev. B: Condens. Matter Mater. Phys.* **13** 5188

[14] Kresse G and Furthmüller J 1996 *Comput. Mater. Sci.* **6** 15–50

[15] Blöchl P E 1994 *Phys. Rev. B:* Condens. *Matter Mater. Phys.* **50**, 17953–79

[16] Momma K, Izumi F and Mineral J 2010 *Petrol. Sci.* **39** 136–45

[17] Jamimovic J, Mettan X, Pisoni A, Gaal R, Katrych S, Demko L, Akrap A, Forro L, Berger H, Bugnon P and Magrez A 2014 Enhanced low-temperature thermoelectrical properties of BiTeCl grown by topotactic method *Scripta Materialia* **76** 69–72

[18] Sans J A, Manjon F J, Pereira A L J, Vilaplana R, Gomis O, Segura A, Munoz A, Rodriguez-Hernandez P, Popescu C, Drasar C and Rulevas P 2016 Structural, vibrational, and electrical study of compressed BiTeBr *Phys Rev B.* **93** 024110.1–11

[19] Shevelkov A V, Dikarev E V, Shpanchenko R V and Popovkin B A 1995 Crystal structures of bismuth telluro-halides BiTeX (X=Cl, Br, I) from X-Ray powder diffraction data *J Solid State Chem.* **114** 379–84

[20] Yandong M, Ying D, Wei W, Xinru L and Baibiao H 2014 Emergence of electric polarity in BiTeX (X = Br and I) monolayers and the giant Rashba spin splitting *Phys. Chem.* **16** 17603

IOP Conf. Series: Materials Science and Engineering **347** (2018) 012018 doi:10.1088/1757-899X/347/1/012018

The study of adsorption of hydrogen onto copper and gold clusters by method of the density functional

A A Kirsankin, N V Dohlikova and S Y Sarvadii

N.N. Semenov Institute of Chemical Physics RAS, Moscow, Russian Federation

E-mail: kirsankin@mail.ru

Abstract. The paper presents the results of quantum-chemical modeling of gold and copper nanoparticles and a comparison of their adsorption properties with the example of interaction with hydrogen atoms. The goals of the work were both a fundamental study of the electronic structure and the elementary act of the heterogeneous catalysis reaction and the comparison of gold and copper nanoparticles as catalysts for the hydrogenation reaction.

1. Introduction

Bulk copper has good electrical conductivity (the best of any metal, except silver), thermal conductivity and plasticity, that determines its value in various areas of practical application. Nanostructured copper also has a high potential for use in catalysis, production of functional materials, medicine and energy. A wide range of applications is due to the high catalytic and chemical activity of copper-based nanostructures, which has been studied and demonstrated in a variety of works, both experimental [1-4], and theoretical [5-8]. In addition, it is important to note that the synthesis of copper nanoparticles is cheaper compared to gold and platinum nanoparticles. All this makes copper nanoparticles extremely attractive for use as catalysts for a variety of reactions. In the present work, the hydrogenation reaction, namely, the interaction of a nanocluster of copper with a hydrogen atom, is considered.

2. DFT-study and discussion

2.1. Calculation method

A numerical experiment is performed using the density functional theory. Software packages OpenMX3.8 и QuantumEspresso 5.1.1 are used for calculations. In both cases, the generalized gradient approximation and the PBE functional were chosen. The relativistic effects of the electrons motion were taken into account in the generation of pseudopotentials. The accuracy of the chosen basis sets roughly corresponded to the double Slater basis function. For the given purposes did not include achievement of experimental values, therefore in order to avoid ambiguities in calculations the overlapping of bases is not considered.

2.2. Cluster models

The models of gold and copper nanoparticles were Au_n and Cu_n (n = 13, 20, 55) clusters. The atomic structure of the clusters is calculated using quasi-Newtonian methods and corresponded to a local minimum of the total energy. The search for a global minimum was not the goal of the work. The obtained stable isomers of clusters, even those with filled icosahedral shells, $n = 13$, 55, have no certain type of symmetry, and that agrees with previous studies [9]. As an example, 13-atoms isomers of gold and copper clusters are shown in figure 1. The symmetry of this clusters is close to icosahedral.

Calculations of binding energy between atoms in clusters have shown the expected dimensional effect: an increase in the binding energy takes place with an increase in the nuclear system, that is explained by a decrease in the fraction of surface atoms in larger clusters (table 1). It is worth pointing out that the binding energy in copper clusters is greater.

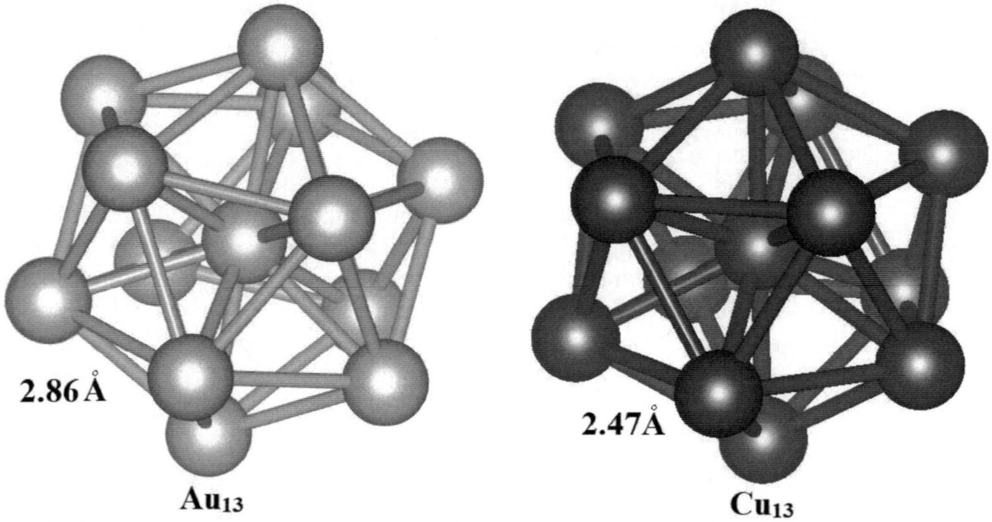

Figure 1. 13-atoms isomers of gold (left) and copper (right) clusters, used in the numerical experiment in this paper.

Table 1. The binding energy between gold atoms in clusters.

Cluster	Au_{13}	Au_{20}	Au_{55}	Cu_{13}	Cu_{20}	Cu_{55}
Binding energy, eV	–1.95	–1.99	–2.19	–2.40	–2.53	2.87

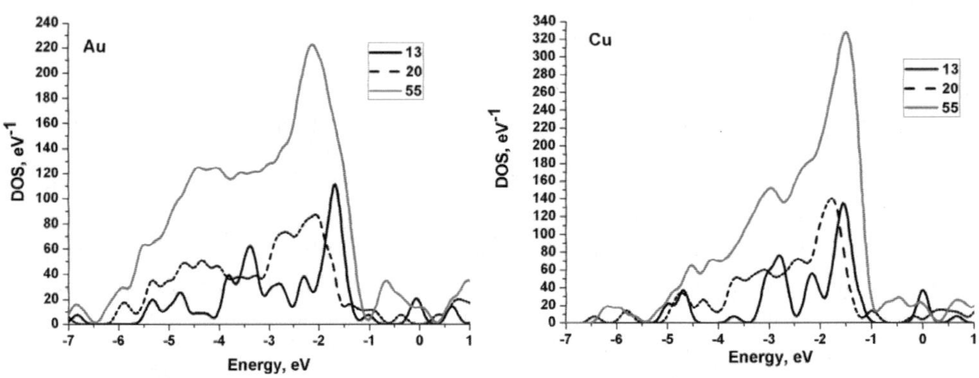

Figure 2. Density of states (DOS) for clusters of gold and copper of various sizes, $n = 13, 20, 55$.

Electronic spectra of gold and copper clusters of various sizes are shown in figure 2. As can be seen, the distribution of the density of states for the cluster Cu_{13} and Au_{13} has separate local maxima,

IOP Conf. Series: Materials Science and Engineering **347** (2018) 012018 doi:10.1088/1757-899X/347/1/012018

which gradually merge with an increase in the cluster size of Cu_{55} and Au_{55}, forming the structure similar to the zone in a solid body. In addition to this demonstration of the size effect, attention should also be paid to the general distribution of the density of states for gold and copper clusters, the main part of which is located below the Fermi level. In general, the features of calculated density of states are in accordance with the detailed results for gold clusters, carried out in [10].

Like most metal clusters, gold and copper clusters of the same size have many almost isoenergetic isomers [11]. In particular, planar isomers of Au_{13} and Cu_{13} clusters are found in this work. It should be noted that the planar isomer of gold is energetically more stable than the icosahedral one (0.25 eV) [12], while in the case of copper the icosahedral isomer is more stable (0.32 eV). The distance between the nearest atoms in the gold cluster is ~ 2.75 Å, in copper ~ 2.36 Å, and the planes are formed by practically regular triangles (figure 3).

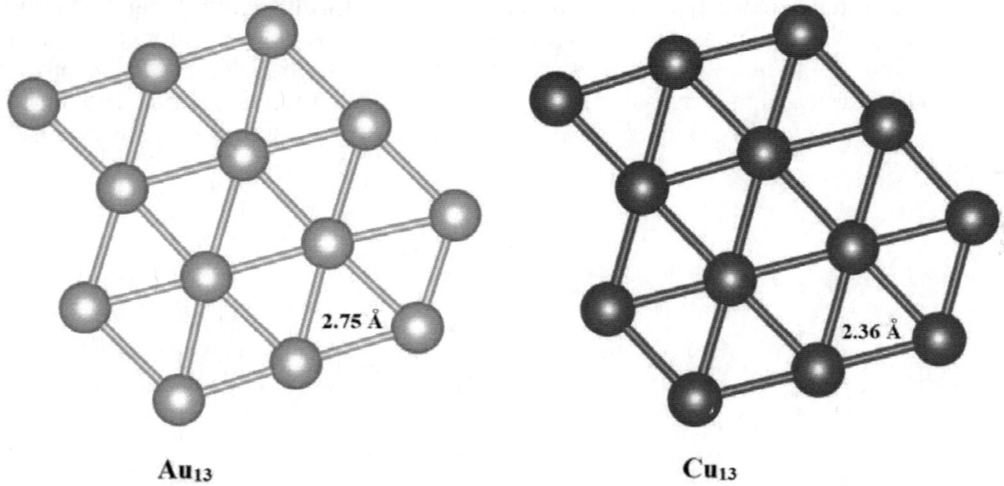

Figure 3. Planar clusters of Au_{13} and Cu_{13}.

2.3. The hydrogen atom position

Table 2 shows the binding energies and lengths of the Au_{13} cluster with hydrogen in different positions relatively to the nearest gold atoms. According to the calculations, the values of the energy and length of the Au-H bond at different positions of the hydrogen atom above the surface of the gold cluster differed insignificantly and averaged as 3.67 eV and 1.77 Å, respectively. However, the bridge-position is the most favorable.

Table 2. The binding energy and bond length of the Au_{13} cluster with the hydrogen atom at different positions relatively to the nearest gold atoms.

Position	atop	bridge	hollow	average
Binding energy, eV	–3.62	–3.75	–3.64	3.67
Bond length, Å	1.6	1.8	1.9	1.77

Table 3. The binding energy and bond length of the Cu13 cluster with the hydrogen atom at different positions relatively to the nearest copper atoms.

Position	atop	bridge	hollow	average
Binding energy, eV	–2.59	–3.13	–3.18	2.97
Bond length, Å	1.5	1.65	1.7	1.62

Carrying out similar calculations for the copper cluster Cu_{13} has shown that hollow-position of the hydrogen atom on the cluster surface is the most stable (table 3). Differences in the magnitude of the binding energy between bridge- and hollow-positions are small. Hydrogen atom binding energy in the atop-position is much smaller. The average binding energy and bond length are 2.97 eV and 1.62 Å, respectively.

2.4. Locality

In order to establish how decidedly the interaction with hydrogen affects the electronic structure of gold and copper clusters, the distribution of the projected density of states on atoms with and without a hydrogen atom is calculated. The local electronic density of d-states on gold and copper atoms interacting with hydrogen decreases with the formation of a bond with hydrogen (figure 4 (a) and (c)). The changes in the distribution of the local density of states for a neighboring atom are much smaller (figure 4 (b) and (d)). Thus, sequential adsorption acts for clusters of gold and copper can be considered independent – in the sense of smallness of the spatial perturbation region determined by the adsorption, that is $E(k) - E(k+1) << (E(k), E(k+1))$, where $E(k)$ is adsorption energy of the k-th hydrogen atom.

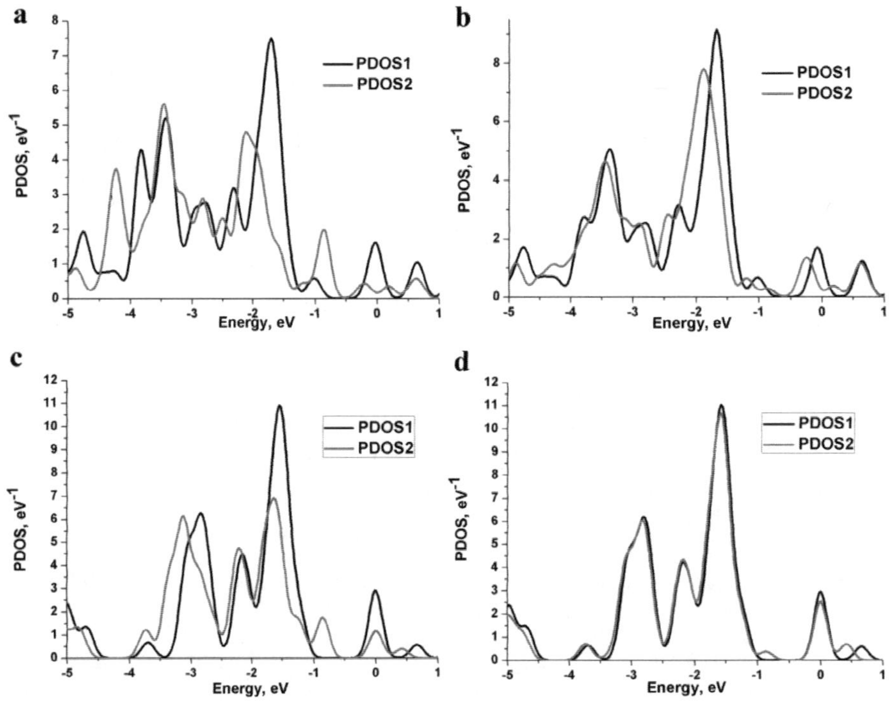

Figure 4. The projected density of states on the gold and copper atoms with a hydrogen atom (a) and (c), on the adjacent atom (b) and (d), in states without (PDOS1) and with (PDOS2) hydrogen.

2.5. Sequential adsorption of hydrogen atoms

Also, to estimate the effect of interaction with hydrogen atoms on the energy structure, the density of states for clusters Au_{13}, Au_{20}, Au_{55}, Cu_{13}, Cu_{20}, Cu_{55} is calculated without hydrogen atoms and with a surface completely filled with hydrogen atoms (figure 5). With an increase in the number of hydrogen atoms, the first moment of density of states distribution (d-band center) shifts from the Fermi level towards negative values of energy in case of gold and also copper clusters, which means a decrease in the chemical activity of the cluster, according to the resonance model of chemisorption. The measure of the chemical activity in this case is the value of the binding energy.

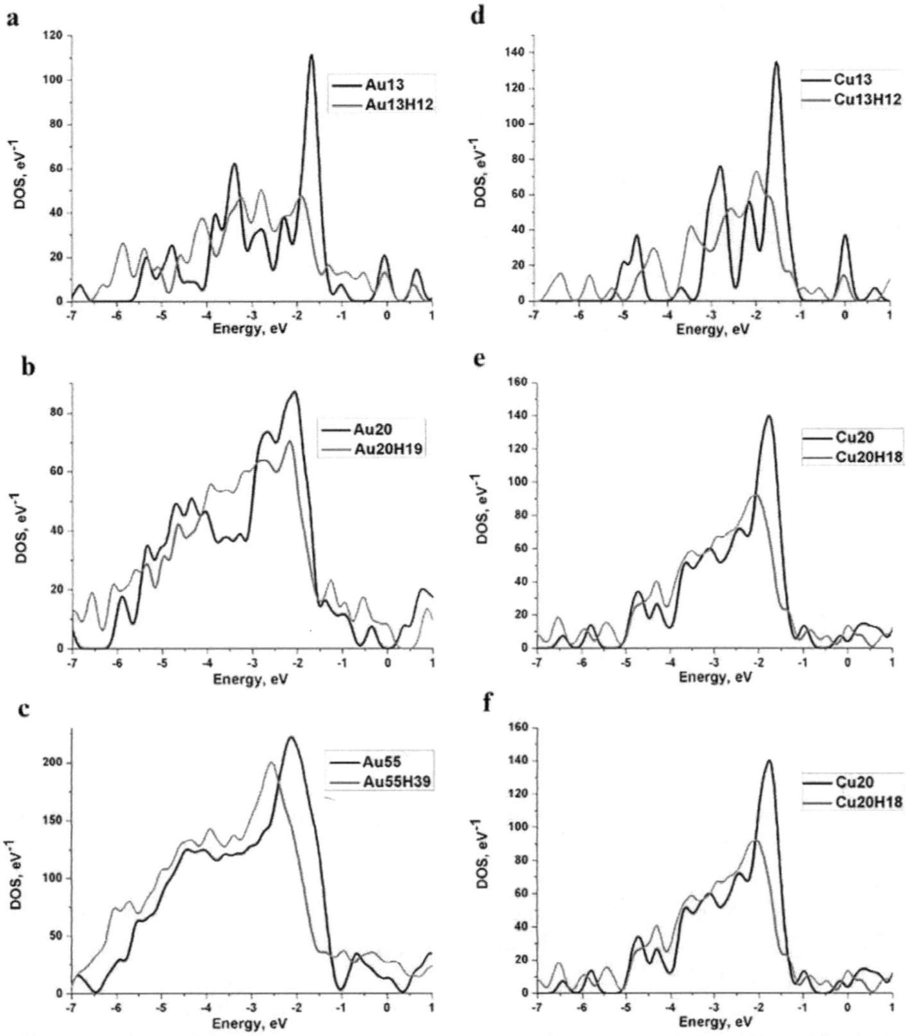

Figure 5. Density of cluster states Au_nH_m, where $n = 13, 20, 55, 100, 147$, $m = 1$–86.

Table 4. The binding energy of hydrogen atoms with gold and copper clusters [13].

m	Au_{13}, eV	Au_{20}, eV	Au_{55}, eV	Cu_{13}, eV	Cu_{20}, eV	Cu_{55}, eV
1	–3.13	–3.22	–3.50	–3.29	–3.35	–3.15
ful	–3.14	–3.26	–3.14	–3.27	–3.23	–3.22

The calculated values of the binding energy of hydrogen atoms in gold and copper clusters for various surface occupancies are shown in table 4: with only one hydrogen atom and with complete filling. A complete correlation with changes in the energy spectra of the clusters does not occur, that indicates the complication of the resonance model of chemisorption. This leads to conclusion that the value of the binding energy in the investigated objects is determined not only by the coordination saturation of the copper and gold atoms, but also by the distribution of the local density of states near the Fermi level. Since the chosen clusters have low nuclearity, their chemical properties can be determined to a considerable extent not only by the local density of d-states below the Fermi level, but also by *sp*-states that are smaller in value but close to the Fermi level.

2.6. Delocalization

As it is known, clusters with highly symmetric geometric structure or clusters with a complete filling of a certain group of electronic states are distinguished by increased stability ("magic" clusters).

The electronic structure of gold ([Xe]4f145d106s1) and copper ([Ar]3d104s1) atoms allows to consider nanosystems based on these elements can exhibit features that correspond to both "geometric" (the main contribution to the interaction of atoms is made by d-electrons, similar to transition metals) and "electronic" (one "external" *s*-electron, as in alkali metals) magic numbers. The calculations did indeed show a nonlinear dependence of the binding energy of gold and copper clusters with hydrogen atoms for certain values of the total number of atoms $n + m$ in clusters Au_nH_m and Cu_nH_m corresponding to the "electronic" series of magic numbers, namely in the vicinity of 18 and 20 (figure 6).

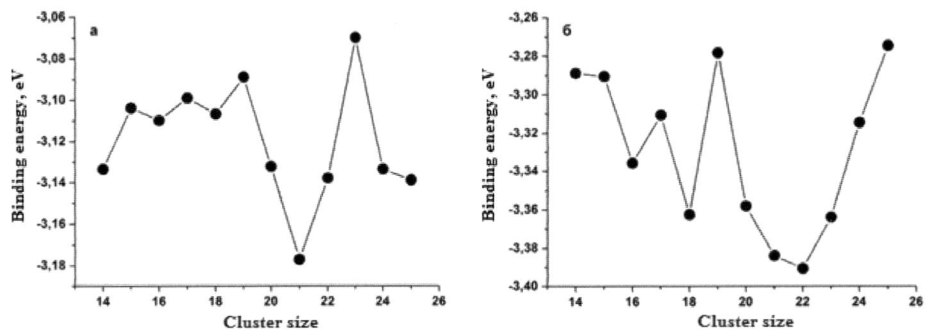

Figure 6. Dependence of the binding energy on the number of adsorbed hydrogen atoms in clusters Au_nH_m (a) and Cu_nH_m (b), where $n = 13$, $m = 1–12$.

Deviations from the exact values of magic numbers arise due to the asymmetric structure of heterogeneous clusters. This means that the replacement of metal atoms by chemisorbed hydrogen atoms in the construction of the outer electron shell of the cluster is inherent both for gold and copper. At the same time, the electrons added by hydrogen are delocalized throughout the volume. However, the general redistribution of the electron density in the heterogeneous metal-hydrogen system is insignificant, that agrees with the previously drawn conclusion about the locality of the introduced perturbations. A similar effect – some kind of the electronic and atomic structure of silicium hydride with the replacement of hydrogen by gold – is detected and investigated by photoelectron spectroscopy in [14].

3. Conclusions

Acts of sequential adsorption in the studied systems can be considered as independent. As the number of adsorbed hydrogen atoms increases, the d-band center shifts toward smaller values of energy.

In the outer electron shell of a heterogeneous metal-hydrogen cluster, the electrons of chemisorbed hydrogen atoms play the same role as delocalized s-electrons of metal atoms.

Acknowledgments

The reported research was funded by Russian Foundation for Basic Research, grant № 16-03-00045.

IOP Conf. Series: Materials Science and Engineering **347** (2018) 012018 doi:10.1088/1757-899X/347/1/012018

References

[1] Yin G, Nishikawa M, Nosaka Y, Srinivasan N, Atarashi D, Sakai E and Miyauchi M 2015 *ACS Nano* **9** 2111–9
[2] Kirsankin A A, Grishin M V, Sarvadii S Yu, Zamota P F and Shub B R 2017 *Russ. J. Phys. Chem. B* **11** 521–5
[3] Nikolaev S A, Dmitriev G S, Zanaveskin K L, Egorova T B and Khadzhiev S N 2017 *Petroleum Chemistry* **57** 1074–80
[4] Nikolaev S A, Chistyakov A V, Zharova P.A, Tsodikov M V, Krotova I N and Ezzhelenko D I 2016 *Petroleum Chemistry* **56** 730–7
[5] Guvelioglu G H, Pingping Ma, Xiaoyi He, Robert C F and Hansong Ch 2006 *Phys. Rev. B* **73** 155436–46
[6] Gogol' V V, Pichugina D A and Kuz'menko N E 2016 *Russ. J. Phys. Chem. A* **90** 2402–7
[7] Li J, Liu Y, Zhang J, Liang X and Duan H 2016 *Chem. Phys. Lett.* **651** 137–43
[8] Gautam S, Dharamvir K and Goel N 2013 *Comput. Theor. Chem.* **1009** 8–16
[9] Michaelian K, Rendon N and Garzon I 1999 *Phys. Rev. B* **60** 2000–10
[10] Larsen A, Kleis J, Thygesen K, Nørskov J and Jacobsen K 2011 *Phys. Rev. B* **84** 245429–32
[11] Li T, Lee S, Han S and Wang G 2002 *Phys. Lett.* **300** 86–92
[12] Bravo-Perez G, Garzyn I and Novaro O 1999 *Chem. Phys. Lett.* **313** 655–64
[13] Gatin A K, Grishin M V, Dokhlikova N V, Kolchenko N N and Shub B R 2016 *Dokl. Phys. Chem.* **470** 125–8
[14] Kiran B, Li X, Zhai H-J and Wang L-S 2006 *J. Chem. Phys.* **125** 133204–11

IOP Conf. Series: Materials Science and Engineering **347** (2018) 012019 doi:10.1088/1757-899X/347/1/012019

Variability of Young's modulus and Poisson's ratio of hexagonal crystals

M A Komarova[1,2], V A Gorodtsov[1] and D S Lisovenko[1,2]

[1] Ishlinsky Institute for Problems in Mechanics RAS, prospect Vernadskogo 101-1, Moscow, 119526, Russia
[2] The Russian Presidential Academy of National Economy and Public Administration, prospect Vernadskogo 82-1, Moscow, 119571, Russia

E-mail: lisovenk@ipmnet.ru

Abstract. In this paper, the variability of elastic characteristics (Young's modulus and Poisson's ratio) of hexagonal crystals has been studied. Analytic expressions for Young's modulus and Poisson's ratio are obtained. Stationary values for these elastic characteristics are found. Young's modulus has three stationary values, and Poisson's ratio has eight stationary values. Numerical analysis of these elastic characteristics for hexagonal crystals is given based on the experimental data from the Landolt-Börnstein handbook. Global extrema of Young's modulus and Poisson's ratio for hexagonal crystals are found. Crystals are found in which the maximum values exceeds the upper limit for isotropic materials.

1. Introduction

A distinctive simplifying feature of the linear elasticity of isotropic materials is that it is completely determined by two elastic constant coefficients. Moreover, the thermodynamic requirement of the positive definiteness of the quadratic form of the strain energy imposes restrictions on them. For example, Young's modulus must be positive, and Poisson's ratio may vary only in the range of from -1 to 0.5 ($-1 \leq \nu \leq 1/2$) [1]. For most isotropic materials we have $\nu > 0$.

In the case of anisotropic materials the complication of elastic behavior takes place. The number of determining elastic coefficients increases. Poisson's ratio, understood as the ratio of transverse compression to longitudinal stretching of the sample, depends in the magnitude and sign of the orientation of the stretched sample. For single-crystal materials, Poisson's ratios can have arbitrary values and signs [2]. Young's modulus remains positive.

The elastic properties of hexagonal crystals are analyzed in [3-10]. Young's modulus, Poisson's ratio and shear modulus for hexagonal Tl and Cd crystals are studied for some particular orientations in [3]. It was shown in [5] that a zinc crystal can have negative Poisson's ratio for some orientations. Only seven crystals (MoS_2, Zn, Be, BeCu (2.4% at. Cu) alloy, TiB_2, C_7H_{12}, MnAs) are known with negative Poisson's ratio among the hexagonal crystals [6-8]. An analysis of average Poisson's ratio have shown that all hexagonal crystals have positive Poisson's ratio [9]. We note that more than four hundred crystals with negative Poisson's ratio (auxetics) are detected for crystals of other crystalline systems [5-8, 11-24]. The least number of auxetics is found among hexagonal crystals, and the largest amount is detected among cubic crystals (more than three hundred).

The extremal values of Young's modulus re investigated using the Lagrange multiplier method for cubic and hexagonal crystals in [4]. Below we will discuss. The extreme values of Young's modulus

IOP Conf. Series: Materials Science and Engineering **347** (2018) 012019 doi:10.1088/1757-899X/347/1/012019

and Poisson's ratio for hexagonal crystals are analyzed below by the extremum of functions of one and two variables.

2. Elastic properties of hexagonal crystals

In modern materials science, nanomaterials that are made of anisotropic single crystals have been actively used. The elastic properties of anisotropic materials depend on the orientation of the samples. Such elastic characteristics of crystals as Young's modulus $E(\mathbf{n})$ and Poisson's ratio $\nu(\mathbf{n},\mathbf{m})$ depend on two unit vectors $\mathbf{n}$ and $\mathbf{m}$ [25]

$$\frac{1}{E(\mathbf{n})} = s_{ijkl} n_i n_j n_k n_l ,$$

$$\frac{\nu(\mathbf{n},\mathbf{m})}{E(\mathbf{n})} = -s_{ijkl} n_i n_j m_k m_l .$$

Here s_{ijkl} are the tensor compliance coefficients. In the case of uniaxial stretching, the unit vector $\mathbf{n}$ is directed along the extension axis, and the vector $\mathbf{m}$ is perpendicular to it.

The elasticity of hexagonal crystals is characterized by five independent matrix compliance coefficients s_{11}, s_{12}, s_{13}, s_{33}, s_{44}. Below we will use the matrix compliance coefficients instead of tensor compliance coefficients. The orientation of a crystalline rod in a crystallographic coordinate system is conveniently described by three Euler's angles φ, θ, ψ. The connection between the unit vectors and Euler's angles has the form

$$\mathbf{n} = \begin{pmatrix} \sin\varphi \sin\theta \\ -\cos\varphi \sin\theta \\ \cos\theta \end{pmatrix},$$

$$\mathbf{m} = \begin{pmatrix} -\sin\varphi\cos\theta\cos\psi - \cos\varphi\sin\psi \\ \cos\varphi\cos\theta\cos\psi - \sin\varphi\sin\psi \\ \sin\theta\cos\psi \end{pmatrix}.$$

Expressions of Young's modulus E and Poisson's ratio ν can be rewritten as follows [26]

$$\frac{1}{s_{11}E} = 1 + \left(\Pi_1 - \Pi_{01}\sin^2\theta\right)\cos^2\theta , \quad (1)$$

$$-\frac{\nu}{s_{13}E} = 1 + \left(\Pi_2 \sin^2\psi + \Pi_{02}\cos^2\theta\cos^2\psi\right)\sin^2\theta , \quad (2)$$

$$\Pi_{01} \equiv \frac{\delta}{s_{11}}, \quad \Pi_{02} \equiv \frac{\delta}{s_{13}},$$

$$\Pi_1 \equiv \frac{s_{33} - s_{11}}{s_{11}}, \quad \Pi_2 \equiv \frac{s_{12} - s_{13}}{s_{13}},$$

$$\delta \equiv s_{11} + s_{33} - 2s_{13} - s_{44} .$$

As can be seen, Young's modulus of hexagonal crystals depends only on one Euler's angle θ (does not depend on the other two angles), and Poisson's ratio is dependent on two Euler's angles θ, ψ, and does not depend on the third angle φ. Dimensionless parameters Π_{0n}, Π_n are characteristics of the degree of anisotropy of hexagonal crystals. Young's modulus and Poisson's ratio are periodic functions with periods $T_\theta = T_\psi = \pi$.

The stationary values of Young's modulus are determined in accordance with the disappearance of the derivative of expression (1) with respect to the angular variable

$$\sin 2\theta\left(2\Pi_{01}\cos^2\theta + \Pi_1 - \Pi_{01}\right) = 0 .$$

Two stationary points

$$E_1 = \frac{1}{s_{33}}, \quad E_2 = \frac{1}{s_{11}}$$

are attained at $\theta = 0$ and $\theta = \pi/2$, respectively. The third stationary point $E_3 > 0$

$$E_3 = \frac{4\Pi_{01}}{4\Pi_{01} - (\Pi_1 - \Pi_{01})^2}\frac{1}{s_{11}}$$

can be attained at $\theta = \theta^*$

$$\cos^2\theta^* = \frac{\Pi_{01} - \Pi_1}{2\Pi_{01}} .$$

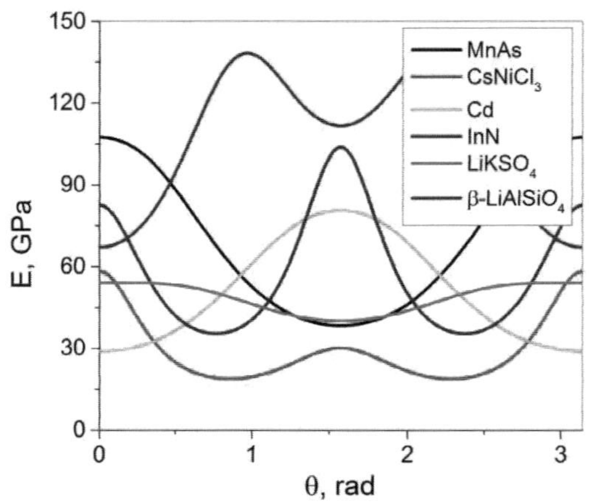

Figure 1. Angular dependences of Young's moduli for some hexagonal crystals.

The angular dependences of Young's modulus for several hexagonal crystals are shown in figure 1. The extreme values of Young's modulus are given for the same crystals in table 1. As can be seen from figure 1 and the data from table 1, the MnAs and Cd crystals have only two extreme values. One value is the maximum (E_1 – the maximum for MnAs and E_2 – the maximum for Cd), and the second value is the minimum (E_1– the minimum for Cd and E_2 – the minimum for MnAs). Four other crystals ($CsNiCl_3$, InN, $LiKSO_4$, -$LiAlSiO_4$) have three extreme values. Two maximum values are observed for $CsNiCl_3$ and InN crystals (CsNiCl3: E_1 – global maximum and E_2 – local maximum; InN: E_1 – local maximum and E_2 – global maximum). The $LiKSO_4$ and β-LiAlSiO4 crystals have two minimum values (β-$LiAlSiO_4$: E_1 – global minimum and E_2 – local minimum; $LiKSO_4$: E_1 – local minimum and E_2 – global minimum).

IOP Conf. Series: Materials Science and Engineering **347** (2018) 012019 doi:10.1088/1757-899X/347/1/012019

Table 1. Extreme values of Young's moduli for some hexagonal crystals. Global maximum and minimum values are shown in bold.

Crystals	E_1		E_2		E_3	
	GPa		GPa		GPa	
MnAs	**108**	max	**38.5**	min	-	-
Cd	**28.9**	min	**80.6**	max	-	-
$CsNiCl_3$	**58.5**	max	30.1	max	**18.8**	min
InN	82.6	max	**104**	max	**35.5**	min
$LiKSO_4$	54.05	min	**40.2**	min	**54.08**	max
β-$LiAlSiO_4$	**67.1**	min	112	min	**138**	max

The extreme values of Poisson's ratio and their positions on the plane of angles θ, ψ are found using formula (2) and solutions of the system of equations

$$\sin^2\theta\sin 2\psi\left(\Pi_2-\Pi_{02}\cos^2\theta\right)=0,$$

$$\sin 2\theta\,[\,\Pi_1+\Pi_{01}\cos 2\theta+\Pi_2\sin^2\psi\left(1+\Pi_1-\Pi_{01}\sin^4\theta\right)+$$
$$+\Pi_{02}\cos^2\psi\left(\cos 2\theta+\Pi_1\cos^4\theta\right)]=0.$$

The value of Poisson's ratio at the stationary point 1

$$\nu_1=-s_{13}/s_{11}$$

is attained on a straight line $\theta=0$, i.e. at an arbitrary angle ψ. The values of Poisson's ratios at the stationary points 2 and 3 have the form

$$\nu_2=-s_{13}/s_{11}$$

and

$$\nu_3=-s_{12}/s_{11}$$

at the points $\theta=\pi/2$, $\psi=0$ and $\theta=\psi=\pi/2$, respectively. The existence of other extrema of Poisson's ratio is possible when certain constraints on the dimensionless complexes of compliance coefficients are fulfilled, i.e. for some specific hexagonal crystals. These extreme values of Poisson's ratio have the following form

$$\nu_{4,5}=-\frac{s_{13}}{s_{11}}\frac{1+0.25\Pi_{02}\sin^2 2\theta}{1+\Pi_1\cos^2\theta-0.25\Pi_{01}\sin^2 2\theta}$$

at the restrictions

$$\psi=0,\;\; 0\le\cos^2\theta_{4,5}=\left(-(\Pi_{01}+\Pi_{02})\pm\sqrt{D_1}\right)/\Pi_1\Pi_{02}\le 1,$$

$$D_1=(\Pi_{01}+\Pi_{02})^2-\Pi_1\Pi_{02}(\Pi_1-\Pi_{01}-\Pi_{02})\ge 0$$

and

$$\nu_{6,7}=-\frac{s_{13}}{s_{11}}\frac{1+\Pi_2\sin^2\theta}{1+\Pi_1\cos^2\theta-0.25\Pi_{01}\sin^2 2\theta}$$

at the restrictions

IOP Conf. Series: Materials Science and Engineering **347** (2018) 012019 doi:10.1088/1757-899X/347/1/012019

$$\psi = \pi/2, \quad 0 \le \sin^2\theta_{6,7} = \left(-\Pi_{01} \pm \sqrt{D_2}\right)/\Pi_{01}\Pi_2 \le 1,$$

$$D_2 = \Pi_{01}^2 + \Pi_{01}\Pi_2(\Pi_1 + \Pi_{01} + \Pi_2 + \Pi_1\Pi_2) \ge 0.$$

The extreme value of Poisson's ratio ν_8 is determined at

$$\cos^2\theta = \Pi_2/\Pi_{02}, \quad \cos^2\psi = A/B,$$

$$A = (\Pi_1 - \Pi_{01})\Pi_{02}^2 + \Pi_2[\Pi_{02}(1 + \Pi_1 + 2\Pi_{01}) - \Pi_1(\Pi_{02}^2 + \Pi_2^2 - 2\Pi_2\Pi_{02})],$$

$$B = \Pi_{02}(\Pi_{02} - \Pi_2)(\Pi_{02} + \Pi_1\Pi_2) + \Pi_{01}\Pi_2(\Pi_{02}^2 + 2\Pi_2\Pi_{02} - \Pi_2^2)$$

The values of Poisson's ratios at eight stationary points for some hexagonal crystals are given in table 2. Global maximum and global minimum values are shown in bold. The global minimum value of Poisson's ratio takes a negative value ($\nu_{min} = -0.04$) for the MnAs crystal. The maximum values of Poisson's ratios for Cd ($\nu_{max} = 0.73$), $CsNiCl_3$ ($\nu_{max} = 0.60$), InN ($\nu_{max} = 0.79$), β-$LiAlSiO_4$ ($\nu_{max} = 0.58$) are more than 0.5 (the upper limit for isotropic materials). Numerical analysis carried out for 143 crystals from [27] showed that the stationary values of Poisson's ratio ν_1 and ν_8 are not extremal values. Surfaces of Poisson's ratio for Cd and InN crystals are shown in figure 2. Poisson's ratios for these crystals take positive values. Analysis carried out in [3] showed that Cd crystal can have negative Poisson's ratio at certain orientations. The results from table 2 and figure 2a indicate that this is not the case. It turns out that the results in [3] contain an error associated with an incorrect recalculation of the compliance coefficients for cadmium.

Table 2. Values of Poisson's ratios at stationary points for some hexagonal crystals. Global maximum and minimum values are shown in bold.

Crystals	ν_1	ν_2	ν_3	ν_4	ν_5	ν_6	ν_7	ν_8
MnAs	0.22	0.08	0.19	-	**-0.04**	**0.24**	-	-
Cd	0.26	**0.73**	**0.10**	-	0.25	0.27	-	-
$CsNiCl_3$	0.20	**0.11**	0.35	**0.60**	-	-	0.13	0.28
InN	0.41	0.52	0.22	**0.79**	-	-	**0.12**	-
$LiKSO_4$	0.28	0.20	0.41	-	**0.17**	**0.41**	-	-
β-$LiAlSiO_4$	0.35	**0.58**	0.14	-	**0.13**	0.42	-	-

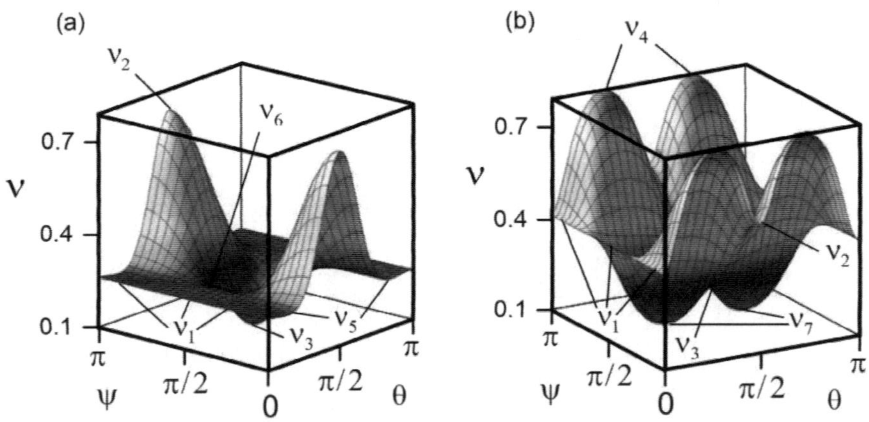

Figure 2. The surfaces of Poisson's ratio for hexagonal Cd (a) and InN (b) crystals.

3. Conclusion

The variability of Young's modulus and Poisson's ratio for hexagonal crystals is studied. Expressions for three stationary values of Young's modulus and eight stationary values of Poisson's ratio are obtained. Numerical analysis of the extrema of these elastic characteristics for some hexagonal crystals is given on the basis of experimental data from the Landolt-Börnstein handbook. For anisotropic materials, elastic characteristics (Young's modulus and Poisson's ratio) are not constant in contrast to isotropic materials. It is shown that global maximum extremes can exceed the value of 0.5, which is the upper limit for isotropic materials. Among the hexagonal crystals, materials for which the global minimum value is less than -1 (the lower limit for isotropic materials) is not found.

Acknowledgments

The study is supported by RFBR, research project 16-01-00325, and the Program for the Basic Research N I.2.32 of the Presidium of the Russian Academy of Sciences.

References

[1] Landau L D and Lifshitz E M 1987 *Theory of Elasticity. Vol. 7 Course of Theoretical Physics*. (Butterworth-Heinemann)
[2] Ting T C T and Chen T 2005 *Quart. J. Mech. Appl. Math.* **58** 73
[3] Li Y 1976 *Phys. Status Solidi A* **38** 171
[4] Cazzani A and Rovati M 2003 *Int. J. Solids Struct.* **40** 1713
[5] Lubarda V A and Meyers M A 1999 *Scripta Mater.* **40** 975
[6] Goldstein R V, Gorodtsov V A and Lisovenko D S 2016 *Phys. Status Solidi B* **253** 1261
[7] Lethbridge Z A D, Walton R I, Marmier A S H, Smith C W and Evans K E 2010 *Acta Mater.* **58** 6444
[8] Goldstein R V, Gorodtsov V A and Lisovenko D S 2010 *Mech. Solids* **45** 529
[9] Goldstein R V, Gorodtsov V A and Lisovenko D S 2013 *Letters on Materials* **3** 7
[10] Goldstein R V, Gorodtsov V A, Komarova M A and Lisovenko D S 2017 *Scripta Mater.* **140** 55
[11] Ballato A. 1996 *IEEE Trans. on Ultrasonics, Ferroelectrics and Frequency Control* **43** 56
[12] Baughman R H, Shacklette J M, Zakhidov A A and Stafstrom S 1998 *Nature* **392** 362
[13] Tokmakova S P 2005 *Phys. Status Solidi B* **242** 721
[14] Branka A C, Heyes D M and Wojciechowski K W 2009 *Phys. Status Solidi B* **246** 2063
[15] Ballato A 2010 *IEEE Trans. on Ultrasonics, Ferroelectrics and Frequency Control* **57** 7
[16] Belomestnykh V N and Soboleva E G 2012 *Letters on Materials* **2** 13
[17] Goldstein R V, Gorodtsov V A and Lisovenko D S 2013 *Phys. Status Solidi B* **250** 2038
[18] Goldstein R V, Gorodtsov V A, Lisovenko D S and Volkov M A 2014 *Phys. Mesomech.* **17** 97
[19] Goldstein R V, Gorodtsov V A and Lisovenko D S 2015 *Phys. Mesomech.* **18** 213
[20] Volkov M A 2014 *Letters on Materials* **4** 167
[21] Krasavin V V and Krasavin A V 2014 *Phys. Status Solidi B* **251** 2314
[22] Goldstein R V, Gorodtsov V A, Lisovenko D S and Volkov M A 2015 *Letters on Materials* **5** 409
[23] Goldstein R V, Gorodtsov V A, Lisovenko D S and Volkov M A 2016 *Letters on Materials* **6** 93
[24] Epishin A I and Lisovenko D S 2016 *Tech. Phys.* **61** 1516
[25] Sirotin Y I and Shaskol'skaya M P 1982 *Foundations of crystal physics* (Moscow, Mir)
[26] Goldstein R V, Gorodtsov V A and Lisovenko D S 2011 *Dokl. Phys.* **56** 602
[27] Nelson D.F. (Ed.) 1992 *Second and Higher Order Elastic Constants, Vol. 29a of Landolt-Börnstein - Group III Condensed Matter* (Springer)

IOP Conf. Series: Materials Science and Engineering **347** (2018) 012020 doi:10.1088/1757-899X/347/1/012020

Experimental analysis of the bearing greases lifetime

Albagachiev A U, Smirnov N I, Prozhega M V and Smirnov N N

Department of Friction, Wear and Lubrication, Mechanical Engineering Research Institute of the Russian Academy of Sciences, 4 Maly Kharitonyevsky Pereulok, Moscow, Russia

E-mail: prmaksim@gmail.com

Abstract. The purpose of the presented work is to increase the lifetime of space mechanisms bearings. The authors reviewed the studies of the bearings and linear mechanisms operating in air and in vacuum. A testing equipment is developed to assess the lubricity and lifetime of greases in the air. Criteria for the evaluation of relative lifetime are defined. Four types of space greases are tested.

1. Introduction

The increased lifetime demanded of state-of-the-art telecommunications satellites has necessitated the development of new greases for space. Modern lubricants should provide the operation of mechanisms for 15 years. Loads in friction units are increasing constantly. When the pointing mechanism for Ion Thruster Alignment Mechanisms (ITAM) used in the telecommunications satellite ARTEMIS was developed in 1995, thermal vacuum life test performed comprised 15000 movements (= 7500 cycles and 12840 degrees total path) of the spindle over 4 revolutions each, performed around a fixed position on one spindle, and performed at –50°C and +70°C. In 2006 the test was performed in two parts, the first part from 0-37500 cycles with a radial screw load of 9 N and around one fixed point, the second part from 37500-61600 cycles with a radial load of 24 N and 1 mm away from the first spindle point [1]. In table 1 presents the comparative characteristics of greases, which were tested for friction in [2-4]. Most often, as a thickener in the greases production, the polytetrafluoroethylene (PTFE) is used. For the greases preparation the different base oils are used. The perfluropolyalkylether (PFPE) oils are characterized by an increased density, which gives an advantage in elasto-hydrodynamic lubrication. A considerable amount of work has recently been devoted to the study of greases based on multiple alkylated cyclopentane (MAC). The kinematic viscosity at 100°C of a MAC base oils varies from 2 to 20 cSt, and the viscosity index ranges from 71 to 178.

The purpose of the presented work is to compare the lifetime and lubricity of the greases for space.

2. Materials

Greases considered in this study are based on a PFPE oil and a tetrafluoroethylene, an organosilicon or a lithium soap as a thickener. They were developed at VNII NP (Russia), the characteristics and composition are summarized in table 2.

The greases #1 are used for small rolling bearings (at speeds of up to 30000 rpm) and low-power reduction gears, especially operating under conditions of deep vacuum (up to 1.3×10^{-4} Pa) in the temperature range of –80°C...+160°C. The grease #2 was developed for lubrication of parts working in contact with aggressive media in the temperature range –110°C ...+200°C and in conditions of deep vacuum. The greases #2 are recommended for use in aerospace and high vacuum. Low-temperature antifriction grease #3 is intended for use in antifriction bearings operating in vacuum at temperatures from –120°C to +250°C. The grease #4 was developed for lubrication of parts working in contact with aggressive environments, including carbon dioxide, in the temperature range of 130°C...+250°C, as well as in conditions of deep vacuum. The grease #4 is composed of soluble PFPE metallo-cladding additive.

Table 1. Properties of the grease for space [2-5].

	Trademark	Base oil	Thickener	Oil separation (ASTM D6184) %, by weigth	Temperature range, °C
1	Krytox 240AC	PFPE (Krytox 143AC)	PTFE	6 (100°C, 30 h)	-34…288
2	MAPLUB PF100	PFPE (Fomblin Z25)	PTFE	2.7 (100°C, 30 h)	-60…130
3	MAPLUB SH050	MAC (NYE 2001A)	PTFE	4.2 (100°C, 30 h)	-60…130
4	Rheolube 2000	MAC (NYE 2001)	Sodium soap	3.3 (100°C, 24 h)	-45…125
5	Braycote 601EF	PFPE (Castrol Brayco 815Z)	---	10.53 (204°C, 30 h using FTM 321)	---

Table 2. Grease composition.

	Base oil	Thickener	Viscosity at -40°…-50°C, Pa*s	Evaporation in air at +150…+200°C (1 hour), %
1	Silicon liquid	Lithium soap	54	N/A
2	PFPE	Organosilicon	210	2,5
3	PFPE	PTFE	79	4
4	PFPE	Organosilicon	100	3

3. Experimental equipment

The authors developed a test method and equipment for bearing lubricant's evaluation, figure 1. Technical parameters of the bearing test machine: shaft velocity up to 200 rpm; vertical force on the bearing - up to 16000 N; temperature of the sample at nominal conditions from –70°C to + 180°C.

The thrust roller bearing is used as a sample for tests. The bearing is loaded with the pneumatic cylinder through the arm. The shaft for load transfer from the arm to bearing is installed inside in the housing using two radial bearings. The housing is movable in a vertical direction along the guides. The friction moment is recorded with a strain sensor which fixed to the shaft. A computer data acquisition system (cDAQ-9184 chassis and the modules NI 9219, NI 9217, NI 9203, NI 9269) constantly monitored bearing load, temperature and friction moment. The test terminated when a coefficient of friction exceeds 0.015.

The thrust roller bearing 84104 from NTN Company with inner diameter 20 mm, outer diameter 35 mm, roller diameter 5.5 mm is used (figure 2). All tests are performed using a maximum Hertzian stress of 0.9 GPa, a lower bearing ring rotational speed of 50 RPM. The temperature of the bearings under the nominal conditions is 25…30°C.

For the lifetime tests, the greases are applied only to the lower ring. The small amount of grease deposited on the ring (25 to 30 mg) is determined using the O-Haus Advanture balance with an accuracy of ± 0.1 mg. All bearings are made of AISI 52100 steel.

Relative lifetime is determined by calculating the number of the lower bearing ring revolutions to a sharp increase of friction coefficient per microgram of used lubricant

$$R = \frac{n}{M}, \tag{1}$$

where n is the number of lower bearing ring revolutions, and M is the mass of grease.

IOP Conf. Series: Materials Science and Engineering **347** (2018) 012020 doi:10.1088/1757-899X/347/1/012020

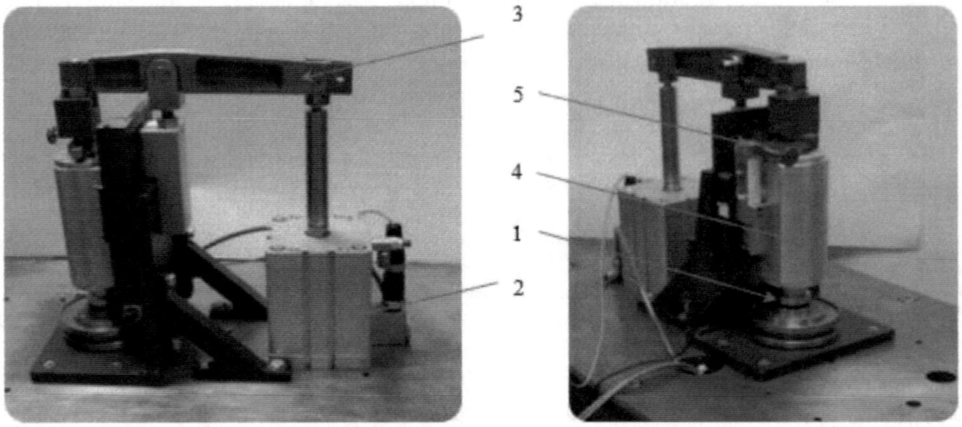

Figure 1. Test equipment for lubricant evaluation in air (1 – thrust roller bearing, 2 – pneumatic cylinder, 3 – arm, 4 – housing, 5 – strain sensor).

Figure 2. Sample for lifetime grease test.

4. Results

The results of grease lifetime tests are shown in figure 3. All greases are tested thrice with the bearing test machine. The data, reported in figure 3, clearly show the lifetimes of the grease #3 to be much greater than samples #1, #2 and #4 by 1.4 … 3.9 times. The presence of PTFE in the grease #3 improve its lifetime.

The initial friction coefficient of the greases is shown graphically in figure 4.

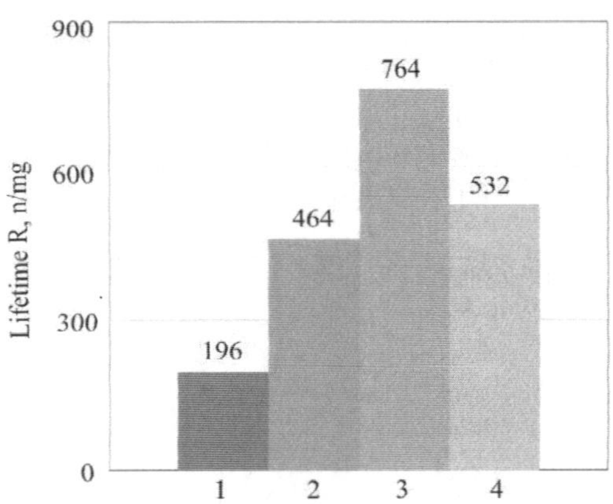

Figure 3. Normalized lifetimes of the various greases.

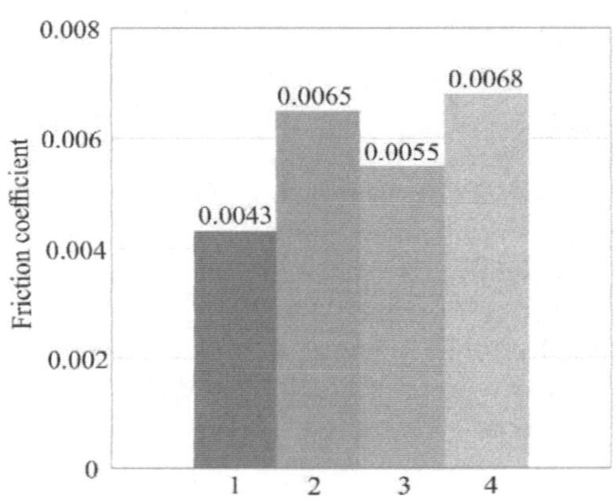

Figure 4. Friction coefficient of the various greases.

IOP Conf. Series: Materials Science and Engineering **347** (2018) 012020 doi:10.1088/1757-899X/347/1/012020

5. Discussion

When evaluating the performance characteristics of grease, the most difficult and time-consuming procedure is to determine the resource characteristics of the lubricant. A sharp increase of the friction coefficient is the result of the lubricant degradation in the contact. The surface of the lower bearing ring is shown in figure 5. An important feature of the developed test methodology is the providing of rolling friction in the contact.

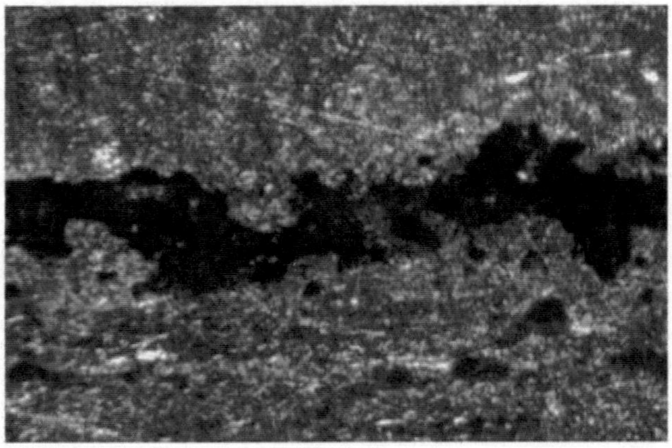

Figure 5. Surface of the lower bearing ring after test with grease #1 (magnification x100).

6. Conclusions

The following conclusions are drawn in relation to the tested greases. These conclusions are specific to greases operating at ambient temperatures.

- The lifetimes of the grease #3 with PTFE outperform another grease.
- The performance of the grease #2 is similar to that of grease #4.
- The friction coefficient of the Silicon liquid base oil grease lower than of PFPE base oil greases.

References

[1] Falkner M, Nitschko T, Supper L, Traxler G, Zemann J and Roberts E 2003 *Proc. 10th Eur. Space Mechanisms and Tribol. Symp.* (San Sebastian, Spain) p 297
[2] Marchetti M et al. 2002 *Tribology letters* **12** 209
[3] Stephen V P 2005 *Proc of WTC2005 World Tribology Congress III* (Washington, USA)
[4] Michael J D et al. 2003 *Tribol. Let.* **15** 3
[5] Jansen M J, Jones W R Jr. and Stephen V P 2003 *Tribol. Let.* **14** 61

IOP Conf. Series: Materials Science and Engineering **347** (2018) 012021 doi:10.1088/1757-899X/347/1/012021

Evaluation of fiber's misorientation effect on compliance and load carry capacity of shaped composite beams

A N Polilov and N A Tatus'

Mechanical Engineering Research Institute named by A.A. Blagonravov of the Russian Academy of Sciences, 4 Maly Kharitonyevsky Pereulok, Moscow, 101990, Russia

E-mail: polilovan@mail.ru, nikalet@mail.ru

Abstract. The goal of this paper is analysis of design methods for composite beams and plates with curvilinear fiber trajectories. The novelty of this approach is determined by the fact that traditional composite materials are typically formed using prepregs with rectilinear fibers only. The results application area is associated with design process for shaped composite structure element by using of biomechanical principles. One of the related problems is the evaluation of fiber's misorientation effect on stiffness and load carry capacity of shaped composite element with curvilinear fiber trajectories. Equistrong beam with constant cross-section area is considered as example, and it can be produced by unidirectional fiber bunch forming, impregnated with polymer matrix. Effective elastic modulus evaluation methods for structures with curvilinear fiber trajectories are validated. Misorientation angle range (up to 5o) when material with required accuracy can be considered as homogeneous, neglecting fiber misorientation, is determined. It is shown that for the beams with height-to-width ratio small enough it is possible to consider 2D misorientation only.

1. Introduction

The international term "bio-inspired method" refers to a methods inspired by Nature. The main slogan of new scientific area named "strength biomechanics" is "Nature study". By analogy with the structure of knots, rational equally stressed curvilinear fiber trajectories "flowing" around the hole can be constructed for the fiber reinforced polymers (FRP) laminates [1-3]. The structure of a bamboo stalk shows us the need to consider different fracture modes of composite tube elements in order to specify rational equistrong dimensions of tube or bamboo construction [4-6].

One of the major advantages of fiber composites is the ease of complex shape components fabrication when considerable loads and high temperatures are not required as for metals. For example, a modern version of pultrusion method – pulforming allows to create some workpiece from unidirectional fibers, impregnated by polymer matrix, and then it is possible to form it into desired shape, in absence of fibers cutting. Due to these features there is an interest to the design of uniform strength beams as elastic elements for effective substitution of steel multi-leaf springs [7]. It is shown in [8] using special FRP strength criteria [9] that it is possible to design the shaped beam which satisfies the strength and stiffness requirements with three times less weight in comparison with rectangular beam. And because of fiberglass lower elastic modulus and density the weight of this beam can be reduced approximately 15 times, compared with steel beam. The analogy of the bending flexibility of shaped beams with constant cross section area and branching composite structures like a

IOP Conf. Series: Materials Science and Engineering **347** (2018) 012021 doi:10.1088/1757-899X/347/1/012021

treetop with a constant sum of branches section areas (Leonardo's rule) [10, 11] has been studied in [12].

This article is aimed to analyze the effects, which should be considered in case of misorientation of fibers, packed in beam body along curvilinear trajectories, corresponding to complex beam shape and stress state.

Constarea beam (beam with constant cross-section area) is chosen as a most effective form for composite design, because of absence of cut fibers and constant volume fraction of fibers. The following problems are considered:

1. Changing of design relations based on reasonable simplifying hypotheses and taking into account fiber misorientation.
2. Determination of the angle range where fiber misorientation can be neglected with required calculation accuracy.
3. Changing of uniform strength beam shape taking into account fiber misorientation.

Standard triangular, parabolic, constarea uniform stressed beams for homogeneous material are known. It is required to determine shape of uniform stressed beam by iteration process to optimal design using fiber uniform stress criterion. Last problem can be solved only by computer modeling of complex spatial fiber trajectories.

2. Shaped constarea beam

Earlier, in [7, 8], equally strong, shaped composite leaf springs (figure 1) are modeled by a cantilever beam, the dimensions of which (width w and height h) are assumed to be variable according to the laws:

$$w(x) = w(0)(1-x^*)^{\alpha},\ h(x) = h(0)(1-x^*)^{\beta},\ x^* = x/L \tag{1}$$

In the rational design of leaf spring it is necessary to find sizes of root section $h(0)$, $w(0)$ and parameters α, β of their variation for the simultaneous implementation of controversial requirements for rigidity and for strength (load carry capacity):

$$C = \frac{P}{v} = \frac{E\,w(0)\,h^3(0)}{4\delta_v L^3}, \tag{2}$$

$$\frac{6P_{\max}L}{w(0)\,h^2(0)} = \sigma^{\max} \le \sigma^*. \tag{3}$$

The contradiction of the requirements (2) and (3) is that, to increase the load carry capacity (to reduce stresses), the thickness of the beam should be increased and, to reduce rigidity (to increase the accumulated elastic energy), the thickness of the beam should be decreased. The best design corresponds to the fulfillment of both (2) and (3) equations.

The additional condition of equal strength is

$$\frac{6P(L-x)}{w(x)\,h^2(x)} = \frac{6PL}{w(0)\,h^2(0)} \tag{4}$$

and leads from equations (1) to the linear ratio

$$\alpha + 2\beta = 1. \tag{5}$$

The deflection v of the shaped beam under load P (figure 1) can be easily defined from the equality of force work $1/2Pv$ and elastic energy accumulated in the beam as follows:

IOP Conf. Series: Materials Science and Engineering **347** (2018) 012021 doi:10.1088/1757-899X/347/1/012021

$$v(L)=\int_0^L \frac{P(L-x)^2\,\mathrm{d}x}{E(x)I(x)}=\frac{PL^3}{3EI(0)}\cdot\frac{1}{1-\alpha/3-\beta}=v_0\delta_v \tag{6}$$

$$I(x)=\frac{w(x)h^3(x)}{12}=\frac{w(0)h^3(0)}{12}\left(1-x^*\right)^{\alpha+3\beta}=I(0)\left(1-x^*\right)^{\alpha+3\beta}$$

The coefficient of the deflection form of equation (6) is:

$$\delta_v=\frac{v(L)}{v_0}=\frac{1}{1-\alpha/3-\beta}. \tag{7}$$

That is equal to the ratio of the maximum deflection of the shaped beam to the deflection of the rectangular beam v_0 with the same dimensions of the root section.

The constarea beam, which is best for the fibrous structure, is designed to retain the cross-section area and the number of fibers in each section, i.e. there are no cut fibers. The equations (1) shows the constant section area should meet the condition:

$$\alpha+\beta=0. \tag{8}$$

So the values α and β are –1 and 1 respectively, based on the equal strength condition (5). For these values, it is clear that a constarea beam has three times higher flexibility than a rectangular beam (see (7)). The necessary dimensions of the spring root section can be found based on simultaneous fulfilling conditions (2) and (3) as follows:

$$h(0)=\frac{2C\sigma^*L^2}{3EP_{\max}}\delta_v=h_0\delta_v,\; w(0)=\frac{27P_{\max}^3E^2}{2\sigma^{*3}C^2L^3\delta_v^2}=\frac{w_0}{\delta_v^2}, \tag{9}$$

where h_0, w_0 are the constant dimensions of a rectangular beam section that meet conditions (2) and (3). The mass of the shaped beam is easily calculated by the integral:

$$m(L)=\int_0^L \rho w(x)h(x)\,\mathrm{d}x=-\rho w(0)h(0)L\int_0^L\left(1-x^*\right)^{\alpha+\beta}\mathrm{d}\left(1-x^*\right)=\frac{\rho w(0)h(0)L}{1+\alpha+\beta}=m(0)\delta_m, \tag{10}$$

$$\delta_m=\frac{1}{1+\alpha+\beta}.$$

A possible mass of the spring is found from equations (9), (10) that fulfill the conditions of rigidity (2) and strength (3):

$$m(L)=\rho w(0)h(0)L\delta_m=\frac{9\rho P_{\max}^2E}{\sigma^{*2}C}\cdot\frac{\delta_m}{\delta_v} \tag{11}$$

where $\delta_m/\delta_v=\delta_\Sigma$ is the coefficient of the mass decrease of the shaped beam compared to a rectangular beam. It can be seen from (11) that the mass of the beam depends on both the material density ρ and the elasticity modulus E. This effect makes GFRP (fiberglass plastic) the most efficient construction material for the elastic elements.

For steel $E^s\approx 210$ GPa, strength $\sigma^{*s}\approx 800$ MPa and density $\rho^s=7800$ kg/m^3. For a unidirectional GFRP $E^s\approx 45$ GPa, strength $\sigma^{*s}\approx 800$ MPa and $\rho^s=2500$ kg/m^3. Thus, according to (11), in the ideal case, the fiberglass spring can be approximately 15 times lighter compared to the steel one.

An interesting conclusion can be drawn from analyzing the influence of an equally strong spring upon reducing its mass:

$$\delta_\Sigma = \frac{3-\alpha-3\beta}{3(1+\alpha+\beta)}, \qquad (12)$$

and resulting from equation (5)

$$\delta_\Sigma = \frac{2-\beta}{3(2-\beta)} = \frac{1}{3}, \qquad (13)$$

i.e. by simultaneously fulfilling the conditions of rigidity and strength, any equally strong ideal beam turns out to be three times lighter than a rectangular beam. The same result is obtained for ideal branching. This is the best limit case; due to the form, no larger reduction in mass can be obtained.

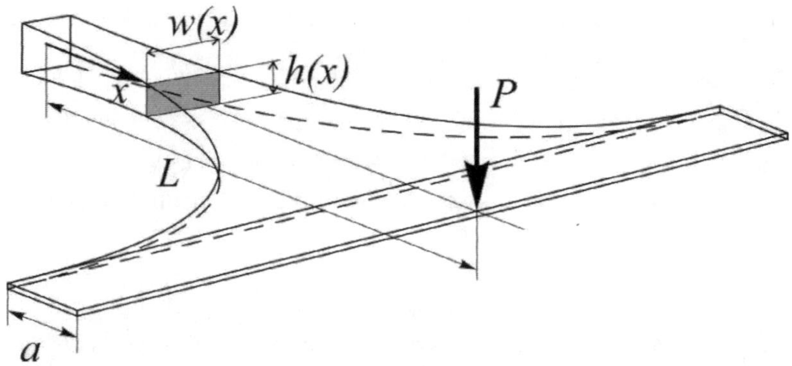

Figure 1. Cantilever constarea beam with constant width-thickness end area.

3. Fibers trajectories in constarea beam

Changing of cross-section dimensions of "ideal" constarea beam is described by power law dependences (1). For taking into account effect of misorientation of fibers on deflection it is necessary to accept some model of "fibers" distribution in the beam.

In this work it is proposed to use "spreading" principle, then in every point direction only is determined as for infinitely thin fiber. Trajectories in this case are conformed with beam shape (1) and are determined by initial coordinates $y(0)$, $z(0)$

$$y(x) = y(0)\left(1-x^*\right)^\alpha, \; z(x) = z(0)\left(1-x^*\right)^\beta, \qquad (14)$$

where $\alpha = -1$, $\beta = 1$. Trajectories slope in planes xy and xz are characterized by derivatives of the functions (14)

$$\frac{\mathrm{d}\,y}{\mathrm{d}\,x} = \frac{y(0)}{\left(1-x^*\right)^2}, \; \frac{\mathrm{d}\,z}{\mathrm{d}\,x} = z(0)^*.$$

Model of misorientation angle calculation is shown in figure 2. "Fiber" element along axis 1 has projections onto axes $\mathrm{d}\,x$, $\mathrm{d}\,y$, $\mathrm{d}\,z$. Angle φ between axis 1 and axis x is determined from apparent dependences:

$$\cos\varphi = (1+A)^{-1/2}, \; \mathrm{tg}\,\varphi = \sqrt{A}, \; A = \left(\frac{\mathrm{d}\,y}{\mathrm{d}\,x}\right)^2 + \left(\frac{\mathrm{d}\,z}{\mathrm{d}\,x}\right)^2. \qquad (15)$$

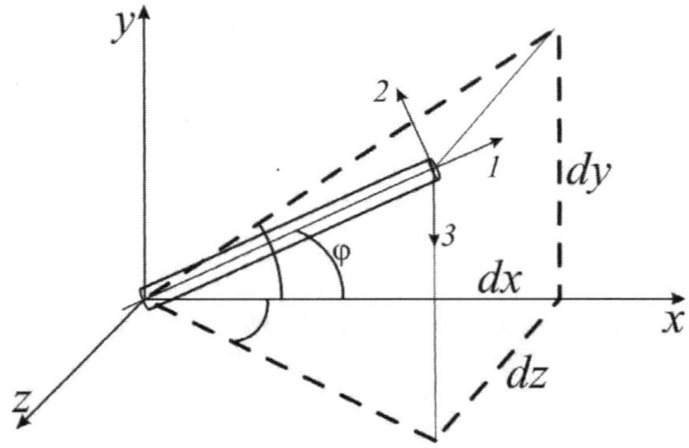

Figure.2. Scheme for calculation of the local misorientation angle of fiber.

If misorientation angle is known, it is possible to calculate local value of longitudinal elastic modulus E_x, which depends on coordinates x, $y(x)$, $z(x)$, connected by equations (14) with initial coordinates of each "fiber" $y(0)$, $z(0)$. The "fiber" tensor modulus E_{11}^0 is much greater than other three moduli E_{22}^0, E_{12}^0, E_{66}^0. Then formula for elastic modulus transformation is extremely simple:

$$E_x(x, y, z, \varphi) = E_1^0 \cos^4 \varphi = E_1^0 / (1 + A)^2 . \tag{16}$$

As effective elastic modulus for layer, section or whole beam the value of modulus for homogeneous element with the same dimensions and the same summary stiffness is understood. In the case of beam tension we can suppose that deformations in each section are homogeneous, i.e. deformations are the same for each structural element and the effective modulus can be estimated by simple averaging:

$$E_x = \frac{1}{N} \sum_{i=1}^{N} E_1^0 \cos^4 \varphi_i , \tag{17}$$

where i – "fiber" number in section, N – total number of "fibers".

More complex bending problem can be solved only based on some kinematic hypotheses:

1. Deformations are changing linearly with height.
2. In each layer (across the width – along y axis) deformations are uniform, so the effective modulus of the layer can be determined by (17).
3. If height of section is small in comparison with width, only 2D misorientation can be considered, and effective modulus in each layer can be assumed constant, thereby averaging (17) can be used for whole section.

For bending of beam with dimensions $w(x)$, $h(x)$ it is possible to estimate the effective modulus:

$$E_x^{\text{eff}}(x) w(x) h^3(x) = 48 \int_{-w/2}^{w/2} \mathrm{d}\, y \int_{-h/2}^{h/2} E(x, y, z) z^2 \, \mathrm{d}\, z . \tag{18}$$

Neglecting small "fibers" misorientation across the height for beams with small height-to-width and height-to-length ratios it is possible to estimate the effective modulus by averaging along width only, and considering that this averaged value is constant along section

$$E_x^{\text{eff}}(x) b(x) = \int_{-b/2}^{b/2} E(x, y) \, \mathrm{d}\, y . \tag{19}$$

Form of homogeneous "uniform strength" constarea beam is chosen and the maximum stress on the surface $\sigma_x = E_x^{\text{eff}} \varepsilon$ reaches the tensile strength $\sigma(0)$ simultaneously in every section. For taking into account the misorientation effect it is necessary to consider the different local "fiber" strength $\sigma(\varphi)$ and reduced local stress $\sigma_x(\varphi)$, associated with a decrease of the local modulus of elasticity $E_x(\varphi)$. The deformation ε is proposed to be uniform along the width. So the local stress change for critical state can be expressed as

$$\sigma_x(\varphi) = \varepsilon E_x(\varphi) = \sigma(0) E_x(\varphi) / E_x^{\text{eff}} .$$

For the tensile strength at an angle φ to the fibers two linear criteria are affirmed in [9] for fibers rupture and for splitting along fiber-matrix interface:

$$\sigma_1 + m_1 \cdot \sigma_6 = \sigma(0), \; \sigma_2 + m_2 \cdot \sigma_6 = \sigma(90) .$$

The parameters of the second criterion are usually determined from tests of unidirectional composite specimens, cut at different angles to fibers. For qualitative estimates only small misorientation angles are considered, and then first criterion for fiber rupture is valid:

$$\sigma(\varphi_i) = \frac{\sigma(0)}{\cos^2 \varphi_i + m_1 \cos\varphi_i \sin\varphi_i} . \tag{20}$$

For a unidirectional GFRP the angle φ_0 at change of fracture modes 1-2 is about $5°$. For clarity, in such small range of angles the trigonometric functions can be represented by power series: $\cos^2\varphi \approx 1-\varphi^2$, $\cos^4\varphi \approx 1-2\varphi^2$, $\cos\varphi_0 \cdot \sin\varphi_0 \approx \varphi - 2\varphi^3/3$. The effective modulus of elasticity is estimated by the average value of $\cos^4\varphi$:

$$\frac{E_x^{\text{eff}}}{E_1^0} = \left(\cos^4\varphi\right)^* = \frac{\int_0^{\varphi_0} \cos^4\varphi \, d\varphi}{\varphi_0} = 1 - \frac{2}{3}\varphi_0^2 .$$

Strength $\sigma(\varphi)$ in the range $\varphi_i \le \varphi_0$ can be estimated in following form:

$$\frac{\sigma(\varphi)}{\sigma_x(\varphi)} = \frac{\sigma(0) E_1^0 (\cos^4\varphi_0)^*}{\sigma(0) E_1^0 (\cos^4\varphi_0)(\cos^2\varphi_0 + m_1 \cdot \cos\varphi_0 \cdot \sin\varphi_0)} \approx 1 - m_1\varphi_0 + 7\frac{\varphi_0^3}{3} - 2m_1\varphi_0^3 + o\left(\varphi_0^3\right) \approx 1.013 ,$$

for accepted values $m_1 \approx \varphi_0 \approx 0.1$. It means that for these composite material properties the "equal strength" condition is satisfied even with a small reserve. This ratio may be slightly less than one, but it is important to note that the strength reduction for small fiber misorientation is insignificant.

In order to "design" beam shape with the "equal strength" condition an iterative computational simulation of trajectories and rebuilding of beam shape by using at least two of the above criteria is necessary, and for more correct results another criterion for interlaminate shear fracture [9] $\sigma_x + m \cdot \tau_{xy} = c$ has to be added.

Decrease of the effective elastic modulus of the material is recommended for elastic elements such as leaf springs, if it does not decrease the strength. Simplified example shows that misalignment can have a positive effect in the elastic elements. It is important to note that a small misalignment is useful to increase of unidirectional composites splitting resistance, which is especially important during cyclic loading.

More rigorous estimates are obtained with use of the finite element method upon misoriented inhomogeneous structure and this method does not require the hypotheses of width and/or height uniform deformation.

4. Bending flexibility increase due to branching

The analogy of the bending flexibility of shaped beams with constant cross section area and branching composite structures like a treetop with a constant sum of branches section areas (Leonardo's rule) is studied. In ideal case shaping or branching provide threefold bend flexibility growth with strength retaining, i.e., threefold increase of accumulated elastic energy for the fixed applied load and the same mass of the elastic element. It is shown that the use of unidirectional GFRP (glass fiber reinforced plastic) in shaped elastic beams makes it possible to reduce mass approximately 15 times in comparison with the steel analog.

An interesting method for increasing the bending flexibility of the composite elastic elements is to observe the structure of the top of an apple tree. In notes of Leonardo da Vinci [11, 12] the following statement is already expressed: "The sum of the squares of the diameters of branches is the same before and after branching".

Let us imagine the simplest branching model, in which the console cylinder beam is loaded at the end with a concentrated force P (figure 3) branched into N similar cylinder rods that preserve the total section area as follows:

$$d^2 = Nd_{N1}^2 \tag{21}$$

where d_{N1} is the diameter of each of the similar branches; the second suffix number indicates the number of the branch point, in this case it is only one. The distance from the seal in which it is reasonable to make the branch is selected based on the condition of equal maximum bending stresses in the root section and at the branch point as follows:

$$\frac{PL}{\pi d^3} = \frac{PL_{N1}}{N\pi d_{N1}^3} \Rightarrow L_{N1} = LN^{-1/2} \Rightarrow \lambda_{N1} = N^{-1/2} \tag{22}$$

So the bending stiffness changes after branching and specify the deflection of the initial cylinder beam is specified as v. After branching into N parts

$$v_{N1} = \frac{P}{E}\left[\frac{1}{I}\int_0^{L-L_{N1}} (L-x)^2\,\mathrm{d}x + \frac{1}{I_{N1}}\int_{L-L_{N1}}^{L} (L-x)^2\,\mathrm{d}x\right] = v\left[1+(N-1)\lambda_{N1}^3\right],$$

$$v = \frac{PL^3}{3EI},\ I = \frac{\pi d^4}{64},\ I_{N1} = N\frac{\pi d_{N1}^4}{64} = \frac{I}{N},\ \lambda_{N1} = \frac{L_{N1}}{L} \tag{23}$$

$$\delta_v = v_{N1}/v = 1+(N-1)N^{-3/2} \xrightarrow[N\to 1;\ N\to\infty]{} 1$$

As a result of branching, the tree becomes more flexible, which means it better resists the wind loads. It is interesting to note that branching does not provide any advantages of flexibility for either the trivial case ($N=1$ in (23)) or a large number of branches ($N\to\infty$). This means that there is some optimum number of branches for the maximum increase in the bending flexibility:

$$\mathrm{d}\delta_v/\mathrm{d}N = 0 \Rightarrow 2N^{3/2} - 3N^{1/2}(N-1) = 0 \Rightarrow N_{\mathrm{opt}} = 3 \tag{24}$$

IOP Conf. Series: Materials Science and Engineering **347** (2018) 012021 doi:10.1088/1757-899X/347/1/012021

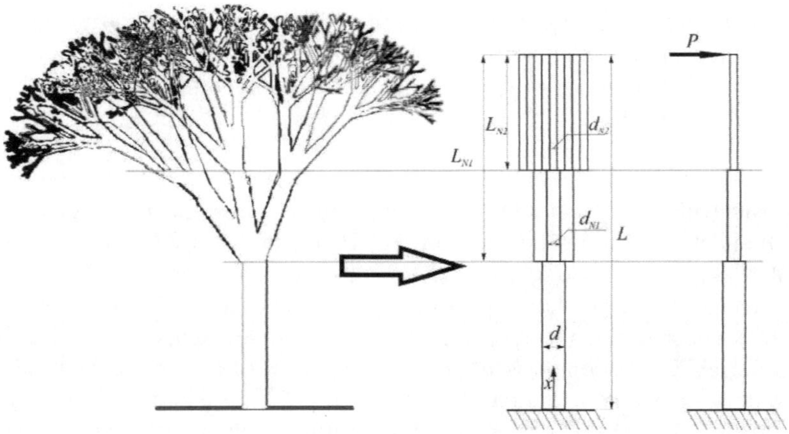

Figure. 3. Calculation model and pattern of variations in the number of branches for increasing flexibility under conditions of equal stress.

For $N=3$: $\lambda_{31}=3^{-1/2} \Rightarrow v_{31}/v=1+2/3^{3/2}\approx 1.38$. This is the maximum possible value of the deflection coefficient for retaining equal strength for one branch.

The flexibility increases with growth of branches number. A general formula can be written for n points of branching into N branches at each point while retaining the condition of equal strength (22) for each subsequent branch as follows:

$$\delta_v=\frac{v_{Nn}}{v}=1+\sum_{i=1}^{n}\frac{N^i-N^{i-1}}{N^{3i/2}}=1+\sum_{i=1}^{n}\frac{1-N^{-1}}{N^{i/2}}. \tag{25}$$

These power series certainly converge, and let us settle on the optimal case of $N=3$. Then, the following expression for additional deflection represents a geometric progression sum with the initial member $a_1=2/3\sqrt{3}$ and $q=1/\sqrt{3}$. The sum of this progression is: $a_1\left(1-q^n\right)/\left(1-q\right)$.

With unlimited number growth ($n\to\infty$) of equally strong branches into three branches ($N=3$), the deflection coefficient

$$\delta_v=\frac{v_{3n}}{v}\xrightarrow[n\to\infty]{}1+\frac{a}{\left(1-1/\sqrt{3}\right)}=\frac{2}{3\left(\sqrt{3}-1\right)}\approx 1.91.$$

This is the maximum coefficient of flexibility growth while retaining the strength that can be obtained during branching into a number (three) of similar branches. The situation is analogous to the case when the constarea beam profile (figure 1) could only be changed step by step, not smoothly.

In order to estimate the limit value of the deflection coefficient for continuous (fractal) branching, it is necessary to present a number N of branches as permanently changing value while retaining the total area $d^2=N(x)d^2(x)$ and equal stress $LN(x)d^3(x)=(L-x)d^3$. So $N(x)=\left(L/(L-x)\right)^2 \Rightarrow I(x)=I/N(x)$ and the deflection

$$v^*_{N(x)}=\frac{P}{EI}\int_0^L (L-x)^2 N(x)\,\mathrm{d}x=\frac{PL^3}{EI}=3v \tag{26}$$

Thus, in the ideal case, the branching cylinder rod with the same strength has three times greater flexibility compared to a homogenous rod (see (13)).

A branching composite beam can be used as an effective elastic element, e.g., in space-based structures (with no dimension limits), while for the fixed weight three times greater elastic energy is accumulated. If the weight efficiency of these elements is compared to the steel springs, then the low-modulus and high-strength GFRP can provide a gain by more than 15 times.

5. Conclusions

Traditional composite technology is based on the application of rectilinear (layout fabrics, prepregs) or spiral (winding bundles, belts) fibers trajectories. Therefore, methods of stiffness and strength engineering estimation of profiled composite components with complex fiber layup trajectories are not well developed. This article is intended to fill this gap. On the basis of the proposed approaches to the estimation of effective elastic modulus and strength, composite elements can be calculated taking into account the local variable misorientation of the fibers. The range of misorientation angles about 5° leads to negligible refinements. Computer simulation of the trajectories of fibers and rather cumbersome calculation of stresses is necessary for strength evaluation in the general case. This will clarify the form of a «uniform strength» composite elements in comparison with the known homogeneous analogs.

Acknowledgments

This work is financially supported by the Russian Foundation for Basic Research (grants no. 18-08-00372, 18-58-53020).

References

[1] Cho H R and Rowlands R E 2009 Optimizing fiber direction in perforated orthotropic media to reduce stress concentration *Journal of Composite Materials* **43** 1177–98

[2] Malakhov A V and Polilov A N 2016 Design of composite structures reinforced curvilinear fibers *Composites: Part A* **87** 23–8

[3] Malakhov A V and Polilov A N 2017 Design algorithm of rational fiber trajectories in arbitrarily loaded composite plate *J. Machinery Manufacture and Reliability* **46** 479–87

[4] Wegst U G K and Ashby M F 2007 The structural efficiency of orthotropic stalks, stems and tubes *J. Mat. Sci.* **42** 9005–14

[5] Polilov A N and Plitov I S 2015 Rational dimensions of segments of bamboo stems and composite tubes subjected to compression, flexure and torsion *J. Machinery Manufacture and Reliability* **44** 239–48

[6] Polilov A N, Malakhov A V, Tatus' N A and Shabalin V V 2011 Biomechanical principles of designing a quasi-unidirectional composite structures and their fasten places *Bul. Nizhny Novgorod Lobachevsky Univ.* **4** 2437–40

[7] Polilov A N, Tatus' N A and Shabalin V V 2011 Peculiarities of constructing elastic elements in the form of shaped composite beams *J. Machinery Manufacture and Reliability* **6** 532–7

[8] Polilov A N, Tatus' N A and Plitov I S 2013 Estimating the effect of misorientation of fibers on stiffness and strength of profiled composite elements *J. Machinery Manufacture and Reliability* **42** 390–7

[9] Polilov A N and Tatus' N A 2008 FRP strength criteria for some experimental effects substantiation *Eng. & Automation Probl.* **3** 103–10

[10] Minamino R and Tateno M 2014 Tree branching: Leonardo da Vinci's rule versus biomechanical models *PlosOne* **9** e9535

[11] Eloy C 2011 Leonardo's rule, self-similarity and wind-induced stresses in trees *Phys. Rev. Lett.* **107** 258101

[12] Polilov A N and Tatus' N A 2017 Designing branching or shaped composite elements by analogy with the structure of treetops *J. Machinery Manufacture and Reliability* **46** 385–93

IOP Conf. Series: Materials Science and Engineering **347** (2018) 012022 doi:10.1088/1757-899X/347/1/012022

Influence of annealing on mechanical properties of TiNi (55.8 mass % of Ni) wire made for medical purposes

M A Kaplan, E O Nasakina, M A Sevostyanov, K V Sergienko, A S Baikin, S V Konushkin, A G Kolmakov, M A Smirnov, A V Izvin and Yu N Ustinova

Baikov Institute of Metallurgy and Materials Science

E-mail: mishakaplan@yandex.ru

Abstract. Mechanical properties of titanium nickelide processed with various heat treatments are investigated. Fractographic, microstructural and X-ray diffraction analysis' were carried out as well as strength and plasticity measurements. It is shown that titanium nickelide (55.8 mass % of Ni) wires treated by two annealing stages, which consist of heating up to 300^{0}C on the first stage and 450^{0}C on the second, with 4 hours and 15 minutes long exposures on each stage respectively, conducted in an air environment, possess the best strength while saving good elasticity figures.

1. Introduction

Alloys possessing shape memory effect are materials, which are capable of shape and mechanical properties alteration in return for temperature change [1]. Nowadays, among all the variety of materials carrying shape memory effect, titanium nickelide (TiNi) appears to be the most widely used material of such kind. It earned such popularity due to its spectacular properties, such as shape memory effect and superelasticity. Within a small range of temperatures, it can exist with low-temperature body-centered cubic lattice (martensite) as well as a high-temperature face-centered cubic lattice (austenite). Transition temperature approximately varies from -50 to 166°C depending on the chemical composition of the material. Due to its unique properties, TiNi is widely used in the manufacturing of bio-medical equipment such as orthopedic arched wires, bone plates etc [2].

Moreover, titanium nickelide is used as a material for the production of medical articles by the type of stent. Stents are mostly used for distension of narrowed cavities in the human body. In the cooled state, stents are plastically deformed before being placed into the catheter to be inserted in the living organism in the future. Inside the organism, stents regain their original form [2-5].

It is essential to use annealing for structure stabilization and shape fixation of medical products made of TiNi [6-13].

This article describes research methods of 280mkm thick wires made of titanium nickelide with 55.8 mass % of Ni and 44.1 mass % of Ti processed by diverse heat treatments in the temperature range of 300 - 500 °C.

IOP Conf. Series: Materials Science and Engineering **347** (2018) 012022 doi:10.1088/1757-899X/347/1/012022

2. Methodology

Researched samples are presented in a form of 280mkm thick wires made of titanium nickelide consisting of 55.8 mass % of Ni and 44.1 mass % of Ti. Detailed information on chemical composition can be obtained from the table 1.

Table 1. Chemical composition of titanium nickelide wires.

Ni(mass %)	Ti(mass %)	C(mass %)	O(mass %)	Other el.(mass %)
55,8	44,1	0,03	0,037	≤0,001*

*Overall quantity of corresponding elements ≤0,027 %

To obtain TiNi wires with optimal mechanical properties, annealing in vacuum heater or high-temperature chamber furnace (ПВК-1,4-А) was used. Rod silicon carbide (SiC) heaters located across the walls of furnace's working chamber were used for heating samples up. Primarily, the furnace's (figure 1) inner empty space was heating up to the temperatures chosen for annealing, only after that samples placed in ceramic cylinders could be placed inside the furnace.

Figure 1. Figure 1High-temperature chamber electric furnace ПВК-1,4-А.

Electromechanical machine INSTRON 3382 was used to carry out static tests of finished samples. The test speed was 1 mm per min. The accuracy of the traverse speed was ± 0,2% of the value of the set speed. The accuracy of load measurement was ± 0.5% of the measured value and up to 1/100 of the maximum value of the load cell. With the help of static tensile tests, the following mechanical properties were determined: yield stress ($\sigma_{0.2}$), tensile strength (σ_t) and elongation (δ). The phase composition was determined using a X-ray diffractometer "Ultima IV" (Cu Kα-radiation) manufactured by "Rigaku", which includes: 1.graphite monochromator intended for use of various polycrystalline inorganic and organic substances, metals and alloys, films and coatings, composites; 2.vertical goniometer; 3.high-speed semiconductor detector "D / teX" with CuKα radiation, which is used to conduct researches using the Bragg-Brentano method, as well as method of oblique shooting with a fixed angle of rotation of the X-ray tube. The phase composition was determined by the PDXL software using the ICDD database.

After static tests, fractographic studies on samples were carried out with help of a scanning electron microscope (SEM) by the name of TESCAN VEGA II SBU equipped with an attachment for the energy-dispersive analysis INCA.

To investigate the microstructure of titanium nickelide wires, its surface was etched for 2-3 minutes with a mixture of the following composition: 50ml C2H5OH + 50ml HCl + 10g CuSO4. After that, each sample was washed with distilled water several times. On the last step of sample preparation, each of them underwent air drying. The studies were carried out on an optical metallographic microscope Olympus GX51. To study the structure each sample was poured by resin to be further pressed-in on the machine for hot pressing Struers CitoPress-20. The final step of preparation, sampled polishing, carried out in the automatic grinding polishing system Struers Tegramin-25.

3. Results and discussion

To obtain stable structure and properties, each sample wire had undergone annealing before the research was carried out. Wires treated by two annealing stages, which consist of heating up to 300^{0}C on the first stage and 450^{0}C on the second, with 4 hours and 15 minutes long exposures on each stage respectively, conducted in an air environment, proved to have the best strength while saving good elasticity figures (table 2).

Table 2. Annealed titanium nickelide resulting mechanical properties obtained by conducting the static stretching tests

Sample	δ(%)	$\sigma_{0,2}$(MPa)	σ_t(MPa)
Original titanium nickelide	9,8	640	1301
Titanium nickelide, annealing 390°C, 30 minutes, vacuum annealing	7,8	347	1830
Titanium nickelide, annealing 450°C, 30 minutes, annealing in air environment	6,3	255	1766
Titanium nickelide, annealing 450°C, 30 minutes, vacuum annealing	4,2	269	1640
Titanium nickelide, annealing 500°C, 15 minutes, annealing in air environment	13,1	349	1585
Titanium nickelide, annealing 350°C, 4hours, then 450°C, 15 minutes (air)	8,8	278	1904
Titanium nickelide, annealing 300°C, 4hours, then 450°C, 15 minutes (air)	12	340	1777

In order to research the surface of wires treated with drawing and heating, X-ray phase and microstructural analysis were carried out. As it is shown on figure, the bulk of the alloy consists of the B2 phase with a slight presence of Ti2Ni intermetallides, which size ranges from 0.1μm to 3μm and does not change with an increase in the annealing temperature. This inclusion is formed during the crystallization of the alloy and is much harder than the matrix [14].

As the result of fractographic studies of samples, figures 3 and 4 were obtained. It is immediately apparent that samples destruction occurred with the formation of the neck, the fracture surface is oriented almost perpendicular to the extension axis and consists of a set of heterogeneous self-similar viscous fracture pores ("cups"). Upon stretching the trunk fracture in wire samples arises on the surface and then propagates to the inner layers of the samples. The surface pattern of fracture of the near-surface layers differs from that for the inner layers, being visually more complex and having directed character of the relief, in addition, the number of "cups" on the near-surface layers exceeds those on inner-surface layers, "cups" located on the near-surface layers also have much smaller dimensions. The spread of the cup sizes for the fracture of the inner layers is approximately 2 times greater than for the fracture of the near-surface layers. kinks differences of the inner and near-surface layers of the samples indicate that the processes of structural self-organization at the stage of propagation of the main transverse crack was proceeding in them differently.

IOP Conf. Series: Materials Science and Engineering **347** (2018) 012022 doi:10.1088/1757-899X/347/1/012022

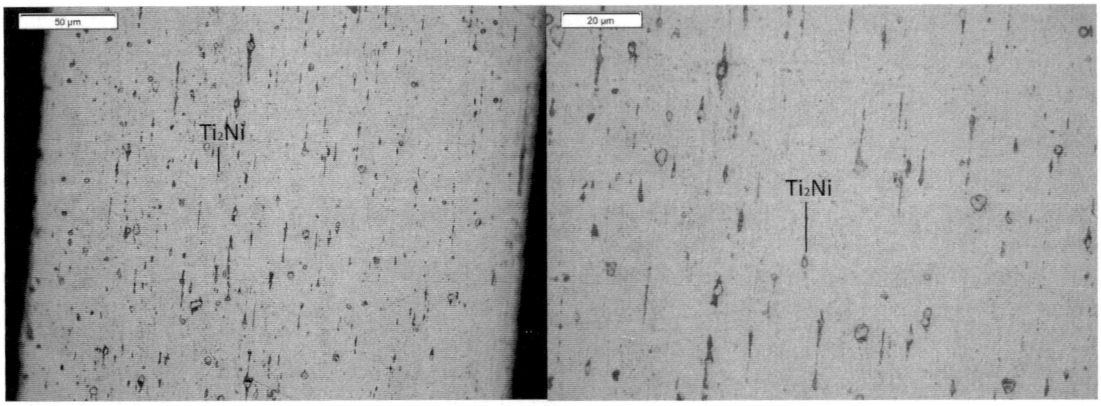

a) b)

Figure 2. Wires microstructure (a – 500x optical magnification, b – 1000x optical magnification).

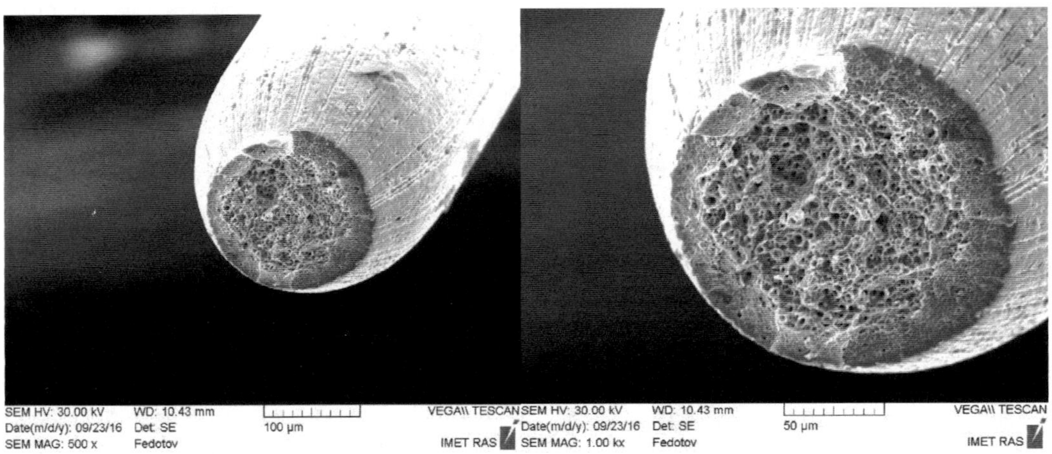

a) b)

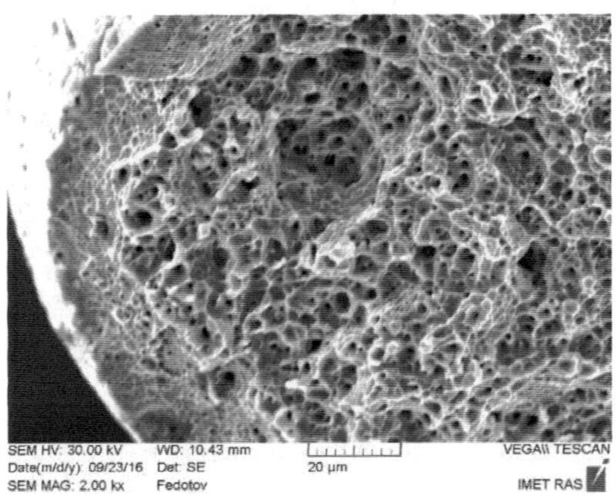

c)

Figure 3. Views on the wire after the destruction (a, b –general view; c - characteristic view on fracture of near-surface layers).

IOP Conf. Series: Materials Science and Engineering **347** (2018) 012022 doi:10.1088/1757-899X/347/1/012022

4. Conclusion

The structure of titanium nickelide samples is heterogeneous. Alloy itself consists of the B2 phase and Ti2Ni intermetallides inclusions.

Proper heat treatment mode, which provides titanium nickelide wires with satisfactory mechanical properties, has been obtained. This heat treatment mode consists of two annealing stages, heating up to 300^0C on the first stage and 450^0C on the second, with 4 hours and 15 minutes long exposures on each stage respectively. The whole heating process should be conducted in an air environment.

Acknowledgments

The work was supported by the grant of the President of the Russian Federation for state support of young Russian scientists - candidates of sciences (MK-4521.2018.8)

References

[1] C M Wayman April 1993 *MRS Bulletin* pp 49-56

[2] Petrini L, Migliavacca F 2011 Biomedical Applications of Shape Memory Alloys *Journal of Metallurgy* **2011** pp 1-15

[3] Duerig T W, Melton K N, Wayman C M, Stockel D 1990 Engineering aspects of shape–memory alloys (Oxford: Butterworth Heinemann Ltd) p 491

[4] Shabalovskaya S 1996 On the nature of the biocompatibility and medical applications of NiTi shape memory and superelastic alloys *Bio–Medical Materials and Engineering* **6** pp 267–289

[5] Tian H, Schryvers D, Shabalovskaya S, Van Humbeeck J 2009 Microstructure of Surface and Subsurface Layers of a Ni-Ti Shape Memory Microwire *Microscopy and Microanalysis* **15** pp 62–70

[6] Nolan M, Tofail S A M 2010 Density functional theory simulation of titanium migration and reaction with oxygen in the early stages of oxidation of equiatomic NiTi alloy *Biomaterials* **31** pp 3439–3448

[7] Hu T, Chu C, Xin Y, Wu S, Yeung K W K, Chu P K 2010 Corrosion products and mechanism on NiTi shape memory alloy in physiological environment *Journal of Materials Research* **25** pp 350–358

[8] Her-Hsiung Huang, Yu-Hui Chiu, Tzu-Hsin Lee, Shih-Ching Wu, Hui-Wen Yang, Kuo-Hsiung Su, Chii-Chih Hsu 2003 Ion release from NiTi orthodontic wires in artificial saliva with various acidities *Biomaterials* **24** pp 3585–3592

[9] Tomić S, Rudolf R, Brunčko M, Anžel I, Savić V, Čolić M 2012 Response of monocyte-derived dendritic cells to rapidly solidified nickel-titanium ribbons with shape memory properties *European cells and materials* **23** pp 58-81

[10] Her-Hsiung Huang 2005 Surface characterizations and corrosion resistance of nickel–titanium orthodontic archwires in artificial saliva of various degrees of acidity *Journal of Biomedical Materials Research* **74A** pp 629–639

[11] Clarke B, Kingshott P, Hou X, Rochev Y, Gorelov A, Carroll W 2007 Effect of nitinol wire surface properties on albumin adsorption *Acta Biomaterialia* **3** pp 103–111

[12] Shabalovskaya S A, He Tian, Anderegg J W, Schryvers D U, Carroll W U , Van Humbeeck J 2009 The influence of surface oxides on the distribution and release of nickel from Nitinol wires *Biomaterials* **30** №30 pp 468–477

[13] Michiardi A, Aparicio C, Planell J A, Gil F J 2006 New oxidation treatment of NiTi shape memory alloys to obtain Ni-free surfaces and to improve biocompatibility *Journal of Biomedical Materials Research Part B* **77B** pp 249–256

[14] Nasakina E O, Baikin A S, Sevost'yanov M A, Kolmakov A G, Zabolotnyi V T, Solntsev K A 2014 Properties of Nanostructured Titanium Nickelideand Composite Based on It *Theoretical Foundations of Chemical Engineering* **48** №4 pp 477–486

IOP Conf. Series: Materials Science and Engineering **347** (2018) 012023 doi:10.1088/1757-899X/347/1/012023

Effect of sintering in a hydrogen atmosphere on the density and coercivity of $(Sm,Zr)(Co,Cu,Fe)_Z$ permanent magnets

G S Burkhanov[1], N A Dormidontov[1,3], N B Kolchugina[1] and A G Dormidontov[2]

[1]IMET RAS, 49, Leninsky pr., 119334 Moscow, Russian Federation
[2]Tekhma ltd., 49, Leninsky pr., 119334 Moscow, Russian Federation
[3] Author to whom any correspondence should be addressed

E-mail: ontip@mail.ru

Abstract. The effect of heat treatments in manufacturing $(Sm,Zr)(Co,Cu,Fe)_Z$-based permanent magnets sintered in a hydrogen atmosphere on their properties has been studied. It was shown that the dynamics of the magnetic hardening of the studied magnets during heat treatments, in whole, corresponds to available concepts of phase transformations in five-component precipitation-hardened SmCo-based alloys. Peculiarities of the studied compositions consist in the fact that the coercive force magnitude of magnets quenched from the isothermal aging temperature is higher by an order of magnitude than those available in the literature. It was noted that, in using the selected manufacturing procedure, the increase in the density of samples does not finish at the sintering stage but continues in the course of solid-solution heat treatment.
Key words: rare-earth permanent magnets, coercivity, precipitation-hardening, heat treatment.

1. Introduction

Permanent magnets based on rare-earth intermetallic phases with transition metals attract the attention of specialists and are widely used in modern industry. The rare-earth permanent magnets (REPMs) exhibit the best characteristics as compared to those of previously known materials when operating in electric and hybrid drives, wind turbines, gyroscopes, contactless bearings and couplings, loudspeakers, MRI machinery, focusing systems of microwave tubes, etc. These advantages are associated primarily with outstanding hysteretic properties of NdFeB-type permanent magnets exhibiting record values of remanence Br > 14 kG, and maximum energy product $(BH)_{max} > 50$ MG×Oe. However, $(Sm,Zr)(Co,Cu,Fe)_Z$-type magnets remain are highly relevant owing to the stability of their characteristics in the wide temperature range (0 – 800 °K), corrosion resistance, in particular, in hydrogen-containing atmosphere.

$(Sm,Zr)(Co,Cu,Fe)_Z$ alloys are complex metallurgical systems, and their hysteretic properties and structural state are closely related to their chemical composition and heat treatment parameters. Moreover, the magnets based on the alloys should be produced using fine micron-sized powders; otherwise, the energy potential of compositions cannot be realized completely. Despite a great number of works related to the study of the structure and properties of these alloys and magnets based on them, the nature of their high-coercivity state and processes of its formation are poor understood [1-6]. High-coercivity state of $(Sm,Zr)(Co,Cu,Fe)_Z$ alloys is achieved by four-stage heat treatment (HT). The purpose of this study is to determine the effect of sintering in a hydrogen atmosphere on the properties of $(Sm,Zr)(Co,Cu,Fe)_Z$-type magnets.

2. Experimental

Samples were prepared by standard powder technology used for sintered magnets; final dimensions of magnets are ∅15×14 mm. The alignment of samples is along the axis rotation.

The $Sm_{0,84}Zr_{0,16}(Co_{0,70}Cu_{0,07}Fe_{0,23})_{6,25}$ composition was prepared by melting in a vacuum induction furnace in a high-purity argon atmosphere using individual components of more than 99,5 wt % purity. The chemical composition was controlled by atomic emission spectrometry with microwave plasma.

The alloy was crushed to a particle size of <500 μm using a mechanical crusher and a nitrogen atmosphere for prevent oxidation. A fine powder with an average particle size of 3.5-4 μm was prepared in a vibratory ball mill; pots were completely filled with absolute isopropyl alcohol used as the oxidation protector. The powders were dried in vacuum and compacted at a pressure of 75 kg/mm^2 in an aligning magnetic field of 16 kOe.

To obtain the high degree of powder particles alignment, the blank-samples were compacted using pseudo-isostatic pressing in a matrix with a polyurethane insert. This scheme ensures uniform compression of powder in all directions. The green blanks were subjected to four step-by-step heat treatments, which included sintering, solid solution heat treatment (SSHT), quenching, and two consecutive aging – isothermal aging (IA), and slow-cooling aging ('step aging' - SA). To clarify the effect of slow cooling aging on the formation of high-coercivity state, two additional heat treatments were performed, namely, heating to the isothermal aging temperature, 15 min holding, and quenching, and second step-aging. Conditions of experimental heat treatment are shown in Figure 1 (a) and Table 1.

Table 1. Conditions of experimental heat treatment (HT) of samples.

HT1	Sintering (T=1190-1200°C, t=1 h)
HT2	HT1 + SSHT (T=1165-1170°C, t=5 h) + Quenching
HT3	HT2 + Isothermal aging (T=760, 800, 840 and 880°C, t=4-20 h) + Quenching
HT4	HT3 + Step aging (from T(IT) to 400°C, 90°/h) + Quenching
HT5	HT4 + "Recovery" (T(IT), t=0,25 h) + Quenching
HT6	HT5+ Isothermal aging (T=760, 800, 840 and 880°C, t=4 h) + Step aging (from T(IT) to 400°C, 90°/h), 90°/h) + Quenching

All heat treatments were carried out in a vacuum resistance furnaces at an excess argon pressure of +0.1 bar and followed by quenching in a helium flow. The average particle size was monitored by measuring the specific surface area of powders. The density of blanks was determined by hydrostatic weighing method. Magnetic properties were investigated in fields of ± 30 kOe using a completely closed magnetic circuit of hysteresis-graph.

3. Results

The main results are presented in Figures 1 – 3. Figure 1(a) shows the general scheme of the experiments - four main steps of heat treatment and two additional pilot stages of heat treatment.

Figure 1 (b, c) shows the effect of the corresponding heat-treatment stages on the density (d) and coercive force (Hcj) of the samples.

The density of the sintered samples was in a range of 8.22-8.25 g/cm^3 and increases to 8.32-8.35 g/cm^3 during solid solution heat treatment (Figure 1(b)). The density of the samples was unchanged during agings.

Schematically, the dynamics of change in the coercive force during step-by-step heat treatments are shown in Figure 1 (c), and hysteretic properties of samples after different stages of heat treatments are presented in Figures 2 and 3. Figure 2(a) shows the magnetization-reversal portions of the major hysteresis loops for the samples subjected to complete heat treatment cycle – HT4 and to heat treatment HT6 (data are shown for isothermal aging at temperature T = 760°C). The samples had the same properties after both heat treatments (HT4 and HT6).

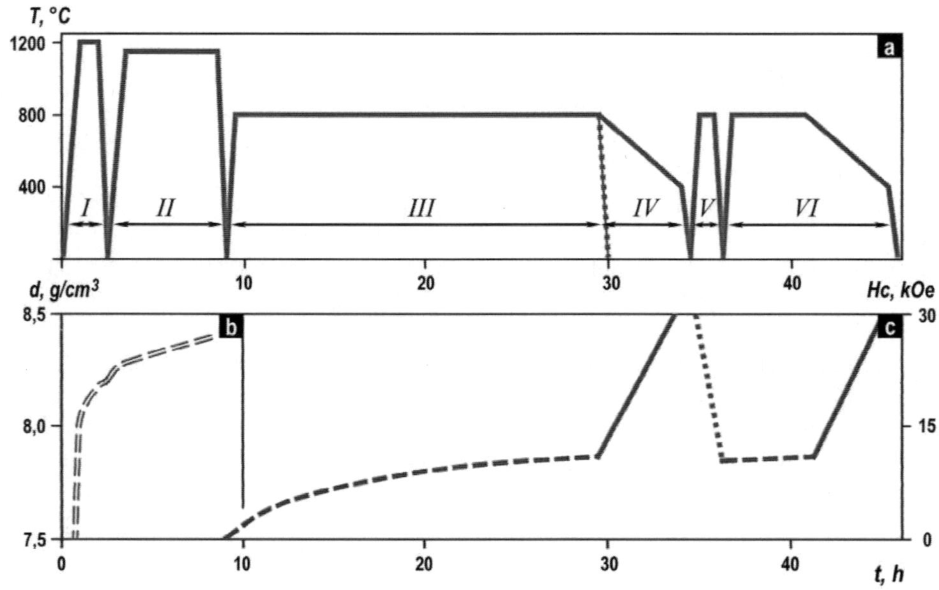

Figure 1. Scheme of heat treatment and experimental results: (a) four stages heat treatment during the manufacturing of magnet samples based on the $Sm_{0.84}Zr_{0.16}(Co_{0.70}Cu_{0.07}Fe_{0.23})_{6.25}$ alloy and two additional experimental stages (see explanations in the text and Table 1), and (b) density (d) and coercive force (H_{cj}) of samples at different stages of heat treatment.

Figure 2 (b) presents the magnetization reversal portion of the major hysteresis loops for the samples subjected to HT3 and HT5; digits at the curves correspond to the isothermal aging time (h). In this case, the heat treatments (HT3 and HT5) lead to identical properties of the samples. The properties of samples heat-treated at other isothermal aging temperatures (T=800, 840 and 880°C) demonstrate qualitatively identical behavior.

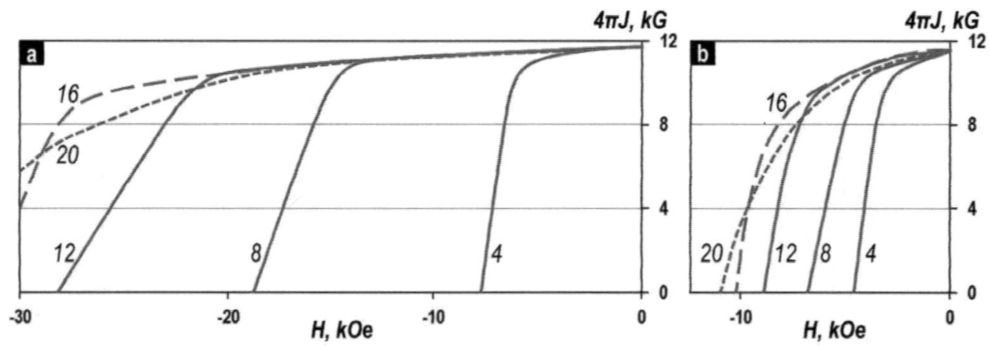

Figure 2. Magnetization reversal portions of the major magnetic hysteresis loops for $Sm_{0.84}Zr_{0.16}(Co_{0.70}Cu_{0.07}Fe_{0.23})_{6.25}$ permanent magnet samples (a) after complete heat-treatment cycle (HT4) and restore of high-coercivity state after a short-time heating to the isothermal aging temperature (HT6); (b) after quenching from the isothermal aging temperature (HT3) and after complete heat-treatment cycle and following short-time heating to the isothermal aging temperature; digits near the curves correspond to isothermal aging time (h) at isothermal aging temperature T = 760°C.

Summary results for all investigated temperatures are shown in Figure 3, where: a) and b) are the dependences of the coercive force (H_{cj}) and the critical field (H_k) on the temperature and time of isothermal aging, respectively, as a result of HT3 and HT5 (hereinafter *1)* – parameters of the samples that were tempered at 760, *2)* – 800, *3)* – 840 and *4)* – 880°C); c) – H_k of the high coercive state samples that were subjected to HT4 and HT6.

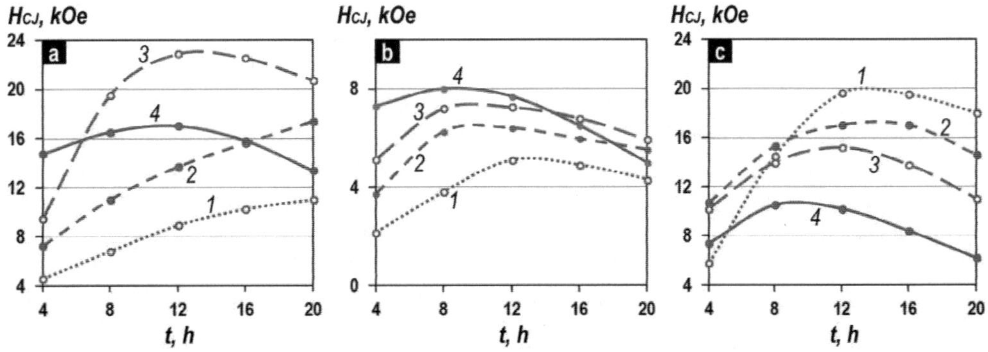

Figure 3. Summary results for heat treatments of $Sm_{0.84}Zr_{0.16}(Co_{0.70}Cu_{0.07}Fe_{0.23})_{6.25}$-based permanent magnet samples: (a) H_{cj} and (b) H_k after quenching from the isothermal aging temperature (HT3) and after complete heat-treatment cycle and following short-time heating to the isothermal aging temperature; and (c) H_k after complete heat-treatment cycle (HT4) and restore high-coercivity state after short heating to the isothermal aging temperature (HT6) (at isothermal aging temperature T = 760°C - *1*, 800°C - *2*, 840°C - *3* and 880°C - *4*).

Since the majority of samples in the high-coercivity state had the H_{cj} values, which are higher than the limit of the used measurement method, the H_k parameter is adopted as a criterion of coercivity (H_k is the magnetization reversal field, which results in the 10-% reduction of the remanence ($4\pi J_k = 0.9\times4\pi J_r$)).

4. Discussion

4.1. Thermal compaction of the samples

The remanence of the magnets is directly proportional to the density of the samples. Therefore, the maximum energy product is directly proportional to the squared density. Thus, the important task is to reach the maximum density of powder samples, which is solved in two sequential technological stages, namely, by mechanical and thermal compactions. We have not been analyzing the mechanical compaction stage in the present paper. The sintering solves the problem of reaching the maximum density by thermal compaction of green blanks. This is the most high-temperature stage among heat treatments. When determining the sintering temperature, we should consider several opposing trends.

The sintering of fine (activated) powders is accompanied by grain-growth processes and formation of a structure characterized by closed porosity at high temperature. A benefit, however, associated with the grain growth process is the improvement in alignment during sintering as compared to that of pressed compacted powder. The improved alignment occurs as a result of the incorporation of small poorly aligned particles into large well-aligned grains during the material compaction under thermally driven forces.

However, too high degree of recrystallization is the consequences of lowering the degree of crystallographic texture (alignment) and reduction of the remanence of the samples. Thus, the sintering conditions (temperature-time) are limited by normal recrystallization processes.

The thermal compaction of the samples ends after sintering when the sintering is performed in an inert gas. All other heat treatments are performed at the lower temperatures and do not result in an increase in the density of samples.

In this work, the density of samples after sintering in a hydrogen atmosphere was in a range of 8.22-8.25 g/cm^3. As is shown in Figure 1 (b), the density continues to increase to 8.32-8.35 g/cm^3 during the solid solution heat treatment (HT2); in some cases, it was 8.45 g/cm3 (when the manufacturing method was used). Note that solid-solution heat treatment temperature was 30-35° below the sintering temperature of the samples. Thus, the densification of samples is not completed at the end of sintering and continues during the first stage of formation of high-coercivity state of samples, namely, during high-temperature supersaturated solid-solution heat treatment at a temperature that is lower than that of previous stage (Figure 1(a), stage II).

The most logical explanation of this fact was made by Kumar [1]. During sintering in an inert gas atmosphere, a pore is filled with an inert gas trapped at the pressure existed during sintering. An expected consequence of the sintering in a hydrogen atmosphere is a dosed pore structure; after subsequent pumping out hydrogen, pores will have a vacuum in them. The subsequent heat treatment (SSHT) completes the evacuation of hydrogen, following the densification to a closed porous structure at a temperature of 1170°C and an argon pressure above the atmospheric one.

Figure 4 shows the difference in the morphology of pores in samples sintered in hydrogen and argon.

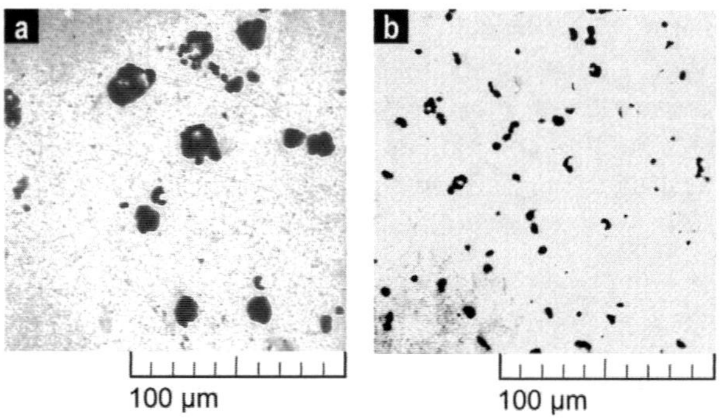

Figure 4. Pores in samples of permanent magnets $Sm_{0.84}Zr_{0.16}(Co_{0.70}Cu_{0.07}Fe_{0.23})_{6.25}$, sintered in (a) argon and (b) hydrogen.

This feature of the hydrogen sintering is especially attractive for manufacturing completely dense magnet with a high degree of magnetic crystallographic texture (alignment).

4.2. Magnetic hardening of the samples

As we noted above, $(Sm,Zr)(Co,Cu,Fe)_Z$-type permanent magnet alloys are precipitation-hardened materials. The main purpose of heat treatment stages II – IV is the formation of crystal structure of alloy, which ensures the effective domain-wall pinning during magnetization reversal.

The structure of the $(Sm,Zr)(Co,Cu,Fe)_Z$ alloys is constructed based on the main intermetallic phases Sm_ACo_B (hereinafter, A:B) of the Sm-Co equilibrium phase diagram. The structure results from the decomposition of the disordered supersaturated 1:7 solid solution into three structural components; these are the main cell R2:17-type phase (R = rhombohedral), the cell boundary structural components, and thin plates (lamellas) of Z-phase (1:3), penetrating cells and cell boundary phase strictly in the basal plane (perpendicular to the c-axis) of the samples [2-6]. Cell boundary is not a phase. This structural component is sequential alternating layers of the phases (2:7 + 5:19 + 1:5) in different volume proportions, which are formed at various stages of aging.

The formation of high-coercivity structural state in the studied samples mainly corresponds to known concepts.

The internal stresses in the structure are induced during quenching of supersaturated solid-solution samples. This initiates the decomposition of solid solution during quenching even at high cooling rates of the alloy [4]. As is shown in the Figure 2 (a, b), the isothermal aging for the first four hours of treatment (HT3) allows one to attain H_{cj} of samples of more than 4.5 kOe, and, after slow cooling (step aging - HT4), – of more than 7.5 kOe.

The long time isothermal aging occurs in coarsening the cellular structure, which is accompanied by increase in H_{cj}. The increase in the isothermal aging temperature causes the similar effect leading to the accelerated increase in the coercive force in the initial period of treatment. The coalescence of precipitates occurs after more prolonged isothermal aging, which causes the statistical irregularity of the cell boundary thickness and, as a result, worsening the squareness of hysteresis loop of the samples (Figure 2 (curves 20) and Figure 3 (curves 4)).

Many investigators have discussed the asymmetry results in a sequential rotation of step aging (HT4 and HT6) and quenching from the isothermal aging temperature (HT3 and HT5). The short-time (5-15 min) heating of high-coercivity samples to the isothermal aging temperature and subsequent quenching (HT5) lead to the drop of H_{cj} by 18 times [3, 5] and even up to 30 times [2]. At the same time, to restore the level of H_{cj} to the previous high-coercivity state of samples, the treatment for 5-15 min is insufficient. The complete step-aging in the course of very slow cooling (HT6) should be repeated, i.e. the heat treatment for about 5-10 h should be performed again. Properties are completely restored after repeated step-aging [2-5, 7, 8].

Unlike data of [2, 3, 5], according to which the "recovery" causes the catastrophic drop of the coercive force of the samples from 25-30 kOe to 1.4-1.6 kOe, in the present study, H_{cj} reaches 11, 17.3, 22.9 and 17 kOe after quenching from an isothermal tempering temperature of 760, 800, 840 and 880°C, respectively (Figure 3(a)).

Differences in the behavior of H_{cj} are likely to be related to the chosen working alloy. Studies [2, 3, 5] were made for magnets with narrow compositional interval: in $Sm(Co,Cu,Fe,Zr)_Y$ alloys, Y = 7.19-7.5; in $(Sm,Zr)(Co,Cu,Fe)_Z$ alloys, Z = 5.48-5.54. In the present work, the alloy is described by the corresponding coefficients equal to Y = 7.63, and Z = 6.25. The differences in the chemical composition of the alloys are likely to produce significant differences in the structural state of samples after isothermal aging and quenching.

It is known that the alloys with Z of more than certain values have the high values of the coercive force H_{cj} immediately after isothermal aging before slow-cooling aging. As in this study, the slow cooling results in only from one and a half to threefold increase in the coercive force of these alloys [7-9].

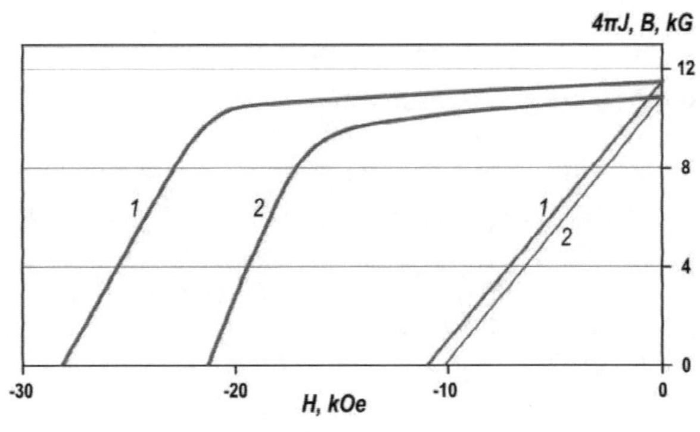

Figure 5. Magnetization reversal portions of the major magnetic hysteresis loops for $Sm_{0.84}Zr_{0.16}(Co_{0.70}Cu_{0.07}Fe_{0.23})_{6.25}$ permanent magnet samples, sintered in (1) hydrogen and (2) argon after complete heat-treatment cycle.

Table 2. The parameters of $Sm_{0.84}Zr_{0.16}(Co_{0.70}Cu_{0.07}Fe_{0.23})_{6.25}$ samples sintered in argon and hydrogen.

Sintering atmosphere	Sintering T, °C	Density g/cm^3	B_r kG	H_{cb} kOe	H_{cj} kOe	$(BH)_{max}$ MG*Oe
Ar	1215	8.20	10.9	10.2	28.2	28.3
H_2	1195	8.35	11.5	11.0	21.3	32.2

The existence of samples (similar to those studied in the present work), the structural state of which ensures the high coercive force after isothermal aging followed by quenching, i.e., without the slow cooling to 400°C, even more sharpens the question about the mechanism of phase transformations upon

IOP Conf. Series: Materials Science and Engineering **347** (2018) 012023 doi:10.1088/1757-899X/347/1/012023

magnetic hardening of the (Sm,Zr)(Co,Cu,Fe)$_Z$ alloys and the processes that generate "return" asymmetric effects during heat treatments.

The magnetization reversal portions of the major hysteresis loops of two samples sintered in hydrogen and argon are shown in Figure 5. Both samples were produce from the same alloy and subjected to optimal treatment for high-coercivity state. The starting and resulting data for the samples are presented in Table 2.

5. Conclusions

1. Four stages of heat treatment of (Sm,Zr)(Co,Cu,Fe)z alloys allow one, at the first stage (I, sintering in hydrogen), to produce dense samples, and after full heat treatment cycle (II, solid solution heat treatment, III, isothermal aging, IV, step aging), to ensure the development of high-coercivity structural state of the alloy with the effective domain-wall pinning.

2. The increase in the density of the samples is not completed at the first stage (I, sintering in hydrogen), and continues at the second stage (II, solid solution heat treatment).

3. Results obtained for the $Sm_{0.84}Zr_{0.16}(Co_{0.70}Cu_{0.07}Fe_{0.23})_{6.25}$ alloy are rare consistent with available data on the formation of phase structure during the heat treatments.

4. The peculiarity of the investigated alloy is the high coercive force Hcj reached after isothermal aging, which is higher than that observed for other alloys of this system.

References

[1] Kumar K 1988 $RETM_5$ and RE_2TM_{17} permanent magnets development *J. Appl. Phys.* **63** R13-57.

[2] Goll D, Kronmüller H and Stadelmaier H H 2004 Micromagnetism and the microstructure of high-temperature permanent magnets *J. Appl. Phys.* **96** 6534-45.

[3] Sepehri-Amin H, Thielsch J, Fischbacher J, Ohkubo T, Schrefl T, Gutfleisch O and Hono K 2017 Correlation of microchemistry of cell boundary phase and interface structure to the coercivity of $Sm(Co_{0,784}Fe_{0,100}Cu_{0,088}Zr_{0,028})_{7,19}$ sintered magnets *Acta Materialia* **126** 1-10.

[4] Maury C, Rabenberg L and Allibert C H 1993 Genesis of the Cell Microstructure in the Sm(Co,Fe,Cu,Zr) Permanent Magnets with 2 : 17 Type *Phys. Stat. Sol.* **140** 57-72.

[5] Yan A, Gutfleisch O, Gemming T and Müller K-H 2003 Microchemistry and magnetization reversal mechanism in melt-spun 2:17-type Sm-Co magnets. *Appl. Phys. Lett.* **83** 2208-10.

[6] Xiong X Y, Ohkubo T, Koyama T, Ohashi K, Tawara Y and Hono K 2004 The microstructure of sintered $Sm(Co_{0,72}Fe_{0,20}Cu_{0,055}Zr_{0,025})_{7,5}$ permanent magnet studied by atom probe *Acta Materialia* **52** 737-48.

[7] Suponev N P, Dormidontov A G, Levandovsky V V, Shamorikova E B and Nekrasova Ye M 1992 Structure and magnetic properties of Sm-Zr-Co-Cu-Fe alloys in the high-coercivity state. 2. Model of structure formation *Physics of magnetic materials (Herald Tver state University)* 78-98 (in Russian).

[8] Dormidontov A G, Suponev N P, Levandovsky V V, Shamorikova E B and Nekrasova Ye M 1991 Alloys for highly coercive permanent magnets of type (Sm,Zr)(Co,Cu,Fe)$_Z$. *Electronic industry* **12** 18-23 (in Russian).

[9] Lyakhova M B, Semenova E M, Sinekop V I and Velichko E S 2010 Effect of nanostructure on the formation of domain structure of (Sm,Zr)(Co,Cu,Fe)$_Z$ permanent magnets *Herald Tver state University* **9** 4-18 (in Russian).

Sintering activation energy $MoSi_2$-WSi_2-Si_3N_4 ceramic

D D Titov[1], A S Lysenkov[1], Yu F Kargin[1], M G Frolova[1], V A Gorshkov[2] and S N Perevislov[3]

[1] Baykov Institute of Metallurgy & Material Science RAS, Moscow, Russia
[2] Institute of Structural Macrokinetic and Material Science RAS, Chernogolovka, Russia
[3] Central Research Institute of Materials, St. Petersburg, Russia

e-mail: mitytitov@gmail.com

Abstract. The activation energy of sintering process was calculated based on dilatometric studies of shrinkage processes (Mo,W)Si_2 + Si_3N_4 composite ceramic. (Mo,W)Si_2 powders was obtained by solid-phase solutions of 70 wt% $MoSi_2$ and 30 wt% WSi_2 by SHS in the ISMAN RAS. The concentration rate Si_3N_4 was from 1 to 15 wt.%. The sintering was carried out to 1850°C in Ar atmosphere the heating rate of 5, 10, 12 and 15°C/min by the way of dilatometer tests. Based on the differential kinetic analysis method (Friedman's method), the sintering process activation energy of (Mo,W)Si_2 + Si_3N_4 were calculated. The two-stage sintering process and the dependence of the activation energy on the Si_3N_4 content was shown. Average value of 370 kJ/mol for Q was obtained.

1.Introduction

$MoSi_2$ intermetallic compound is one of the candidate materials for high temperature applications because of its moderate density (ϱ = 6,23 г/см^3), high melting point (2030 °C), excellent oxidation resistance at high temperatures (≥1000 °C) and low electrical resistivity (σ = 21.6 $\mu\Omega$/cm^2) [1]. $MoSi_2$ has the highest oxidation resistance above 1000°C among all disilicides. The resistance to oxidation at high temperatures is due to the formation silicon oxide (SiO_2) coating on the surface, it prevents diffusion of oxygen.

$MoSi_2$-based materials are commonly used as a heating element for atmospheric high-temperature furnaces and protective coatings because of a unique properties: resistance to oxidation at high temperatures and electrical conductivity. The pure $MoSi_2$ has not been widely spread as a structural ceramic (for example, for the production of turbine blade) [2] due to brittleness at room temperature, creep at high temperatures, and low oxidation resistance at 500 to 800 °C. Molybdenum disilicide is subject to catastrophic oxidation ("pesting") in this temperature range [3] and is completely destroyed forming MoO_3 and SiO_2.

Since the discovery of molybdenum disilicide many different additives have been investigated to improve the properties of $MoSi_2$ (for example, WSi_2, TiB_2, ZrO_2, Cr_2Si_3, Y_2O_3, Mo_5Si_3, BN, NbB, Si_3N_4, etc.) [4,5,6 and 7]. The recent articles are devoted to the study of the silicon nitride (Si_3N_4) additive. Si_3N_4 ceramic samples have phenomenal strength, according to some data, their flexural strength are 1000-1200 MPa [8], Si_3N_4 ceramic has good resistance to oxidation up to 1500 °C [9]. The main drawback of this material, which prevents its use in its pure form, is the energy-consuming process for obtaining dense ceramics, usually hot isostatic pressing (HIP) or hot pressing (HP) is used together with sintering additives to lower the firing temperature [3].

Since the earliest studies of sintering processes many models have been developed to relate the sintering rate to particle characteristics, density, atmosphere and sintering temperature. The kinetic parameters determination is necessary to predicting the thermal ageing process development: at what temperature and for how long can a given substance be used without impairing its properties. Computer data processing allows modeling and predicting even overlapping multi-stage processes. There are three main modifications of the iso-conversion method:

- differential (Friedman) [10];
- integral(Flynn-Ozawa-Wall) [11-12];
- improved integral, based on non-linear procedures (Vyazovkin) [13].

The determining kinetic parameters by Friedman method is based on the iso-conversion principle, that the reaction rate at a constant conversion degree α is only a function of temperature. This method allows us to find the activation energy (E) (or the dependence of E on α) without knowing the exact form of the dependence $f(\alpha)$.

In previous work [14-15], it was shown that composites based on molybdenum disilicide with 30 and 50 wt% WSi_2 addictive are characterised by a decrease in porosity and an increase in the flexural strength. The influence of tungsten disilicide in the $MoSi_2$ - WSi_2 system on the properties of ceramic samples $Mo_xW_{1-x}Si_2$, (where x varied from 1 to 0.3) was studied in detail in the paper [14]. It has been established that during solid-phase sintering of components mechanical mixtures at 1650-1800 °C for 30 minutes, complete interaction between disilicides of molybdenum and tungsten with formation a solid solution does not occur. Ceramic samples have greater strength and low porosity sintered from $Mo_{0.7}W_{0.3}Si_2$ powder (obtained by comminuting of cast sample and synthesised by SHS).

This work is aimed at dilatometric properties studying of $Mo_{0.7}W_{0.3}Si_2$ solid solution depending on the content of α-Si_3N_4 reinforcing additive in $Mo_{0.7}W_{0.3}Si_2$ + Si_3N_4 composite. Conclusions are made about the sintering rate influence on the sintering start point displacement, relative shrinkage. The activation energy of the sintering process for $Mo_{0.7}W_{0.3}Si_2$ + Si_3N_4 (from 1 to 15 wt.%) are calculated.

2.Materials and methods of researches

The cast mixture 70 wt% $MoSi_2$ and 30 wt% WSi_2 ($Mo_{0.7}W_{0.3}Si_2$) [14], was obtained by the SHS method at the Institute of Structural Macrokinetics and problems of materials science of the Russian Academy of Sciences (ISMAN RAS) in Chernogolovka Russia and characterised by SEM, X-ray and granulometric analyzes. The dominant fraction in powder $Mo_{0.7}W_{0.3}Si_2$ is 10÷20 μm particles.

The Si_3N_4 powder was presented by well-crystallized silicon nitride whiskers, with an average length about 2 μm and a thickness up to 200 nm. The α-Si_3N_4 content is not less than 95%, the specific surface area of the powder is 8.2 m^2/g. The impurities content are O2 = 1.64 wt%, Fe = 0.023 wt%. The powder was obtained by SHS method at the ISMAN RAS in Chernogolovka.

The $Mo_{0.7}W_{0.3}Si_2$ + α-Si_3N_4 mixtures containing 1; 2.5; 5; 10 and 15 wt% (respectively 2.3, 5.6, 10.8, 20.3 and 28.8 vol%) of Si_3N_4 were prepared in a planetary mill (Pulverisette 5 from Fritsch GMBH, Germany).

The mixtures were formed in the cylinders about 6x10 mm at 200 MPa uniaxial pressing on a hydraulic press in a metal form at room temperature. The weight of each sample was about 1 g.

The microstructure and phase composition were examined on an electron microscope (Supra 50 VP (LEO, Germany) with an INCA Energy + Oxford microanalysis system with a prefix for local X-ray spectroscopy) and X-ray (Rigaku D/MAX 2500 with a rotating anode, Japan).

Dilatometric studies were carried out on a dilatometer (DIL 402C Netzsch, Germany), capable of recording up to 2000°C [16]. The sample was placed in a horizontal graphite holder equipped with a graphite pusher with a high-precision displacement transducer (measuring range 500-5000 μm). The thermocouple (tungsten-rhenium alloy) was in close proximity to the sample and accurately recorded its temperature, the second thermocouple (tungsten-rhenium alloy) is in the chamber with a heater. This chamber has an independent of the working chamber argon atmosphere. The flow of argon introduced into the furnace was 70 ml/min, the heating rate was 5, 10, 12 and 15 °C/min, the heating was continued to 1850 °C, the isothermal section was 60 minutes, it was cooled to room temperature at a rate of 20 °C/min. The samples obtained after dilatometric studies density determined by archimedean method. Distilled water was used as a liquid.

Assuming isotropy in densification of all the specimens, the relative density of the sintered specimen (ϱs) was calculated using the following equation [17]

$$\rho_s = [\frac{1}{1 + dL/L_0 - \alpha(T - T_0)}]^3 \rho_g \quad (1)$$

where dL/L_0 is instantaneous linear shrinkage obtained by the dilatometer test, L_0 is the initial length of the specimen, T is the measured temperature, T_0 is the room temperature, ϱ_g is the green density, and α is the coefficient of thermal expansion. From a practical point of view, α is determined from the cooling steps of the dilatometer run [18,19,20]. An average value α as a function of temperature T by the following estimation equations was determined from the cooling steps of the dilatometer runs performed with the different heating rate adopted during our investigations [17].

$$\alpha(K^{-1}) = 9.1861 \times 10^{-6}(445 < T < 2030K)$$

Young and Cutler [21] have derived the sintering-rate equations that are applicable to quantitative analysis for shrinkage data measured under constant rates of heating (CRH) and estimated the activation energy to induce diffusion at the initial sintering stage. Wang and Raj [22, 23] have derived the sintering- rate equation used for estimation of the activation energy at intermediate sintering stage by CRH technique. Matsui et al. [24, 25] derived two sintering equations for constant heating rate as follows.

$$Ln[TC\frac{d\rho}{dT}] = -\frac{Q}{RT} + Ln[f(\rho)] + Ln[\frac{K\gamma\Omega D_0}{k_b}] - pLn(a) \quad (2)$$

$$\frac{d(dL/L_0)}{dT} = Ln(\frac{K\gamma\Omega D_0 R}{ka^pCQ})^n exp(\frac{nQ}{RT^{2-n}})exp(-\frac{nQ}{RT}) \quad (3)$$

where T is absolute temperature, C the heating rate, dϱ/dT the densification rate, Q the activation energy, R the universal constant of gases, $f(\varrho)$ the function of density, K the numerical constant, γ the surface energy, Ω the atomic volume, D_0 the frequency factor, k_b the Boltzmann's constant, a the particle radius, $dL(=L_0-L)$ the change in length of the compacts, L_0 the initial length of the compacts, and the parameters n and p the order depending on the diffusion mechanism. Eqs. (2) and (3) is applicable to the fractional shrinkages of <4%, which fulfill the initial or early stage sintering condition.

Using CRH experiments, the analysis method that is able to determine the diffusion mechanism at the initial sintering step is derived as follows: in Eq. (2), using the slope S_1 of the Arrhenius - type plot of $Ln[TC\frac{d\rho}{dT}]$ against $\frac{1}{T}$ at same density, the activation energy is expressed as:

$$Q = -RS_1 \quad (4)$$

In equation (3), using the slope S_2 of the Arrhenius - type plot of $Ln[T\frac{d(dL/L_0)}{dT}]$ against $\frac{1}{T}$, the apparent activation energy is expressed as:

$$nQ = -RS_2 \quad (5)$$

Equations (3)–(6) were used to determine the values of activation energy. Combining equations (4) and (5) we have:

$$n = \frac{nQ}{Q} = \frac{S_2}{S_1} \quad (6)$$

3.Results and discussion

In this paper cast $Mo_{0.7}W_{0.3}Si_2$ solid solution dilatometric studies were carried out depending on the α-Si_3N_4 reinforcing additive content in the composite. The dilatometry was carried out at rate of 5, 10, 12 and 15 °C/min to 1800 °C. Figures 1 and 2 highlight the change of the shrinkage $Mo_{0.7}W_{0.3}Si_2$ with Si_3N_4 additive (from 1 to 15 wt.%).

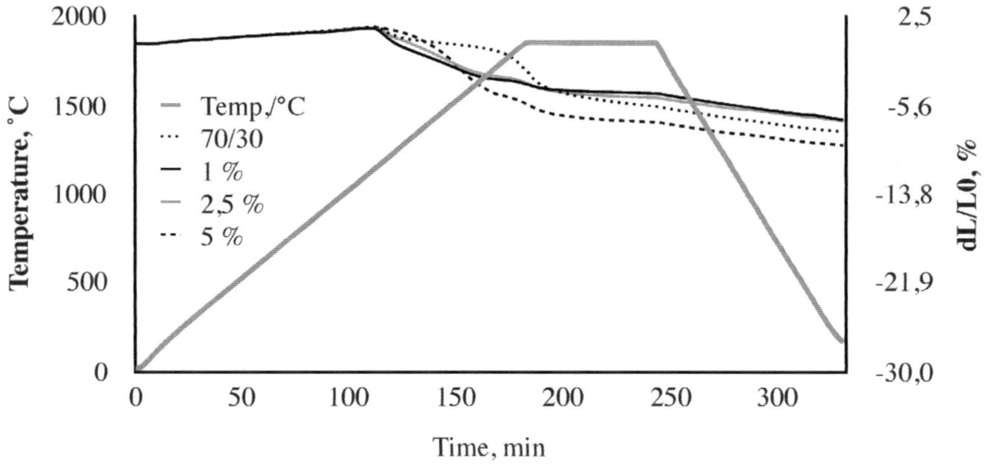

Figure 1. The curve of the continuous shrinkage of $Mo_{0.7}W_{0.3}Si_2$ and $Mo_{0.7}W_{0.3}Si_2$ + 1; 2.5 and 5 wt.% Si_3N_4, hearting rate 10 °C/min

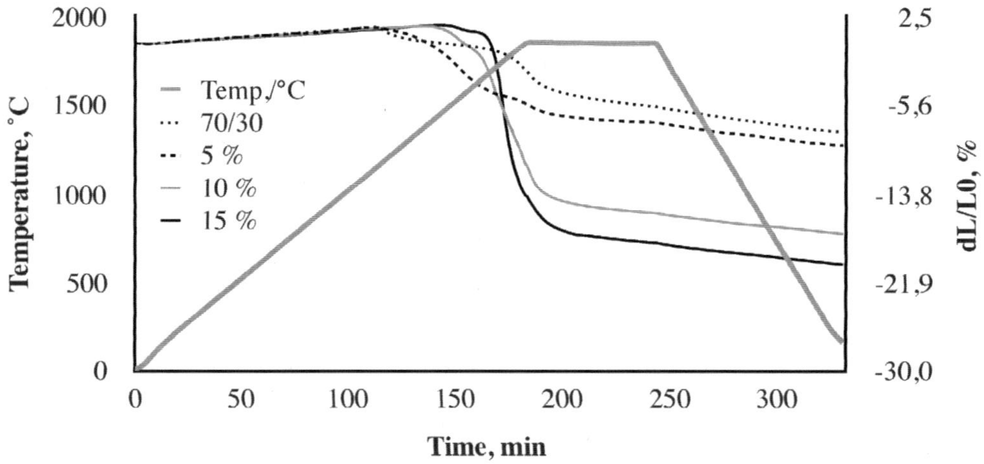

Figure 2. The curve of the continuous shrinkage of $Mo_{0.7}W_{0.3}Si_2$ and $Mo_{0.7}W_{0.3}Si_2$ + 5; 10 и 15 wt.% Si_3N_4, hearting rate 10 °C/min

The figure below shows the change of the shrinkage and shrinkage rate of $Mo_{0.7}W_{0.3}Si_2$ + 1 and 15 wt.% Si_3N_4 compact during non-isothermal sintering at three different heating rates of 5, 10 and 15 °C/min, respectively (figure 3). In the previous work [14-15] it was shown that Si_3N_4 additive concentration increase led to shift the composite sintering beginning point from 1150 °C to 1697 °C and the composite shrinkage changed from 6.75% to 20.21% (respectively, at 1 and 15 wt.% Si_3N_4). The comparative data of the sintering response is summarised in Table 1.

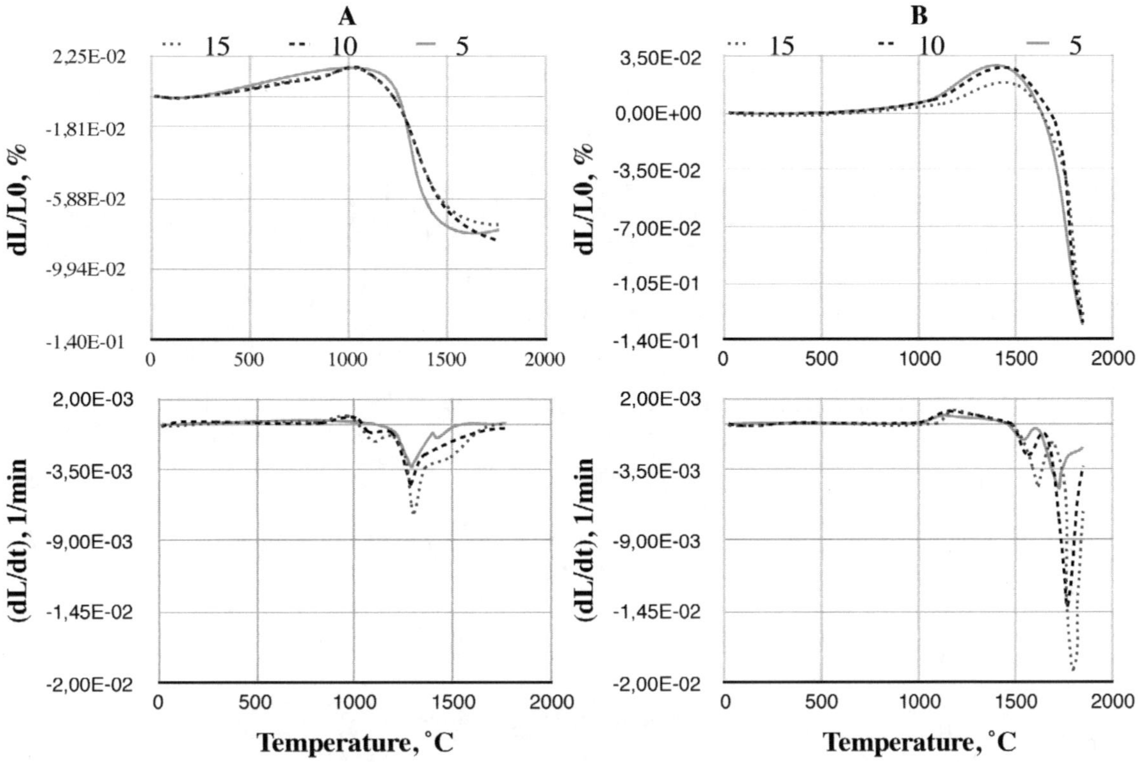

Figure 3. Effect of heating rate on the shrinkage and shrinkage range of (**A**) $Mo_{0.7}W_{0.3}Si_2$ + 1 wt.% Si_3N_4 and (**B**) $Mo_{0,7}W_{0,3}Si_2$ + 15 wt.% Si_3N_4 compact as a function of temperature

Table 1. Dilatometry data for $Mo_{0.7}W_{0.3}Si_2$ + 1;5;10 and 15 wt.% Si_3N_4 powder compact sintered at different heating rates (5,10 and 15 °C/min).

Heating rate (°C min^{-1})	**$Mo_{0,7}W_{0,3}Si_2$ + 1 wt.% Si_3N_4**		**$Mo_{0,7}W_{0,3}Si_2$ + 5 wt.% Si_3N_4**		**$Mo_{0,7}W_{0,3}Si_2$ + 10 wt.% Si_3N_4**		**$Mo_{0,7}W_{0,3}Si_2$ + 15 wt.% Si_3N_4**	
	T_{onset} (°C)	T_{max} (°C)	T_{onset} (°C)	T_{max} (°C)	T_{onset} (°C)	T_{max} (°C)	T_{onset} (°C)	T_{max} (°C)
5	1234	1282	1297	1531	1561	1675	1662	1721
10	1240	1284	1322	1560	1614	1729	1723	1766
15	1261	1299	1360	1588	1629	1750	1749	1772

Increasing the heating rate increased the temperatures at which sintering starts (arbitrary set to 0.5% shrinkage, T_{Onset}) and proceeds. For example, T_{Onset} 1234 and 1261 °C for $Mo_{0.7}W_{0.3}Si_2$ + 1 wt.% Si_3N_4 when heating rates increased from 5 to 15 °C/min. It is a usual behaviour for this systems [17]. For the same reason, the shrinkage rate curves of specimens shifted to higher temperature, as the heating rate increased. The maximum shrinkage rate was found to be dependent on heating rate, i.e. the higher the heating rate, the higher is the instantaneous shrinkage rate whatever is the temperature.

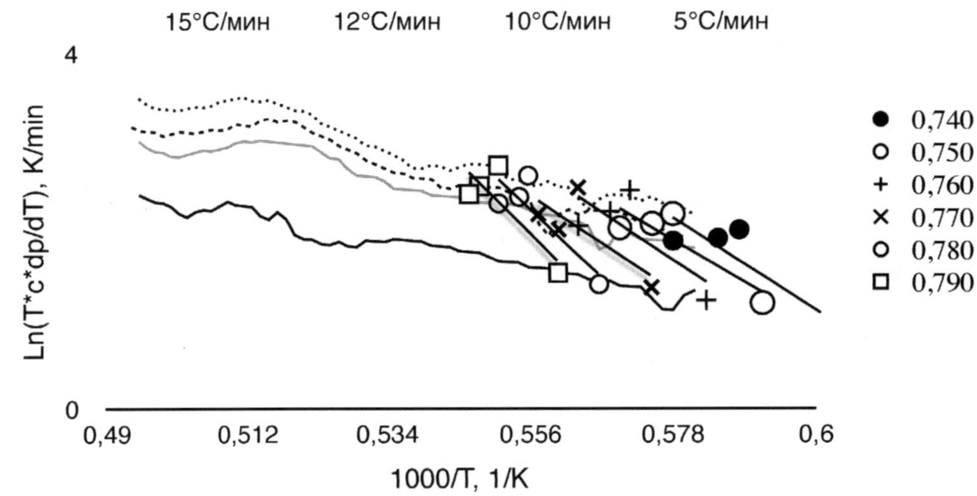

Figure 4. Arrhenius type plots for estimate of activation energies of sintering $Mo_{0.7}W_{0.3}Si_2$+1wt.% Si_3N_4

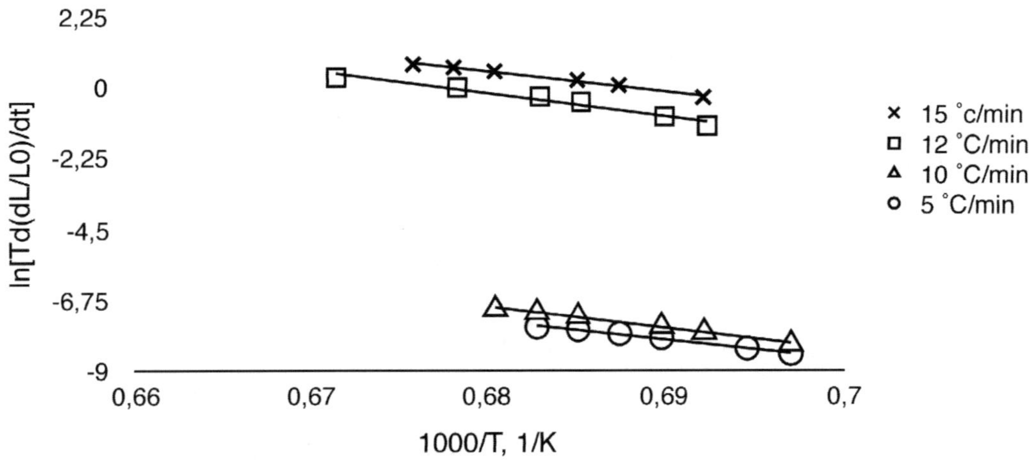

Figure 5. Plot of ln[Td(dL/L_0)/dt] against 1/T of $Mo_{0.7}W_{0.3}Si_2$ + 1 wt.% Si_3N_4 at different heating rates (5; 10; 12 and 15 °C/min)

The activation energy (Q) and the apparent activation energy (nQ) of diffusion, and order on diffusion mechanism (n) can be appraised by applying Eqs. (2), (3) and (6) to the results in figure 3. Eq. (2) is applied in the following way. For each heating rate (dT/dt)=c, both T and (dT/dt) at the same relative density were determined, and their values were plotted as ln[Tc(dϱ/dT)] against 1/T (figure 4). The Q at each relative density was determined from the slope of the straight line and as summarised in Table 2. On the other hand, Eq. (3) is applied in the following way. In the fractional shrinkage range of <4%, the nQ was determined from the slope of the straight line in the plot of ln[Td(dL/L_0)/dt] versus using the shrinkage curve of each heating rates. The proper plots at different heating rate are shown in figure 5. As summarised in Table 2.

The activation energy for each concentration of the composites was calculated. It was shown that activation energy decreased to when the concentration of Si_3N_4 increase. Average value of 370 kJ/mol for Q was obtained. All values of the activation energy are given in Table 2. The average value of 440 kJ/mol for nQ was obtained.

Table 2. Activation energy, apparent activation energy, and order on diffusion mechanism at the initial sintering stage of $Mo_{0.7}W_{0.3}Si_2$ + 1; 2.5; 5; 10 and 15 wt.% Si_3N_4

	1 wt.% Si_3N_4	2.5wt.% Si_3N_4	5 wt.% Si_3N_4	10 wt.% Si_3N_4	15 wt.% Si_3N_4
Q, kJ/mol	396÷51	402÷54	389÷36	344÷23	319÷36
nQ, kJ/mol	550÷88	540÷64	400÷64	373÷83	349÷76

4.Conclusion

Sintering in Ar atmosphere of $Mo_{0.7}W_{0.3}Si_2$ + 1; 2.5; 5; 10 and 15 wt.% Si_3N_4 powder has been investigated by the way of dilatometer tests. Using the constant rates of heating technique and applying the analytical method, the activation energy value as Q = 370 kJ mol. The activation energy decrease when the concentration of Si_3N_4 increase in the composite.

Acknowledgements

The reported study was funded by RFBR project № 18-38-00327. The methodical part of the work (X-ray and SEM) was carried out according to the state task № 007-00129-18-00.

References

[1] Stoloff N S 1999 *J. Mat Sci. and Eng.* **A261** 169
[2] Khanra A K 2005 *J. Mat. Sci* **40** 3027
[3] Melsheimer S 1996 Doctoral Dissertation
[4] Lohfeld S and Schuetze M 2005 *J. Mat. and Corrosion* **56** 93
[5] Zhu G, Lu Q, Wu G and Feng P 2015 *J. App. Surface Sci.* **343** 41
[6] Titov D D, Lysenkov A S, Kargin Y F, Gorshkov V A, Goldberg M A and Petrakova N V 2016 *J. Inorg. Mat.: Applied Research* **7** 624
[7] Titov D D, Kargin Yu F, Lysenkov A S, Popova N A and Gorshkov V A 2013 *J. Inorg. Mat.: Applied Research* **4** 66
[8] Perevislov S N, Lysenkov A S, Titov D D, Tomkovich M V 2017 *J. Inorg. Mat.* **53** 220
[9] Lysenkov A S, Timoshkin I A, Kargin Y F, Titov D D, Fedotov A Y, Ashmarin A A, Baranchikov A E 2016 *J. Inorg. Mat.: Applied Research* **7** 517
[10] Burnham A 2000 *Thermochim Acta* **355** 165
[11] Roduit B 2000 *Thermochim Acta* **355** 171
[12] Liu C [et al.] 2003 *Polymer Degrad. and Stab.* **81** 197
[13] Kang Y [et al.] 2004 *J. the Korean Ind. and Eng. Chem.* **15**, N2 188
[14] Titov D D, Akhrorov A Yu, Lysenkov A S, Kargin Yu F, Gorshkov V A, Baikin A S 2016 *Forum New Materials (Sochi)* (Moscow: Intercontact Science) p 25
[15] Titov D D 2015 *Symp. New mat. and tech. (Kazan)* vol 2 (Moscow: Intercontact Nauka) p 896
[16] Trusova E A, Khrushchev A A, Vokhmintcev K V, Titov D D 2013 *J. Eur. Ceram. Soc.* **33** 2327
[17] Abbasian A R, Rahimipour M R and Hamnabard Z 2013 *Iranian J. Mat. Sci. & Eng.* **10** N 3 44
[18] Bernard-Granger G, Guizard C 2007 *J. Am. Ceram. Soc.***90** 1246
[19] Prosvirnin D V et al 2017 *J. Mat. Sci.* **10** 45
[20] Kolmakov A G, Oparina I B, Sevoctyanov M A, Arzhakov M S, Lukovkin G M 2015 *J. Eurasian Phys. Tech.* **12** 33
[21] Young W S, Cutler I B 1970 *J. Am. Ceram. Soc.* **53** 659
[22] Wang J and Raj R 1990 *J. Am. Ceram. Soc.* **73** 1172
[23] Wang J and Raj R 1991 *J. Am. Ceram. Soc.* **74** 1959
[24] Matsui K, Ohmichi N, Ohgai M, Enomoto N, Hojo J 2005 *J. Am. Ceram. Soc.* **88** 3346
[25] Matsui K, Tanaka K, Enomoto N, Hojo J 2006 *J. Ceram. Soc. Jpn.* **114** 763

IOP Conf. Series: Materials Science and Engineering **347** (2018) 012025 doi:10.1088/1757-899X/347/1/012025

The structure of biocoats based on TiO_2 doped with nitrogen study

E L Boytsova, L A Leonova, V F Pichugin
Tomsk Polytechnic University, Institute of Physics and Technology,
Lenin Avenue 30, Tomsk, 634050, Russia

E-mail: boi5@list.ru

Abstract. Nitrogen-doped titanium dioxide (N-TiO_2) nanofilms were deposited by reactive magnetron sputtering under different bias voltage. The mode of sputtering influences to formation and properties of titanium films. X-ray diffraction (XRD) was used to study the phase transition and crystallinity of the nanofilms. A technique of layer-by-layer measurement of Raman scattering from nanostructured titanium dioxide films based on a preliminary sputtering of the films by argon beam under an angle of 45° and less has been developed. Experimentally confirmed low dissolution rate of the coating in NaCl saline (0.9%).

1. Introduction

Films of titanium dioxide, doped with different elements, found their application in various fields of technology, including medicine. Films for medical purposes are used for many types of implants, including vascular stents, should improve the quality of bar metal stents. The coatings of the stents should have the following properties: biocompatibility, stability of properties and composition, corrosion resistance, that is, reduce the effect on the surrounding tissue [1]. Complex N-TiO_2 films are one of the most promising coatings for coronary stents [2, 3]. It is known that such crystalline modifications of TiO_2 as anatase and rutile have bioactive properties, manifested in their faster (in comparison with amorphous natural titanium oxide) osteointegration, with preference given to titanium oxide, with anatase structure. Indeed, it has been found that the anatase satisfies, first of all, the crystallographic criteria in the selection of materials for bioactive coatings [4]. The bioactivity of the rutile phase of TiO_2 is significantly inferior to the bioactivity of the anatase phase. The necessity of modification of the anatase form of titanium dioxide is due to its use in a number of areas of medicine, where the content of the rutile form is regulated.

In this work we studied N-doped titanium dioxide (N-TiO_2) nanofilms obtained by reactive magnetron sputtering by using UVN-200MI laboratory system [5, 6]. The purpose of this work is to investigate the influence of the sputtering conditions, in particular, the bias voltage U_b; the influence of the liquid phase, in particular, prolonged contact with the physiological fluid (NaCl saline, 0.9%) on the surface structure of the coating, composition and morphology of the surface of N- TiO_2 nanofilms.

2. Experimental part

In the present work the N- TiO_2 nanofilms were deposited on the NaCl crystals (plates 10×10×1 mm) and stainless steel substrates using the reactive magnetron sputtering method. The description of laboratory system and using method were given in [5, 6]. The following parameters are used for the nanofilms deposition: cathode material is Ti, operating pressure in the chamber is 0.1 Pa, power is 1.2 kW, current is 3A, working gas leakage rate is 5 ml/min, bias voltage is 0 V and -100 V. The partial pressure ratio of pure plasma-forming gases N_2 and/or O_2 are $p(O_2)/p(N_2)$ =1/1, $p(O_2)/p(N_2)$ =1/3, O_2 deposition time is 90 minutes. The X-ray diffraction (XRD) was used to study the structural changes and phase transitions in the nanofilms using XRD-7000 diffractometer (Shimadzu) with grazing angle mode (α=1°) under Cu K-α radiation. The average crystallite size was calculated using Scherrer's equation:

$$D = \frac{\lambda}{\beta \cdot cos\theta} \quad (1)$$

where D is the crystallite size, λ is the X-ray wavelength (1.5418 Å), β is the full width at half maximum, and θ is the diffraction angle. PDF-4 database of International Center for Diffraction Data (ICDD) was used for the phase analysis (ICDD: anatase #21-1272, rutile #21-1276).

For studying of chemical properties stability the samples were exposed of prolonged interaction with saline (0.9% NaCl), which simulates approximate blood composition of a living organism [7]. The required volume of NaCl solution (4 ml) was determined according to ISO 10993-12 [8]. After 600 days, the samples were removed from solution. The structure of the coatings before and after contact with the solution was investigated by Raman spectroscopy on the Centaur U HR complex [9]. The spectra were measured along the slope of a crater obtained by sputtering an 6 keV argon ion beam with a cross section of 1 mm under an angle of 45° from the bottom of the crater to the surface of the sample. The Raman image of the films was obtained using the Olimpus 100 microscope.

3. Results and Discussion

3.1 XRD Analysis

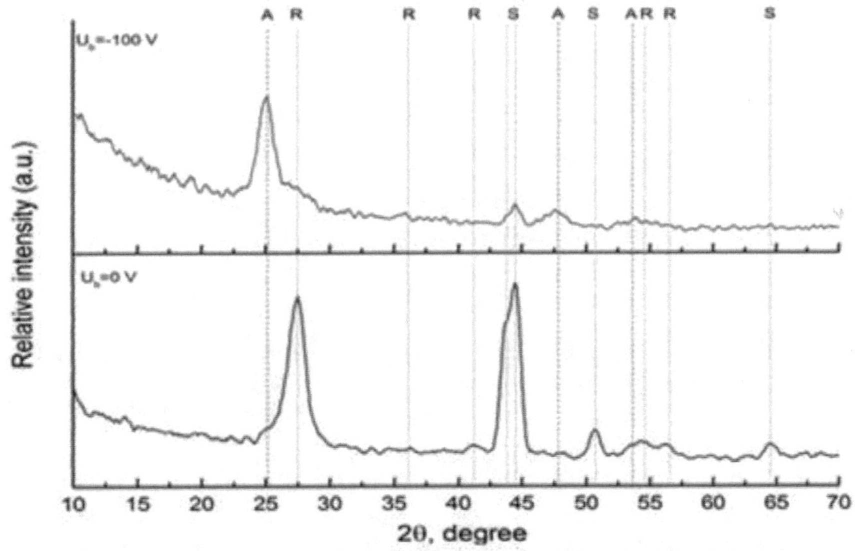

Figure 1. XRD patterns of N-TiO2 nanofilms deposited at different conditions: grounded substrate (Ub = 0 V) and negative bias voltage (Ub = -100 V). A - anatase; R - rutile.

Figure 1 shows XRD patterns of the N-TiO_2 nanofilms obtained at different bias voltage values. As observed in the XRD patterns, the nanofilms deposited at U_b = 0 V consist of 75% rutile and 12% anatase phase mixture. When the negative bias voltage -100 V)\ is applied, the films structure changes to a low amorphous/crystalline phase and consists of 78% anatase and 18% rutile mixture including amorphous part. There are some peaks from steel substrate in XRD patterns. The crystallite size (characterized by A (101) and R (110) reflections) is about 14 nm for all nanofilms. As XRD analysis

IOP Conf. Series: Materials Science and Engineering **347** (2018) 012025 doi:10.1088/1757-899X/347/1/012025

showed the rutile-anatase transition occurs at the bias voltage mode. The rutile phase volume has decreased from 75% to 18%, while the anatase one has increased up to 78%.

3.2 Raman spectroscopy

The structural and phase characteristics and transitions of TiO_2-based films were studied using Raman spectroscopy. Raman spectra of TiO_2 are characterized by a set of Ti—O stretching lines in the range (140-800) cm^{-1}. In the Raman spectrum, there are six characteristic lines of the anatase phase and four modes of the rutile phase.

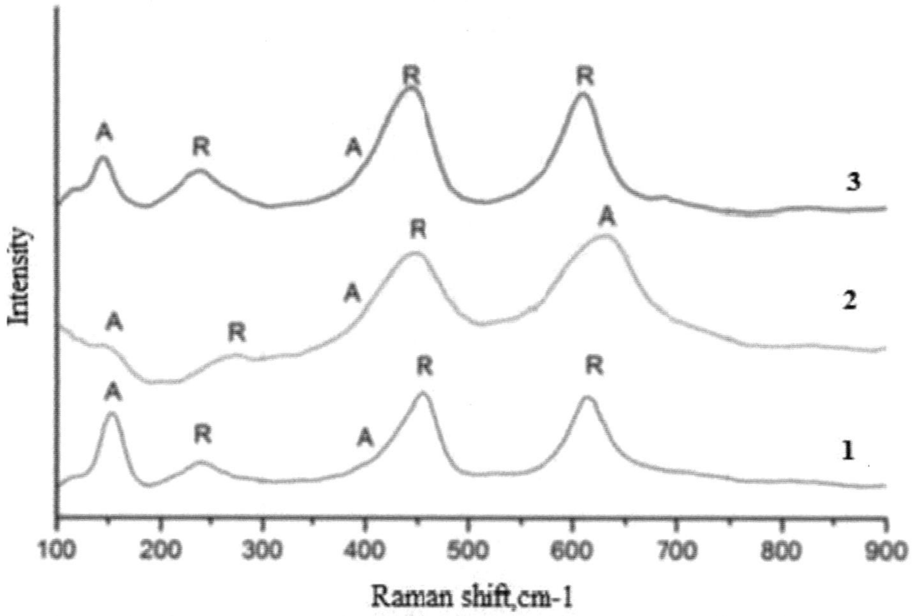

Figure 2. Raman spectra from films deposited in the atmosphere: 1) O2, 2) N2 + O2 and 3) 3N2 + O2 at a bias potential *Ub* = 0 V.

Figure 2 shows the Raman spectra of N-TiO_2 films formed in the zero-bias potential mode. The films demonstrate a two-phase structure of titanium dioxide (A-anatase, R-rutile). In the Raman spectra, characteristic lines of the anatase phase are observed: 154 cm^{-1} and 405 cm^{-1}, and also for the rutile phase: 240 cm^{-1} (two-phonon scattering mode, 2-background). The main anatase peak, observed at 154 cm^{-1}, is shifted toward higher wave numbers compared to the single-crystal anatase phase (144 cm^{-1}), which may indicate nanoscale structural components of the films. The broadening and decrease in the intensity of the low-frequency maximum also indicates a change in the crystallite size and the degree of crystallinity of the phases.

Figure 3 shows Raman spectra from N-TiO_2 films formed at the negative bias potential mode. The spectra contain bands characterized by a mixed structure of titanium dioxide. The blurring of Raman lines may indicate the nanocrystalline structure of the films or their partial amorphization (quasi-morphic structure) [10].

IOP Conf. Series: Materials Science and Engineering **347** (2018) 012025 doi:10.1088/1757-899X/347/1/012025

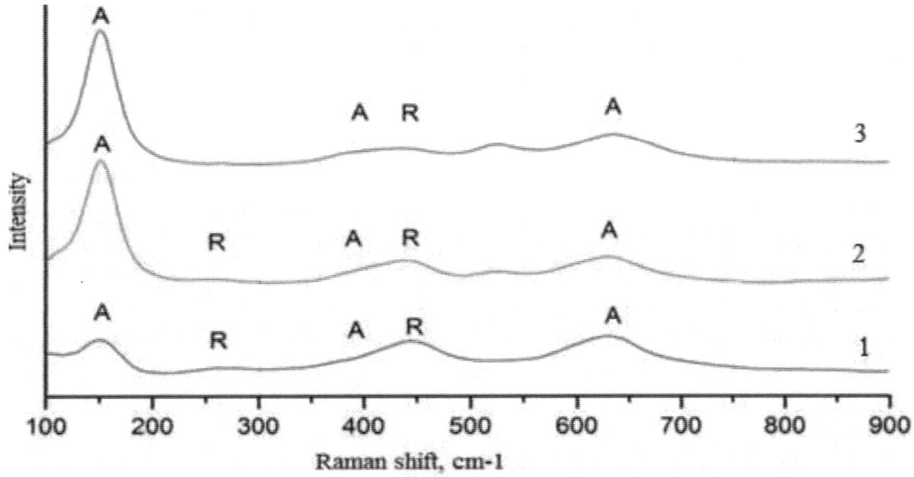

Figure 3. Raman spectra from films deposited in the atmosphere: 1) O_2, 2) $N_2 + O_2$ and 3) $3N_2 + O_2$ at a bias potenial Ub = -100

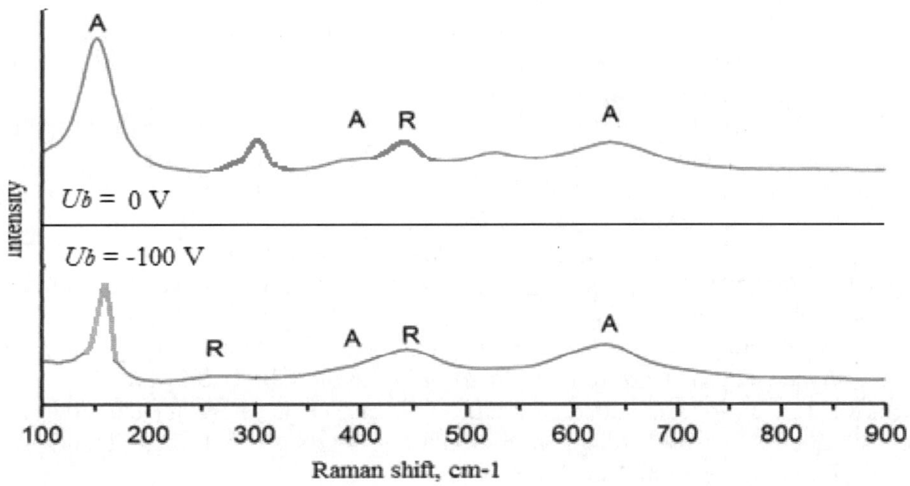

Figure 4. Raman spectra from N-TiO_2 films retained in a model biological fluid (NaCl solution).

In the Figure 4 the Raman scattering spectrum of anatase, sustained in physiological solution, one can observe two peaks that are located at 144, 640 cm^{-1}, belonging to doubly degenerate modes of optical phonons, and two peaks at 387 cm^{-1} and 520 cm^{-1} belonging to nondegenerate modes of optical phonons. The peak of 144 cm^{-1} is the most informative and most intense.

The oscillation frequency (144 cm^{-1}) of samples fabricated at a negative bias (-100 V) mode on a substrate is less intense than for samples deposited in a grounded substrate mode. This type of scattering is related to several reasons, one of which is the quantum mechanical effect of Raman scattering. For the small grain sizes, the phonons of the entire Brullian zone contribute to the formation of the Raman spectrum, ie, boundary phonons [11]. This can be explained by the fact that a negative bias voltage U_b= -100 V leads to an increase in the amorphous phase, in contrast to the nanocrystalline structure at U_b = 0 V. Although the anatase form of titanium dioxide formed with a negative bias voltage applied to the substrate is less stable and can release into solution [7].

We made the Raman scan of the sample surface coated by the N-TiO2 films before and after a long exposure at the solution. The investigation revealed the change of the color and structure of the films, which is due to the change in the morphology and modification of the bio-coating.

4. Conclusions

Nitrogen-doped titanium dioxide films with a dominant phase of anatase were obtained by the method of reactive magnetron sputtering. The influence of the bias voltage on the composition change, the morphology of the surface in NaCl solution of N-TiO_2 nanofilms was observed. Negative bias voltage leads to an increase in the anatase phase and the appearance of an amorphous part. The prevalence of anatase in the coating was revealed during prolonged contact with the liquid phase, which confirm of the films corrosion-resistant properties. This property makes the obtained films promising for their use in medicine.

Acknowledgement

The authors wish to thank their colleague M. Konishchev for the help in preparing the samples.

References

[1] López-Huerta F, Cervantes B, González O, Hernández-Torres J, García-González L, Vega R, Soto E 2014 Biocompatibility and Surface Properties of TiO2 *Thin Films Deposited by DC Magnetron Sputtering Materials* **7(6)** 4105–4117

[2] Rizzo A, Signore M A, Tapfer L, Piscopiello E, Cappello A, Bemporad E, & Sebastiani M 2009 Graded selective coatings based on zirconium and titanium oxynitride *Journal of Physics D: Applied Physics* **42(11)** 115406

[3] Yin Z F, Wu L, Yang H G, & Su Y H 2013 Recent progress in biomedical applications of titanium dioxide *Physical chemistry chemical physics: PCCP* **15(14)** 4844–58

[4] Masaki Uchida, Hyun-Min Kim, Tadashi Kokubo, Shunsuke Fujibayashi, Takashi Nakamura 2003 Structural dependence of apatite formation on titania gels in a simulated body fluid *Journal of Biomedical Materials Research* **Part A Volume 64A Issue 1 1** 164-170

[5] Pichugin V P, Pustovalova A A, Konishchev M E, Khlusov I A, Ivanova N M, Sun Zhilei 2016 In-Vitro Dissolusion and Structural and Electrokinetic Characteristics of Titanium-Oxynitride Coatings Formed via Reactive Magnetron Sputtering *Journal of Surface Investigation X-ray, Synchrotron and Neutron Techniques* **10(2)** 282–291

[6] Pustovalova A, Ivanova N 2016 Structural Changes of Titanium Dioxide Thin Films Deposited by Reactive Magnetron Sputtering through Nitrogen Incorporation *Key Engineering Materials* **683** 383–388

[7] Leonova L A, Boytsova E L, Pustovalova A A 2016 The study of titanium oxynitride coatings solubility deposited by reactive magnetron sputtering *IOP Conference Series: Materials Science and Engineering* **Vol. 135** 5

[8] ISO 10993-12:2012(en) 2012 *Biological evaluation of medical devices* 20

[9] Kelly S, Pollak F H, Tomkiewicz M 1997 *J. Phys. Chem.* **101** 2730 – 2734

[10] Pustovalova A A , Pichugin V F 2017 Structural features of N-containing titaniumdioxide thin films deposited by magnetron sputtering *Thin Solid Films* **627** 9–16

[11] Kiselyova E S, Sypchenko V S, Nikitenkov N N, Pozdeeva E V, Zeylun V 2017 Study of the composition and structure of titanium dioxide based coatings deposited by the method of reactive magnetic sputtering *Letters on materials* **7(2)** 117-119

IOP Conf. Series: Materials Science and Engineering **347** (2018) 012026 doi:10.1088/1757-899X/347/1/012026

Investigation of the influence of the composition on mechanical properties of polylactide

A S Baikin, M A Sevostyanov, E O Nasakina, K V Sergienko, M A Kaplan, S V Konushkin, A A Kolmakova, A D Yakubov, A G Kolmakov

A.A.Baikov Institute of Metallurgy and Materials Science, Russian Academy of Sciences

Email: baikinas@mail.ru

Abstract. In this paper we describe the creation of films from polylactide. Studied the mechanical properties of developed polymer films of polylactide. The effect of the molecular weight of polylactide on the mechanical properties of the resulting polymer films is shown. The dependence of the mechanical properties of polylactide films on the polymer concentration in chloroform was studied. The possibility of creating biodegradable films with specified mechanical properties is shown.

1. Introduction

Treatment of many diseases is very often inextricably linked with the negative impact on healthy human organs. Modern medicine has long sought to minimize such possible side effects, including the use of various medications [1-3].

One of the variants of solving this problem is the scientific society sees in the creation of systems of controlled drug delivery. Such systems provide therapeutic effects directly in the desired area, reducing or completely eliminating the effect on the rest of the body [4-6].

The promising basis for the creation of these systems are biodegradable polymeric materials of synthetic and natural origin, among which polyesters: polylactides, polylactide glycolides, polyethylene glycol, polyanhydrides, polyorthoesters, polysaccharides (starch, dextran, chitosan) are the most developed [7-8].

Films based on polylactide of different molecular weights are considered in this paper. Polylactide is an aliphatic polyester whose monomer is lactic acid. It is biodegradable, biocompatible and thermoplastic. An obvious advantage is the fact that the raw materials for the production of polylactide are renewable resources, namely silo-forming agricultural crops. Polylactide is already used in medicine, for the manufacture of surgical threads and the processing of pins. [9].

Thus, polymers can be obtained with different molecular weights and structures, and can also encapsulate molecules of almost any size. Using the proposed molecular weight of the polymer, different concentrations of polymer solutions, it is possible to provide a wide range of coatings with different properties, with further selection of the required characteristics. [10]

Polylactide is soluble in most solvents, including chlorinated solvents, tetrahydrofuran, acetone, ethyl acetate. In the work as a solvent, chloroform was selected. [11]

2. Materials and methods

To create the films, we prepared samples of polymers of mass 2; 6 and 10 g (± 0.01 g).

Chloroform, 200 ml in volume, was placed in a 500 ml flask and heated to 80° C with a magnetic stirrer.

Further, the polymer samples weighed were dissolved to a homogeneous state in chloroform at 80° C for 1 hour with constant stirring by means of an electronic overhead stirrer.

At the end of dissolution of the polymer, the volume of the solution was adjusted with chloroform to 200 ml. The resulting solution was aged for 5 minutes at 80° C and spread over glass pallets.

Drying was carried out for 2 days in air at 37°C in a thermostat.

At the end of the drying, the resulting films were removed.

The obtained samples were assigned ciphers used for independent research in the future (Table 1):

Table 1. Code samples of films

Code	Polymer	Molecular weight, kDa	Concentration, g per 100 ml of chloroform
1.1	Polylactide	45	1
1.2	Polylactide	45	3
1.3	Polylactide	45	5
2.1	Polylactide	90	1
2.2	Polylactide	90	3
2.3	Polylactide	90	5
3.1	Polylactide	180	1
3.2	Polylactide	180	3
3.3	Polylactide	180	5

Tensile strength studies of polymeric films from polylactide were carried out on a universal test machine INSTRON 3382 at a loading rate of 10 mm / min. Samples of polymer films for testing were made in accordance with GOST 14236-81, in the form of a double blade. The sample was fixed in the grippers of the testing machine, which was tightened evenly to ensure that the sample did not slip during the test. Tests of polymer films with determination of the relative elongation, yield strength and tensile strength were carried out in accordance with GOST 14236-81. The processing of the test results in determining the characteristics of mechanical properties was carried out using the software INSTRON Bluehill 2.0. The measurement error of the test machine is less than 1%. Five samples were tested for one experimental point. The values of yield strength σ_y, tensile strength σ_B and deformation δ were determined.

3. Results and discussion

Table 2 presents the averaged results of mechanical testing of polymer films for each composition based on polylactide (PLA) with different molecular weights with different dissolution concentrations in chloroform.

From the results obtained, graphs of the dependence were plotted, along which it is possible to trace the tendency of the change in mechanical characteristics (Figures 1-3).

As can be seen from the graphs shown in Figures 1-3, the thickness of polylactide films, their plasticity and strength is directly related to the concentration of the solution and the molecular mass of the substance. Increasing the concentration of the polymer in the solution promotes thickening of the polymer layer, thereby increasing plasticity, but decreasing the strength of the polymer film.

The change in the molecular weight of the polylactide showed that films based on polylactide 90 kDa and a strength of 45 kDa have a better plasticity. For lower concentrations in this study (1 and 3 g per 100 ml of chloroform), the change in the molecular weight of the polymer does not significantly affect the thickness of the resulting film. For a concentration of 5 g per 100 ml, the film thickness is reduced to 45% with an increase in molecular weight from 45 kDa to 180 kDa.

Table 2. Mechanical properties of polymer films.

Code	Polymer and concentration	Deformation, %	Yield strength, MPa	Ultimate Strength, MPa	Thickness, µm
1.1	PLA 45kDa 1x100	7,08	22,90	35,63	42
1.2	PLA 45kDa 3x100	60,21	17,39	21,68	107
1.3	PLA 45kDa 5x100	209,69	2,86	5,73	339
2.1	PLA 90kDa 1x100	58,09	17,27	20,12	38
2.2	PLA 90kDa 3x100	225,95	3,10	4,41	125
2.3	PLA 90kDa 5x100	305,37	0,63	2,05	262
3.1	PLA 180kDa 1x100	43,86	16,77	19,63	39
3.2	PLA 180kDa 3x100	140,88	5,37	6,64	82
3.3	PLA 180kDa 5x100	304,71	1,87	3,28	152

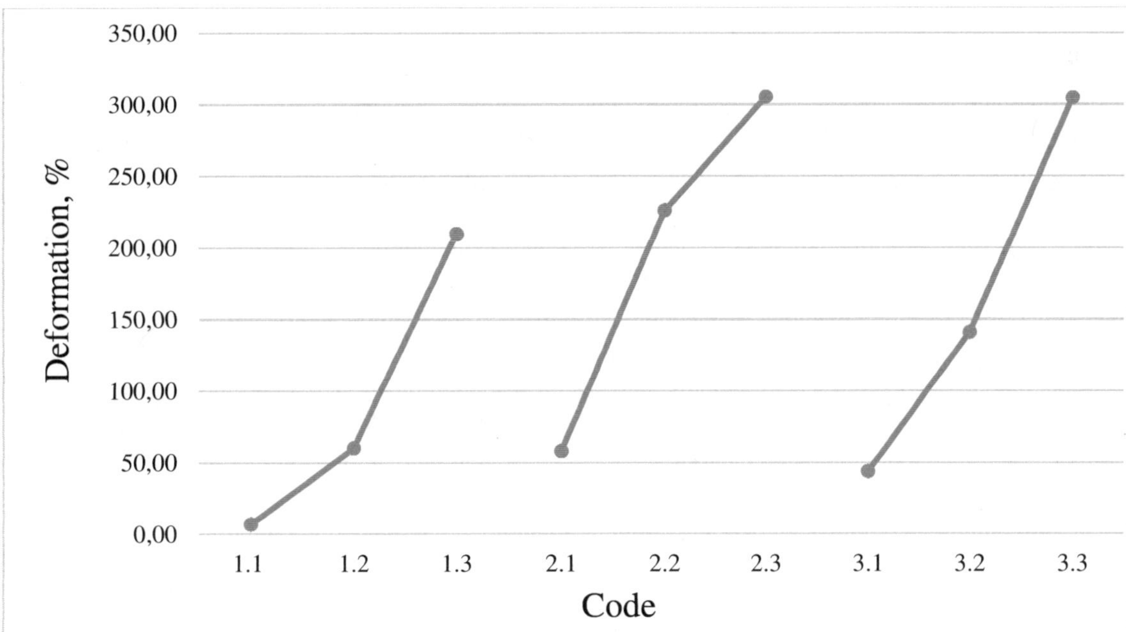

Figure 1. Diagram of deformation depending on the concentration of the solution for PLA

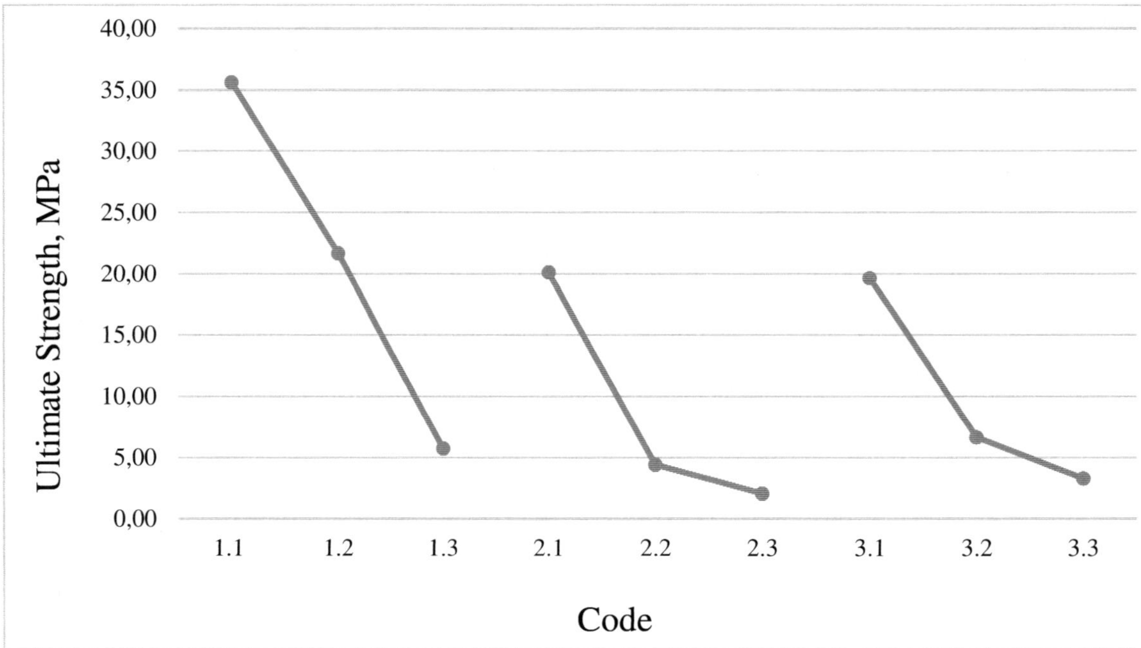

Figure 2. Diagram of the dependence of tensile strength on the concentration of solution for PLA

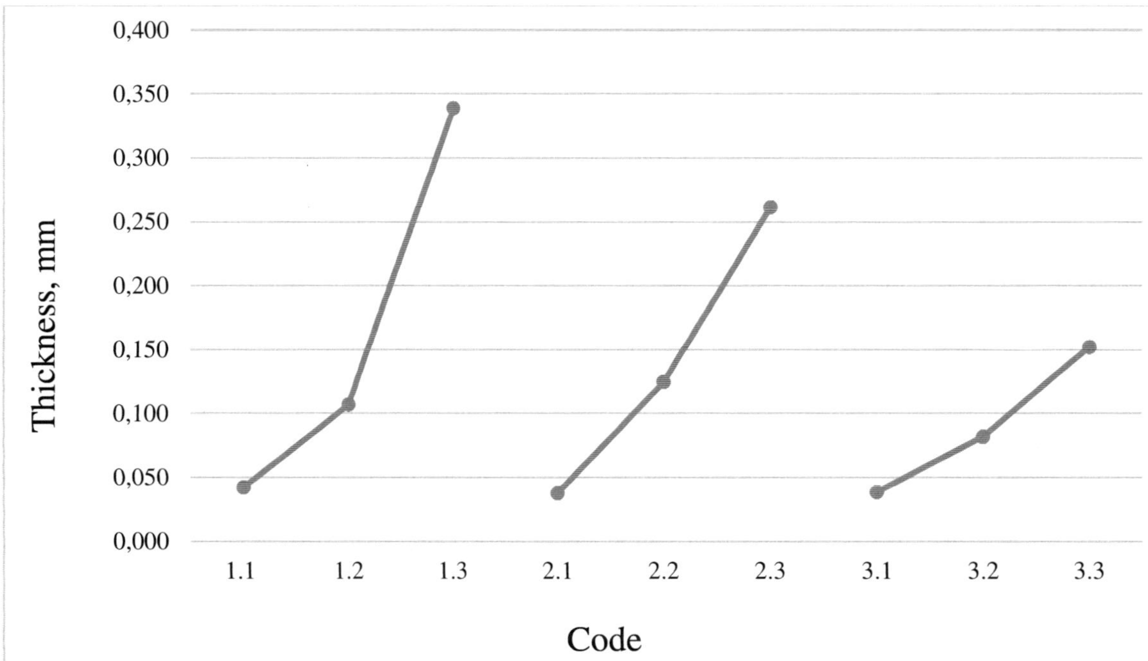

Figure 3. The diagram of the dependence of the thickness of polymer films on the concentration of solution for PLA

4. Conclusions

Investigations of the mechanical properties of the developed polymer films of polylactide were carried out and the possibility of creating biodegradable films with specified mechanical properties was demonstrated.

Acknowledgements

The work was supported by the Ministry of Education and Science of Russia (grant identifier RFMEFI60417X0196).

References

[1] Giessen WV, Lincoff A, Schwartz R, Beusekom HV, Serruys P, Holmes D, et al. Marked inflammatory sequelae to implantation of biodegradable and nonbiodegradable polymers in porcine coronary arteries. *Circulation* 1996;**94(7)**:1690–7.

[2] Nasakina E.O., Baikin A.S., Sevost'yanov M.A., Kolmakov A.G., Zabolotnyi V.T., Solntsev K.A. Properties of nanostructured titanium nickelide and composite based on it *Theoretical Foundations of Chemical Engineering*, 2014. - **V.48. №.4.** - P.477–486.

[3] Nasakina E.O., Sevost'yanov M.A., Mikhailova A.B., Gol'dberg M.A., Demin K.Yu., Kolmakov A.G., Zabolotnyi V.T. Preparation of a nanostructured shape memory composite material for biomedical applications *Inorganic Materials*, 2015, **Vol. 51, No. 4**, pp. 400-404

[4] Cooper SL, Visser SA, Hergenrother RW, Lamba NMK. Polymers.In: Ratner BD, Hoffman AS, Schoen FJ, Lemons JE, editors. *Biomaterials science an introduction to materials in medicine*. 2nd ed. San Diego: Elsevier Academic Press; 2004. p. 67–79.

[5] Stack R, Califf R, Phillips H, Pryor D, Quigley P, Bauman R, et al. *Interventional cardiac catheterization at Duke medical center*. Am J Cardiol 1988;**62**:3F–24F.

[6] Lincoff AM, Schwartz RS, Giessen WJVD, Beusekom HMMV, Serruys PW, Holmes DR, et al. Biodegradable polymers can evoke a unique inflammatory response when implanted in the coronary artery. *Circulation* 1992;**86(Suppl.4)**:

[7] Supper S., Anton N., Seidel N., Riemenschnitter M., Curdy C., Vandamme T. Thermosensitive chitosan/glycerophosphate-based hydrogel and its derivatives in pharmaceutical and biomedical applications. *Expert Opin. Drug Deliv.* 2014;**11**:249–267. doi: 10.1517/17425247.2014.867326.

[8] Sevost'yanov M.A., Fedotov A.Yu., Nasakina E.O., Teterina A.Yu., Baikin A.S., Sergienko K.V., Kolmakov A.G., Komlev V.S., Ivanov V.E., Karp O.E., Gudkov S.V., Barinov S.M. Kinetics of the release of antibiotics from chitosan-based biodegradable biopolymer membranes *Doklady Chemistry*, 2015. **V.465. Part.1.** P.278-280.

[9] Timchenko TV, Shcherbakova LI, Kompantsev VA Poly-D, L-lactide-co-glycolide: methods of preparation, properties and use for the development of drugs with micro- and nanodelivery means. *Modern problems of science and education*. 2015. **№ 4**. P.559-570 (in russian)

[10] Kedik SA, Zhavoronok ES, Sedishev IP, Panov AV, Polymers for delayed drug delivery systems (review) (May 3, 2013) (in russian)

[11] Glotova VN, Novikov VT, Izhenbina TN, Titova NG Solubility of lactide and glycolide in organic solvents *Polzunovskii vestnik*. 2014. **No. 3.** P. 145-147. (in russian)

IOP Conf. Series: Materials Science and Engineering **347** (2018) 012027 doi:10.1088/1757-899X/347/1/012027

Ceramics based on zirconium dioxide stabilized with indium oxide and praseodymium oxide

M G Frolova [1], Yu F Kargin[1], E S Lukin[2], N A Popova[2], A S Lysenkov[1], D D Titov[1] and S N Ivicheva[1]
[1]Baykov Institute of Metallurgy & Material Science RAS, Moscow, Russia
[2]Mendeleyev University of Chemical Technology of Russia, Moscow, Russia

E-mail: frolovamarianna@bk.ru

Abstract. The solid solutions based on Zirconia doped with indium oxide and praseodymium oxide with different concentration was obtained by the method of inverse heterophase co-deposition.The physico-chemical properties of obtained samples were examined. The highest properties have been achieved in samples of the composition $Zr_{1-x}R_xO_{2-\delta}$ (R=23 mol .% In). This composition has a density of 5.38 g/cm3; open porosity is 0 %, the bending strength is 203 MPa.

1. Introduction

Modern technology requires the development of new materials which possess ideal structures with high-performance characteristics. Currently, research works are aimed at the improvement of materials structures in order to enhance properties such as physical-mechanical and thermo-mechanics [1]. On the other hand, there is a need to decrease the sintering temperature. Additionally, the improved properties can promote the expansion of the application area.

One of the leading materials for the development of new structures is ceramics. It is the third most prevalent industrial material after metals and plastics.

The combination of unique properties determines an increase in the area of application of ceramic-based materials in the industry. It can be explained by the limit of some characteristics of present metals and alloys which does not allow their use for engineering of new constructions and new products. These characteristics include heat durability, heat resistance, corrosion resistance, etc.

The main ceramic disadvantages are fragility, sensitivity to thermal shocks (especially overcooling), and the complexity of mechanical restoration. These features set a task for the further development of technology and a properties improvement.

Recent achievements have shown a possibility of formation of new ceramic materials based on oxides and their derivatives which have high mechanical strength, resistance to abrasion, to simultaneous exposure to high temperatures and to stresses.

The current ceramic-based materials progress can be explained by the success in powder technology, the new methods development for products formation of highly disperse powders, the use of more advanced principles of the additives selection, the sintering theory development. All these factors promote the existing technologies modernisation and the properties improvement, as a result, it expands the application area.

The present study purpose was the preparation of solid solutions based on zirconia doped with indium oxide and praseodymium oxide and subsequent investigation of their physicochemical properties.

In order to achieve this goal, the main areas of research include:

- A zirconium dioxide cubic solid solution with additions of indium oxides and praseodymium synthesis;
- Study of the physicochemical properties of solid solutions of zirconia with different concentrations of the doping component;
- Study of the influence of the firing temperature and the concentration of the doping component on the ceramic properties of the samples.

2. Materials and Method

The monoclinic phase (m) ZrO_2 is thermodynamically stable below 1170 °C. It is tetragonal (t) in the range 1170 – 2370 °C and cubic (c) above 2370 °C. The transition t-$ZrO_2 \leftrightarrow$ c-ZrO_2 has a diffusion nature and plays a very important role in the production of so-called partially stabilised zirconia. The m-$ZrO_2 \leftrightarrow$ t-ZrO_2 transformation proceeds according to the martensitic mechanism and accompanies by volumetric changes of 5-9 %. Such a significant expansion of the material upon cooling, accompanied by cracking, does not allow the production of compact products of unalloyed ZrO_2. For this reason, only solid solutions of various ZrO_2-based oxides are of practical value, or composites with a rigid matrix capable of stabilizing the high-temperature phases of ZrO_2 at low temperature.

The majority of researchers explain the stabilization of solid solutions upon doping of zirconium dioxide with metal oxides in the tetragonal and cubic phases due to the presence of oxygen vacancies that ensure electroneutrality for the heterovalent character of the Zr^{4+} substitution in the trivalent or divalent cation in the alloying oxide, rather than the actual M^{3+}, M^{2+} cations [2]. With the formation of only oxygen vacancies in the crystal lattice of ZrO_2, the formation of t-ZrO_2, rather than m-ZrO_2, becomes energetically favorable. However, this mechanism alone is not enough to describe the stabilization of the tetragonal and cubic phases of zirconium dioxide when the oxides of elements are doped with oxidation states of 2+ and 3+. This is indicated by the discrepancy between the phase diagrams of the ZrO_2-stabilizer, constructed in the coordinates of the concentration of oxygen vacancies [3]. It is obvious that for all the attractiveness of the mechanism of stabilization by oxygen vacancies, some influence on phase stability is exerted by the alloying cations themselves, differing in valence and ionic radius.

The systematic research of the role of various ions in the stability of the tetragonal and cubic phases was carried out using X-ray absorption spectroscopy, in which the influence of trivalent and quadrivalent cations was investigated, as well as the effect of stabilizers with the ionic radius larger (Y^{3+}, Gd^{3+}, Ce^{4+}) and smaller (Fe^{3+}, Ga^{3+}, Ge^{4+}) of the ionic radius Zr^{4+}. In the case of doping with trivalent metal oxides, the oxygen vacancies arise in order to compensate the charge. In the case of the ions with the large radii, the oxygen vacancies appear to be associated with Zr^{4+}, whereas, in the case of the ions with the small radii the vacancies are bound to two ions of the addition. Both configurations are advantageous for the sevenfold coordination of the Zr^{4+} cations (CN = 7) by oxygen anions, which leads to the stabilization of the tetragonal and cubic phases of zirconia dioxide.

Apart from the achievement of a stable fluorite structure, these additives are also resistant to evaporation and it leads to a significant expansion of the homogeneity region of the solid solution of ZrO_2.

A zirconium oxychloride stabilized with different compounds was used as initial components for the preparation of ultradispersed powders of solid solutions based on zirconia dioxide. The first option is praseodymium oxide, whereas, the second option is indium oxide. The choice of stabilizer oxides was based on the proximity of the ionic radius (according to Belov and Bockiy), which corresponds with substituting elements (Zr^{4+} = 0.82 Å, In^{3+} = 0.92 Å, Y^{3+} = 0.97 Å, Pr^{3+} = 1.00 Å) [4].

The salt of zirconium oxychloride with oxides of indium and praseodymium was dissolved in distilled water at 105 °C. The resulting solution was evaporated to 90% concentration of the limiting solubility at the given temperature. The solution concentration was controlled by density. The salt solution at 100±5 °C was sprayed with compressed nitrogen under pressure (1÷2) 10^4 Pa through a glass capillary with a diameter of 1.0÷1.5 mm into a saturated cooled ammonia solution. The resulting suspension was transferred to the Buchner funnel and washed with distilled water to pH 6.0 in the filtrate. The drying of the precipitate was carried out in natural conditions (Figure 1).

IOP Conf. Series: Materials Science and Engineering **347** (2018) 012027 doi:10.1088/1757-899X/347/1/012027

Figure 1. SEM micrographs ZrO_2 powder with: a) 10 mol.% Pr_2O_3;b) 15 mol.% Pr_2O_3; c) 23 mol.% In_2O_3; d) 25 mol.% In_2O_3

The 5 % polyvinyl alcohol aqueous solution (PVA, TU 10779-99) was used as the temporary technological binder for semi-dry pressing, it was introduced in an amount of 15 % by weight of the powder. The pressed powder was twice wiped by a No. 01 sieve in order to homogenize the polyvinyl alcohol aqueous solution. The samples with the size of 40×6×6 mm were formed by a method of double-sided uniaxial semi-dry pressing at the pressure of 100 MPa. The samples were fired in the air at 1550 °C, with the holding time at the final temperature of 3 hours.

3. Results and discussion

In determining the main parameters of the ceramic samples which were sintered at different temperatures with the different content of the additive, it has been established that sintering of the ceramic improves with a rise in the sintering temperature, as well as, with an increase in the content of the doping additive. Such observations as the density increase, the significant linear shrinkage and the decrease in porosity were made. The samples obtained were examined for physical and mechanical properties (Table 1). According to the data, samples of the composition $Zr_{1-x}R_xO_{2-\delta}$ (R = 23 mol% In) have demonstrated the most enhanced characteristics. These samples have the density of 5.38 g/cm^3, the open porosity of 0 % and the flexural strength of 203 MPa.

Table 1. Properties of the ceramics based on ZrO_2 doped with praseodymium oxide and indium oxide

Doping additive	The amount of additive added mol %	The average density of the compact, g/cm^3	The average density of the burned product, g/cm^3	The linear shrinkage, %	The open porosity, %	The flexural strength, MPa
Pr_2O_3	10	2.82	5.21	20	0	155±5
	15	3.08	5.33	21	0	176±5
	20	2.92	5.15	21	0	192±5
In_2O_3	23	3.01	5.38	22	0	203±5
	25	3.23	5.67	25	0	230±5

4. Conclusion

Traditionally, the ceramics based on zirconium oxide has been used in the metallurgical industry for the manufacture of crucibles for metals smelting. Nowadays, this type of ceramics is one of the most promising ceramic materials for structural and instrumental use and it has been utilized in the production of gas turbine's parts, diesel engines, friction units, sealing rings of pumps, elements of stop valves, nozzles of spray chambers, wire drawing dies and cutting tools. Moreover, the ceramics based on zirconium oxide is used in the medical application for the manufacture of bone tissue implants.

During this research, the solid zirconium dioxide-based solutions doped with indium oxide and praseodymium were obtained by the method of inverse heterophase co-deposition. The most enhanced properties were achieved in the samples of the composition $Zr_{1-x}R_xO_{2-\delta}$ (R = 23 mol% In). The resulting ceramic samples have the density of 5.38 g/cm^3, the open porosity of 0 % and the flexural strength which equals 203 MPa.

Acknowledgments
The work was carried out according to the state task № 007-00129-18-00.

References

[1] Lipilin A S 2007 Doctoral Dissertation
[2] Pezzotti G, Munisso M C, Porporati A A, Lessnau 2010 Biomaterialy **31** p 6901
[3] Li P, Chen I W, Penner-Hahn J E 1994 J. Am. Ceram **77** p 118
[4] Matrenin S V, Slosman A I 2004 Technical Ceramics (Tomsk TPU) p 39

IOP Conf. Series: Materials Science and Engineering **347** (2018) 012028 doi:10.1088/1757-899X/347/1/012028

Structure and properties of ZrB_2, ZrSiB and ZrAlSiB cathode materials and coatings obtained by their magnetron sputtering

I V Iatsyuk, M V Lemesheva, Ph V Kiryukhantsev-Korneev and E A Levashov

National University of Science and Technology 'MISIS', 4 Leninsky Prospekt, Moscow 119049, Russia

Emails: lemescheva.margarita@yandex.ru

Abstract. The ceramic ZrB_2, ZrSiB, and ZrAlSiB cathodes were manufactured by means of self-propagating high-temperature synthesis (SHS). The parameters of SHS process including dependence of the combustion temperature and rate on the initial temperature of the reaction mixtures, as well as values of effective activation energy were estimated. Cathodes were subjected to the magnetron sputtering in the argon atmosphere. The structure and properties of cathodes and coatings were studied by means of X-ray diffraction, scanning electron microscopy, energy-dispersive and glow discharge optical emission spectroscopy. Bulk ceramic samples and coatings were characterised in terms of their hardness, elastic modulus, elastic recovery, density, and residual porosity. Results obtained shows that cathodes posses homogeneous structure with low porosity level in range 2-6% and hardness between 10 and 17 GPa. Coatings demonstrate dense defect-free structure and contain nanocrystallites of h-ZrB_2 phase. The grain size and hardness decrease from 8 down to 2 nm and from 37 down to 16 GPa with the addition of the silicon and aluminum dopes.

1. Introduction

Recently, ultra-high-temperature ceramics (UHTC) is of great interest in various branches of science and technology. Among the UHTC, the leading material is zirconium diboride due to its high melting point of 3245°C, high thermal conductivity of 57.9 W/(m·K), low thermal expansion coefficient of 5.9×10^{-6} °C^{-1} [1], hardness up to 22 GPa [2] and strength up to 565 MPa [3], high heat resistance up to 2000°C [4]. As a result, ceramics based on ZrB_2 are used as materials for heaters, crucibles, covers and electrodes of high-temperature thermocouples, refractory components, high-speed bearing, as well as for functional coatings deposited onto cutting tools, parts of hypersonic aerospace vehicles, high-temperature sensors etc. Bulk ZrB_2 can be used also as a cathode material for the magnetron sputtering and deposition of protective coatings. Addition of Si and Al is expected to improve the oxidation resistance and thermal stability of the bulk metal borides and coatings [5-8].

The bulk ZrB_2-based ceramics can be manufactured by powder metallurgy methods, such as spark-plasma sintering [9-11] or hot pressing [10]. However these processes are not economically advantageous, because for their implementation the large expenditures of electricity and expensive equipment (furnaces) are required. In connection with this, self-propagating high-temperature synthesis (SHS) for the production of boride ceramics has become increasingly widespread [9,12]. One of the important advantages of the SHS method is the self-cleaning effect of the combustion products during the synthesis process [13,14].

The magnetron sputtering is one of the most common methods of coating deposition in vacuum. The main advantage of magnetron sputtering is that a coating of the desired composition can easily be deposited onto the surface of different substrates [15-19]. Coatings deposited by magnetron sputtering do not significantly change the substrate geometry [20,21], posses low defect density, and a dense homogeneous structure with uniform elemental distribution through the coating thickness [22]. Method is useful in the deposition of multicomponent coatings [23,24].

This work is devoted to the study of the ZrB_2, ZrSiB, and ZrAlSiB bulk ceramic electrodes and coatings produced by their sputtering in the magnetron deposition system. Special attention was paid to the investigation of kinetics and mechanism of the combustion during SHS, mechanical properties of the deposited coatings.

2. Experimental part

Cathodes were manufactured by SHS technology using powders of zirconium (grade PTsrK-1) with an average particle size in range of 10-15 μm, aluminium (grade ASD-1) with a dispersion of about 50 microns, amorphous boron (grade B-99A) with an average particle size ~ 0.1 μm, and silicon (powder was prepared by grinding KEF-4.5 (100) monocrystals) with grain size < 45 μm. The powders were mixed in a stainless steel ball mill using the hard-alloy grinding bodies. To prevent self-ignition of zirconium powder, mixing was performed in isopropyl alcohol medium. Drying of the finished mixtures was carried out in a vacuum oven at a temperature of 50 °C. Table 1 shows the test compositions of powder mixtures.

Table 1. Compositions of powder mixtures.

Composition	Concentration							
	Zr		Si		B		Al	
	Weit.%	At.%	Weit.%	At.%	Weit.%	At.%	Weit.%	At.%
1	81.0	33.3	-	-	19.0	66.7	-	-
2	64.7	20.4	20.0	38.7	15.3	40.9	-	-
3	75.9	40.0	11.7	20.0	6.8	30.0	5.6	10

Composition №1 was calculated to form a 100% ZrB_2 phase. Composition №2 was calculated assuming complete chemical interaction of zirconium with boron and the formation of 80% ZrB_2, as well as 20% of free silicon. Composition №3 was calculated assuming the formation of ZrB_2 diboride and ZrSi silicide in the 1:1 ratio and the addition of aluminium (10 at.%) in analogy to [23]. Calculation of the adiabatic combustion temperatures T_{ad} of these mixtures and the equilibrium phase compositions of the combustion products at the initial temperature $T_0 = T_r$ (293 K) was carried out using "THERMO" software under the assumption of adiabatic process [14]. The macrokinetic characteristics of the combustion process were studied in a home-made SHS reactor using the methods described in [25,26]. Cylindrical briquettes with a diameter of 10 mm and a height of 18 mm with a relative density of 55-60% were formed from the powder mixtures. The combustion temperature T_c was measured with W-Re thermocouples installed in the holes of a cylindrical sample. The combustion rate Uc was measured by the method of high-speed video recording using a "Panasonic" WV-BL600 video camera at a 15-fold magnification of the image. Based on the experimental dependences T_c vs T_0 and U_c vs T_0, the values of the effective activation energy E_{eff} of self-sustained reactions were calculated by the method described in [15], which allowed us to assume the limiting stage of the combustion process. Bulk ceramic were obtained by forced SHS-pressing technology [14]. Hydraulic press DA-1532B and "sand" reaction mold were used. Synthesized samples were subjected to polishing and electroerosion cutting to get cathodes 120 mm in diameter and 8-10 mm in height.

The phase composition of the SHS products was determined by X-ray diffraction (XRD) analysis using monochrome Cu-Kα radiation. Vickers hardness of bulk ceramic samples was measured on an HVS-50 instrument at a load of 10 kg. The density was evaluated by the hydrostatic weighing method on the "AND1 GR-202" analytical balance. The true density of the compact samples was determined using "AccuPyc 1340" Micromeritics helium pycnometer. The residual porosity was calculated from the relative density values, which was found to be the ratio of the hydrostatic (experimental) to the true density of the bulk material.

The ceramic ZrB_2, ZrSiB, and ZrAlSiB cathodes were subjected to the DC magnetron sputtering (current 2 A, voltage 500 V) in pure argon (99.993 %) using a UVN-2M vacuum unit. The principal scheme of the vacuum chamber configuration has been shown elsewhere [27]. Polished plates of polycrystalline alumina (VK-100-1 trademark, JSC Polykor), 2x3 cm^2 in size, were used as a substrate material. Samples were mounted on a 2-axis rotating table equipped with bias voltage supply system, heater, and thermo-couple for temperature control. Before coating, the substrates were ultrasonically cleaned in isopropyl alcohol for 5 min using an UZDN-2T machine operating at a power of about 100 W and a frequency of 22 kHz, after which they were etched for 20 min in a vacuum chamber by Ar^+ ions with an average energy of about 2 keV using slit-type ion source, 180 mm in length. The distances form the target and ion source to substrate were 9 and 13 cm, respectively. The total pressure was maintained at 0.1-0.2 Pa. During deposition, the negative substrate bias voltage was kept constant at -250 V, the temperature was controlled in the range of 350-400 °C, and the deposition time was kept constant at 40 min.

The coating elemental compositions were determined from the elemental depth profiles obtained by glow discharge optical emission spectroscopy (GDOES) using a PROFILER 2 instrument (Horiba Jobin Yvon) [28]. The microstructure of coatings was examined by a S-3400N Hitachi scanning electron microscope equipped with a Noran 7 Thermo energy dispersive X-ray spectrometer. X-ray diffraction (XRD) patterns were recorded using a D8 Advance X-ray diffractometer (Bruker) using CuKα monochromatic radiation. The coatings were characterized in terms of their hardness (H),Young's Modulus (E) and elastic recovery (W) using a Nano-Hardness Tester (CSM Instruments) equipped with a Berkovich diamond indenter tip under an applied load of 5 mN. The detail description of method is presented in [27].

3. Results and Discussion

T_0 study of kinetics and mechanism of combustion the experimental dependences of T_c and U_c on the T_0 were obtained (Figure 1). While the graphs for compositions №2 and №3 were successfully constructed the data for compound №1 was not obtained because the value of T_c at T_0 = 298 K was 2900 K, which exceeds the recommended measuring range of the W-Re thermocouple (2773 K). According to the literature data the values a T_{ad} of ZrB_2 formation are in the range 3200-3300 T_0 [14,29]. It is worth considering that often the experimental T_c values are below the calculated values of T_{ad} in average 150 T_0 [14], which is associated with thermal losses the heating of the environment and the local heterogeneity of the starting mixture. Thus experimental values of T_c can be in the range 3000-3200 K.

An increase in the initial temperature T_0 for compositions №2 and №3 leads to linear growth of T_c and U_c. This indicates that the stage of chemical transformations in the combustion wave does not change with increasing T_0. For composition №2, an increase in T_0 by 300 K leads to an insignificant increase in T_c of about 100 K (Figure 1a). In this case, the most sensitive characteristic to the change in T_0 is U_c [15], the values increase by an order of magnitude (Figure 1b). This is probably due to the largest amount of liquid phase due to the melting of all the initial reagents (Zr, Si and B) in the temperature range 2268÷2403 K. Therefore, it can be assumed that the motive power of the combustion process in this multicomponent reaction mixture is the formation of the Zr-Si-B melt.

For composition №3, an increase in T_0 by 400 K leads to a linear growth of T_c by approximately 200 K (Figure 1a), and U_c increases by a factor of three. In this case, like the driving force of the combustion process is the formation of the melt, while all the initial reactants (Zr, Si, B and Al) also melt, liquid-phase interaction occurs. According to the calculation results, the E_{eff} were 63 and 225 kJ/mol for compositions №2 and №3, respectively. The values obtained indicate the limiting role of liquid-phase interaction processes. Note that according to the literature data the E_{eff} value for ZrB_2 is 150 kJ/mol [14] in the temperature range of 2600-2800 K or 300 kJ/mol for 2100-3100 K [14].

The results of XRD analysis of compact ceramic samples are presented in Table 2.

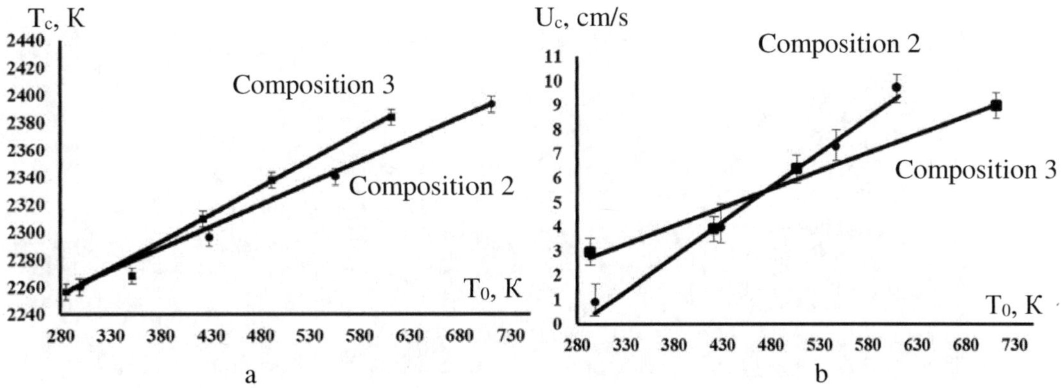

Figure 1. Dependences of the combustion temperature T_c (a) and rate U_c (b) of compositions №2 and №3 on the initial process temperature T_0

Table 2. Results of XRD analysis of the bulk samples

Composition	Concentration, Weit.%							
	ZrB_2	Si	$ZrSi_2$	ZrO_2	ZrSi	$ZrSiAl_2$	Al	ZrB_{12}
1	94	-	-	3	-	-	-	3
2	66	6	26	2	-	-	-	-
3	39	-	5	-	46	7	3	-

The sample №1 contains 94 Weit.% of ZrB_2 phase. Formation of ZrB_{12} can be due to the presence of a fraction bound to ZrO_2 in the initial Zr powder, i.e. locally zirconium interacts with an excess of boron to form ZrB_{12}. In sample №2 the stable $ZrSi_2$ phase is formed, whereas metastable ZrSi phases are not detected. The phase composition of sample №3 includes ZrB_2, ZrSi, $ZrSi_2$ and the ternary compound $ZrSiAl_2$. At the same time, their concentrations are 39, 46, 5 and 7%, respectively. Also in the structure there is 3% free Al. The presence of zirconium silicide $ZrSi_2$ is probably associated with a high conversion depth. When the sample is cooled slowly after synthesis, the monosilicide ZrSi is saturated with silicon from the zirconium depleted melt.

The SEM and EDS data correlate well with the XRD results (Figure 2). The main structural component in sample №1 is ZrB_2 phase, represented by rounded grains of the order of 10-20 μm in size (Figure 2a). On the microstructures of the sample №2 (Figure 2b), light faceted ZrB_2 grains 1-5 μm in size and a gray matrix of $ZrSi_2$ and silicon are distinguishable. Zirconium oxide ZrO_2 is presumably concentrated mainly in the pores. Figure 2c shows the microstructure of sample №3. The ZrB_2 grains with an average size of 2-3 μm are distributed in a matrix of monosilicide ZrSi. Additional $ZrSi_2$ and $ZrSiAl_2$ phases were detected.

Table 3 presents some of the physical and mechanical properties of synthesized bulk ceramic samples.

Table 3. The physical and mechanical properties of the bulk samples

Composition	Density, g/cm^3	Residual porosity, %	Hardness, GPa
1	4,5	5,8	9,6
2	4,6	2,7	14,2
3	5,3	3,0	17,1

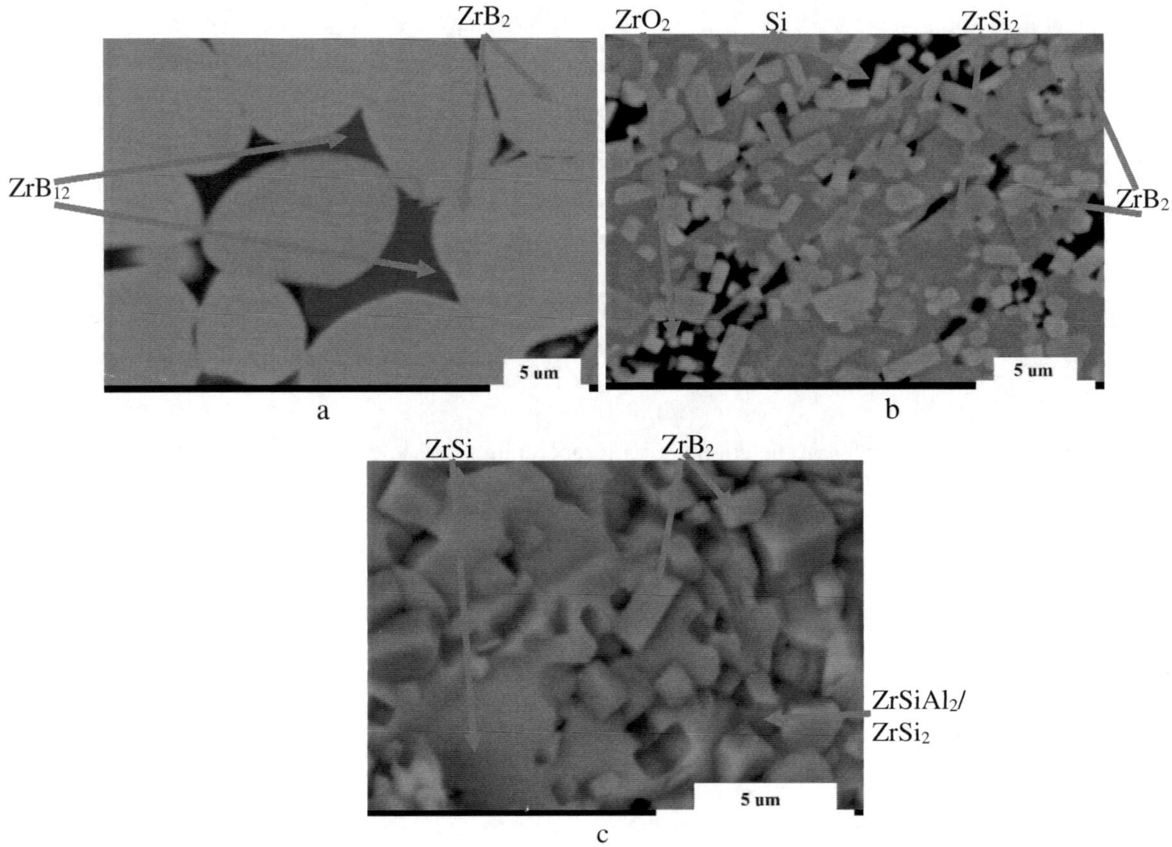

Figure 2. Microstructure of the bulk samples №1 (a), №2 (b), №3 (c)

Samples have low residual porosity and are characterized by a sufficiently high hardness comparable to the hardness of structural ceramics. Note that in early work [30] the 100 %ZrB_2 cathodes were obtained by cold isostatic pressing with subsequent sintering at a temperature of 1500°C. Density of products was limited by 3.96 g/cm^3, and a residual porosity was about 35 %. In [31] compact ceramics ZrB_2- $ZrSi_2$ obtained by hot pressing (30 MPa, 1850 °C) showed a residual porosity in range 1-6 %.

The SEM-images of microstructure of coatings is shown in Figure 3. It's possible to note a formation of very dense structure without coarse columnar elements. The coating thicknesses estimated from the SEM-images are 3, 5, and 4 μm for ZrB_2, ZrSiB, and ZrAlSiB, respectively.

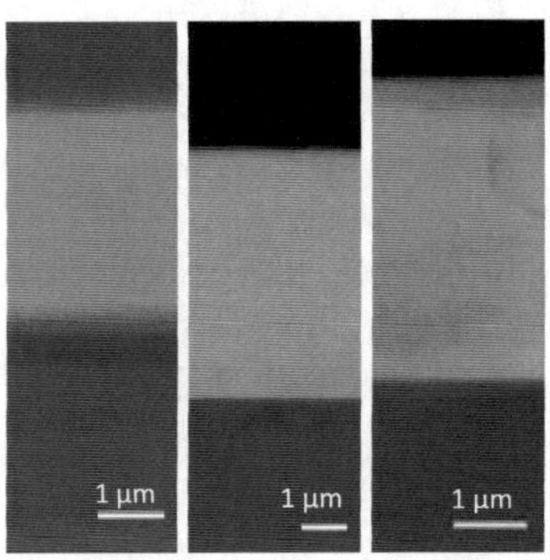

(a) (b) (c)

Figure 3. Microstructure of coatings №1 (a), №2 (b) and №3 (c)

The main structural component of ZrB_2, ZrSiB and ZrAlSiB coatings is the hexagonal phase of ZrB_2 (Figure 4). Some lines of ZrB_2 phase coincide with the lines from the Al_2O_3 substrate, but the positions at angles 2θ: 25.2; 32.5; 41.5, 51.5 (JCPDS 89-3930), accurately indicate the presence of the h-ZrB_2 phase in these coatings. The preferrent orientation of ZrB_2 grains corresponds to the crystallographic direction (101), which is typical for coatings of similar systems [32]. In the ZrB_2 coating, the crystallite size of h-ZrB_2, calculated by the Debye-Scherrer formula, is 8 nm. It is known that silicon is a strong amorphizer. As a result of silicon doping, the size of crystallites of the h-ZrB_2 phase in the ZrSiB and ZrAlSiB coating decreases down to 2 nm.

The hardness of the ZrB_2 coating is 37.4 GPa (Table 4), which exceeds the previously known values from the literature data [2,33,34]. Also, the resistance to elastic and plastic deformation of fracture in the ZrB_2 coating proved to be the highest. The hardness of ZrSiB and ZrAlSiB coatings is almost 2 times lower than that of ZrB_2 (see Table 4). Usually the addition of silicon to the coatings increases their hardness [35], but there is some value of the silicon content in the coating, after exceeding which the reverse effect is observed. Reduction of hardness with increasing content of silicon in the coating can be explained by a reduction in the size of the grains. There is a critical grain size d_c of 10 nm, at which the maximum hardness of the material is achieved, and any deviations in either direction ($d<d_c$ or $d>d_c$) lead to a decrease. This fact is called the Hall-Pitch effect, which explains the low hardness, in comparison with ZrB_2, silicon containing coatings. Decreasing of the mechanical properties after Al-alloying was described in [23] dedicated to the MoSiB and MoAlSiB coatings. Calculation of H/E and H^3/E^2 values (Table 4) may be useful for predicting the behavior of coatings under conditions of tribological contact and establishing inclination to inhomogeneous deformation 36-38].

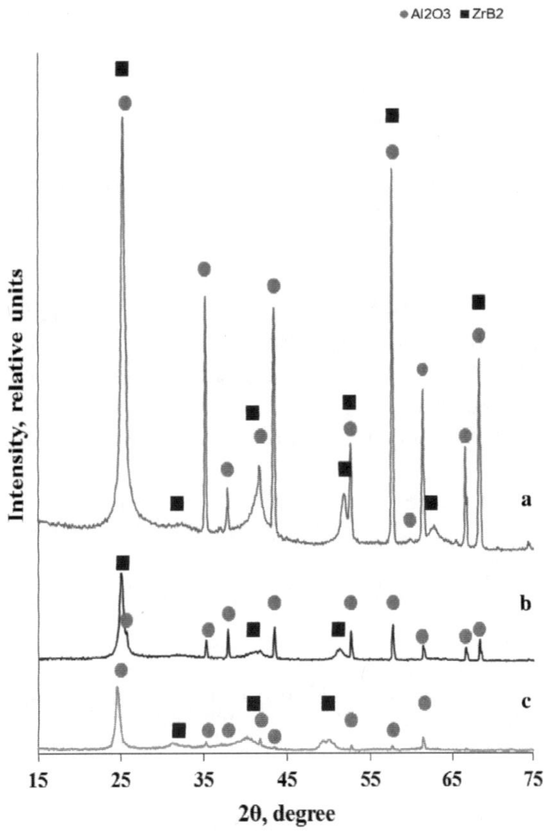

Figure 4. X-ray diffraction pattern of coatings №1 (a), №2 (b), and №3(c)

Table 4. Mechanical properties of coatings.

Composition	Mechanical properties				
	H, GPa	E, GPa	W, %	H/E	H^3/E^2, GPa
1	37.4	399.4	73	0.094	0.33
2	22.2	255.8	60	0.090	0.17
3	16.0	227.0	47	0.070	0.08

4. Conclusion

The kinetics and mechanisms of the combustion process in the Zr+B, Zr+Si+B, and Zr+Al+Si+B powder mixtures have been studied. The dependence of the combustion temperature and rate on the initial temperature of the reaction mixtures, as well as values of effective activation energy were determined.

The main structural components of the bulk ceramic sample №1 are ZrB_2 and an insignificant amount of ZrB_{12} boride. The sample №2 contains the phases of ZrB_2, $ZrSi_2$ and Si. The ZrB_2, ZrSi, $ZrSi_2$ and the ternary $ZrSiAl_2$ compound were detected in the sample №3. The obtained bulk ceramic samples have a low residual porosity (2.7-5.8%) and high hardness (10-17 GPa).

The ceramic ZrB_2, ZrSiB, and ZrAlSiB cathodes obtained using force SHS-compacting technology were successfully used for the magnetron sputtering. The main component of the deposited coating was hexagonal ZrB_2 phase. The addition of silicon or silicon and aluminum to the coatings resulted in a decrease in the crystallite size of the ZrB_2 phase from 8 to 2 nm. All coatings had a dense and homogeneous structure. The greatest mechanical properties (hardness of 37.4 GPa) were shown by the ZrB_2 coating.

Acknowledgments

The authors gratefully acknowledge the financial support from the Federal Target Program "Investigations and Implementations over Priority Directions of the Development of the Scientific and Technology Complex of Russia for 2014–2020" of Ministry of Education and Science of the Russian Federation (Agreement No. 14.578.21.0227, unique project identifier RFMEFI57817X0227).

References

[1] Sonber J.K., Murthy T.S.R. Ch., Subramanian C., Kumar Sunil, Fotedar R.K., Suri A.K. 2011 *Int. J. Refract. Met. Hard Mater.* **29** 21-30

[2] Wang Tie-Gang, Liu Yanmei, Zhang Tengfei, Kim Doo-In, Kim Kwang Ho. 2012 J. *Mater Sci. Technol.* **28 (11)** 981-991

[3] Neuman Eric W., Hilmas Gregory E., Fahrenholtz William G.2017 *Ceram. Int.* **43 (9)** 6942-6948

[4] Saunders Theo, Grasso Salvatore, Reece Michael J. 2015 *J. Alloys Compd.* **653** 629-635

[5] Zeman P., Zuzjaková Š., Mareš P., Čerstvý R., Zhang M., Jiang J., Meletis E.I., Vlček J. 2016 *Ceram. Int.* **42 (4)** 4853–4859

[6] Grančič B., Mikula M., Roch T., Zeman P., Satrapinskyy L., Gregor M., Plecenik T., Dobročka, E., Hájovská, Z., Mičušík M., Šatka A., Zahoran M., Plecenik A., Kúš P. 2014 *Surf. Coat. Technol.* **240** 48–54

[7] Paternoster Carlo, Fabrizi Alberto, CecchiniRaimondo, Spigarelli S., Kiryukhantsev-Korneev Ph.V., Sheveyko A. 2008 *Surf. Coat. Technol.* **203 (5-7)** 736-740

[8] DeJun L.I., YuanBin KANG, Lei DON G, XiangYun DENG, Hao LIU, XueLiang SUN. 2010 *Sci. Chin.a Tech. Sci.* **53 (3)** 772-775

[9] Musa C., Licheri R., Orrù R., Cao G. 2013 *Eurasian Chemico-Technological J.* **15 (2)** 117-126

[10] Zhang X., Xu L., Du S., Liu Ch., Han J., Han W. *2008 J. Alloys Compd.* **466** 241-245

[11] Locci, A.M., Orrù, R., Cao, G., Munir, Z.A. 2006 *J. Am. Ceram. Soc.* **89 (3)** 848-855

[12] "Concise Encyclopedia of Combustion Synthesis: History, Theory, Technology, and Products", Eds I. Borovinskaya, A. Gromov, E. Levashov, Yu. Maksimov, A. Mukasyan, A. Rogachev, *Elsevier*, 2017, 466 p.
[13] Levashov E.A., Rogachev A.S., Kurbatkina V.V., Maksimov M., Yukhvid V.I. 2011 *Perspective materials and technologies of self-propagating high-temperature synthesis (textbook).* p 377
[14] Rogachev A.S., Mukasyan A.S. 2015 *(New York: Taylor and Francis)*
[15] Levashov E.A., Mukasyan A.S., Rogachev A.S., Shtansky D.V. 2017 *Intern. Materials Reviews. **62** (4) 203-239*
[16] Fu Wenbin et al. 2016 *Rare Metal Mater. and Eng.* **45(10)** 2543-2548
[17] Mário Vančo et al. 2016 *Procedia Engineering* **136** 341-345
[18] Sukhorukova I. et al. 2017 *Biomed. Mater.* **12(3)**:035014
[19] Shtansky et al. 2011 *Surf. Coat. Technol.* **206** 1188–1195
[20] Wang Ming'e et al. 2016 *Rare Metal Materials and Engineering* **45 (12)** 3080-308
[21] Chu C.W.et al. 2009 *Thin Sold Films* **517** 5197-5201
[22] Ph.V. Kiryukhantsev-Korneev et al. *Russian Journal of Non_Ferrous Metals*, 55
[23] Ph.V. Kiryukhantsev-Korneev et al. 2017 *Corros. Sci.* **123** 319–327
[24] Shtansky et al. 2011 *Surf. Coat. Technol.* **205** 4640
[25] Levashov E.A., Pogozhev Yu.S., Potanin A.Yu., Kochetov N.A., Kovalev D.Yu., Shvyndina N.V., Sviridova T.A. 2014 *Ceram. Int.* **40** 6541-6552
[26] Pogozhev Yu.S., Potanin A.Yu., Levashov E.A., Kovalev D.Yu. 2014 *Ceram. Int.* **40** 16299–16308
[27] Levashov E.A. et al. 2010 *Russian Metallurgy (Metally)* **10** 917–935
[28] Kiryukhantsev-Korneev Ph.V. et al. 2012 *Protection of Metals and Physical Chemistry of Surfaces* **48** 585
[29] Erdem Camurlu H., Maglia Filippo. 2009 *J. of the Europ. Cer. Soc.* **29** 1501–1506
[30] Liu Pengchuang, Zhang Pengcheng, Pang Xiaoxuan, Wang Qingfu, Liu Tingting. 2012 *Procedia Engineering. 27 1305 – 1312*
[31] Guo S.Q., Kagawa Y., Nishimura T., Tanaka H. 2008 *Scripta Mater.* **58** 579–582
[32] Niu Yaran , Wang Hongyan, Huang Liping , Li Hong , Liu Xuanyong , Zheng Xuebin, Ding Chuanxian. 2014 *J. Therm. Spray Technol.* **23(3)** 479-476
[33] Mitterer C., Losbichler P., Werner W.S.M., Störi H., Barounig J. 1992 *Surf. Coat. Technol.* **54/55** 329-334
[34] Tengdelius Lina, Lu Jun, Forsberg Urban, LiXun, Hultman Lars, Janzén Erik, Högberg Hans. 2016 *J. Cryst. Growth* **453** 71-76
[35] Nanostructure Coatings edited by Cavaleiro Albano and Hosson Jeff T. De *(Moscow, Technosphere)* p 752
[36] Leyland A., Matthews A. 2000 *Wear.* **246** 1–11
[37] Levashov E.A. et al. 2013 *Mater. Sci. Eng.* **A 570** 51-62
[38] Shtansky D.V. et al. 2002 *Thin Solid Films.* **420/421** 330-337

IOP Conf. Series: Materials Science and Engineering **347** (2018) 012029 doi:10.1088/1757-899X/347/1/012029

Environmentally safe oil-field reagents for development and operation of oil-gas deposits

A V Fakhreeva[1,2], D A Manaure[1,3], V A Dokichev[1,2], A I Voloshin[4], A G Telin[5], Yu V Tomilov[6] and N E Nifantiev[6]

[1]Ufa State Aviation Technical University, Karl Marks st. 12, Ufa, Bashkortostan, 450077 Russia
[2]Ufa Institute of Chemistry, Russian Academy of Sciences, Oktyabrya pr. 71, Ufa, Bashkortostan, 450054 Russia
[3]Venezuelan Institute of Scientific Researches
[4]OOO RN-UfaNIPIneft',. Sochinskaya st 12, Ufa, Bashkortostan, 450103 Russia
[5]OOO Ufa-NTC, Aksakov st. 59, Ufa, Bashkortostan, 450076 Russia
[6]Zelinskii Institute of Organic Chemistry, Leninskii pr. 47, Russian Academy of Sciences, Moscow, 119991 Russia

Email: alsu.allagulova@mail.ru

Abstract. Sodium-carboxymethylcellulose and arabinogalactane inhibits the crystallization of calcium carbonate from a supersaturated aqueous solution at 80°C. The sizes of formed crystals $CaCO_3$ in the presence of arabinogalactane, sodium-carboxymethylcellulose and neonol AF 9-10 decrease on an average 7-12 μm and a change of their structure. It is expected, that the mechanism of inhibitionis in specific adsorption polysaccharides and neonol on occurring crystalline surface of the calcium carbonate, both at the expense of electrostatic interaction of functional groups with Ca^{2+} ions, located on the surface of the crystal, and due to coordination and hydrogen bonds with oxygen atoms and HO-groups of additives. Oil-water emulsion rheology in the presence of neonol AF 9-10 has been studied. It is shown that neonol AF 9-10 decrease viscosity natural water-oil emulsion by 25 times. Addition of 5% neonol to water-oil emulsion leads to formation more than 20 stable emulsion forms of different density and composition. New highly effective "green" oilfield reagents have been developed on the basis of neonol and natural polysaccharides.

1. Introduction

The world's explored oil reserves contain approximately 70% of heavy and extra-heavy oils, what is about 810 billion tonnes [1,2]. Extraction and transportation of high viscosity hard-to-recover oil requires development of new economically feasible technologies and reagents ensuring multi-phase flow flotation and crude oil viscosity reduction which represent water-oil emulsion of a w/o type, which are stabilized by lipophilic natural surfactants which are present in oil – asphaltenes, resins, naphtenic acids, etc. [3]. Among lots of reagents regulating the rheological properties of high viscosity oils, the most widely used are surfactants of various structure [4]. One type of perspective for this purpose surfactants is polyoxyethylated alkylphenols (neonols) whose effectiveness is not affected by mineralization of water [5-8].

Compatibility with other reagents, interaction with formation water and influence on the properties of oil is essential in using oilfield reagents. Scaling on downhole oilfield equipment and bottom-hole formation zone is one of the most common types of complications in the technological processes of development and operation of oil and gas deposits [9,10]. Formation of slightly-soluble salts in bulk is the reason for stabilization of water-oil emulsions and decline in efficiency of demulsifiers used in oil treatment. The most common salt deposits include the natural mineral $CaCO_3$.

Results on the effect of neonol AF 9-10 on the rheology of natural water-oil emulsion Severo-Komsomolskoye oilfield was used to develop methods of controlling its viscosity. We investigated the properties of carboxymethyl cellulose, arabinogalactan and neonol on the scaling of calcium carbonate. The obtained results are used to create new "green" chemicals for oilfield chemistry.

2. Experimental part

A sample of water-oil emulsion of the Severo-Komsomolskoye oilfield was investigated. Anhydrous crude oil and water, separated from the emulsion, had the following physical and chemical characteristics: oil density – 0.9464 g/cm^3 at 20°C; density of separated water – 1.003 g/cm^3; water cut determined by the Dean and Stark method – 31.4%. Oil blend composition: wax – 7.0%, asphaltenes – 0.7%, resins – 4.3%. Reservoir water of the Severo-Komsomolskoye oilfield belongs to the calcium-chloride type and has the following characteristics: salinity – 13.2 g/l;; pH – 8.1; ionic composition – Ca^{2+} – 210.4; Mg^{2+} – 79.0; $Na^+ + K^+$ – 4662.9; Cl^- – 7927.9; HCO_3^- – 292.8; SO_4^{2-} – 7.4 mg/l.

Neonol AF 9-10 ("Nizhnekamskneftehim" OJSC), arabinogalactan ("Ametis" JSC) and sodium carboxymethylcellulose (Sigma-Aldrich) were used in this study. Water of the following ionic compositions was used as a reservoir water model: Ca^{2+} – 270.55; Mg^{2+} – 30.35; Na^+ – 3120.25; Cl^- – 4492.10; HCO_3^- – 1525.50 mg/l, prepared from analytically pure salts $CaCl_2$, $MgCl_2{\cdot}6H_2O$, NaCl, $NaHCO_3$.

The viscosity and rheological characteristics were determined with the rotary viscometer Haake Viscotester iQ. For the explored samples a shear test was used with the shear rate varying from 1 to 300 s^{-1}, at which the dependence of shear stress and viscosity on shear rate was determined. The influence of neonol AF 9-10 on the water-oil emulsion stability was studied using the integrated tool set for dispersion stability measuring "Turbiscan Tower" (Formulaction SA). The efficiency of the neonol AF 9-10, arabinogalactan and sodium carboxymethylcellulose as scale inhibitors was determined at a temperature of 80°C and concentrations of 10, 30 and 50 mg/l by the method, based on blocking of a 0.8 mm thick and 1.5 m long steel capillary with the occurring calcium carbonate deposits during pumping of the reservoir water model [11]. The sizes of resulting $CaCO_3$ crystals were recorded using the laser particle size analyzer Analysette 22 NanoTec plus.

3. Results and discussion

The reviews [12–14] consider interrelation between composition and structure, structural features of emulsions, which determine their rheological properties.

The object of our study was the natural water-oil emulsion of the Severo-Komsomolskoye field with viscosity of 1234 mPa·s at water cut 31,4%. Anhydrous crude oil has the API value of 19.5, calculated with the formula $141.5\times\rho_w/\rho_o$ - 131.5, where ρ_w = 1.003 g/cm^3 - reservoir water density, and belongs to heavy oils according to classification of the American Petroleum Institute [15]. Microscopic analysis of the water-oil emulsion has shown that it is a "packed" in membranes emulsion of a water-in-oil type, containing water drops mainly of sizes less than 135 μm (Figure 1). This emulsion is stable, obviously, the stabilization due to the formation of the "armoring" layers of molecules of asphaltenes, resins and paraffin crystals at the interface of water and oil phases. It is well known that these oil components increase the stability of emulsions, creating a protective barrier around the water droplets [3, 16,17].

IOP Conf. Series: Materials Science and Engineering **347** (2018) 012029 doi:10.1088/1757-899X/347/1/012029

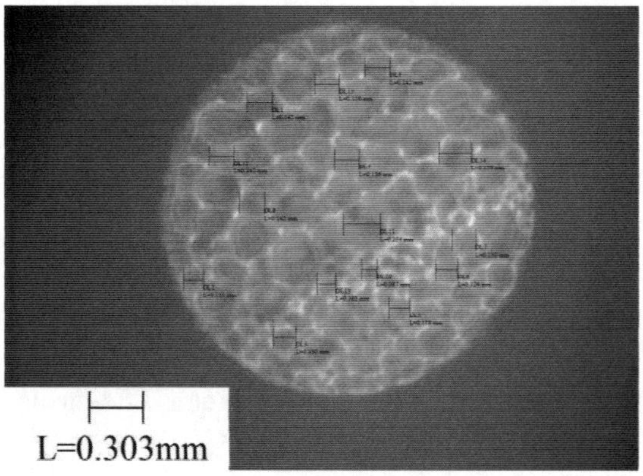

Figure 1. The Micrograph of crude oil

The conducted analysis of effective viscosity curves of shear rate has shown that in the temperature range of -10 – +50 °C flowing of the crude oil obeys the Newton's law, i.e. the emulsion behaves like a Newtonian fluid (Figure 2).

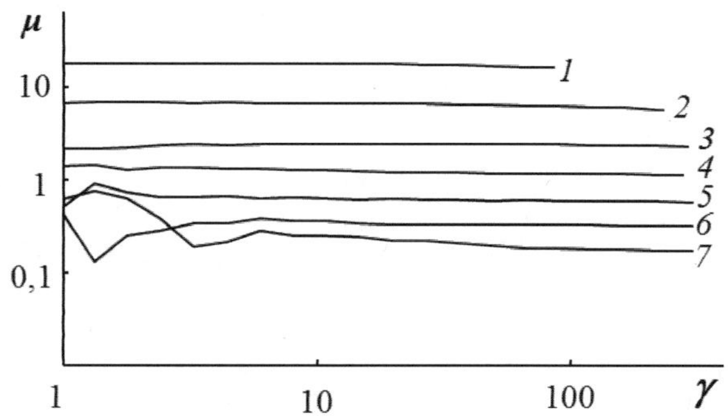

Figure 2. Dependence of crude oil effective viscosity on shear rate at T = -10 (*1*), 0 (*2*), 10 (*3*), 20 (*4*), 30 (*5*), 40 (*6*) and 50°C (7). μ – effective viscosity (Pa·s), γ – shear rate (1/s).

It should be noted that at low shear rates and temperature above 30°C a non-monotonic change in viscosity is observed, what is apparently connected with the breakdown of crystal structure of paraffins (melting) and structural changes in asphaltenes "armoring" layer at the oil/water phase interface, since asphaltenes, being diphilic substances, can form hard 2-10 nm thick films on the surface of water droplets, consisting of resins and asphaltenes [3,18]. At temperature above 40°C and shear rate of 9 s^{-1} effective viscosity does not depend on shear rate and approaches to an asymptote, what is connected with the breakdown of spatial coagulation-crystallization structure of high-molecular oil compounds.

The effect of neonol AF 9-10 on crude oil viscosity was studied at concentrations of 0.5, 1, 3 and 5%. Based on the results of the conducted rheological studies of oil with neonol AF 9-10 addition it was established that the best viscosity decline is observed in samples with surfactant concentration of 3%. Effective viscosity at 20°C in the explored samples decreases by 25 times (Figure 3).

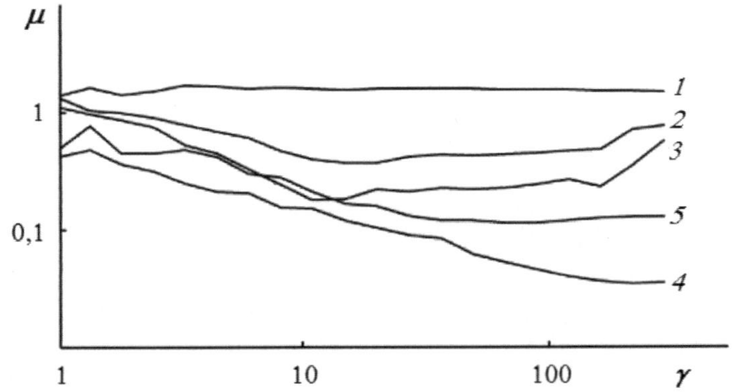

Figure 3. The influence of neonol AF 9-10 on dependence of crude oil effective viscosity on shear rate at concentrations of 0 (*1*), 0,5 (*2*), 1 (*3*), 3 (*4*) and 5% (*5*) at 20°C. μ – effective viscosity (Pa·s), γ – shear rate (1/s).

Containing 5% surfactant the oil viscosity increases what is connected with possible emulsification and stabilization of water-oil emulsion by neonol. Strongly pronounced non-monotonic change in oil effective viscosity in the presence of neonol AF 9-10 is due to, apparently structural changes in interfacial layer at the water/oil phase interface as a result of introduction of surfactant molecules into structures formed by asphaltenes and resins molecules and formation of various emulsion forms, including micelles, since critical concentration of micelle formation of neonol AF 9-10 is 0.082 mmol/l [19].

The curve of dependence of water-oil emulsion viscosity logarithm in the presence of 3% neonol AF 9-10 on reciprocal absolute temperature in coordinates lnμ – *1/T* shows breaks at temperatures of 10 and 23.6 °C (Figure 4). Temperature dependencies in the temperature ranges 0 – +10°C, +10 – +20°C and +25 – +50°C obey the Arrhenius equations $\ln(\mu) = 2461.3 \cdot T^{-1} - 10.70$, $\ln(\mu) = 15100 \cdot T^{-1} - 53.25$ and $\ln(\mu) = 7427 \cdot T^{-1} - 26.18$, respectively. Activation energy of viscous flow in the temperature range 0 to +10°C is 62.0 kJ/mole, from +10 to +20 °C – 126 kJ/mole and from +25 to +40 °C – 20.5 kJ/mole. At temperatures of 10 and 23.6 °C the water-oil emulsion structure changes from less ordered to ordered (high activation energy) to the least ordered. Low value of effective activation energy is apparently connected with a parallel process of destruction of hydrocarbon bonds of water with oil components (the process ends at temperature of 35 – 40°C). Viscosity values at 35 and 40°C are corrected from approximation of the flow curve by the Bingham equation $= K\gamma^n$

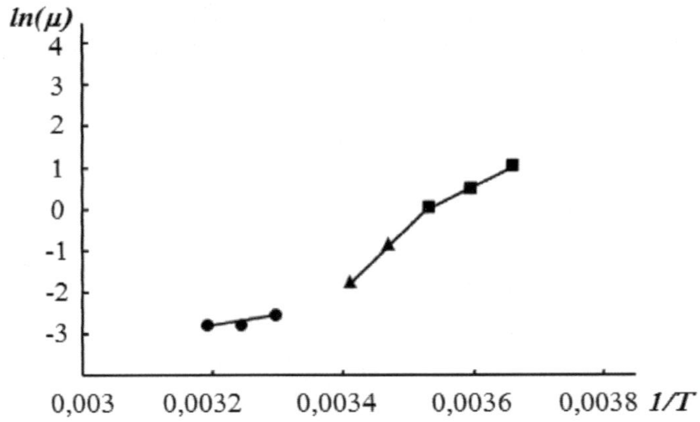

Figure 4. Dependence of crude oil effective viscosity on temperature in the presence of 3% neonol AF 9-10. ln(μ) – effective viscosity logarithm, *1/T* – reciprocal temperature (1/K).

The results of studying the water-oil emulsion stability dynamics in gravitational field, obtained by laser scanning with Turbiscan Tower device [20] have shown that at surfactant concentration of 3-5% during 20-30 h more than 20 stable emulsion forms of various densities are formed in water-oil emulsion without separation of water phase. The variety of emulsion forms is due to formation of various packings of water droplets in oil as a result of neonol AF 9-10 adsorption on the interface and its interaction with oil components during formation of solvate layers of the dispersed phase.

The study of the effect of neonol AF 9-10, arabinogalactan and sodium carboxymethylcellulose (NaCMC) on the scaling process (calcium carbonate crystallization) by capillary blocking at temperature of 80°C has shown that at concentrations of 50, 20 and 30 mg/l, respectively, the reagents inhibit $CaCO_3$ scaling processes and their efficiency makes ~90-98%. The capillary does not occur deposition of calcium carbonate crystals by passing saline water during the experiment.

Data on the formed $CaCO_3$ crystals size distribution by the action of neonol, arabinogalactan and sodium carboxymethylcellulose, obtained with the laser particle size analyzer has shown that in their presence a decrease in the average size of crystals by ~7-12 μm is observed.

Electron micrographs of $CaCO_3$, formed in the presence of NaCMC, indicate polymorphism of calcium carbonate (Figure 5). The sample, obtained in the absence of polysaccharide, contains elongated aragonite crystals, which is the most thermodynamically stable phase under the specified conditions. During $CaCO_3$ crystallization in the presence of NaCMC formation of aragonite is practically not observed. The crystals are irregular-shaped particles with rounded faces.

a) b)

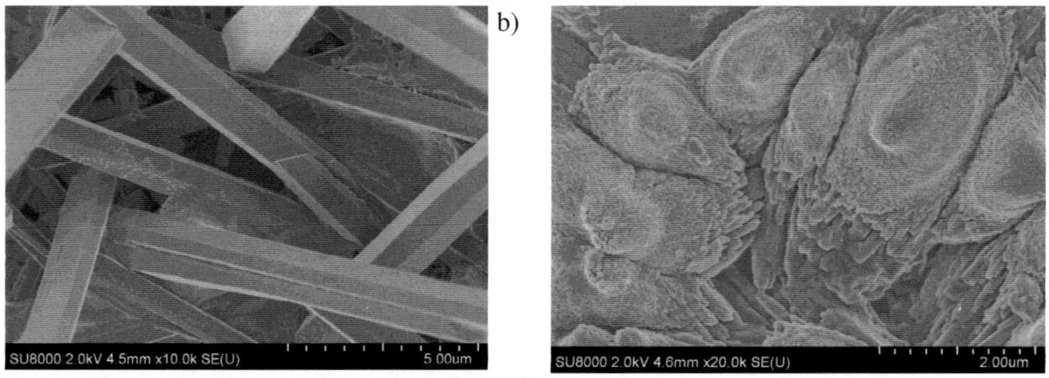

Figure 5. Electron micrograph of $CaCO_3$ crystals, obtained without (a) and in the presence of (b) NaCMC

Sedimentation stability of the formed aqueous dispersed $CaCO_3$-NaCMC systems was studied at 80°C and NaCMC concentrations of 10, 20, 30, 50 and 100 mg/l using the Turbiscan Tower device. In the absence of NaCMC a finely dispersed suspension of crystals, which settle as a result of aggregation and agglomeration [21], is formed and clarification of the solution is observed (Figure 6). In this case stability index of the TSI dispersed system has the highest value for the entire observation time.

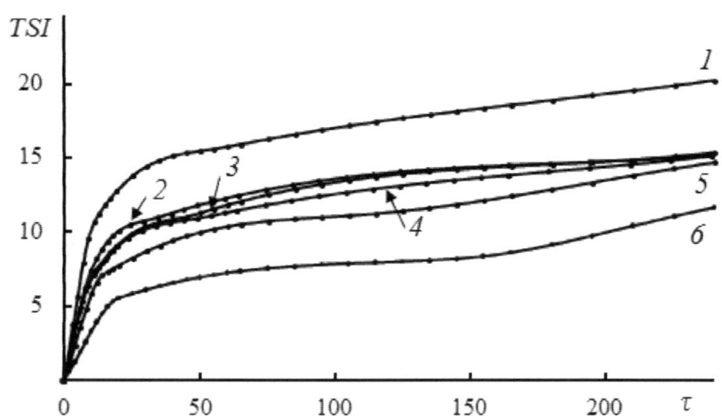

Figure 6. The effect of NaCMC on the $CaCO_3$ scaling process at concentrations of 0 (*1*), 10 (*2*), 20 (*3*), 30 (*4*), 50 (*5*), 100 (*6*) mg/l at 80°C. TSI –Turbiscan stability index, τ – time (min).

At NaCMC concentrations more than 10 mg/l the TSI value decreases, indicating an increase in stability of NaCMC-$CaCO_3$ and NaCMC-$CaSO_4$ dispersed systems. Within three hours formation and precipitation of $CaCO_3$ crystals did not occur, which illustrate high efficiency of NaCMC as a scale inhibitor.

The main reason for the effect of neonol, arabinogalactan and NaCMC on crystal formulation of calcium carbonate may be considered to be their specific adsorption on the emerging faces of calcium carbonate both due to electrostatic interaction of ionized carboxyl groups with Ca^{2+} ions, located on the crystal surface, and due to coordinate and hydrogen bonds with oxygen atoms and HO-groups of carbohydrate fragments [22,23]. In particular, NaCMC macromolecules in aqueous medium have conformation of elongated coils with sizes of hypothetically straight-line sections of 13-18 nm (approximately 25 glucose cellulose monomers), corresponding to sizes of the formed calcium salts nanoparticles [24]. Such interaction leads to formation of polymer forms without distinct morphological features and to changes in calcium carbonate particles size. According to the mode of action, these reagents can be referred to the threshold-action scale inhibitors, since their specific adsorption on the emerging faces of microcrystalline calcium carbonate nucleus, which leads to slowing of the crystals growth and their suspension in solution at concentration above the deposition level.

The study of the effects of neonol, arabinogalactan and sodium carboxymethylcellulose on the process of calcium carbonate crystallization inhibition has shown that they are promising as environmentally safe inhibitors of calcium carbonate scaling.

4. Conclusion

It was stated that neonol AF 9-10 leads to a significant decrease in effective viscosity of the natural water-oil emulsion of the Severo-Komsomolskoye oilfield, inhibition of calcium carbonate crystallization and, along with the “green” scale inhibitors – arabinogalactan and sodium carboxymethylcellulose, is an effective oilfield reagent.

Acknowledgement

The results were obtained within the framework of the state task of the Russian Ministry of Education and Science (project No. 4.2703.2017/ПЧ).

References

[1] Alboudwarej H, Felix J and Taylor S 2006 *J. Oilfield Rev.* **18 №2** 34-53
[2] Danilova E 2008 *The Chem.J.* **12** 34–37
[3] Zadymova N M, Skvortsova Z N, Traskin V Yu et al. 2016 *Colloid J.* **78 №6** 735–46
[4] Zhang J, Chen X, Zhang D and Xu J 2017 *J of Pet. Sci.and Eng* **156 №7** 563–74
[5] Langevin D, Poteau S, Henaut I and Argillier J F 2004 *J. Oil Gas Sci. Technol.* **59 №5** 511–24
[6] Lakatos I and Lakatos-Szabo J 1997 *J. Colloid and Polymer Science* **275 №5** 493–501
[7] Mosayebi A, Angaji M T and Khadiv-Parsi P 2016 Recovery, Utilization, and Environmental Effects *J. Energy Sources* A **38. №19** 2891–97
[8] Romero N, Cardenas A, Henriquez M and Rivas H 2002 Physicochemical and Engineering Aspects *J. Colloids and Surfaces* A **204 №1-3** 271–84
[9] Crabtree M, Eslinger D, Fletcher P et al. 2002 *J. Oilfield Rev.* **11 № 3** 30–45
[10] Ishmiyarov E R, Rakhimova N T, Latypova D R et al. 2015 *Russ. J. of Applied Chem.* **88 №7** 1174–77
[11] Ishmuratov F G, Koptyaeva E I, Voloshin A et al 2015 *Russ. J. of Applied Chem.* **88 №12** 1920–22
[12] Derkach S R 2009 *Adv. Colloid Interface Sci.* **151 №1-2** 1–23
[13] Foudazi R, Qavi S, Masalova I and Malkin A Ya 2015 *J. Adv. Colloid Interface Sci.* **220** 78–91
[14] Malkin A Ya and Kulichikhin V G 2015 *J. Russ. Chem. Rev.* **84 №8** 803–25

[15] Conaway C 1999 *The Petroleum Industry: A Nomenclature Guide* (Tulsa: Pennwell Publishing Company) p 289
[16] McLean J D and Kilpatrick P K 1997 *J. of Colloid and Interface Sci.* **196** 23-34
[17] Li M, Xua M, Ma Y et al. 2002 Physicochemical and Engineering Aspects *J. Colloids Surfaces* A **197** 193–201
[18] Alvarado V, Wang X and Moradi M 2011 *J. Energies* **4** 1058–86
[19] Tikhova A A, Glukhareva N A and Kolesnikova E N 2010 *J. Sci. Bulletin Nat.Sci.* **№ 21 (92) 13** 127–130
[20] Mengual O, Meunier G, Cayre I et al 1999 Physicochemical and Engineering Aspects *J. Colloids and Surfaces* A **152** 111–23
[21] Linnikov O D 2014 *J. Russ. Chem. Rev.* **83** 343–64
[22] Backfolka K, Lagergeb S, Rosenholmc J and Eklundd D 2002 *J. Colloid Interface Sci.* **248** 5–12
[23] Shestak I V, Vorob'ev P D, Cherednichenko D V et al. 2011 *Russ. J. Inorg. Chem.* **56 №2** 176–80
[24] Bikles N and Segal L 1971 *Cellulose and Cellulose Derivatives* (New York: Wiley)

IOP Conf. Series: Materials Science and Engineering **347** (2018) 012030 doi:10.1088/1757-899X/347/1/012030

Effect of sintering methods and temperatures on porosity of the ceramics from aluminum oxinitride

D V Prosvirnin[1], A G Kolmakov[1], M D Larionov[1], M E Prutskov[1], A S Alikhanyan[2], A V Samokhin [1], A S Lysenkov[1] and D D Titov[1]

[1] Baikov Institute of Metallurgy & Materials Science, Moscow, Leninsky prospekt 49, Russia
[2] Kurnakov Institute of General and Inorganic Chemistry of the Russian Academy of Sciences, Moscow, Leninsky prospekt 31, Russia

Abstarct. The paper presents the results of studies of the effect of temperature regimes and time on porosity in ceramic samples made of aluminum oxynitride. Getting rid of the porous structure allows reducing the scattering of rays and, as a result, achieving the required optical characteristics.

1. Introduction
Ceramic materials have a wide range of applications from the construction industry to the aviation industry. However, not many ceramic materials have optical properties. Optimum combination of optical and high strength properties is possessed by ceramics based on aluminum oxynitride $Al_{23}O_{27}N_5$ [1]. The preparation of such ceramics consists of several stages. The first stage is synthesis of powder materials [2]. The second stage is compacting (pressing), specifies the form of future samples. Next, the compacted preform is sintered. The final stage is mechanical processing, grinding and polishing. However, in each stage, precise technological parameters must be observed to obtain the required qualities of the finished ceramic product. Requirements are given for the chemical purity of the powders, as well as for the particle size, of 1-10 μm. Sintering of the powder is required to be carried out in a vacuum, at temperatures about 2000-2100 ° C, for 10-12 hours [3,4].

To reduce the sintering temperature and porosity, a variety of sintering additives, such as yttrium oxide Y_2O_3 [6], are used. Also, hot isostatic pressing (HIP) method is used to combine two processes, sintering and pressing. This technology also reduces the porosity and thus increases the optical and mechanical properties [7].

In this paper, tests of samples from a ceramic material, aluminum oxynitride $Al_{23}O_{27}N_5$ are presented and an analysis of the data is presented.

2. Materials and techniques
The material for creating ceramic samples was powdered materials Al_2O_3 and AlN, obtained by the method of plasma-chemical synthesis. The parameters of powders are given in Table. 1.

Table 1. Characteristics of the starting powders

Initial powder	Chemical purity,%	Average particle size, μm
Al_2O_3	98,0	0,1
AlN	98,0	10,0

The powders were mixed in planetary micro furniture in a ratio of 40:60 for 50-60 minutes. The process was carried out in isopropyl alcohol to prevent the agglomeration of the powder particles and the abrasion of the drum walls by grinding bodies made of zirconia. The resulting mixture was subjected to uniaxial pressing to obtain blanks for further sintering.

The sintering process was carried out according to two schemes. Samples of type I were obtained by uniaxial pressing of a mixture of powders followed by reaction sintering. The pressing was carried out in a mold at a pressure of 50 MPa. Sintering was carried out in a vacuum chamber for 30-120 minutes at temperatures of 1700-1800 ° C. In this case, the chamber was subjected to two consecutive cycles of "nitrogen purge" - evacuation to a residual pressure of 10^{-2} ... 10^{-3} mm of mercury. The pressed billets were placed in a boron nitride crucible.

Samples of type II were obtained by hot pressing a mixture of powders at temperatures of 1700-1900 ° C and pressure of 50 MPa for 12 minutes. The mixture of powders was placed in a graphite form, whose walls were covered with a coating of boron nitride on an alcohol base to prevent the effect of carbon. The obtained samples of both types were subjected to grinding and polishing.

Tescan Vega scanning microscope was used for testing and research of compacts and powders. The XRD method was used to analyze the phase composition on a Bruker D8 ADVANCE diffractometer. The microhardness test was carried out on a Wolpert Wilson instruments 402mvd device at a load of 500 g and a test time of 10 seconds. Strength characteristics were determined on a universal test machine INSTRON 3382 in a three-point bending scheme. Measurements of the velocity of longitudinal ultrasonic waves were carried out according to standard procedure using an ultrasonic flaw detector from Panametrics, EPOCH-4, at frequency of 10 MHz using a scheme of combined transducers. The velocity of the longitudinal ultrasonic waves V was calculated from the formula V = 2d / t, where d is thickness of the sample, t is propagation time of ultrasonic signal at the measurement point, and the attenuation coefficient of longitudinal ultrasonic waves was calculated from the relation (i.e., (1)), where A_1 and A_2 are the amplitudes of two consecutive ultrasonic pulses.

$$\alpha = \left[\frac{1}{2d}\right] ln\left(\frac{A_1}{A_2}\right) \quad (1)$$

3. Results and its discussion

The results of the described technological operations are samples of ceramics measuring 65 mm in diameter and 10-15 mm in height (Figure. 1).

Figures 1b and 1c show images of samples obtained by hot pressing technology. The sample has light spots of 1-3 mm in size. To eliminate the identified shortcomings, a number of additional sinterings were carried out at elevated temperatures.

Unfortunately, the samples did not show the expected optical properties. Such results are associated with insufficient sintering time and low temperature. These factors did not allow to get rid of porosity. The surface of the broken sample shows the presence of pores and the size difference of some crystals, 1-5 μm (Figure 2).

It was suggested that this result is related to the sintering temperature. At temperature of 1900 ° C, only the homogeneity region is achieved, while for the healing of pores, a liquid phase is required, which correlates with the phase diagram of this system (Figure. 3).

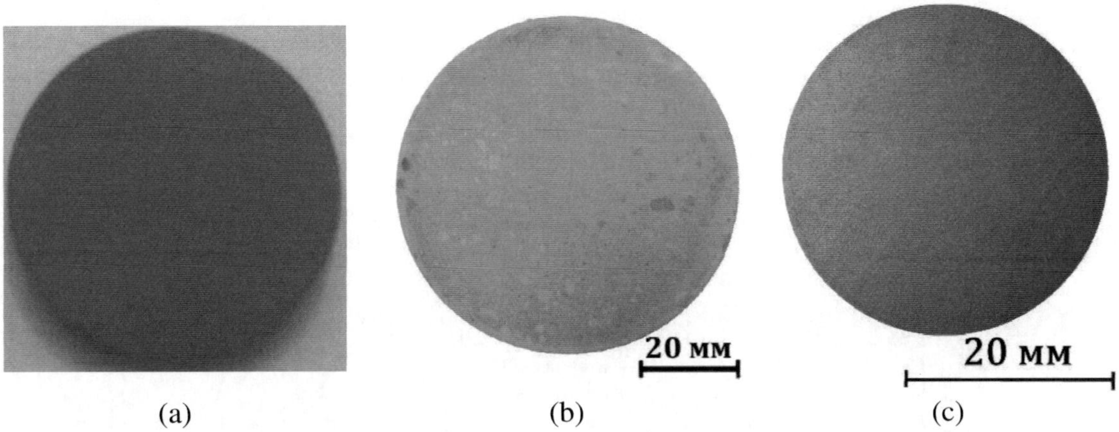

Figure 1(a)**.** A sample obtained by the technology of reaction sintering. Type I.
(b). A sample obtained by hot pressing technology at a temperature of 1700 ° C. Type II.
(c). A sample obtained by hot pressing technology at a temperature of 1900 ° C. Type II.

Figure 2. Surface of a broken specimen with 3-point bending

A possible solution to this problem is addition of yttrium oxide Y_2O_3, of the order of 1.5% by weight or increase in the sintering temperature and holding time.

X-ray phase analysis of samples after sintering of mixture of aluminum oxide and aluminum nitride powders by hot pressing technology at temperatures 1700°-1900°C (samples type II) showed the presence of aluminum oxynitride and alumina in the samples (Figure 4).

The analysis showed the presence of the $Al_{23}O_{27}N_5$ phase and not a large number of Al_2O_3 phases, as well as aluminum oxycarbide and aluminum carbonitride. These data indicate that the conversion process of oxynitride phase was not completed. The effect of carbon impurities from the coating was also affected.

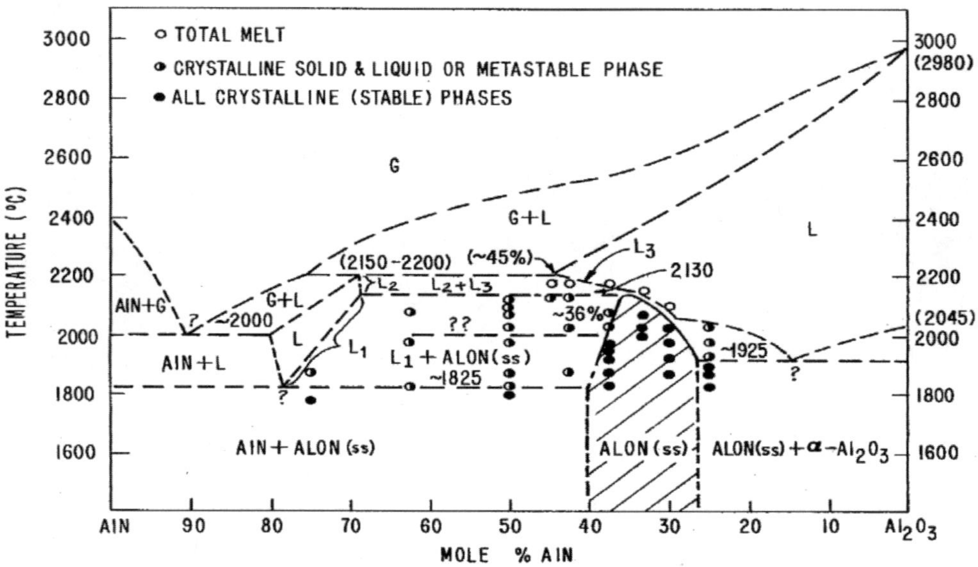

Figure 3. Phase diagram AlN-Al_2O_3 [8]

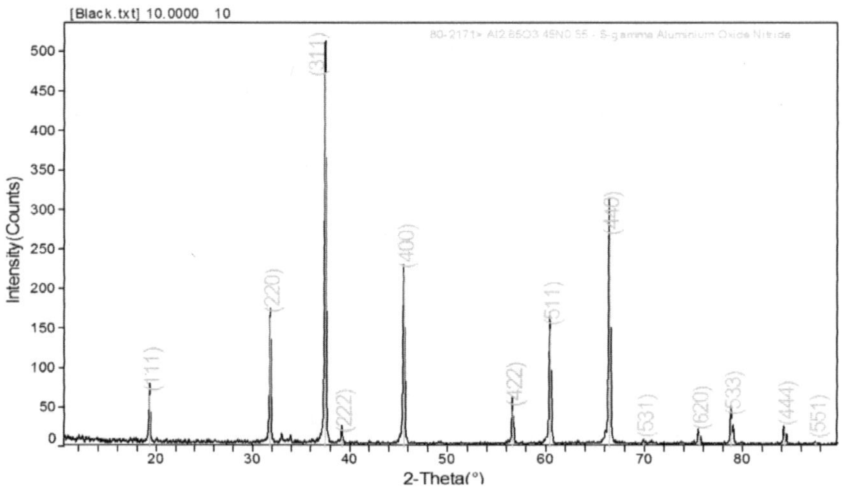

Figure 4. Diffractogram of the sample type II

Ultrasonic method was used to study the speed of sound in the samples. Measurements of the velocity of longitudinal ultrasonic waves were carried out according to a standard procedure using an ultrasonic flaw detector from Panametrics, EPOCH-4, at a frequency of 10 MHz using a scheme of combined transducers. The velocity of the longitudinal ultrasonic waves V was calculated from the formula $V = \frac{2d}{t}$, where d is thickness of the sample, t is propagation time of the ultrasonic signal at the measurement point, and the attenuation coefficient of the longitudinal ultrasonic waves was calculated from the relation (i.e., (1)), where A_1 and A_2 are the amplitudes of the two consecutive ultrasonic pulses. The results are shown in Table 2.

Because of that, the higher attenuation coefficient and low sound propagation velocity in the sample, it can be assumed that in samples of type I there is larger number of inhomogeneous regions (pores, defects, impurities). The values of the sound velocity and the attenuation coefficient obtained for Type II samples are close to those given in the literature.

Table 2. Physical and mechanical characteristics of samples

	$V_{average}$,m/s	$\alpha_{average}$, dB/mm	$HV_{0,3}$ MPa	σ_i, MPa	$K_{IC,}$ MPa/m^2
Type I:	9304±46	0,6±0,03	958±98,5	122,13±13,37	4,02±0,37
Type II:	10127±506	0,18±0,01	1335,8±144,4	138,52±11,17	4,51±0,21

The obtained samples showed characteristics similar to those given in the literature, such as hardness and fracture toughness, but the flexural strength showed lower values. Such indicators are associated with structural heterogeneity and defects in the sample [9].

4. Conclusions

The influence of sintering additives and temperature regimes on the formation of mechanical and optical properties of ceramics should be studied to achieve transparency. These data will help reduce the porosity in the preparation of the following samples and, accordingly, achieve optical properties.

Obtained samples by the methods of reaction sintering (I) and hot pressing (II) showed presence of the $Al_{23}O_{27}N_5$ phase, a small amount of phases of Al_2O_3, aluminum oxycarbide and aluminum carbonitride. The presence of these phases shows that the process of conversion of the oxynitride phase was not completed and as a result did not lead to the expected optical properties. Obtained microstructure and strength data indicate the presence of an inhomogeneous structure and structural defects and is possibly associated with an insufficient sintering temperature of 1700 -1900 ° C. An increase in temperature would lead to the appearance of a liquid phase and the healing of pores.

The way to achieve the required characteristics can be a method of liquid-phase sintering. The sintering is carried out in two stages, heating up to the temperature of existence of the liquid phase and sintering with a temperature of 50-100 °C below the liquid phase existence temperature, in the homogeneity region.

Hot pressing technology is more efficient, as fewer defects are achieved in the sample (pores, defects, impurities).

Acknowledgments

This work was partially supported by a grant from RFBR № 16-08-00815 A
The work was carried out according to the state task № 007-00129-18-00

References

[1]. Arzhakov M.S., Zhirnov A.E., Arzhakov S.A., Lukovkin G.M., Kolmakov A.G., Zabolotnyi V T 2015 Glass ceramic and polymer impact-resistant materials and protective constructions based on them *Russian Metallurgy (Metally),*. №10, Moscow pp.800-4.

[2]. Takao Y and Sando M 2005. Al- System Non- Oxide Spherical Powder Synthesis by Liquefied Petroleum Gas Firing. *Journal of the American Ceramic Society*, *88*(2). pp 450-2.

[3]. Bakas M P, Lillo T M and Chu H S 2010. U.S. Patent No. 7,833,922. *U.S. Patent and Trademark Office*. Washington, DC

[4]. Huang S et al. 2016 AlON phase formation in hot-pressing sintering Al2O3/AlN composites and their oxidation behavior. *Journal of Alloys and Compounds*. Vol. 685. – pp. 309- 15.

IOP Conf. Series: Materials Science and Engineering **347** (2018) 012030 doi:10.1088/1757-899X/347/1/012030

[5]. McCauley J W et al 2009 AlON: a brief history of its emergence and evolution *Journal of the European Ceramic Society.. – V. 29. – №. 2.* – pp. 223-36.
[6]. Jiang Hua Wei, et al. 2010. "Influence of Y2O3 additive on transparent of AlON ceramics." *Advanced Materials Research*. Vol. 105. Trans Tech Publications,
[7]. Chen, Feng, et al 2014 Microstructure and optical properties of transparent aluminum oxynitride ceramics by hot isostatic pressing. *Scripta Materialia 81* pp. 20-23.
[8]. McCauley J W and Corbin N D 1980. U.S. Patent No. 4,241,000.: *U.S. Patent and Trademark Office*. Washington, DC
[9]. Lysenkov A S, Timoshkin I A , Kargin, Y F, Titov D D, Fedotov A Y, Ashmarin A A and Baranchikov A E 2016 «Synthesis of aluminum oxynitride (AlON) and study of the properties of ceramics based on it» *Inorganic Materials: Applied Research,* Vol 7 pp. 517-19

IOP Conf. Series: Materials Science and Engineering **347** (2018) 012031 doi:10.1088/1757-899X/347/1/012031

Features of the structure of a TiB_2-based powder material obtained under self-propagating high-temperature synthesis and shear deformation

P M Bazhin[1], A M Stolin[1], A S Konstantinov[1], N I Mukhina[1] and A Pazniak[2]

[1] Merzhanov Institute of Structural Macrokinetics and Materials Science Academician Osipyan str. 8, Chernogolovka, Moscow Region, 142432, Russia
[2] National university of science and technology «MISiS», Leninskii ave. 4, Moscow, 119049, Russia

olimp@ism.ac.ru

Abstract. The results of an experimental study of TiB_2-based powder material obtained under the combination of SHS processes with shear deformation are presented. The effects of the rotor velocity and the delay time before shear deformation application upon the structure of the synthesized powder are studied. The grain structure of titanium diboride is shown to become predominantly round with particles size of 1-5 μm with increasing the rotor velocity from 120 to 600 rpm. At the same time, particles of 200-400 nm size can be observed on the surface of the agglomerates.

1. Introduction

Self-propagating high-temperature synthesis (SHS) is one of the most promising areas of modern materials science [1-3]. Products synthesized by this method are characterized by a complex of effective properties that are achieved by presence in the product structure phases with complementary physicochemical, mechanical and other characteristics. The main goal of SHS is traditionally to produce powder composite materials for manufacturing products of various functional purposes in powder metallurgy [4-6], such as protective coatings using methods of plasma, arc or detonation spraying, and others [7-9].

The wide range of possibilities to create new composite materials is opened by the use of the SHS process in combination with shear high-temperature deformation of the initial components and the resulting combustion products during the synthesis [10]. These conditions are realized in the SHS-grinding method. The essence of the method consists in grinding of the material obtained in a combustion wave under conditions of shear high-temperature deformation and application of external pressure. The implementation of the method is based on the ability of the hot mass of the synthesized product to macroscopic deformation.

It is shown that the deformation parameters, in particular the deformation rate, strongly influence both the size and morphology of the synthesized particles and powder phase composition [11]. This paper demonstrates that the intensity of shear deformation and time delay before deformation influence the structure of the synthesized material.

2. Objects and methods of research

Materials based on TiB_2 are chosen as the objects of research. The choice stemmed from well-known properties of titanium diboride such as high melting point, hardness, thermal conductivity, resistance to abrasion and aggressive media [12, 13]. Titanium diboride is widely used in modern technology, chemical equipment, and as a part of cermets for nuclear power engineering [14, 15]. Commercial powders of 68.6 wt% PTM titanium (>99.1%, 45 μm) and 31.4 wt% amorphous black B-99B boron (99%, 5 μm) are used by authors. The mixture is premixed in ball mill and dried in drying cabin at a temperature of 100 °C for 6 hours.

The experiments are carried out in a closed reactor without rotation of the rotor and at rotational velocitiess of 120, 240 and 600 rpm [11]. The rotor has conic form with the apex angle of 140°. After the combustion wave propagation in the SHS mode, the rotor with a specified frequency of rotation is dropped to the bottom base of the graphite beaker for 40-60 s until the structure transformations in the synthesized charge are complete.

For the structure studies of synthesized powder materials the equipment of the Distributed Center for Collective Use of ISMAN is used: the high-resolution field-scanning electron microscope Carl Zeiss Ultraplus (Germany), the powder X-ray diffractometer ARL X'TRA, and other certified equipment, methods and techniques.

3. Discussion

The results of scanning electron microscopy showed that the particles obtained by the SHS method without shear deformation are agglomerates of 1-3 mm size. The presence of this agglomerates of 1-3 mm in size require more time for the subsequent grinding process. The agglomerates contain titanium diboride fragments sintered together (figure 1). Most of the titanium diboride particles are crystallites from 1-3 μm to 9-30 μm. Rounded particles ranging from 1-3 μm to 25 μm were also found. The surface of the agglomerated particles is coated with a thin oxide film (figure 1b). The presence of oxide inclusions in the material, even in the amount that does not exceed 5% (since oxides are not observed on XRD), will adversely influence on further use of the material in powder metallurgy to produce nonporous products.

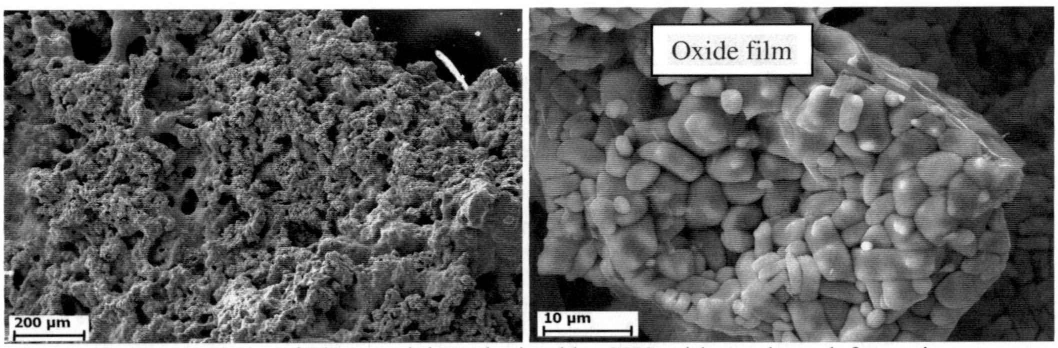

Figure 1. SEM of TiB_2 particles, obtained by SHS without shear deformation

The macrostructure of the material synthesized during SHS and shear deformation, regardless of the rotational velocity of the rotor, is analogous, and presented by agglomerated sintered titanium diboride particles. Material obtained at the rotor rotational velocities of 120 and 240 rpm is characterized by particles with comminuted structure (figure 2). The particle sizes are predominantly 5-10 μm. The particles of 200-400 nm can also be observed on the surface of the agglomerates, but the oxide films were not detected.

For the material obtained by SHS with shear deformation at rotational velocity of 600 rpm (figure 3), the grain structure is predominantly round. The grain sizes are approximately the same (1-5 microns), compared with the powder material obtained at low rotational frequencies of 120-240 rpm. The presence of carbon in the synthesized powder is associated with sample preparation procedure.

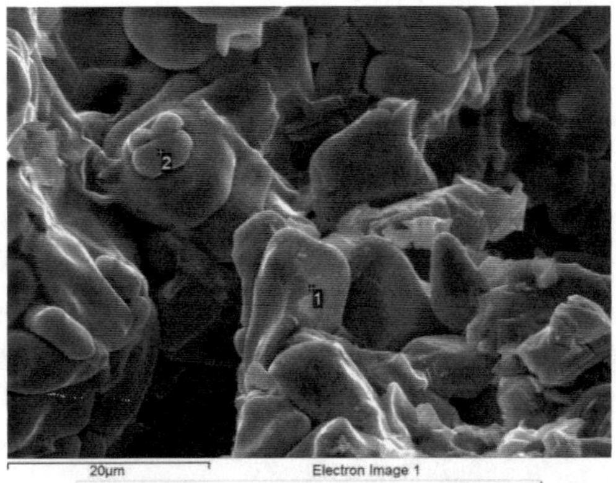

Spectrum	B	C	Ti
1	14.55	5.11	80.34
2	27.82	5.11	67.07

Figure 2. Microstructure and energy dispersive analysis of TiB_2 particles obtained by SHS method with shear deformation at 120 rpm

Spectrum	B	C	O	Ti
1	33.09	4.7	6.58	48.73

Figure 3. Microstructure and energy dispersive analysis of TiB_2 particles obtained by SHS method with shear deformation at 600 rpm

The influence of the time delay before the shear deformation implementation on the synthesized material microstructure at a rotor velocity of 240 rpm is studied (figure 4). The time delay is varied between 0 and 15 s with 5 s step. The countdown begins at the moment of the combustion wave passage and ends at the start of shear deformation. It is found that applying of shear deformation immediately after the combustion wave passage results in the formation of rounded particles with size up to 5-10 µm. With an increase of time delay, transformation of the titanium diboride grains occurs during cooling. At the 10 s time delay, the TiB_2 grains are characterized by comminuted structure, and at 15 s delay the grain form of the titanium diboride particles becomes predominantly parallelepiped. At long time delays before external effects implementation the grains grow dominantly in one direction and become mostly parallelepiped.

Shear deformation in SHS and subsequent cooling of material prevents the agglomeration of the formed particles with each other, and also leads to the destruction of the sintered titanium diboride particles. Intensive shear deformation of the powder prevents the enlargement of TiB_2 particles during cooling of the synthesized material.

Figure 4. The microstructure of TiB_2 particles obtained by the SHS method with mechanical influences at a rotor rotation velocity of 240 rpm with time delay: a-c 5, 10, 15 s respectively.

IOP Conf. Series: Materials Science and Engineering **347** (2018) 012031 doi:10.1088/1757-899X/347/1/012031

4. Conclusion

The structure of a powder material based on titanium diboride obtained under SHS conditions and shear deformation in a closed type reactor depends on the intensity and time delay before deformation. For the material obtained by the SHS method without deformation, titanium diboride particles are crystallites with sizes from 1-3 μm to 9-30 μm. With increase of the rotor rotation velocity from 120 to 600 rpm, the grain structure becomes predominantly round with a size of 1-5 μm. At low rotational velocities (120-240 rpm) particles have a comminuted structure with particle dominant size of 5-10 μm. At the same time, particles of 200-400 nm can also be observed on the surface of the agglomerates.

It is found that the shear deformation application immediately after the passage of the combustion wave leads to the formation of rounded particles with sizes up to 5-10 μm. With the increase in the delay time, transformation of the titanium diboride grains occurs in the process of cooling with onset of parallelepiped grain shape. At long delay times before implementation of external load, the grains grow in one direction and become predominantly parallelepiped.

Acknowledgements

The study is performed within the framework of the grant of the President of the Russian Federation No. MK-3213.2017.8.

References

[1] Levashov E A, Mukasyan A S, Rogachev A S and Shtansky D V 2017 Self-propagating high-temperature synthesis of advanced materials and coatings *Int. Mat. Rev.* **62** 203-39

[2] Merzhanov A G, Munir Z and Holt J 1990 B Self-propagating high-temperature synthesis: Twenty years of search and findings *Combust. and Plasma Synth. of High-Temp. Mater.* Eds. Z.A. Munir, J.B. Holt. N.Y., etc.: VCH Publ. 1-53

[3] Amosov A P, Luts A P, Latukhin E I and Ermoshkin A A 2016 Application of SHS processes for in situ preparation of alumomatrix composite materials discretely reinforced by nanodimensional titanium carbide particles *Russ. J. Non-Ferrous Met.* **57** 106-12

[4] Merzhanov A G and Rogachev A S 1992 Structural macrokinetics of SHS processes. *Pure and Appl. Chem.* **64** 941

[5] Borovinskaya I P, Akopdzhanyan T G, Ignatieva T I, and Chemagina E A 2017 Structure and phase formation of combustion products during the synthesis of gamma-AlON in self-propagating high-temperature synthesis *Russ. J. Non-Ferrous Met* **58** 405-10

[6] Bazhin P M, Stolin A M, Konstantinov A S, Kostitsyna E V and Ignatov A S 2016 Ceramic Ti—B Composites Synthesized by Combustion Followed by High-Temperature Deformation *Mater.* **9** 1-7

[7] Bazhin P M, Stolin A M, Chizhikov A P, Alymov M I and Kuznetsov D V 2016 Structure, Properties, and Use of Protective Cermet Coatings Prepared By Electric-Spark Alloying and Electric-Arc Hardfacing *Refract. Ind. Ceram.* **57** 1–6

[8] Panteleenko F I , Sarantsev V V and Stolin A M. 2011 Formation of Composite Coatings Based on Titanium Carbide via Electrospark Alloying *Surf. Eng. Appl. Electrochem.* **47** 328-37

[9] Kudryashov A E, Potanin A Yu, Lebedev D N, Sukhorukova I V, Shtansky D V and Levashov E A 2016 Structure and properties of Cr-Al-Si-B coatings produced by pulsed electrospark deposition on a nickel alloy *Surf. Coat. Tech.* **285** 278-88

[10] Bazhin P M, Stolin A M, Mikheev M V and Alymov M I 2017 Self-Propagating High-Temperature Synthesis under the Combined Action of Pressure and Shear *Dokl. Chem.* **473** 95–97.

[11] Bazhin P M, Stolin A M and Konstantinov A S 2017 The Impact of Mechanical Effects on Granulometric Composition of TiB_2 – Based Materials *Adv. Mater. Tech.* **3** 40-43

[12] Liu B, Liu Y, Xe X Y, Tang H P, Chen L F and Huang B Y 2007 Preparation and mechanical properties of particulate-reinforced powder metallurgy titanium matrix composites *J. Metall. Mater. Trans* A **38** 2825–31

[13] Leyens C A, Leyens C A and Peters M L 2003 Titanium and titanium alloys. *Fundamentals and Applications* (Weinheim: Wiley-VCH Verlag GmbH)
[14] Mroz C 1995 Titanium Diboride *Am. Ceram. Soc. Bull.* **74** 158-59
[15] Abkowitz S C 2004 Ti Discontinuously Reinforced Ti-Matrix Composites: Manufacturing, Properties, and Applications *J. Miner. Met. Mater. Soc.* **56** 38-41

IOP Conf. Series: Materials Science and Engineering **347** (2018) 012032 doi:10.1088/1757-899X/347/1/012032

Composite fibres based on cellulose and vinyltriethoxysilane: preparation, properties and carbonization

I S Makarov, L K Golova, M V Mironova, M I Vinogradov and V G Kulichikhin

A.V.Topchiev Institute of Petrochemical Synthesis, RAS, 119991, Leninsky prospekt, 29, Moscow, Russia

E-mail: makarov@ips.ac.ru

Abstract. For the first time the composite fibers based on cellulose with additives of vinyltriethoxysilane (VTEOS) have been obtained. The choice of the additive was justified by the chemical structure of the VTEOS, namely the Si-C links content and the low C/O ratio. Composite fibers were prepared from solid phase pre-solutions of cellulose with VTEOS in N-methylmorpholine-N-oxide (NMMO). An investigation of the rheological behavior of the filled cellulose solutions with VTEOS showed a slight effect of the additive on the viscosity properties of the system. Introduction of 5% of VTEOS to cellulose does not lead to significant structural changes and, as a result, mechanical properties of the fibers. The thermal behavior of composite fibers differs from cellulose fibers.

1. Introduction

Carbon fibers (CF) based on cellulose, remain irreplaceable to date due to a number of their unique functional properties, for example, small values of the thermal expansion coefficient. Improvement of the carbonization process includes using as additives chemical reagents of various types (organic, inorganic and silicon- or halogen-containing compounds, metal salts, etc.) affecting the carbonization process, i.e., playing a role of catalyst [1].

Along with the change in the mechanism of carbonization process, which manifests in the change of the DSC traces, the additives influence (as a rule, increase) the value of the carbon residue at producing CF.

From this point of view the additives work as fire retardants, and silicon-containing compounds are popular for this aim [2-4]. Silicon flame retardants are applied to the precursor fibers by impregnation. Unfortunately, this method does not allow to reach a uniform distribution of the additive in fibers – the surface is enriched, but the volume is depleted with active substance. As an alternative method to achieve a good distribution of the additive in the fiber volume is it introducing into the spinning solution (the dope) [5].

Since the rayon production in many countries is diminished because of ecological problems, the NMMO process, where the of cellulose dissolves in direct solvent - N-methylmorpholine-N-oxide (NMMO) becomes more and more popular [6]. That is why we have used these solutions for spinning cellulose fibers of Lyocell type.

Introduction of tetraethoxysilane (TEOS) into cellulose solutions in NMMO was proposed by our group and described in [7]. The viscosity of such solutions is slightly differed from that, typical for the neat cellulose solutions. This renders it possible to obtain composite fibers in conditions, previously defined for cellulose solutions. When introducing up to 10% TEOS into a cellulose matrix strength and elongation at break of fibers do not change practically, while the modulus of elasticity increases on 20-30%. It allowed us to conduct the further thermal treatment of composite precursors and to obtain carbon-silicon carbides fibers [8].

Compared with TEOS, vinyltriethoxysilane (VTEOS) is characterized by the presence of one Si-C bond and only three Si-O bonds that allows us to expect that lower oxygen content in precursor fibers could lead to increase of carbon yield at thermolysis of composite fibers due to formation of silicon carbide. Thus, the aim of this work consists in preparing composite fibers based on cellulose and VTEOS solutions in NMMO and studying their structure and properties.

2. Experimental

2.1.Materials

Cellulose with DP = 600 was received from Baikal Pulp and Paper Mill (Russia) (H_2O ~ 8%, alpha cellulose content not less than 94% (GOST 6840-78). N-methylmorpholine-N-oxide (MMO) ($T_{m.p.}$ 120-160°C, H_2O <10%) were supplied by Demochem (China). Triethoxyvinylsilane (VTEOS) ($H_2C = CHSi(OC_2H_5)_3$) (ID 24850510) and Propylgallate (ID 329757396) were from Sigma-Aldrich (USA).

2.2.Preparation of composite fibers

Composite fibers based on cellulose with VTEOS were spun by a dry-wet method on laboratory unit Rheoscope 1000 capillary viscometer (Ceast, Italy) from solutions prepared via solid-phase treatment of components according to the procedure described in [7]. The diameter of the capillary used was 1 mm, and the length-to-diameter ratio was 40.

2.3. Characterization

X-ray diffraction studies of the obtained fibers were carried out at room temperature in an apparatus with rotating copper anode (Rigaku, Japan) in the transmission mode. The beam radiation was CuKα with a wavelength $\lambda = 0.1542$ nm. To produce equatorial X-ray diffraction patterns, the parallel folded filaments (~ 100) were used. The morphology of the fibers was studied by the scanning electron microscope JSM U-3 (JEOL, Japan). The mechanical properties of the fibers were determined in Instron 1122 tensile machine (UK) on a 10 mm basis at a strain rate of 10 mm/min. Thermal behavior of the fibers was investigated on a TGA/DSC1 thermal analysis instrument, Mettler Toledo (Switzerland). The measurements were carried out in alumina crucibles of 70 μl in the temperature range from 30 to 1000°C at a heating rate of 10 K/min. The inert gas consumption (argon) was 10 cm^3/min. Carbonization of filaments was carried out in the operating cell of the TMA 402 F1 Hyperion device for thermal mechanical analysis (Netzsch, Germany) up to 1200°C.

Content of silicon in the samples was determined using a multielement CHNS/O analyzer "Thermo Scientific Flash 2000" (USA) and transmission electron microscopy Osiris (FEI, USA).

3. Results and discussion

The equatorial diffractograms of cellulose and composite fibers obtained in the "transmission" mode at room temperature are shown in Figure 1.

On both diffractograms, peaks are observed in the regions 2θ ~ 12-13° and ~ 20.5°, i.e. inherent for cellulose II. Change in the intensity ratio of the peaks belonging to cellulose and composite fibers indicates on of a less ordered structure in composite fibers.

The morphology of composite fibers does not differ from cellulose fibers (Figure 2).

IOP Conf. Series: Materials Science and Engineering **347** (2018) 012032 doi:10.1088/1757-899X/347/1/012032

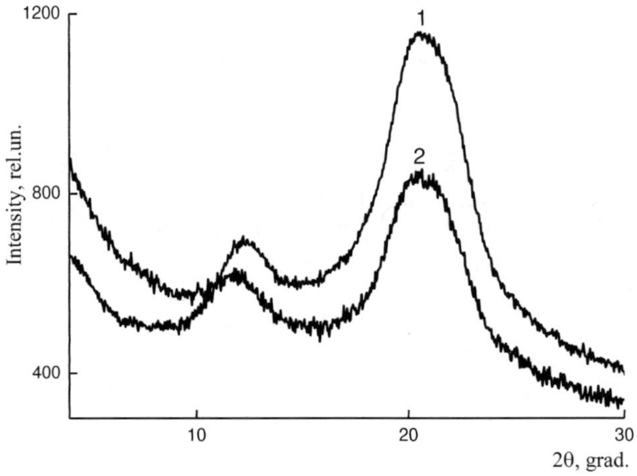

Figure 1. Equatorial diffractograms of cellulose (1) and composite fibers (2).

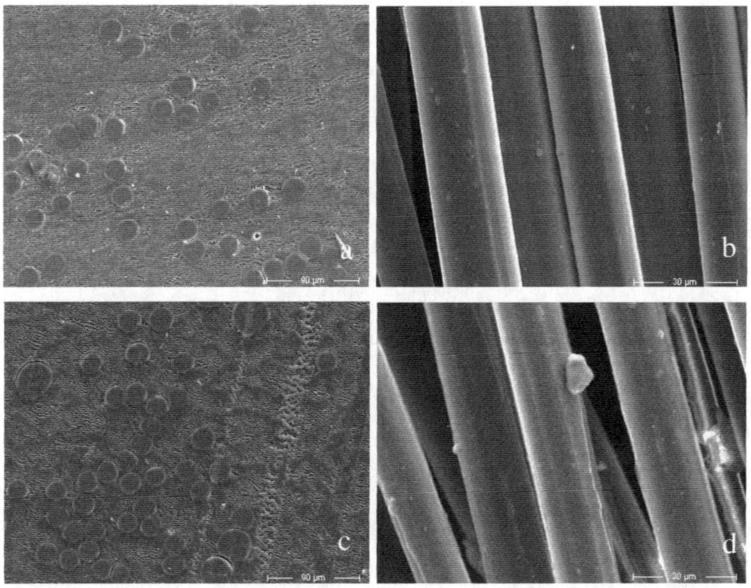

Figure 2. Micrographs of cellulose (a, b) and composite fibers (c, d) containing 90% cellulose and 10% VTEOS, cross-sections (a, c) and surface (b, d).

As can be seen from micrographs, the addition up to 10% VTEOS leads to preparing fibers with a circular cross-section and a smooth defect-free surface of fibers. The average fiber diameter is 15-20 μm.

Mechanical properties of composite fibers revealed that with increase in the VTEOS content to 10%, the strength is slightly decreased (~ 5%), and the values of the elongation at break fell by half, and values of the elastic modulus did not change (Table 1).

IOP Conf. Series: Materials Science and Engineering **347** (2018) 012032 doi:10.1088/1757-899X/347/1/012032

Table 1. Mechanical properties of cellulose and composite fibers.

	Ø, μm	σ, MPa	ε, %	*E*, GPa
100% cellulose	14-20	570-610	10-12	11-12
95% cellulose + 5% VTEOS	11-13	460-510	6-8	11-12
90% cellulose + 10% VTEOS	9-11	460-550	6-8	11-12

The composite precursor fibers were processed into carbon fibers. The morphology of the obtained carbon fibers obtained at a temperature treatment up to 1000°C is shown in Figure 3.

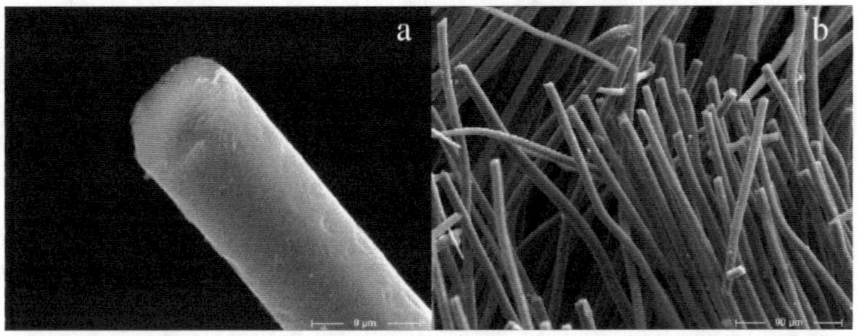

Figure 3. Micrographs of carbon fibers obtained from precursors containing 90% cellulose and 10% VTEOS, obtained by heat treatment in argon atmosphere up to 1000°C at different magnification.

The surface of carbon composite fibers is relatively smooth, but there exists "craters" of a small size (Figure 3). The cross-sectional pattern is uniform. The average diameter of the carbon fibers is ~ 6-9 μm. The distribution of the chemical elements in the resulting carbon fibers is shown in Figure 4.

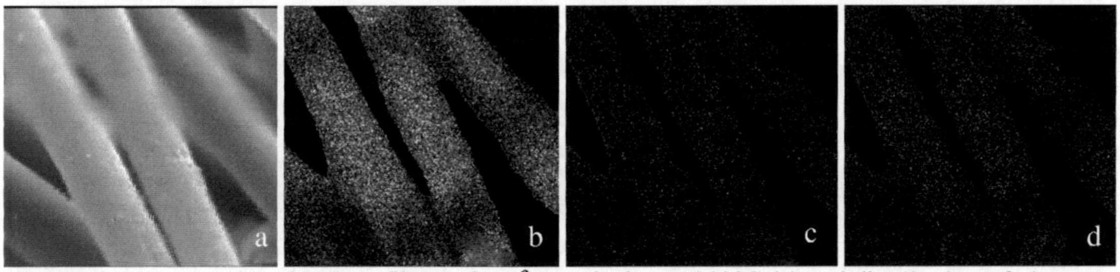

Figure 4. Micrographs of carbon fibers after thermolysis at 1000°C (a) and distribution of elements: C (b), O (c), Si (d) in these fibers.

It is seen from the micrographs (Figure 4) that C, O, Si are evenly distributed, and their absolute values were determined by TEM method and shown in the histogram (Figure 5).

Char yield obtained by heat treatment of composite precursor fibers up to 1000°C in 2-3 times higher than for neat cellulose fibers, meanwhile it first increases at small additive content, but further decreases. The observed decrease in the carbon residue at high level of additive is possibly associated with an increase in the oxygen concentration in the precursor fibers.

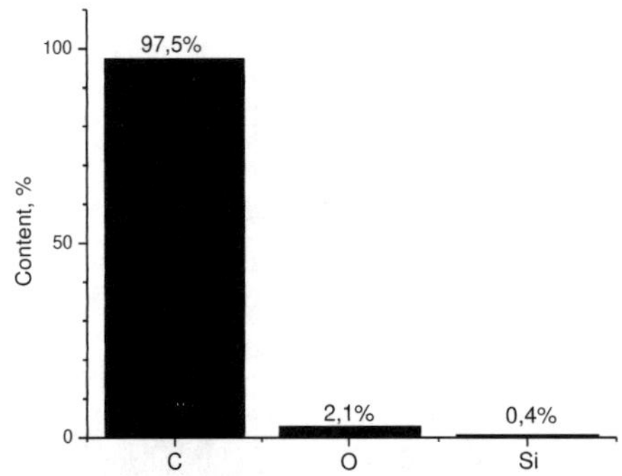

Figure 5. The content of C, O, Si in carbon fibers obtained from a precursor containing 95% cellulose and 5% VTEOS.

5. Conclusion

The introduction of VTEOS into the cellulose matrix at the stage of solution preparation renders it possible to obtain composite precursors with a good distribution of silicon in the volume of the fiber. The mechanical properties of the obtained composite fibers and their morphological quality are slightly differ from cellulose fibers, that allows to process them into carbon fibers with uniform distribution of chemical elements.

Acknowledgments
The reported study was funded by RFBR according to the research project № 16-33-60218 mol_a_dk.

References

[1] Efimova O S and Khokhlova G P 2009 *Chemistry for Sustainable Development* **17** 79-85
[2] Kazakov M E et.al. *Patent RF 2045472*, 1992.
[3] Trushnikov A M et.al. *Patent RF 2047674*, 1993.
[4] Olri P et.al. *Patent RF 2258773*, 2005.
[5] Wizon I, Robertson J A 1967 *J. Polym. Sci.* **19** 267-281.
[6] Golova L K, Makarov I S, Matukhina E V, Kulichikhin V G 2010 *Polym Sci Ser A+* **52 (11)** 1209-1219.
[7] Makarov I S, Golova L K, Kuznetsova L K, Bondarenko G N, Skvortsov I Yu, Mironova M V, and Bermeshev M V 2017 *Fibre Chem.* **49(2)** 101-107.
[8] Makarov I S, Golova L K, Bondarenko G N, Skvortsov I Yu, Berkovich A K, Bermeshev M V, Mironova M V 2017 *Fibre Chem.* **49(4)** in press.

IOP Conf. Series: Materials Science and Engineering **347** (2018) 012033 doi:10.1088/1757-899X/347/1/012033

Receiving finely divided metal powder by inert gas atomization

M A Smirnov, M A Kaplan and M A Sevostyanov

A.A. Baikov Institute of Metallurgy and Materials Science, Russian Academy of Sciences, Leninsky Prospect 49, 119991 Moscow, Russia

Email: mexfexpex@gmail.com

Abstract: In this study, the inert gas atomization method to obtain finely divided metal powder was studied. Two different configurations of inert gas atomizers were compared to identify the most appropriate one to obtain high spherical AM powder. Some problems of free-fall atomizer were discussed.

The method of obtaining small and medium-sized spherical shape metal powders using dispersing of melts called atomization [1]. This process occurs in special installations called atomizers. Spraying (atomization) is widely used in the production of powders of multicomponent alloys, with an amorphous structure that allows to achieve a uniform chemical composition, even with a content of doping components above their solubility limit in the main component of the alloy. In addition, the powders obtained using the methods for dispersing melts have a particle shape that is close to spherical. More than 90% of all powders used in additive technologies are obtained by melt dispersion methods. One of the main methods is gas atomization [2].

The main reasons for using this spraying technology are:

- The possibility of high bandwidth and the disintegration of high mass costs
- Greater amount of heat exchange between the gas and the particles for rapid partial cooling of the particles
- Direct delivery of kinetic energy to accelerate particles towards the substrate / deposition for compaction
- Minimizing the risks of oxidizing sputtered materials during spraying using inert gases.

According to this technology, the metal is melted in the melting chamber (usually in a vacuum or inert medium) and then drained in controlled mode through a special device - a sprayer, where the liquid metal flow is destroyed by an inert gas jet under pressure. Spraying can occur in water or in an inert gas stream. Water has the advantage of rapid heat transfer, but there is a high risk of hydrogen and oxygen generation during spraying [3]. During the flight phase, the particles undergo a spherical shape under the action of surface tension forces. To obtain fine powders d = 10-40 microns, most often used in additive technologies, the VIM-atomizers (Vacuum Induction Melting) are used in which the melting chamber is evacuated to minimize melt contact with oxygen and nitrogen. And the technology itself for obtaining powders using vacuum melting machines is called VIGA-Vacuum Induction Melt Inert Gas

IOP Conf. Series: Materials Science and Engineering **347** (2018) 012033 doi:10.1088/1757-899X/347/1/012033

Atomization, that is, the "technology of gas atomization of a metal melted in a vacuum chamber by induction heating".

VIGA atomizers are used to produce the following powders:

- Nickel refractory alloys (Inconel 718, Rene 88, etc.) for parts of aircraft and stationary turbines;
- Alloys based on cobalt for use in medicine, dentistry and production of ion sputtering targets;
- Powders for plasma deposition (NiCrAlY, CoCrAlY, etc.) for plasma spraying of protective coatings on parts made of heat-resistant alloys;
- Powders for granule metallurgy (17-4 PH, 316L) for automotive parts of mass production;
- Sintering compositions in the powder layer (eg cobalt alloys, precious metals) for use in AM machines;
- High-alloy steels (tool steel, quick-cutters) with very high carbide content;
- Non-ferrous metals (copper or tin alloys for various applications).

The design of the classic free fall atomizer is simple and very reliable. Moreover, such sprayers have several applications. They are used, for example, in spray dryers, in powder manufacturing devices or in spray forming processes [4]. Figure 1 shows a sketch of a conventional external atomizer of free mixing. The design is based on the combination of two systems of gas nozzles, called the primary and secondary gas nozzles [5]. Secondary nozzle is the main spraying unit. Concentric jets of gas from the secondary nozzle fall on the central jet of the liquid, which decays due to instabilities from the shear action of the secondary gas stream and its relative velocity [6]. The process of disintegration of the liquid jet in the spraying zone is under the secondary gas nozzle [7]. Due to the splitting of the alloy jet into droplets already inside the spray gun body, this design is mainly used for spraying viscous or high-temperature liquids and melts.

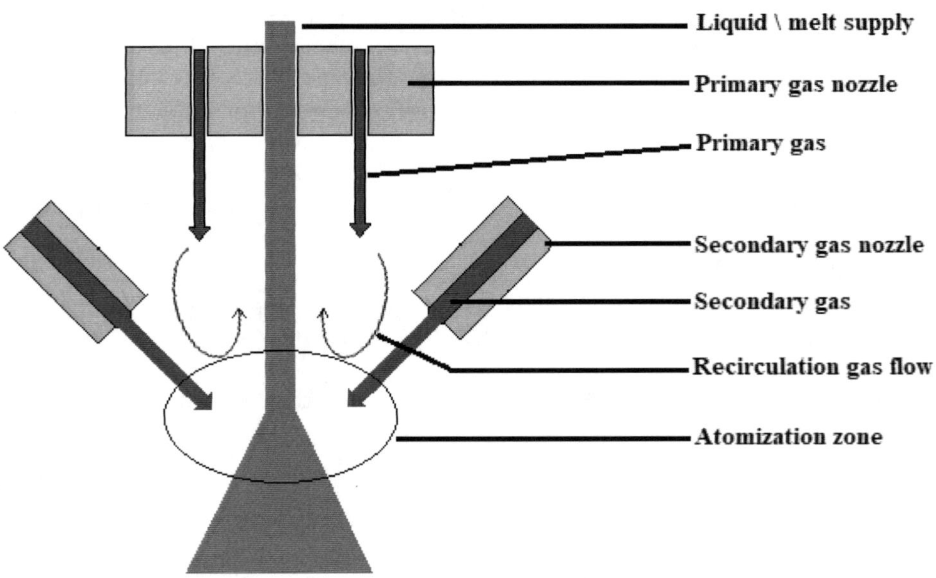

Figure 1. Main areas of the free fall atomizer

In this method, both rotary crucibles and crucibles with bottom draining are used. For melting, ceramic or graphite crucibles can be used. Depending on the material of the crucible, the melting point can reach 1900 °C. The molten metal is poured into a special receiver, to which an inert gas is supplied

IOP Conf. Series: Materials Science and Engineering **347** (2018) 012033 doi:10.1088/1757-899X/347/1/012033

under pressure (usually argon, sometimes nitrogen). The process of metal spraying has three phases - initial, working and final. In the initial phase, the system enters the operating mode: the metal drain valve is opened (it takes some time to stabilize the flow), the supply of the spray gas is switched on, in a precisely defined ratio between the amount of metal and gas. This phase lasts a few seconds.

A common characteristic of the two spray nozzle variants used for sputtering molten metal is the vertical outlet of the melt jet from the casting device through (in most cases cylindrical) molten nozzle in the direction of gravity. Also, in most cases, the central jet stream of the melt is surrounded by a gas flow from one (slotted) configuration of the jet or a set of discrete gas jets that flow parallel to the direction of melt flow or within the angle of inclination of the attack to the melt stream. A coaxial gas atomizer usually leaves the atomizer at high pressures with high kinetic energy.

In the process of spraying molten metal, two basic configurations and types of two-fluid nozzles must be distinguished. The first type is called a closed or closely-coupled atomizer, and the second type is called a free fall atomizer. Both concepts are illustrated in Figure 2.

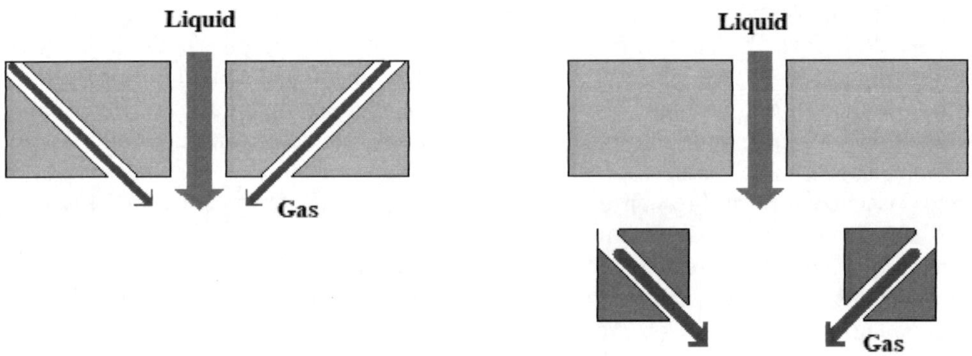

a b

Figure 2. Principles of atomization of a gas of melts: atomizer with a closed connection (a) and a free fall atomizer (b)

Then the working phase begins, in which the metal drainage process is stabilized and the required ratio of metal and argon consumption in the atomizer is achieved. At the end of the working process, the rate of metal exit from the crucible decreases, the flow parameters change, and the balance between the mass ratio of the metal and the gas is disturbed. This process also lasts a few seconds. In the first and final phases, the powder turns out to be substandard. Therefore, to increase the productivity and efficiency of the system, it is necessary to increase the proportion of the working phase in the overall balance of the atomization cycle time. The atomization unit, where the process of metal sputtering actually takes place, is a very complex device that, for structural and technological reasons, can not be performed with dimensions that are less than certain. For qualitative spraying, a certain ratio between the metal consumption and the gas consumption must be maintained [8].

Also, due to the slope of the secondary gas stream, a recirculation gas flow region may occur under the secondary gas nozzle, depending on the design of the atomizer and the operating conditions. If the intensity of the recirculation flow is large enough, liquid bundles or droplets are transported back to the spray gun body. Here, these adhering fragments can clog the gas and / or liquid openings and adversely affect or even stop the spraying process. In conventional free-fall atomizer designs, the recycle gas stream is suppressed by the primary gas nozzle. The primary flow of gas merges with the liquid stream, thereby directing the liquid into the spray zone without recirculation.

The flow of gas in a atomaizer with close coupling immediately closes the outlet of the melt. Within the confined atomizer, the distance between the gas outlet and the melt flow is much less than in the free fall device, where the melt stream moves a certain distance in the direction of gravity before the gas flow falls onto the central jet of the melt. A close-coupled configuration typically tends to give a higher spray efficiency (in terms of smaller particles with the same energy consumption) because of the lower distance between the gas and melt yields. But a closed type of atomizer is more susceptible to problems of melt freezing on the tip of the nozzle. This effect is due to the wide cooling of the melt by the expanding gas flow, which leaves in a close connection near the melt flow. During the expansion of the isentropic gas, the temperature of the atomization gas decreases (sometimes well below 0 ° C). Because of the close spatial relationship between the gas and melt fields, this facilitates the rapid cooling of the melt at the tip of the molten nozzle.

The problem of freezing is relevant, in particular, for spraying. In this process, a discontinuous periodic operation is performed in all technical applications (for example, due to the periodic preparation of the melt or the limited preform that extends to the formation of the spray). The operating time of spray forming processes ranges from a few minutes to about one hour. The problem of the thermal dependence of freezing is most important in the initial phase of the process itself, when the melt flow leaves the nozzle for the first time. At this point, the nozzle tip is still cold and must first be heated, for example by a hot melt flow. This heating process lasts a certain time. Therefore, in the first few seconds of the melt spraying process, there are often problems with freezing associated with heat treatment.

In addition, there are often problems associated with chemical or metallurgical processes in melt delivery systems. Not all these problems have been solved at the moment. A number of problems arise due to a possible change in the composition of the melt material or material of the intermediate ladle or nozzle, due to possible melt / spill response or melt segregation effects from diffusion. This kinetics of the process is somewhat slower than the kinetics of the thermal freezing process mentioned above and can contribute to operational problems in the next stages of the process. [2]

Free fall atomizers are much less problematic than atomizers with close coupling, because the flow of melt streams and the gas stream are well separated at the outlet from the tundish. Therefore, the cooling of the melt with cold gas occurs later than in atomizers with a close bond. As an additional advantage, there is the possibility of controlled mechanical or pneumatic scanning and, consequently, oscillate the gas atomizer with respect to one axis. Also, for the spraying, the concepts of spatial / temporal distribution of the gas jet segments within the nozzle can be used. Thus, the free fall atomizer gives the operator an additional degree of freedom in controlling and controlling the distribution of the mass flow of droplets in the atomizer. This is the most important physical property, since in other systems the nozzle can only affect the pressure of the atomizer gas (during the working process). With controlled scanning of the nozzle, the mass flow can also be distributed over a certain area (necessary, for example, to form flat products). [10]

Acknowledgments
The work was carried out according to the state task No. 007-00129-18-00

References

[1] Lujia L, Yingjie L and Lipeng L 2015 Atomization mechanism of a charged viscoelastic liquid sheet *Chinese Journal of Aeronautics* **28(2)** 403–9.

[2] Introduction to additive manufacturing technology in *Erasteel* Retrieved January 24 2018 from http://www.erasteel.com/content/additive

[3] Busch G. Metal powder - the raw material of future production

[4] Yule A J and Dunkley J J 1994 Atomization of melts *Clarendon Press* Oxford UK

[5] Fritsching U and Bauckhage K 1992 Investigations on the atomization of molten metals: the coaxial jet and the gas flow in the nozzle near field *J. Comp. Fluid Dynamics* **5(1)** 81-9

[6] Markus S, Fritsching U and Bauckhage K 2002 Jet break up of liquid metals in twin fluid atomization *Materials Sci. & Engng.* A **326** 122 – 33

IOP Conf. Series: Materials Science and Engineering **347** (2018) 012033 doi:10.1088/1757-899X/347/1/012033

[7] Lohner H, Czisch C and Fritsching U. 2003 Impact of gas nozzle arrangement on the flow field of a twin fluid atomizer with erxternal mixing *Int. Conf. on Liquid Atomization and Spray Systems*
[8] Fritsching U and Uhlenwinkel V 2012 Hybrid gas atomization for powder production *University of Bremen* 99-124
[9] Hohmann M, Brooks G and Spiegelhauer C 2005 Production methods and applications for high quality metal powders and sprayformed products *Stahl und Eisen* **4** 125
[10] Boulos M 2004 Plasma power can make better powders *Metal powder report* **59** **(5)** 16–21

IOP Conf. Series: Materials Science and Engineering **347** (2018) 012034 doi:10.1088/1757-899X/347/1/012034

Tricosane ($C_{23}H_{48}$) and Octacosane ($C_{28}H_{58}$) mixture phase transition insight via Light scattering techniques

V N Kuryakov[1], P G De Sanctis Lucentini[2] and D D Ivanova[3]
[1] Oil and Gas Research Institute of RAS (OGRI RAS),
Gubkina Street, 3, 119333, Moscow, Russia.
[2] Gubkin Russian State University of Oil and Gas (National Research University),
Department of Physics, Leninskii prospect, 65, Moscow, 119991 Russia.
[3] D. Mendeleev University of Chemical Technology of Russia
Miusskaya pl., 9, Moskva, 125047 Russia.

Abstract. In this study we analyse several emulsion samples of pure Tricosane ($C_{23}H_{48}$) and Octacosane ($C_{28}H_{58}$) paraffins as well as their mixtures with different component concentrations, prepared by means of ultrasonic dispersion without the addition of surfactants. We show that from the measurements of Static Light Scattering temperature dependences it is possible to determine the phase transition temperatures of the paraffins emulsion during the heating and cooling cycles (melting, crystallization and rotator phases). The results for the pure paraffin are in good agreement with the literature data. We produce the outcomes of the Dynamic Light Scattering technique to determine the cluster size of the obtained emulsions (radius 70-120 nm). Those emulsions proved to remain stable during several months.

1. Introduction

The n-Alkanes melting and crystallization temperatures take great importance in many research and applied works as, for example, in the Oil and Gas or in the phase change material world. Furthermore we have an unusual family of plastic-crystalline bulk phases enriching the phase diagrams in-between the liquid and the lower crystal phases temperature, often called rotator or *Cristal-Rotator* phases.

As evidence of the wide interest aroused, we can find experiment performed with different methods with details on the different phase diagram zones. Among the methods commonly used for studying the phase transitions, we would like to mention the result of the adiabatic scanning calorimetry and the differential one [1–3]. There are indeed other stimulating results from the study of the paraffins rotator phases via the well established X-ray diffraction methods [2–4], from the Dynamic mechanical spectroscopy (DMS) [3] and from the recently proposed use of atomic force microscope cantilever [5].

In this paper we propose a technique for determining the phase transitions temperatures of the paraffins and of their mixtures based on optical methods. The studied paraffins are nanoemulsions in water prepared by ultrasonic dispersion without the addition of surfactants. A similar technique was already successfully used to determine the phase transitions in the DODAB vesicles [6] and the light scattering technique is a well established method for the critical phenomena investigation [7, 8].

IOP Conf. Series: Materials Science and Engineering **347** (2018) 012034 doi:10.1088/1757-899X/347/1/012034

2. Materials and methods

For this experiment we prepared emulsions in water from pure paraffins ($C_{23}H_{48}$, $C_{28}H_{58}$) and from their mixtures with 17, 33, 50, 67 and 83 weight percent of the components. The heating and cooling rate in all the experiments was fixed at $2\,K\,h^{-1}$ and it was explored the range of temperatures in which the water remain in its liquid state, from the temperature at which the paraffins are completely solid to the one at which they are completely liquid.

Materials and instruments. We prepared the samples with Tricosane ($C_{23}H_{48}$, 98% pure), and Octacosane ($C_{28}H_{58}$, > 99% pure), purchased from Acros Organics and bi-distilled water purified by reverse osmosis at production time (mean conductivity $10\,mS\,cm^{-1}$). To prepare new emulsion samples and to dilute the highly concentrated ones, we used each time only water from new ampoules. We did not make use of any surfactants.

To prepare the nanoemulsion of paraffin in water, we used an ultrasonic disperser model UZDN-A (USSR, 22 kHz, 75 W) with an immersion transducer. We weight the components of the prepared mixture on a Sartorius BP301S balance (max weight 303 g, $\sigma = 0.1$ mg).
The measurements of light scattering temperature dependence, the hydrodynamic radius (by DLS) and the zeta potential of the emulsions were performed on particle size and zeta potential analyzers Photocor Complex and Photocor Compact (Russia).

Samples preparation. We used the ultrasonic dispersion method to prepare the emulsions of the pure paraffins and their mixtures in water. The dispersing was carried out at a temperature substantially higher than the melting point of the paraffin in use. Before starting the dispersion operations, we visually verified that the paraffin on the surface of the hot water was completely passed into the liquid state. During the dispersion process, the temperature of the water-paraffin mixture did not drop below the crystallization temperature of the used n-Alkane.

To prepare emulsions from a mixture of two paraffins, we preliminarily took samples of the pure species in the right proportions, melted them in one vial, stirred intensively in the liquid state and after cooling the necessary amount of such a mixture was used for the *mixture-sample* preparation.

To carry out the emulsification, we used the ultrasonic disperser up to the power of 75 W. There were prepared samples of about 10 ml each time. The initial paraffin concentration in water was about 0.01% in weight.

3. Results and discussion

We performed Dynamic Light Scattering (DLS) measurements at 25 °C to determine the hydrodynamic radius R_h of the paraffin clusters. For each sample it was in the range $R_h = 70 - 120\,nm$. The zeta potential in emulsions of the single paraffins $C_{23}H_{48}$ and $C_{28}H_{58}$ was close to $-35\,mV$.

We present, in figure 1, the scanning electron microscope (SEM) image of a sample obtained drying the sample studied with the DLS. We chose the paraffin with the higher melting point ($C_{28}H_{58}$) in order to avoid any accidental melting of the clusters during the SEM measurement.

It should be noted that on the dried sample there are no separate particle clusters. Indeed, being the solvent removed, all the clusters stuck together.

The visual hindrance caused by the three-dimensionality of the resulting structure, the impossibility of focusing of different plans and measuring the sizes only on such plans with the images at our disposal, the drying procedure itself that lead to clump together the different clusters of the solution, are all elements that impair the possibility to have a precise and complete reconstruction for the particle size distribution and the relative average.

Taking for example the minimum and the maximum radius of the more evident and peripheral structures, we obtain an average radius of respectively $R^{SEM}_{min} = 55 \pm 20\,nm$ and $R^{SEM}_{Max} = 78 \pm 19\,nm$. Both results are quite in agreement with the DLS one, considering

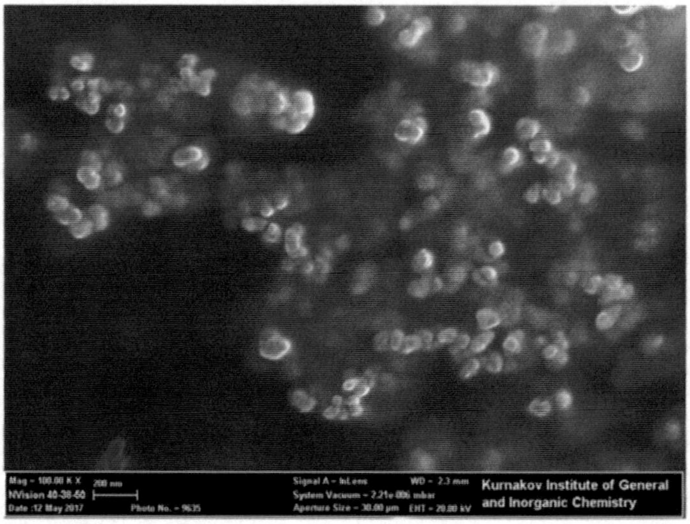

Figure 1. SEM image of paraffin ($C_{28}H_{58}$) particles dried from the studied emulsion.

that the manual selection procedure of the ellipses for the SEM image is prone to introduce a systematic error avoiding to count the bigger structures below the peripheral ones.

Further study with different preparation methods of the SEM specimens have to be implemented to avoid that all the clusters clump together and to allow more informative particle counts.

For each sample we studied the temperature dependence of the scattered light intensity $I_s = I_s(T)$ upon heating and cooling. In figure 2 we show the results for the $C_{28}H_{58}$ emulsion. The intensity $I_s(T)$ shows a linear behaviour and a weak dependence in some temperature ranges, remaining practically constant. Then at special temperatures, or temperature ranges, a significant change occurs.

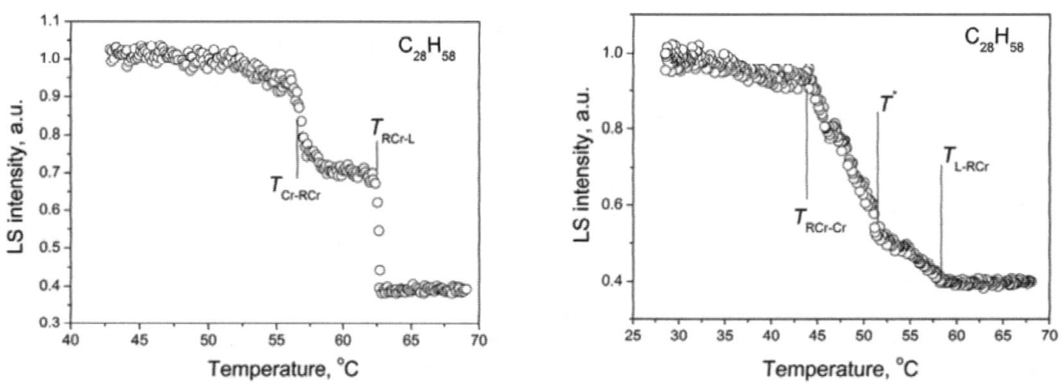

(a) (b)

Figure 2. Temperature dependence of the scattered light intensity I_s, upon (a) heating and (b) cooling, for the paraffin emulsion $C_{28}H_{58}$.

IOP Conf. Series: Materials Science and Engineering **347** (2018) 012034 doi:10.1088/1757-899X/347/1/012034

In general the intensity I_s in an emulsion may depend on the concentration of the droplets, on their size, their shape, on the surface properties and on the difference in the refractive indices between the dispersion medium and the dispersed phase.

We assume that the marked change of I_s in our samples is due to the phase transitions of the paraffin in the emulsion.

The low-temperature intensity gap corresponds to the phase transitions from the *Crystal* (Cr) to the *Crystal-rotator* (RCr) phase and *vice versa* upon heating and cooling, respectively. In figure 2, these temperatures are marked as T_{Cr-RCr} and T_{RCr-Cr}.

The changes in the scattering intensity I_s at higher temperatures correspond instead to phase transitions phase *Crystal-rotator* to liquid (L) (melting point) and *vice versa*, upon heating and cooling respectively. In the figure these temperatures are denoted as T_{RCr-L} and T_{L-RCr}.

Upon cooling, b-panel on the right of the figure 2, it is possible to note another characteristic temperature corresponding to a change in the slope of I_s (marked with T^*) in the temperature interval between the *Liquid* and the *Crystal-Rotator* phases.

The temperature T^* may be correlated to a phase transition from the rotator phase IV to the rotator phase III [2] or the temperature range $T_{L-RCr} - T^*$ may be connected to the range in which a surface crystallization happens in the paraffin emulsion droplets [9]. Further analysis are needed to investigate this zone.

We observed similar temperature dependences of I_s for each of the studied emulsions. From these dependences, we determined the characteristic temperature assumed as the phase transition ones.

The figure 3 shows the temperatures of phase transitions for the mixtures of two n-Alkanes, determined according to the procedure described above.

A number of studies in the field of n-Alkanes binary mixture for bulk samples [10–12] presented analogous forms of the phase transition temperature dependence on the concentration.

It should be noted that in all samples in our study we were able to measure the melting temperature with a good reproducibility. Indeed during this phase transition the refractive index (RI) of the paraffin in emulsion changes greatly, which causes a strong change in the scattered light intensity I_s within a little temperature increment.

For the paraffin mixture emulsion samples, not all the phase transitions have a clearly determined temperature: the intensity of the scattered light during heating and cooling are blurred in a certain temperature ranges. When we prepare an emulsion from a mixture of paraffins we are fixing the overall percentage of each species on the total. Nonetheless it is not possible to obtain the same concentration of paraffins in each drop of the emulsion, which leads to phase transitions in different drops at different temperatures and cause the blurred data reading.

The temperatures of all the phase transitions of the pure paraffin emulsion samples were also determined with good reproducibility.

In the table 1 we report the experimental data of the phase transition temperatures taken from [9] for the pure paraffins, $C_{23}H_{48}$ and $C_{28}H_{58}$, and the ones determined in this study. The used symbols are T^*_m for the melting temperature (solid to liquid phase transition), T^*_{xr} for the temperature of *Crystal* to *Crystal-rotator* phase transition in the emulsion samples. T^*_s is the temperatures of solidification for the bulk, non-emulsified samples. With the symbols T_{RCr-Cr}, T_{Cr-RCr}, T_{RCr-L}, are indicated the phase transitions between crystal (Cr), rotator crystal (RCr) and liquid (L) phases determined in this work and introduced above.

4. Conclusion

We prepared by means of ultrasonic dispersion several samples nanoemulsion in water from two pure n-Alkanes ($C_{23}H_{48}$ and $C_{28}H_{58}$) and their mixtures with different relative percentage, without the addition of surfactants.

IOP Conf. Series: Materials Science and Engineering **347** (2018) 012034 doi:10.1088/1757-899X/347/1/012034

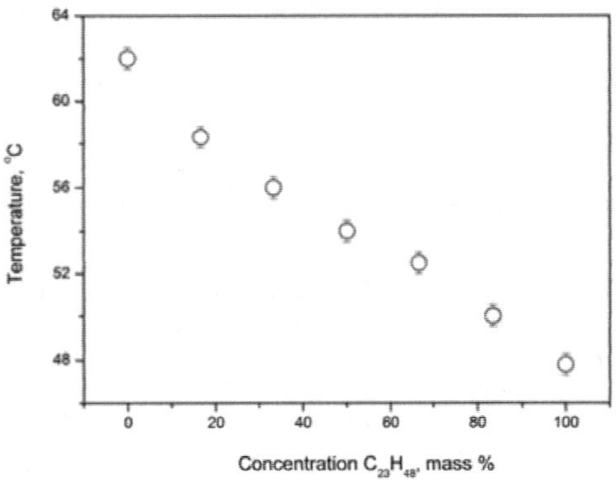

(a) $RCr \rightarrow L$ (Melting point) upon heating.

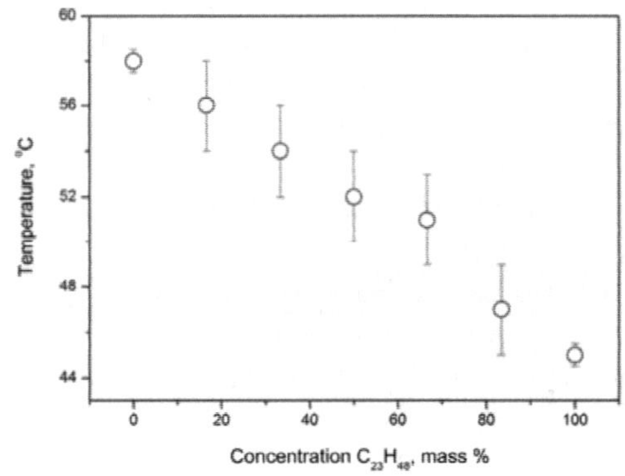

(b) $L \rightarrow RCr$ upon cooling.

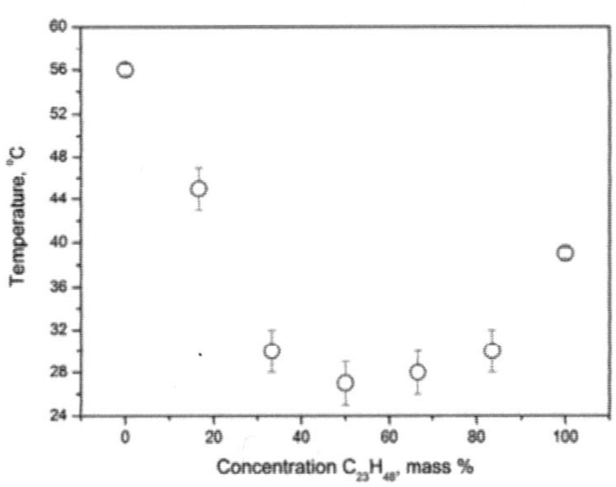

(c) $Cr \rightarrow RCr$ upon heating.

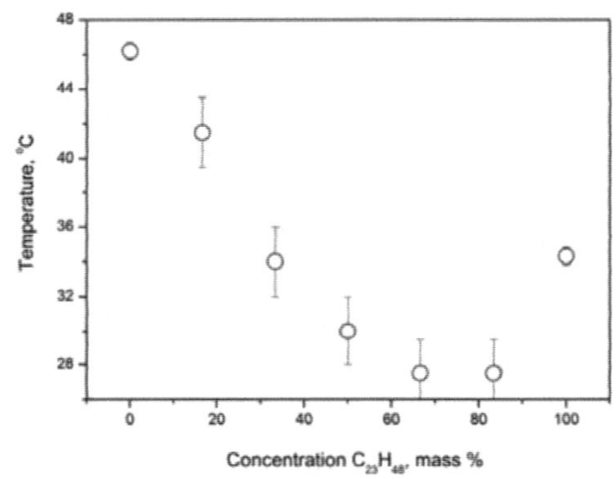

(d) $RCr \rightarrow Cr$ upon cooling.

Figure 3. Phase transitions temperatures of the binary n-alkane mixture ($C_{23}H_{48}$,$C_{28}H_{58}$) versus the concentration of the components as determined from the LS measurements for the nanoemulsions samples. *Crystal* (Cr),*Crystal-rotator* (RCr) and *Liquid*,(L) phases.

We propose a method to determine the phase transition temperatures of those samples observing the scattered light intensity temperature dependence.

Table 1. The comparison of the phase transition temperature of pure n-Alkanes from [9] (with an asterisk) with the ones determined in this work. All the reported temperatures are in Celsius degrees.

n	T^*_m	T_{RCr-L}	T^*_{xr}	T_{Cr-RCr}	T^*_s	T_{RCr-Cr}
$C_{23}H_{48}$	47.2	48	40.6	39	33.4	34
$C_{28}H_{58}$	60.9	62	57.7	56	45.4	45

The phase transition temperatures of the two pure paraffins so determined are in a good agreement with the previous literature data.

The n-Alkanes Crystal-rotator phase transition presents an intrinsic complexity. We believe that the percentage fluctuation of the two types of paraffin in the binary mixture drops may cause a different transition temperature for each drop and a complex pattern of the scattered light temperature dependence of the whole emulsion. The eventual influences of the heating and cooling rates on the metastable phases, moreover, may play an additional role in the *Rotator-Crystal* phase.

The previous ones are all elements that drive us to consider interesting further future studies, applying new methods for the sample preparation, comparing the results of other instruments measurements on the same samples, and applying different methods of data analysis and data acquisition in order to obtain a deeper knowledge of the applicability of this technique for a broader phase transitions comprehension.

Acknowledgments
Authors are grateful to Alexey D.Yapryntsev from Kurnakov Institute of General and Inorganic Chemistry Russian Academy of Sciences for the SEM analysis of samples.

References
[1] Voronov V P 2000 *Journal of Experimental and Theoretical Physics* **91** 144–149 ISSN 1090-6509 URL https://doi.org/10.1134/1.1307242
[2] Sirota E B and Singer D M 1994 *The Journal of Chemical Physics* **101** 10873–10882 URL http://doi.org/10.1063/1.467837
[3] Sirota E B, Jr H E K, Singer D M and Shao H H 1993 *The Journal of Chemical Physics* **98** 5809–5824 (*Preprint* https://doi.org/10.1063/1.464874) URL https://doi.org/10.1063/1.464874
[4] Craievich A F, Denicolo I and Doucet J 1984 *Phys. Rev. B* **30**(8) 4782–4787 URL https://link.aps.org/doi/10.1103/PhysRevB.30.4782
[5] Jung N, Yun M and Jeon S 2012 *The Journal of Chemical Physics* **136** 104903 (*Preprint* https://doi.org/10.1063/1.3692296) URL https://doi.org/10.1063/1.3692296
[6] Voronov V P, Kuryakov V N and Muratov A R 2012 *Journal of Experimental and Theoretical Physics* **115** 1105–1110 ISSN 1090-6509 URL https://doi.org/10.1134/S106377611211012X
[7] Robertson A E, Phan D H, Macaluso J E, Kuryakov V N, Jouravleva E V, Bertrand C E, Yudin I K and Anisimov M A 2016 *Fluid Phase Equilibria* **407** 243 – 254 ISSN 0378-3812 aqueous Solutions URL http://www.sciencedirect.com/science/article/pii/S0378381215003490
[8] Belyakov M, Gorodetskii E, Kulikov V, Kuryakov V and Yudin I 2011 *Chemical Physics* **379** 123 – 127 ISSN 0301-0104 URL http://www.sciencedirect.com/science/article/pii/S0301010410005240
[9] Kraack H, Sirota E B and Deutsch M 2000 *The Journal of Chemical Physics* **112** 6873–6885 (*Preprint* https://doi.org/10.1063/1.481263) URL https://doi.org/10.1063/1.481263
[10] Ungar G and Masic N 1985 *The Journal of Physical Chemistry* **89** 1036–1042 (*Preprint* http://dx.doi.org/10.1021/j100252a030) URL http://dx.doi.org/10.1021/j100252a030
[11] Hammami A and Mehrotra A K 1995 *Fluid Phase Equilibria* **111** 253 – 272 ISSN 0378-3812 URL http://www.sciencedirect.com/science/article/pii/037838129502777C
[12] Mazee W 1957 *Analytica Chimica Acta* **17** 97 – 106 ISSN 0003-2670 URL http://www.sciencedirect.com/science/article/pii/S0003267000870026

IOP Conf. Series: Materials Science and Engineering **347** (2018) 012035 doi:10.1088/1757-899X/347/1/012035

Phase transitions and structural formation of PEG-PCL-PEG copolymer in the processes of fused deposition 3D printing

A Dunaev[1], A Mariyanac[1], A Mironov[1, 3], O Mironova[1], V Popov[1] and M Syachina[1,2]

[1]Institute of Photonic Technologies, Federal Research Centre Crystallography and Photonics', RAS, Moscow, Russia
[2] I.M. Sechenov First Moscow State Medical University, 8/2 Trubetskaya St., Moscow, Russian Federation.
[3]Author to whom any correspondence should be addressed

E-mail: scftlab@gmail.com

Abstract. In present work the analysis of thermal field distribution and thermal analysis were used to study phase and structural transformations in the block copolymer of polycaprolactone and polyethylene glycol in the process of scaffolds fabrication for tissue engineering using fused deposition modeling. It was shown that the intact polymer has a noticeable thermal history and formed degree of crystallinity which is close to its equilibrium value, while the microstructure of the polymer stays unchanged.

1. Introduction

Nowadays one of the actual task of regenerative medicine is the creation of tissue engineering constructs (TEC) using three-dimensional printing. Main concept of tissue engineering was published in 1993 [1], where one of the basic elements of TEC - artificial bioresorbable (gradually dissolving in the body) volumetric porous matrix-frames of certain sizes, shapes and internal structure where described. In general, TECs are a system consisting of living functional cells, artificial scaffolds and biologically active compounds. Currently, there are many research works is focused on different 3D printing methods adaption to create TEC, designed and targeted to repair damaged tissue. They are used for partial or complete restoration of the functions of damaged or lost tissues. The choice of material and technology for these scaffold manufacturing is determined by the set of desired physical, chemical and mechanical properties for TEC. In addition, there are very stringent requirements for scaffold materials. They should neither provoke any negative cellular response nor exhibit immunogenicity or toxicity. These materials should provide mechanical strength and stability of the TEC, both during its implantation and in subsequent processes of tissue regeneration [2-5].

Biocompatible aliphatic polyesters (polylactides, polylactoglycolides, and poly-ε-caprolactones) and its copolymers are widely used both in biomedical research and clinical practice. The physicochemical properties of these polymers allow their processing in various ways, including 3D printing via Fused Deposition Modeling (FDM) [6-8]. However, the process of fused deposition printing is able to influence the properties of the selected material. Thus, there are detailed studies of the physicochemical processes during the FDM printing is needed. The present work is devoted to the study of phase transitions and structure formation processes during fabrication of the polyethylene glycol-poly-ε-caprolactone-polyethylene glycol copolymer (PEG-PCL-PEG) scaffolds by FMD 3D printing. This block copolymer has a number of advantages for biomedical applications in comparison with poly-ε-caprolactone alone [9-11]. PEG-PCL-PEG also has significant crystallinity, which greatly simplifies the studies using thermal analysis methods.

2. Materials and methods

The combined method including studies of the dynamics of the thermal fields distribution during the process of three-dimensional printing of polymer melt and the subsequent model study of thermal effects was applied to study phase transitions and structure formation in the process of FDM printing.

2.1. Materials

In experiments a polyethylene glycol-poly-ε-caprolactone-polyethylene glycol (PEG-PCL-PEG) block copolymer was used as a model compound. PEG-PCL-PEG with four different molecular weights (M_n) 10 kDa, 20 kDa, 30 kDa and 40 kDa was synthesized by the method described in [10]. In all cases, the molecular weight of the polyethylene glycol end blocks was 550 Da. The molecular weight of synthesized polymers was characterized by gel permeation chromatography (GPC) using a polystyrene standard.

2.2. 3D printing

The study of the three-dimensional printing was carried out using a specially designed 3D printer based on polymer pellet melt deposition technology (figure 1).

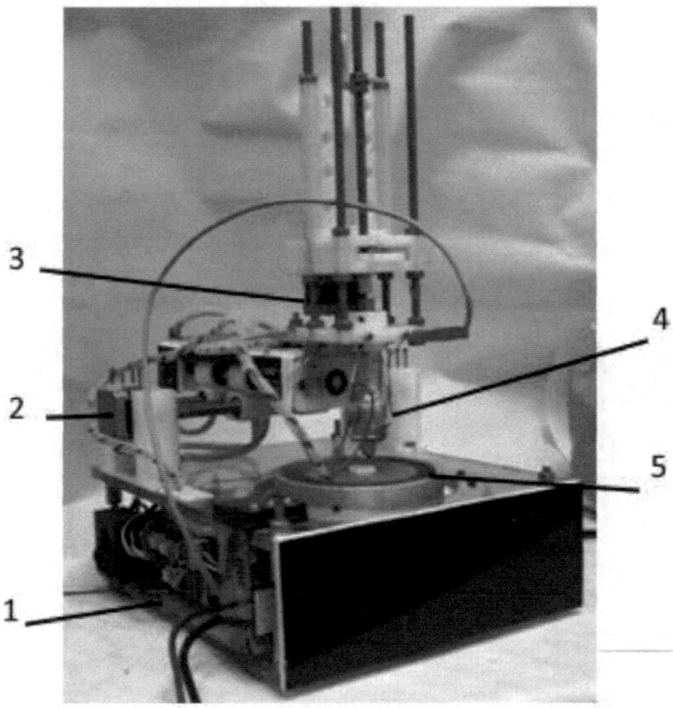

Figure 1. 3D printer for thermal fields distribution analysis:
1 - main frame;
2 - positioning system;
3 - drive of the piston extruder;
4 - extruder with heater and nozzle;
5 - cooling table.

The melting zone of the printer extruder is limited to a small (about 10 mm^3) volume immediately before the nozzle to reduce the negative effect of prolonged thermal stress on thermo labile polyesters. The powder is fed into the melting zone in a solid state at a pressure of up to 1.5 MPa by means of a piston driven by a precision stepping motor. In order to improve the adhesion of the sample and increase the thermal contrast, the microporous steel substrate with a mass of 300 g was used as a working table of the printer.

As a test sample a hollow single-walled cylinder with a diameter of 10 mm was used during experiments. The generatrix of cylinder was printed layer by layer in the spiral way. This provided sufficient time for the transition of the polymer from melt to solid at any point of the sample.

2.3. Thermal imaging analyses

In order to study the dynamics of the of thermal process distribution, a three-dimensional printer was equipped with the Flir A655sc high-speed thermal imaging camera (FLIR, Sweden) and the compact thermal imaging module Thermal Expert TE-Q1 (I3System, Taiwan, ROC). The temperature profiles of the samples which described the temperature distribution in the layers of the printed object were gained directly during the printing process in 8-14 mkm radiation wavelength range.

2.4. TGA/DSC modeling

A combined differential thermal and gravimetric analysis (TGA/DSC Linesys PT1000, Liseis, Germany) was used to model the thermal effects of phase transitions during 3D printing process. To create a temperature profile (the curve of the temperature versus time, according to which the sample is heated and cooled) the data obtained in the analysis of thermal fields by the thermal visualization method were used. The conditions of DTA analysis were chosen as close as possible to the conditions of three-dimensional printing.

2.5. Structural analyses

The microstructure of the resulting samples was examined without further processing with a Phenom Scanning Electron Microscope (Phenom, The Netherlands).

3. Results and discussion

In order to obtain modeling parameters, a thermal imaging of a real 3D printing process was used. Each temperature profile represented a curve with peaks where width increased, till the height of sample did not exceed 1 – 1.2 mm. After that a new layers cooling process was not influenced by heat transfer to substrate and their temperature profiles becomes uniformal (figure 2). For investigated PEG-PCL-PEG samples the thermal profiles were nearly similar and were not affected by polymer molecular weight. The subsequent modeling differential scanning calorimetry (DSC) and thermogravimetry (TG) experiments were performed using measured temperature curves.

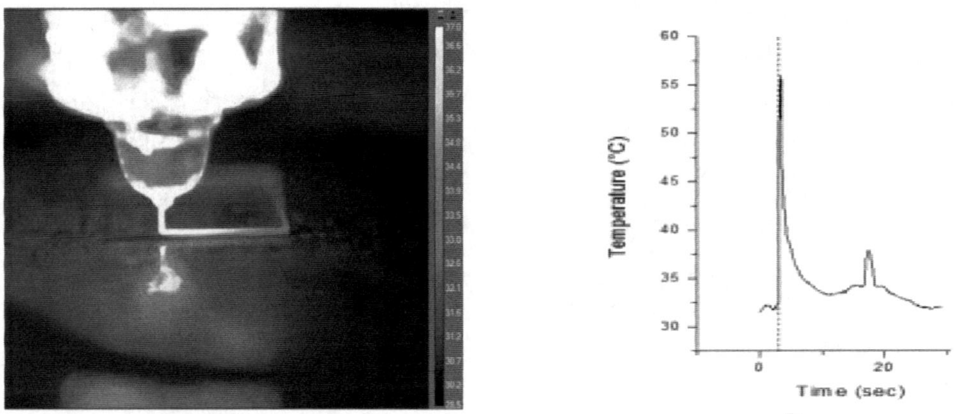

A B

Figure 2. Thermal imaging and trace during extrusion of melt of PEG-PCL-PEG copolymer (A) and single layer temperature profile (B).

The thermal analysis results (figure 3) indicate the presence of significant thermal history of the original polymer because in the process of heating-cooling, the area of the crystallization peak decreases markedly. For example, for PEG-PCL-PEG with a molecular weight M_n = 30000, in the first heating-cooling cycle, the heat of fusion was 88.74 J/g, while the heat of crystallization was 62.53 J/g. In the second cycle, the heat of melting and crystallization had values of 61.36 J/g and 64.20 J/g, respectively.

For all investigated samples in the temperature interval from 30°C to 200°C two cycles of heating-cooling thermogravimetry analysis did not reveal any anomality. The obtained data show that during short-term heating, studied polymers are not subject to significant thermal degradation and do not react with atmospheric components.

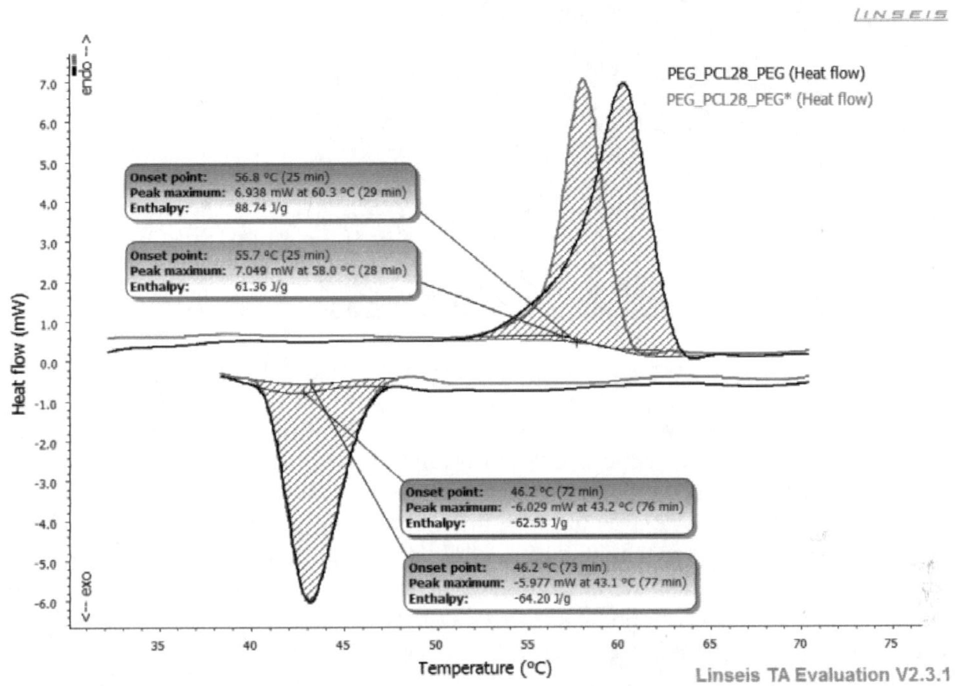

Figure 3. Typical thermogram of successive heating-cooling cycles of PEG-PCL-PEG. a - first cycle (intact PEG-PCL-PEG), b-second cycle (re-crystallized PEG-PCL-PEG).

Combining the data of thermal field distribution (figure 2A) and PEG-PCL-PEG DCS curve (figure 3), shows that each printed layer completely passes into a crystallized state even for a small model sample. The cooling rate of any layer is generally identical and described by exponential law, so significant differences were not found in the thermograms of individual layers. This indicates close degrees of crystallinity, but the real degree of crystallinity of the model triblock-co-polymer was not calculated, because no data were obtained about the heat of fusion of 100% crystalline phase of PEG-PCL-PEG. Also thermal field distribution shows that the following passages of the nozzle re-melting underlying layers into a depth of from 200 to 400 microns. This effect was not observed for layers lying in the immediate vicinity (up to 1 mm) from the metal substrate, which is associated with a more intense heat transfer. As a result, of experimental modeling a picture of the phase and aggregate states of the model triblock-co-polymer during the fused deposition printing process was obtained.

There is no evidence of fundamental changes in the microstructure of the PEG-PCL-PEG in depend of different molecular weight and or layers position of the formed sample. The structure of all the samples is formed by spherulites with characteristic dimensions of 20-50 mkm (figure 4), which agrees well with literature data for block copolymers based on polyesters [12].

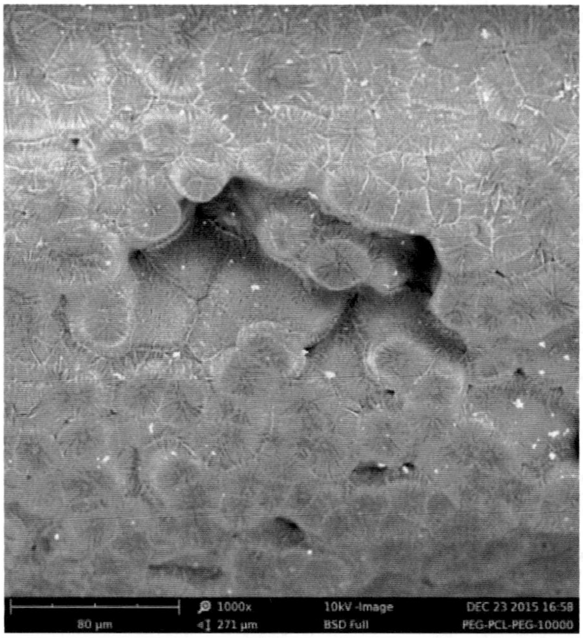

Figure 4. A characteristic image of the microstructure of a PEG-PCL-PEG sample obtained by fused deposition modelling.

4. Conclusions

For thermo-extrusion printing with a thermal imager, the data of thermal fields dynamics were obtained, as well as on peak temperatures during the layer-by-layer creation of a three-dimensional object from the melt of a crystallizing copolymer. Based on thermal modeling data, the DTA method established that: 1) the polymer used has a significant thermal prehistory and the heat of fusion of the starting polymer is 25% higher than that of the polymer subjected to fused extrusion, whether this is a feature of the synthesis of a PEG-PCL-PEG copolymer or cold crystallization processes were not established; 2) in the heating-cooling cycles, the identity of the heat of fusion and crystallization of the polymer from cycle to cycle was manifested.

5. Acknowledgments

This work was supported by the Federal Agency of Scientific Organizations (Agreement No 007-ГЗ/Ч3363/26) in part of thermal analyses and by the Russian Foundation for Basic Research (grant №16-29-11722 ofi_m) in part of analysis of 3D printing and thermal imaging.

Reference

[1] Langer R and Vacanti J P 1993 Tissue engineering Science 260 920-26
[2] An J, Teoh J E M, Suntornnond R and Chua Ch K 2015 Design and 3d printing of scaffolds and tissues *Engineering* **1** 261-68
[3] Mota C, Puppi D, Chiellini F, Chiellini E, 2015 Additive manufacturing techniques for the production of tissue engineering constructs *J. Tissue Eng. Regen. Med.* **9** 174–90
[4] Pedde R et. al. 2017 Emerging Biofabrication strategies for engineering complex tissue constructs *Adv. Mater.* **29** ID1606061
[5] Melchels F P W, Domingos M A N, Klein T J, Malda J, Bartolo P J and Hutmacherad D W 2012 Additive manufacturing of tissues and organs *Progress in Polymer Science* **37** 1079-104
[6] Martina M and Hutmacher D W 2007 Biodegradable polymers applied in tissue engineering research: a review Polymer International 56 145–57

IOP Conf. Series: Materials Science and Engineering **347** (2018) 012035 doi:10.1088/1757-899X/347/1/012035

[7] Chia H N and Wu M W 2015 Recent advances in 3D printing of biomaterials *Journal of Biological Engineering* **9** 4-14

[8] Chiulan I, Frone A N, Brandaburv C and Panaitescu M D 2018 Recent advances in 3d printing of aliphatic polyesters *Bioengineering* **5** 2-15

[9] Cuong N V, Hsieh M F, Chen Y T and Liau I. 2010 Synthesis and characterization of PEG–PCL–PEG triblock copolymers as carriers of doxorubicin for the treatment of breast cancer *J.of App. Pol. Sci.* **117** 3695-3703

[10] Liao H T, Chen Y Y, Lai Y T, Hsieh M F, and Jiang C P 2014 The Osteogenesis of bone marrow stem cells on mPEG-PCL-mPEG/hydroxyapatite composite scaffold via solid freeform fabrication *BioMed Research International* **2014** ID321549

[11] Hwang M J, Suh J M, Bae Y H, Kim S W and Jeong B 2005 Caprolactonic Poloxamer Analog: PEG-PCL-PEG *Biomacromolecules* **6** 885-890

[12] Xue F and Jiang S 2014 Crystallization cehaviors and structure transitions of biocompatible and biodegradable diblock copolymers *Polymers* **6** 2116-45

The obtaining and properties of asymmetric ion transport membrane for separating of oxygen from air

A A Solovieva[1,2] and I V Kulbakin[2]

[1] Bauman Moscow State Technical University (BMSTU), 5, 2nd Bauman Street, Moscow, 105005, Russian Federation
[2] Baikov Institute of Metallurgy & Materials Science (IMET RAS), 49, Leninsky Prospekt, Moscow, 119334, Russian Federation

E-mail: ivkulbakin@mail.ru

Abstract. The bilayer oxygen-permeable membrane, consisting of a thin-film dense composite based on Co_3O_4 - 36 wt. % Bi_2O_3, and of a porous ceramic substrate of Co_2SiO_4, was synthesized and characterized. The way for obtaining of porous ceramic based on cobalt silicate was found, while the microstructure and the mechanical properties of porous ceramic were studied. Layered casting with post-pressing was used to cover the surface of porous support of Co_2SiO_4 by the Co_3O_4 - 36 wt. % Bi_2O_3 - based film. Transport properties of the asymmetric membrane have been studied, the kinetic features of oxygen transport have been established, and the characteristic thickness of the membrane has been estimated. The methods to prevent the high-temperature creep of ion transport membranes based on solid/molten oxides, which are the promising ones for obtaining of pure oxygen from air, are proposed and discussed.

1. Introduction

One of the urgent problems of modern membrane science is the production of high-purity oxygen, which is necessary for needs of nanoelectronics, medicine and pharmaceuticals [1]. For the obtaining of oxygen by using the ion transport membranes many people typically apply inorganic mixed-conducting oxide materials, such as BSCF [2] or LSCF [3], operating at a temperature of above 850 °C. Despite of the many-years investigations, the question of creating of the membrane material, operating in the intermediate-temperature region, with stable transport characteristics during 40,000 hours of exploitation is still remaining the relevant [4]. It is evident that technological efforts for lowering operating temperatures of the membranes to 500-850 °C while maintaining the same values of transport properties as for known for the high-temperature systems are required [5]. One of such decisions is the creating of new more advanced membrane materials, which possess high oxygen permeability and selectivity at separation of an air.

Nowadays the direction, related to the creating of solid/molten composite membrane materials with high mixed oxygen-ion and electron conductivity in the intermediate-temperature range below 850 °C, widely develops and is being investigated [6, 7]. The indisputable advantage of such membranes is the high selectivity of the process of concentrating of oxygen and possibility to generate a product of high purity [8]. It is possible due to the presence of wetting interlayers at grain boundaries of the solid skeleton and forming of so-called "liquid-channel grain-boundary structure" [9]. Predominantly, in such composites the solid component has an electronic conductivity, while the melt is predominantly an oxygen-ion conductor [10]. Thus, the membrane material of the above-mentioned type based on Co_3O_4 - 36 wt. % Bi_2O_3 showed high oxygen permeability with a record selectivity of oxygen over nitrogen (> 100,000) at 790-850 °C [11]. In addition, it was found that the transport of oxygen through the given membrane material of 1,5-3,3 mm thick was carried out in a diffusion regime according to Wagner's mass transfer law (the flux was inversely proportional to the thickness of membrane) [12]. It is assumed that such membrane material may replace the high-temperature membrane materials having the various disadvantages. However, for the practical use of such material in oxygen separators it is necessary to increase its productivity by minimizing the membrane thickness, which may be achieved by creating a thin-film solid/molten oxide membrane onto a porous ceramic substrate. In this case the thin-film ion transport membrane should possess high mixed ionic-electronic conductivity and provide an accelerated ambipolar transport of charge carriers, and the porous substrate should operate as a high-performance track system to penetrate and remove of pure oxygen. When developing the asymmetric "porous substrate / dense ion transport membrane" system, the chemical and thermal compatibility of the membrane material with this substrate, as well as their stability under operating conditions, are required. Besides, the substrate must have big open porosity (about 40 %), but acceptable mechanical properties (strength must be not less than 30-50 MPa), thus the development of such system should take into account all these necessary aspects. In the present work, Co_2SiO_4 with a structure of olivine [13] has been chosen as the porous substrate because it is chemically and thermally compatible with the solid/molten composite thin-film membrane of Co_3O_4 - 36 wt. % Bi_2O_3. This article is devoted to the development of an asymmetric oxygen-permeable membrane based on the above-mentioned materials.

2. Experimental

The initial components to synthesize Co_2SiO_4 phase were Co_3O_4 (99,9% purity) and amorphous SiO_2 with a specific surface area of 350 m^2/g. Amorphous SiO_2 was taken in excess (3%) due to removal of water during the crystallization, while Co_3O_4 was taken in excess (7%) due to decomposition to CoO at heating of above 900 °C. The powder was premixed in a planetary mill (Pulverisette 5, rotation speed of 200 rpm, duration of 5 hours) with additive of ethyl alcohol (1:1 by volume) and held for 3 hours at 120 °C in a drying camera for removal alcohol, then was sintered on zirconia crucible in air at 1200 °C for 40 hours with an intermediate grinding and homogenization. The resulting single-phased product was grinded and used as a starting component to obtain the porous ceramics.

To obtain the porous ceramics of Co_2SiO_4, the powder mixture of particles size of 20-30 μm (60%), of 0.1-1 μm (25%), and of 1-10 μm (15%) was used. Such size distribution, as shown on a Figure 1, was chosen a priori for production of porous but strong enough ceramics, since it is known that the growth of average grains and the formation of isthmuses in the porous skeleton are due to dissolution of small particles with the recrystallization of average particles. A carbon black of 5-10 μm was used as a technological burning out additive. The Co_2SiO_4: carbon black (mass ratio of 3:1) composition was obtained by mixing in ethyl alcohol (1:1 by volume) in a planetary mill for 30 minutes at a rotation speed of 120 rpm with drying to remove an alcohol. The resulting powder was then cold uniaxially pressed (100 MPa) into tablets (diameter of 30 mm, height 4 mm for the subsequent studying of transport properties, and diameter of 6 mm, height of 20 mm for the strength measurements). These tablets were sintered in air on zirconia substrates at 1200 °C for 20 hours with a slow heating of the furnace (2 °C/min) to burn out the carbon black and to form pores in ceramics.

The forming of a thin-film material based on Co_3O_4 - 36 wt. % Bi_2O_3 onto a porous substrate of Co_2SiO_4 was carried out by layered (100 μm) covering by the suspension, which was prepared in an agate mortar by homogenization and was consisted of Co_3O_4 (99,9% purity), Bi_2O_3 (99,9% purity) and 5% polyvinyl alcohol aqueous solution (10% by weight). The suspension was covered onto a pre-polished porous Co_2SiO_4 substrate with a Teflon spatula, then was dried at 120 °C to remove water. Then the next layer was applied, and the procedure was repeated to form the controlled coatings with the thickness of 0,7 mm, 0,5 mm and 0,3 mm. The finish forming of the coatings was carried out by its cold uniaxially pressing (10 MPa) using vacuum-tight rubber, protected with a Teflon film. The obtaining material was annealed at 750 °C for 5 hours with a slow heating of the furnace (2 °C/min).

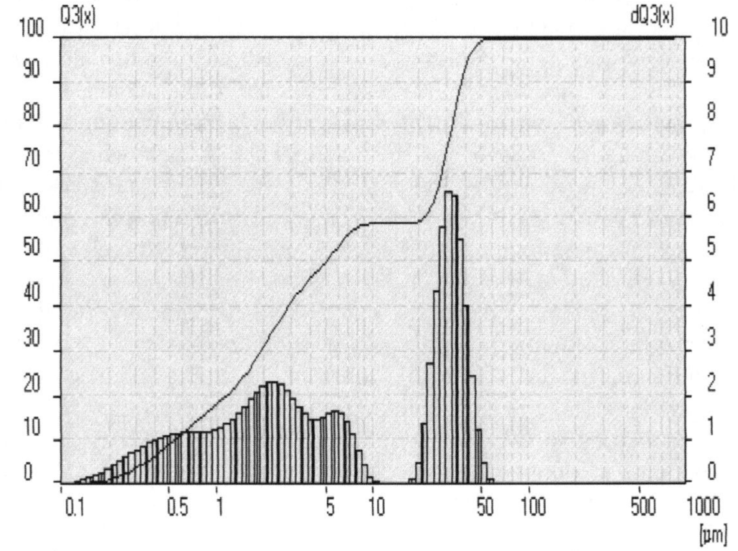

Figure 1. Chosen particle size distribution of Co_2SiO_4 powder

X-ray analysis (XRD) was performed on Shimadzu XRD-6000 diffractometer, the interval of angles was 10-90 ° with the steps of 0.1 ° (CuKα radiation, wavelength λ = 1,54178 Å). The decoding of the initial X-ray diffraction patterns was carried out using ICDD PDF-2.

The manufacturing of polished samples for the subsequent studying of their microstructure was carried out with using STRUERS equipment. The samples were placed in automatic hot press on the bottom of a cylindrical shape (diameter of 30 mm, depth of 10 mm). Then the shape was filled by a conductive resin, which solidified after heat treatment. Both SiC papers (grades of 80 to 4000) and diamond pastes of various granularity were used to polish the samples. The microstructure of polished sections was studied on Olympus GX-51 optical microscope.

The studying of burn-up of carbon black from the initial powder for synthesis of porous ceramics was carried out by thermogravimetry (TG) and differential thermal analysis (DTA) using NETZSCH STA 449 F3 Jupiter thermal analyzer. The powder of Co_2SiO_4 + C was placed in a YSZ crucible and heated in an air atmosphere at a rate of 20 °C/min from room temperature to 1300 °C.

The strength on three-point bending was studied using Instron 3382 testing machine. The measurements were carried out at a room temperature using five cylindrical samples (the length of 15 mm, the diameter of 5 mm, the load of 0 - 100 kg, the distance between supports of 12 mm). The obtained strength results were averaged, and the error of measurements was evaluated.

The oxygen permeation of the composite membranes was measured using gas chromatography. The testing samples were polished to the diameter of 25 mm, and their edges were covered by silver paste with the average grain size of Ag of 1 µm. Then the samples were annealed at 700 °C for 5 hours to form a dense edge coatings and to eliminate of possible air leaks. The resulting samples of 3-4 mm thick (the thicknesses of dense coating of Co_3O_4 - 36 wt. % Bi_2O_3 were of 0,2 mm, of 0,4 mm, and of 0,6 mm) were hermetically sealed at the end of the quartz tube, while the other end of this tube was connected to inlet of a gas chromatograph. The collected cells were placed in a tube furnace and heated to the temperature of 790-850 °C. The samples were in contact with air in the furnace from the one side, while the other side of samples was blown by sweep gas (helium) to removal a penetrated oxygen. The obtaining gas mixture (oxygen and helium) was directed to a chromatograph to determine the concentration of oxygen. The rate of helium was fixed to be 5-20 mL/min.

IOP Conf. Series: Materials Science and Engineering **347** (2018) 012036 doi:10.1088/1757-899X/347/1/012036

3. Results and discussion

According to X-ray analysis data, it was shown that the yield of Co_2SiO_4 olivine (JCPDS No. 84-1298) was 100%. Basing to Zhuravlev's model [14], the full transformation of amorphous SiO_2 to crystalline tridymite completes at 1200 °C, after that, according to CoO - SiO_2 phase diagram [15], the Co_2SiO_4 formation is taken place. Phase analysis of the coating deposited on the surface of a porous ceramic based on Co_2SiO_4 showed the formation of two phases - Co_3O_4 with a spinel structure (JCPDS No. 42-1467) and $Bi_{88}Co_3O_{136}$, which was isostructural to sillenite γ-Bi_2O_3 (JCPDS No. 45-1344). This is in a good agreement to phase diagram of Co_3O_4 - Bi_2O_3 system [16]. The quantitative composition of the coating corresponds to Co_3O_4 - 36 wt. % Bi_2O_3, all peaks have been indexed and determined.

The thermal analysis data for Co_2SiO_4 + C powder indicates slight increasing in mass up to 850 °C, which is apparently due to the sorption of air oxygen by carbon black. At the temperature range of 850 °C to 970 °C the significant decrease in mass is observed, which is probably due to the desorption of CO_2. Since the mass value does not change up to a temperature of 1300 °C, one can be spoken about the full burning out the carbon black at 970 °C. Such a high burning out temperature corresponds to about 0,7 of melting temperature of Co_2SiO_4 (1415 °C). One may conclude about the good choice of technological burning out additive, because, according to the Tamman's semi-empirical rule, the ability of the sintering of particles and of the formation of a strong skeleton in porous ceramics appears near the temperature of 0,6-0,7 by a melting point. Basing on this conclusion, the partial sintering of skeleton firstly appears, then the pores formation becomes due to burning out the technological additive, so the porous but strong ceramics may be obtained. The microstructure studies, then confirmed by hydrostatic weighting, indicated that ceramics with an open porosity of about 40 % was formed. The mechanical strength on the three-point bending of porous ceramic was 35±6 MPa, which was a good and predictable result. One can be concluded that such porous ceramic should be used as a substrate for an asymmetric ion transport membrane for air oxygen separation.

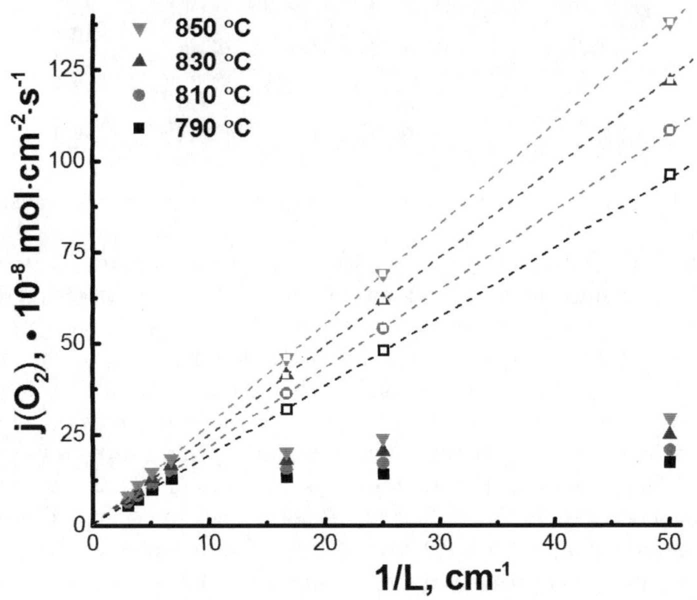

Figure 2. The oxygen flow of Co_3O_4 - 36 wt. % Bi_2O_3 membranes vs their inverse thickness

Figure 2 illustrates the dependences of oxygen flux $j(O_2)$ of Co_3O_4 - 36 wt. % Bi_2O_3 membranes deposited on a porous substrate vs their inverse thicknesses L^{-1} at the temperatures of 790-850 °C (log $P'(O_2)/P''(O_2) = 0,5$). The first four series of points located in the lower left corner of the Figure 2 are the data taken from [11], while the second three series of points are the experimental data obtained by us on membranes with covered thicknesses of 0,6 mm, 0,4 mm, and 0,2 mm, respectively. It is shown

IOP Conf. Series: Materials Science and Engineering **347** (2018) 012036 doi:10.1088/1757-899X/347/1/012036

that the oxygen flux rises at an increase in the temperature and in inverse thickness, but the tendency of such rising describes according to Burggraaf's model [17]. If the oxygen transport was described by the Wagner's model [12], the values of oxygen flux would be much higher, as shown in Figure 2 (the hypothetic values are marked as hollow geometric figures). Because the oxygen transport through the investigated composite system satisfies the Burggraaf's law, the process is carried out in a mixed diffusion-kinetic regime. Hence, the estimations of the characteristic membrane thicknesses indicate the values of 0,9 mm at 790 °C, of 0,85 mm at 810 °C, of 0,8 mm at 830 °C, and of 0,75 mm at 850 °C, respectively. These values are in good agreement with the literature data, since for the known SCF membranes the values of characteristic thickness are of 0,7 to 2 mm [18]. A similar characteristic thickness of 0,5 mm is known for BSCF membrane material [19]. At an increase in the temperature, the characteristic thickness of the membrane decreases, which is quite logical, since the rate of the oxygen surface exchange processes on interfaces is accelerated. It can be concluded that the optimal thickness of membrane material of Co_3O_4 - 36 wt. % Bi_2O_3 is approximately of 0,8 - 1 mm.

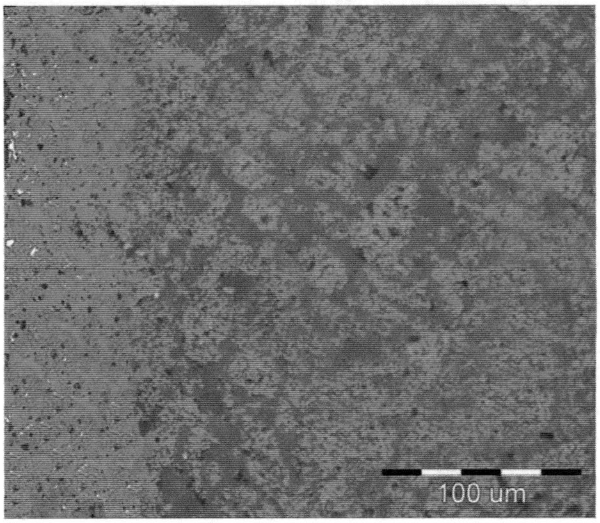

Figure 3. The micrograph of asymmetric membrane after oxygen transport tests

Figure 3 shows the microstructure of polished section of asymmetric "dense Co_3O_4 - 36 wt. % Bi_2O_3 / porous Co_2SiO_4" membrane after 10 hours of oxygen transport investigations at 850 °C. The microstructure indicates good adhesion of dense membrane to a porous support. But, unfortunately, one can be seen the penetration of Bi_2O_3 - based melt (light-grey fraction) on the depth of 50 μm into a porous support. Sadly, we can conclude that such melt interpenetration and its spreading along the membrane body will may subsequently lead to a decrease in the transport properties of the membrane. However, at now this conclusion is difficult to make, more long-term tests of oxygen transport of this membrane are needed. In our opinion, the solving of the problem of impregnation a porous membrane body by a melt can be a decrease in the surface energy of the substrate. In addition, the solving of the problem of impregnation a porous body by a melt can be a decrease in the surface energy of the substrate. It may be achieved by partial amorphization of support and by reduction of its average particle size to submicron, since it is known that surface energy of the material decreases at a decrease in particle size. Perhaps the dramatic creep problem solution which will also allowed to improve the mechanical properties of the substrate is to form the hierarchical porous microstructure of the substrate without a lot of mechanical stress concentrators. The solving of above-mentioned problems is not an easy task, since in our case we are dealing with a bismuth oxide-based melt that interacts with most oxide materials having good mechanical properties (Al_2O_3, ZrO_2, etc.). Possibly it is worthwhile to develop of the membrane material of other composition, operating at lower temperatures and not reacting with known constructional oxide materials, this may be a follow-up of our current researches.

4. Conclusions

As can be seen from all the presented results, the main goal of the research is solved. However, it should be noted that a significant increase in the oxygen flux of composite membranes is difficult because of the rather large value of the characteristic thickness of the membrane, which describes the rate of surface-exchange processes at the interfaces. It should also be understood that the increase in specific oxygen fluxes is difficult to realize due to a significant difference in the total pressure on both sides of the membrane because of the possible creep of membrane and interpenetrating of melt into the porous substrate. In addition, since it is known that Co_3O_4 is not enough thermodynamically stable at a low partial oxygen pressure [20], it is also no possibility of significantly increase in the oxygen flux due to the large oxygen partial pressure difference on both sides of the membrane. Therefore, the record oxygen selectivity with a comparable oxygen permeability of Co_3O_4 - 36 wt. % Bi_2O_3 [11], apparently, entails the subsequently lower membrane productivity, which is the dramatically restraining and hindering factor for the large-scale use of such materials in ion transport membranes.

Acknowledgements

This work was supported by the Russian Science Foundation (Project No. 16-19-10608).

References

[1] Kerry F G 2007 *Industrial gas handbook: gas separation and purification* (Boca Raton, FL: CRS Press, Taylor & Francis Group) 550
[2] Zeng P, Chen Z, Zhou W, Gu H, Shao Z and Liu S 2007 *Journal of Membrane Science* **291** 148
[3] Teraoka Y, Zhang H M, Furukawa S and Yamazoe N 1985 *Chemistry Letters* **11** 1743
[4] Zhu X and Yang W 2017 Mixed conducting ceramic membranes: fundamentals, materials and applications (Berlin: Springer) 367
[5] Yaroslavtsev A B 2016 *Russian Chemical Reviews* **85** 1255
[6] Belousov V V 2017 *Accounts of Chemical Research* **50** 273
[7] Belousov V V 2017 *Russian Chemical Reviews* **86** 934
[8] Belousov V V and Fedorov S V 2012 *Russian Chemical Reviews* **81** 44
[9] Belousov V V 1996 Journal of the American Ceramic Society **79** 1703
[10] Belousov V V 2016 *Ionics* **22** 451
[11] Belousov V V, Kulbakin I V, Fedorov S V and Klimashin A A 2016 *ACS Applied Materials & Interfaces* **8** 22324
[12] Wagner C 1933 Zeitschrift für Physikalische Chemie **21** 25
[13] Morimoto N, Tokonami M, Watanabe M and Koto K 1974 *American Mineralogist* **59** 475
[14] Zhuravlev L T 2000 Colloids and Surfaces A: Physicochemical and Engineering Aspects **17** 1
[15] Zabdyr L A, Garzel G and Fabrichnaya O B 2003 *Computer Coupling of Phase Diagrams and Thermochemistry* **27** 127
[16] Kargin Yu F, Voevodskii V Yu and Skorikov V M 1998 *Russian Journal of Inorganic Chemistry* **43** 1946
[17] Bouwmeester H J M, Kruidhof H and Burggraaf A J 1994 *Solid State Ionics* **72** 185
[18] Wiik K, Aasland S, Hansen H L, Tangen I L and Odegard R 2002 *Solid State Ionics* **152-153** 675
[19] Hong W K and Choi G M 2010 *Journal of Membrane Science* **346** 353
[20] Chen M, Hallstedt B and Gauckler L J 2003 *Journal of Phase Equilibria* **24** 212

IOP Conf. Series: Materials Science and Engineering **347** (2018) 012037 doi:10.1088/1757-899X/347/1/012037

Comparison of mechanical and friction properties of composite materials based on AlMg2 containing nano-dimensional particles of crystalline graphite and nanofibers of gamma oxide of aluminum

A V Aborkin, D M Babin and A V Sobol`kov

Vladimir State University named after Alexander and Nikolay Stoletovs, Vladimir, Russia

E-mail: aborkin@vlsu.ru

Abstract. The method of mechanical synthesis in a planetary ball mill was used for production of composite powders based on the AlMg2 alloy containing 1 wt. % of nanosized particles of crystalline graphite or γ-Al_2O_3. The resulting powders are consolidated by the sintering under pressure. Using the methods of X-ray diffraction analysis, scanning and transmission electron microscopy, the structural-phase composition of bulk composite materials was studied. Comparative analysis of the microhardness, the conditional yield stress at compression, and the friction coefficient of bulk composite materials is carried out. It has been found out that the mechanical properties of composites reinforced with γ-Al_2O_3 nanofibers are higher than when reinforcing with nanoscale particles of crystalline graphite.

1. Introduction

In the world scientific and technical literature there is a rather large number of works showing that dispersed hardening by nanosized reinforcement can significantly improve the structural and functional properties of aluminum and its alloys [1, 2].

The first and one of the most important stages in the production of composite materials by the method of powder metallurgy is the obtaining and preparation of powders. Their properties are largely determined by the physical-chemical conditions of the synthesis. The basis of mechanical synthesis is the mechanical treatment of solid mixtures (for example, in planetary ball mills), which results in grinding and plastic deformation of substances, accelerates mass transfer and mixing of the components down to the atomic level, and chemical interaction of solid reagents is activated [3].

Methods of thermoforming treatment of powder compositions are an effective means of obtaining high-density compacts (bulk materials). Both innovative technologies [4-7] and classical methods of pressure processing can be used for consolidation [8-11].

The aim of this work is investigation and comparison of the structural-phase composition, mechanical and frictional properties of composite materials based on the nanocrystalline matrix of the AlMg2 alloy, reinforced by additions of nanosized particles of crystalline graphite and nano-fibers of gamma-alumina.

2. Experimental methods and equipment

In the first case, the initial charge consisted of globular granules with a diameter of 1-2 mm of an aluminum alloy AlMg2 with addition of 1 wt. % graphite powder (average particle size less than 20 μm). In the second case, the matrix material remained the same, and 1 wt. % γ-Al_2O_3 nanofibers (diameter 10-40 nm) was used as the reinforcement. The SEM images of the reinforcements used in experiments are shown in Figure 1.

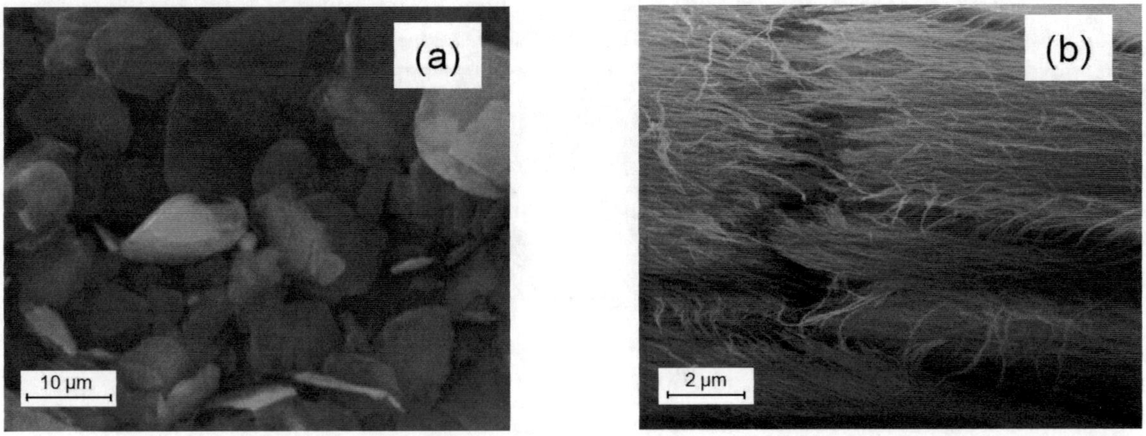

Figure 1. SEM images graphite (a) and nanofibers of γ-Al_2O_3 (b)

Mechanical treatment of the initial charge was carried out in a Fritsch Pulverisette 6 planetary ball mill using surfactants (1 wt.% stearic acid). For grinding, steel balls with a diameter of 8 mm were used at a ratio of the mass of the loaded components to the mass of the grinding bodies equal to 1:15. The treatment was carried out at 600 rpm for 4.5 hours.

The morphology of the resulting powders was studied using an ultra-high resolution field-emission scanning electron microscope Zeiss Ultra Plus based on Ultra 55. The granulometric composition of the obtained powders was determined on a Microsaser-201C analyzer. The powders were previously sieved through a 300 μm sieve.

Consolidation of the obtained powders was carried out by the method of sintering under pressure at temperature of 400 °C, pressure of 600 MPa and a holding under pressure of 30 minutes.

The fine structure of the consolidated samples was studied using the transmission electron microscopy JEM-2010. The microhardness was measured by the Vickers method using a Shimadzu HMV-2 tester at load of 10 N and exposure time of 12 s. The mechanical properties at the compression of the obtained samples were determined using a mechanical testing machine WDW-100E.

The frictional properties were determined on a CSM Instruments tribometer under dry friction conditions in contact with a disc of heat-treated 4Kh5MFS steel. During the tests, the pin-disk scheme was used. The load was 5 N, the friction path length was 200 m.

3. Results

Figure 2 a,b shows typical SEM images of synthesized powders characterizing their morphology. The powder particles have an irregular shape, close in geometry to fragmental. The particles of the reinforcement were not fixed on the surface of the powder particles. This is due to their low concentration and to the regularities of the powder particles formation under conditions of mechanical synthesis [12].

The results of granulometric analysis show that despite the type of reinforcement the average size of the d50 powder particles formed is fairly close and is 59 μm when graphite used as a reinforcement and 62 μm at using γ-Al_2O_3 nanofibers.

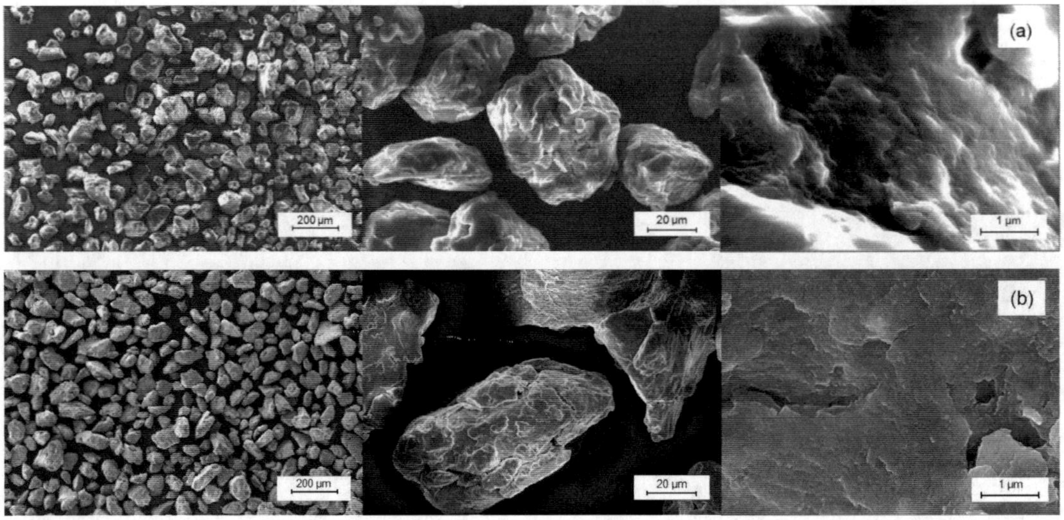

Figure 2. SEM images of mechanically synthesized powders: a – AlMg2+1 wt.% graphite; b – AlMg2+1 wt.% γ-Al_2O_3

Analysis of the results of X-ray diffraction analysis of the obtained powders (see Figure 3a) shows that regardless of the reinforcement type the diffractograms have a qualitatively similar character.

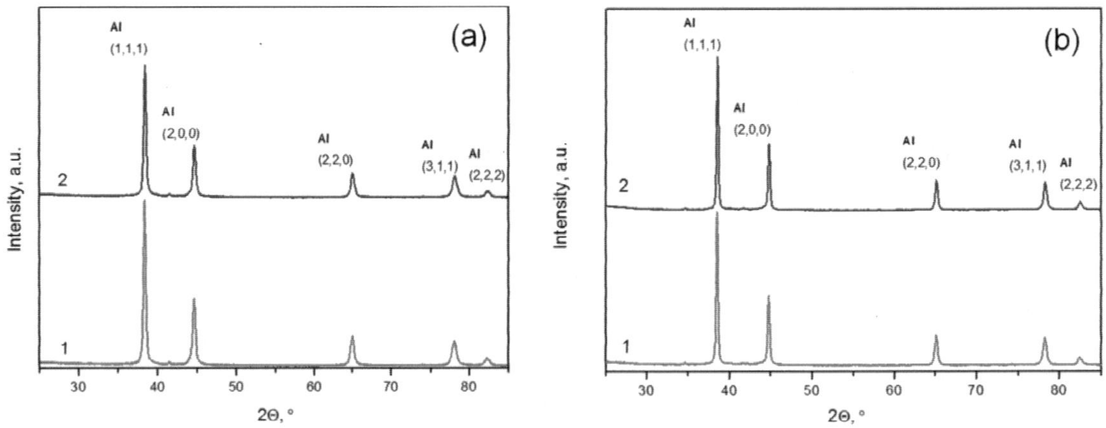

Figure 3. X-ray diffraction analysis of composite powders (a) and consolidated samples (b). The numbers are: 1 – AlMg2+1 wt.% graphite; 2 – AlMg2+1 wt.% γ-Al_2O_3

The presence of peaks corresponding to aluminum was noted [13]. However, there are no peaks corresponding to the reinforcements, which on the one hand can be due to their high dispersion achieved in mechanical synthesis, on the other hand, by the low sensitivity of the X-ray diffraction method to identifying phases with a content of less than 3%. Also, according to the Selyakov-Scherrer formula, the size of the coherent scattering regions was determined, according to which the size of crystallites of the matrix material can be judged with satisfactory accuracy. The volume-averaged sizes of the coherent scattering regions were calculated on the basis of the assumption of the spherical form of the crystallites. The results of the calculations show that for a composite material containing 1 wt.% graphite, the size of the coherent scattering regions is 72 nm, and for 1 wt.% γ-Al_2O_3 is 92 nm. This is due to the influence of graphite on the processes during mechanical synthesis, which consist both in changing the conditions of contact interaction between grinding bodies due to a change in the coefficient of friction, and in the fact that graphite, distributed over the surface of aluminum particles (at least at the initial stage of synthesis), reduces the number of juvenile surfaces, making plastic welding difficult and contributing to the dispersion of the powder.

After consolidation of the powders by the sintering under pressure, bulk samples with a diameter of 5 mm and a height of 5 mm were obtained. The results of X-ray analysis of consolidated samples are shown in Figure 3b. We can note a decrease in the width of the diffraction peaks of aluminum. The calculation of the size of the coherent scattering regions shows an increase in their values to 138 and 196 nm when graphite and γ-Al_2O_3 nanofibers are used as reinforcement, respectively.

Thus, in the process of aging during sintering under pressure, the regenerative processes take place in the consolidated material and the size of the crystallites passes from nano- to submicron scale.

In addition, the samples containing graphite as the strengthening additive were studied by Raman spectroscopy. This method has a high sensitivity to low concentrations of phases formed in the sample and makes it possible to more accurately identify the presence of Al_4C_3. The results of Raman spectroscopy show the presence of replicas at 492 cm^{-1} and 857 cm^{-1}, corresponding to the crystalline phase of Al_4C_3. Those the selected temperature-time consolidation conditions lead to the formation of the dispersion carbide phase in the material AlMg2 + 1 wt.% graphite.

Using the transmission electron microscopy, the structure of consolidated samples was studied. Analysis of the structure indicates that the obtained compact materials correspond to a crystalline structure with a crystallite size from tens to hundreds of nanometers. It can be noted that the calculated values of the average sizes of the coherent scattering regions are consistent with the crystallite size from the transmission microscopy data.

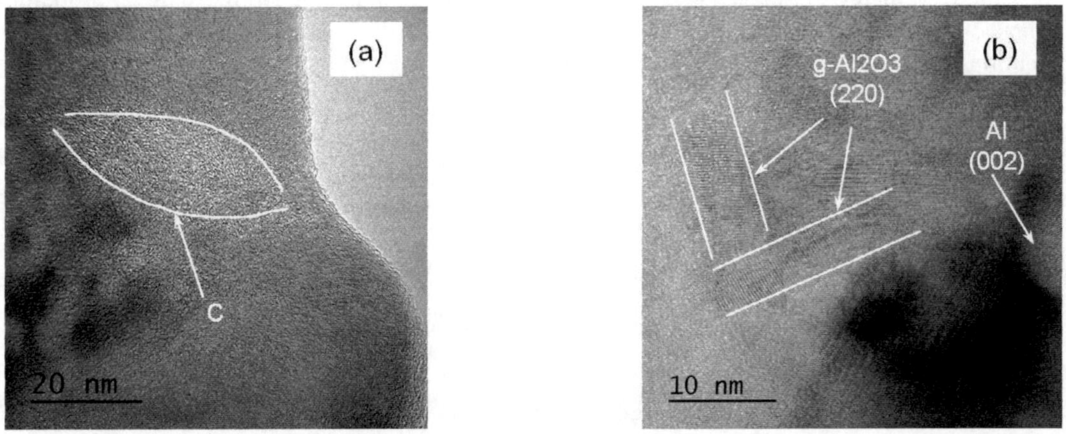

Figure 4. HRTEM images of consolidated composite material: a – AlMg2+1 wt.% graphite; b – AlMg2+1 wt.% γ-Al_2O_3

Figure 4 shows HRTEM images characterizing the fine structure of the resulting composite materials. An analysis of the obtained HRTEM images shows that during the mechanical grinding process a mechanical splitting of the microdimensional graphite particles occurs to form particles smaller than 50 nm. Also, the agglomerated γ-Al_2O_3 nanofibers are dispersed.

In the course of measuring the microhardness, it was established that the composite materials obtained have microhardness exceeding the matrix material in the initial state by an average of 3.4 times. For AlMg2 + 1 wt.% graphite and AlMg2 + 1 wt.% γ-Al_2O_3 microhardness was 1.9 and 2 GPa, respectively.

Figure 5 shows the results of the compression and friction tests of the obtained composite samples. Analysis of the results of compression tests shows a significant increase in the mechanical properties of the composites obtained as compared to the matrix material. Thus, for example, the conditional yield strength increases in comparison with the matrix by 5.6 and 6.4 times for AlMg2 + 1 wt% graphite and AlMg2 + 1 wt% γ- Al_2O_3, respectively. A comparison of the mechanical properties of the obtained composites (see Figure 5b) shows that the use of γ-Al_2O_3 nanofibers as a reinforcement makes it possible to obtain a material having not only high strength properties, but also a relatively high plasticity. While for the material reinforced with dispersed nanosized particles of graphite and dispersion particles Al_4C_3, despite the increase in strength properties, plasticity is significantly reduced, since the presence of the carbide phase promotes embrittlement of the material.

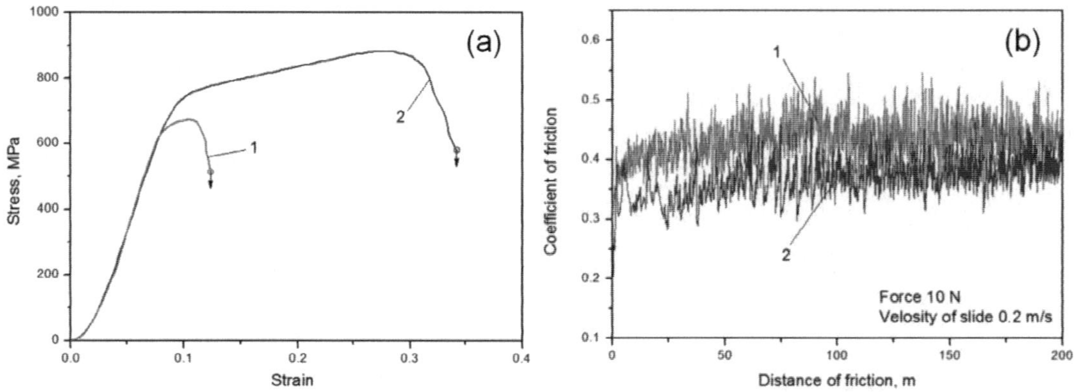

Figure 5. The results of compression tests (a) and friction (b) of the resulting composite samples. The numbers are: 1 – AlMg2+1 wt.% graphite; 2 – AlMg2+1 wt.% γ-Al_2O_3

The summary analysis of the data on the structural-phase composition and mechanical properties of the composite materials obtained allows us to conclude that hardening occurs with the additive contribution of the nanostructuring of matrix material and the dispersive hardening by the reinforcing particles. Moreover, in the case of composite AlMg2 + 1 wt.% graphite, additional dispersion hardening takes place due to the precipitation of the Al_4C_3 phase.

Friction tests show that the composite material reinforced with γ-Al_2O_3 nanofibers have a lower coefficient of friction 0.38. The use of nanosized particles of graphite as a reinforcement led to an increase in the coefficient of friction to 0.44.

Acknowledgments
The reported study was funded by RFBR according to the research project № 17-38-50109.

References

[1] Moghadam A D, Schultz B F, Ferguson J B, Omrani E, Rohatgi P K and Gupta N 2014 *JOM* **66** 872
[2] Baig Z, Mamat O and Mustapha M 2018 *Solid State and Materials Sciences* **43** 1
[3] Alymov M I 2007 *Powder Metallurgy of Nanocrystalline Materials* (Moscow: Science) p 169
[4] Aborkin A V, Elkin A I and Babin D M 2016 *Russian Journal of Non-Ferrous Metals* **57** 14
[5] Morsi K, Esawi A M K, Lanka S, Sayed A, Taher M 2010 *Composites A* **41** 322
[6] Vaganov V E, Aborkin A V, Alymov M I and Berbentsev V D 2015 *Russian Metallurgy (Metally)* **2015** 732
[7] Morsi K, El-Desouky A, Johnson B, Mar A and Lanka S 2009 *Scripta Materialia* **61** 395
[8] Choi H J, Kwon G B, Lee G Y and Bae D H 2008 *Scripta Materialia* **59** 360
[9] Choi H J, Shin J H, Min B H and Bae D H 2010 *Composites A* **41** 327
[10] Zhang H, Xu C, Xiao W, Ameyama K and Ma C 2016 *Mater. Sci. Eng.* **658** 8
[11] Majid M, Majzoobi G H, Noozad G A, Reihani A, Mortazavi S Z and Gorji M S 2012 *Rare Met.* **31** 372
[12] Aborkin A V, Evdokimov I A , Vaganov V E , Alymov M I., Abramov D V and Khor'kov K S 2016 *Nanotechnologies in Russia* **11** 297
[13] Aborkin A V, Alymov M I, Kireev A V, Elkin A I and Sobol'kov A V 2017 *Nanotechnologies in Russia* **12** 395

Study of absorption and IR-emission of Er^{3+}, Dy^{3+}, Tm^{3+} doped high-purity tellurite glasses

S E Motorin[1,2], V V Dorofeev[1,2], B I Galagan[2,3], S E Sverchkov[2,3], V V Koltashev[4] and B I Denker[2,3]

[1] G.G. Devyatykh Institute of Chemistry of High-Purity Substances of the Russian Academy of Sciences, 603950, Nizhny Novgorod, Russia
[2] Center of Laser Technology and Material Science, 119991, Moscow, Russia
[3] A.M. Prokhorov General Physics Institute of the Russian Academy of Sciences, 119991, Moscow, Russia
[4] Fiber Optics Research Center of the Russian Academy of Sciences, 119991, Moscow, Russia

E-mail: semotorin@gmail.com

Abstract. A study of high-purity TeO_2-ZnO based tellurite glasses doped with Er^{3+}, Dy^{3+} or Tm^{3+} that could be used as laser media in the 2-3 μm spectral range is presented. The glasses are prepared by melting the oxides mixture inside a silica glass reactor in an atmosphere of purified oxygen. The low level of hydroxyl groups absorption allowed to measure correctly the luminescence decay characteristics of the dopants. The rare-earth ions absorption bands, the luminescence spectra and kinetic characteristics of emission from the levels $^4I_{11/2}$, $^4I_{13/2}$ of Er^{3+}, $^6H_{13/2}$ of Dy^{3+} and $^3H_4$, $^3H_5$, $^3F_4$ of Tm^{3+} ions are investigated. The results confirm the high potential of tellurite glasses as an active media for bulk, planar waveguide and fiber lasers.

1. Introduction

Laser systems operating at 2-3 μm have important applications in both science and industry. This spectral range is characterized by a very high absorption of hydroxyl groups, so it is very attractive for surgical applications. There are several types of coherent optical sources beyond 2 μm. In recent years significant progress is demonstrated in supercontinuum generation and soliton self-frequency shift using tellurite, fluoride, germanate, and chalcogenide glass fibers [1-7]. However, the use of rare-earth ions (REI) doped fibers remains a promising way to create powerful mid-IR fiber lasers of simple design. At present, the technologies for creating continuous wave and pulsed fiber lasers operating in the range of 2-3 μm are well developed for ZBLAN fluoride glass fibers (see reviews [8-9] with references therein). Nevertheless, along with the advantages of high transparency in the mid-IR region and substantial solubility of REI, ZBLAN glass has pronounced drawbacks. It is mechanically weak, tends to crystallize and to react with atmospheric moisture.

Tellurium dioxide based glasses could be an alternative to fluoride glasses as a medium for the laser generation in the 2-3 μm range. Tellurite glasses containing oxides of tungsten, zinc, molybdenum, bismuth, alkali metals and rare-earth elements are promising materials as an active medium for the bulk, planar and fiber coherent light sources in the near and mid-IR ranges. They are characterized by a high transparency in the near and mid-IR [1], by highly nonlinear optical properties [1-3], by sufficient stability against crystallization [10, 11], and by excellent rare-earth oxides solubility [12]. The transparency range of TeO_2-based glasses in the IR region is much wider than that for silicate glasses (~2.1 μm [13, 14]) and extends up to 5-6 μm for short bulk and planar devices and up to 3.5-4.5 μm for lengthy optical fibers [1, 2]. The main obstacle for their extensive applications is the difficulty to obtain the glasses with low enough concentration of 3d-transition metals impurities and OH-groups. A related problem is the lack of the adequate luminescent measurements, which can be realized only using well-dehydrated and high-purity glasses. Nevertheless the possibility to produce optical fibers with low losses up to 3-3.5 μm [15, 16] and efficient luminescence of REI up to 2.7-2.9 μm are successfully demonstrated [17-19].

One of the interesting tellurite glass compositions is the zinc-tellurite glass containing up to 30 % mol. ZnO and some modifying components such as alkali and rare-earth metals oxides. Zinc-tellurite glasses have noticeably lower phonon energy than tellurium dioxide - heavy metal oxide based systems, like tungstate-tellurite glasses. Transparency range of zinc tellurite glasses is also much wider than for tellurite glasses based on heavy metal oxides (W, Mo). The minimal level of optical losses in these glasses is estimated to be ~0.5–1 dB/m [20] and it is located in the spectral area, where silica glass optical fibers are not transparent (at ~3 μm). Like other tellurite glasses, TeO_2-ZnO based compositions are characterized by high (~2) refractive indices, sufficient chemical resistance, and significant solubility of rare-earth elements. All these properties make zinc-tellurite glasses interesting for mid-IR lasers.

Zinc-tellurite glasses doped with rare-earth ions having radiative transitions that are suitable for the mid-IR range are being actively studied at the present time. Several research groups made Er-doped TeO_2-ZnO based glasses and fibers and studied their luminescent properties, including spectroscopic parameters for the $^4I_{11/2} \rightarrow {}^4I_{13/2}$ energy transition near 2.7 μm [21-23].

There is a number of papers [24-26] devoted to spectroscopic analysis of Tm^{3+}-doped zinc-tellurite glasses of different compositions. Despite the small branching ratio for 2.3 μm transition, the cross-section of this transition in a zinc-tellurite glass can be expected rather high due to high refractive indices and to relatively long wavelength of the transition. The estimated cross-section value is at least 3×10^{-21} cm^2 [26].

In the present investigation we have produced and studied high-purity zinc-tellurite glasses doped with Er^{3+}, Dy^{3+} or Tm^{3+} because these REI have appropriate radiative transitions in the range of our interest (2-3 μm) and their absorption bands are suitable for pumping by commercially available diodes operating at 0.98-μm and at ~0.8 μm as shown in figure 1.

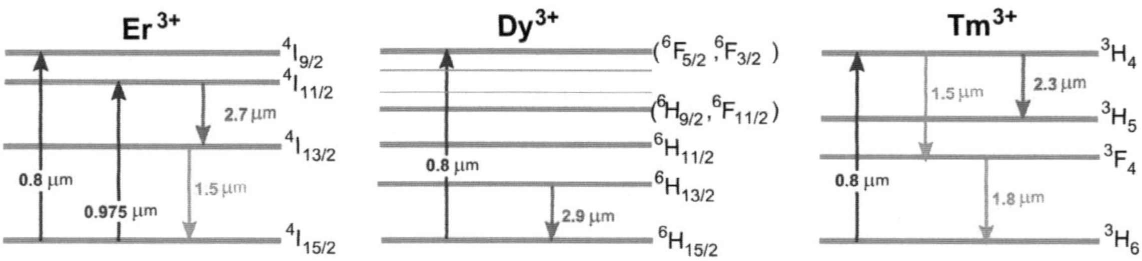

Figure 1. Simplified energy levels schemes of REI.

2. Glass preparation

Tellurite glasses are prepared by melting a batch in crucibles of gold or platinum inside sealed a silica chamber in purified oxygen atmosphere. Crystallization resistant zinc-tellurite glass composition modified with lanthanum and sodium oxides is chosen as a host glass matrix.

High-purity tellurium dioxide TeO_2 that is used in the synthesis is produced by the original patented method [27]. Ultrapure zinc oxide ZnO is produced by diethyl zinc oxidation. Commercially available superpurity grade lanthanum oxide La_2O_3 and sodium carbonate Na_2CO_3 are used for the glasses preparation. The total content of 3d-transition metals impurities in the mixture of initial oxides is determined to be 0.2–2 ppm wt and the total concentration of undesired rare-earth elements does not exceed 1–2 ppm wt same as in [28, 29].

The glass-forming melt is held at 800 °C for several hours with recurrent stirring. Then the melt is poured into silica glass moulds and annealed for several hours at a glass transition temperature. After annealing the obtained samples are extracted from the moulds and mechanically treated for further investigations. The prepared high-purity glasses are characterized by extremely low OH-group absorption less than 0.01 cm^{-1} at the peak at about 3 μm. This absorption is controlled by the procedure described in [30]. The samples with different concentrations of REI have been prepared for spectroscopic investigation.

3. Transmission spectra

The visible spectra are recorded by the spectrophotometer Lambda 900 and the IR spectra are recorded by the IR Nicolet 6700 Fourier spectrometer. The transmission spectra of glass samples doped with Er^{3+}, Dy^{3+} and Tm^{3+} are shown in figure 2 a,b,c respectively. The absorption coefficients are calculated from transmission spectra by the Beer–Lambert–Bouguer equation: the volume absorption coefficient $\alpha = \ln(1/T)/L$, where T is the transmittance; L - sample length [31].

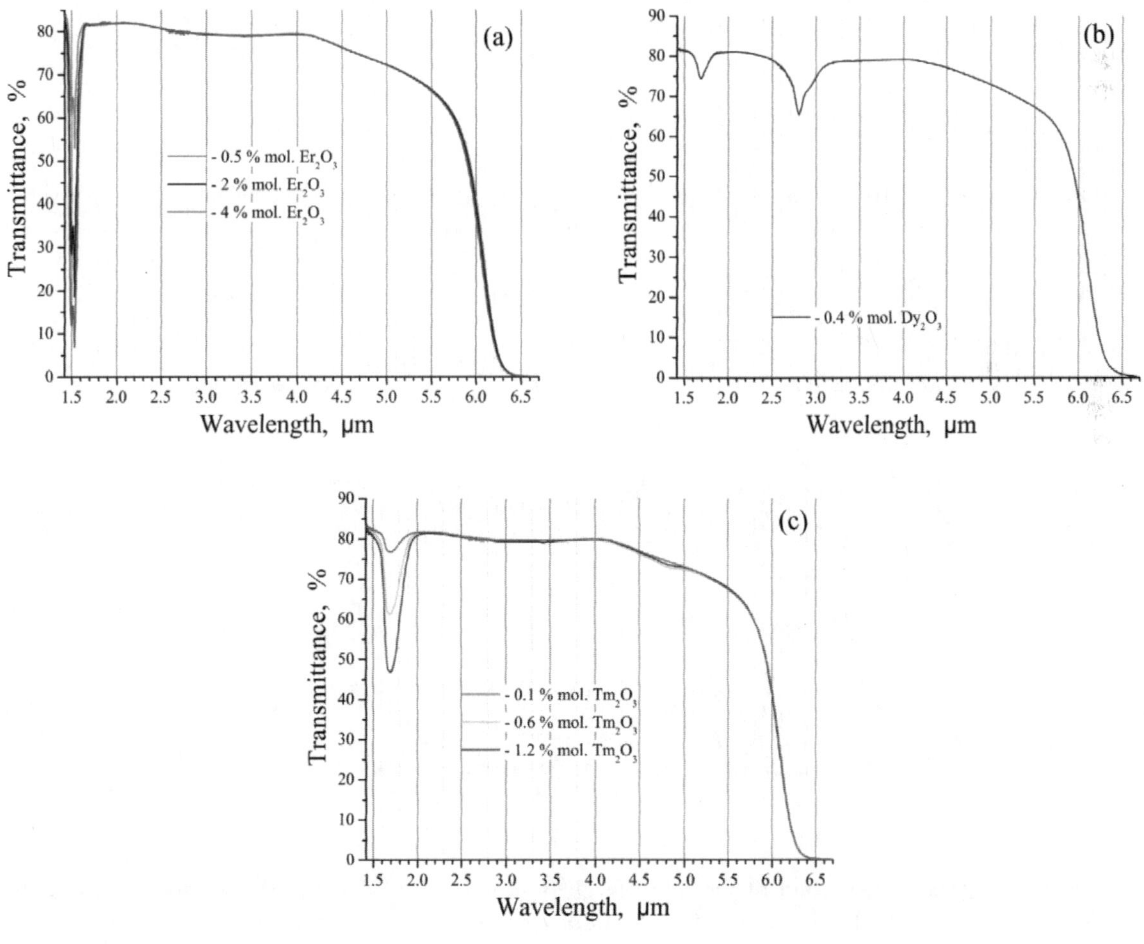

Figure 2. Transmittance spectra of the tellurite glass samples of 2-2.3 mm thick doped with different content of Er_2O_3 (a), Dy_2O_3 (b) or Tm_2O_3 (c). Hydroxyl groups absorption in the vicinity of 3 μm is less than ~0.005 cm^{-1} in all samples.

Rare-earth doped glasses are transparent in the 0.4-5.2 μm wavelength region. The absorption bands of REI with maxima at about 1.5 μm, at both 1.7 and 2.8 μm or at 1.7 μm are observed in transmittance spectra of the tellurite glasses samples doped with different content of Er_2O_3, Dy_2O_3 or Tm_2O_3 accordingly. Thanks to the use of the original procedure of melt dehydration, the absorption bands of hydroxyl groups at around 1.5, 2.3 and 3.3 μm are not observed at the spectra of thin samples.

4. Luminescent properties of glasses

The luminescent properties of REI in the near and mid-IR ranges, including decay times under various excitation wavelengths are investigated as well. Low OH-absorption level allowed to investigate correctly the emission properties of the rare-earth dopants inside the spectral regions of hydroxyl groups absorption bands.

While excited at 0.98 μm, broad luminescence bands with maxima at ~1.5 and ~2.7 μm corresponding to electronic transitions $^4I_{13/2}\rightarrow{}^4I_{15/2}$ and $^4I_{11/2}\rightarrow{}^4I_{13/2}$ of Er^{3+} are recorded (figure 3(a)). To measure lifetimes of excited levels, we used the similar technique described in the paper [18]. This technique in cludes short pulse excitation t the required wavelength and excludes the reabsorption.

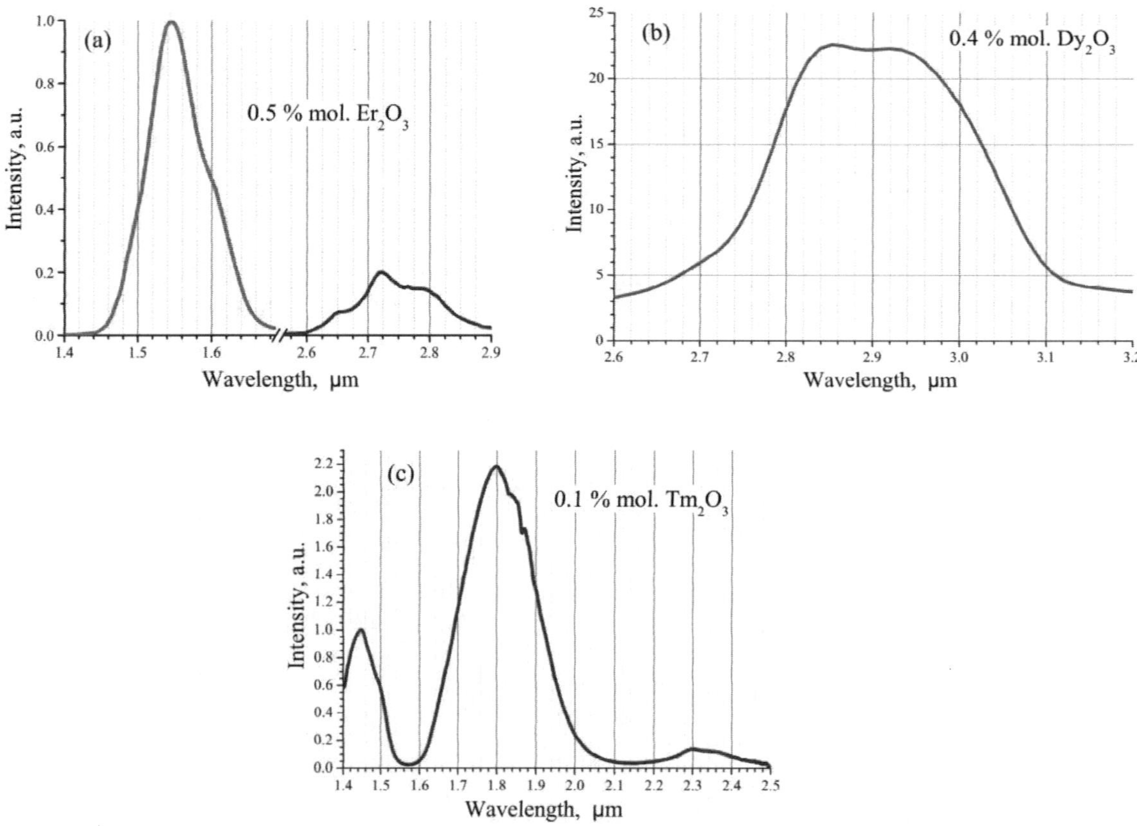

Figure 3. Luminescence spectra of the tellurite glass samples of 2-2.3 mm thick doped singly with Er_2O_3 (a), Dy_2O_3 (b) and Tm_2O_3 (c) under excitation of 0.975, 0.8 and 0.8 μm, respectively.

The luminescence decay times at ~1.5 and ~0.98 μm are at the level of 3.5 ms and 250 μs, respectively, for the best samples. The results of the study showed that zinc-tellurite glasses doped with erbium ions are of considerable interest for erbium laser generation at ~2.7 μm, mostly because the upper laser lifetime in them is twice longer than in heavy metal tellurite glasses.

Luminescence band of Dy^{3+} with maximum at 2.9 μm corresponding to electronic transition $^6H_{13/2}\rightarrow{}^6H_{15/2}$ is registered for the Dy_2O_3 doped glass sample under 0.8 μm excitation (figure 3(b)). The lifetime of the level $^6H_{13/2}$ is equal to ~13 μs. This lifetime is determined mostly by multiphonon luminescence quenching rate.

Three emission bands of Tm^{3+} with maxima at 1.46, 1.8, and 2.3 μm corresponding to electronic transitions $^3H_4\rightarrow{}^3F_4$, $^3F_4\rightarrow{}^3H_6$ and $^3H_4\rightarrow{}^3H_5$, respectively, are recorded under 0.8 μm excitation of the tellurite glass samples doped with Tm_2O_3 (figure 3(c)). The lifetimes of the $^3H_4$, $^3H_5$, $^3F_4$ levels have been determined experimentally. They are 300, 0.13 and 2660 μs, respectively for the glass with 0.1% Tm_2O_3.

5. Conclusion

We have prepared high-purity zinc-tellurite glasses doped with Er^{3+}, Dy^{3+} or Tm^{3+} and experimentally studied their absorption and luminescent properties in the near and mid-IR. Hydroxyl groups absorption in the all samples prepared is less than 0.01 cm^{-1} at the maximum of the main absorption band near 3 μm. The luminescence spectra near 1.5 and 2.7 μm for Er-doped glass corresponding to $^4I_{13/2}\rightarrow{}^4I_{15/2}$ and $^4I_{11/2}\rightarrow{}^4I_{13/2}$ laser transitions; near 2.9 μm for Dy-doped glass corresponding to $^6H_{13/2}\rightarrow{}^6H_{15/2}$ laser transition; and near 1.46, 1.8 and 2.3 μm for Tm-doped glass corresponding to $^3H_4\rightarrow{}^3F_4$, $^3F_4\rightarrow{}^3H_6$ and $^3H_4\rightarrow{}^3H_5$ laser transitions are recorded and analyzed. The measured lifetimes of excited levels are 3.5 ms and 250 μs for $^4I_{13/2}$ and $^4I_{11/2}$ of Er^{3+}; 13 μs for $^6H_{13/2}$ of Dy^{3+}; and 300, 0.13 and 2660 μs for $^3H_4$, $^3H_5$, and $^3F_4$ of Tm^{3+}, respectively. This work may provide promising active materials for lasers emitting at 1.45, 2.3 and 2.7 μm.

Acknowledgments

This work was supported by Russian Science Foundation (grant 17-12-01324).

References

[1] Tao G, Ebendorff-Heidepriem H, Stolyarov A M, Danto S, Badding J V, Fink Y, Ballato J, and Abouraddy A F 2015 *Adv. Opt. Photon.* **7** 379

[2] Kedenburg S, Strutynski C, Kibler B, Froidevaux P, Désévédavy F, Gadret G, Jules J-C, Steinle T, Mörz F, Steinmann A, Giessen H, and Smektala F 2017 *J. Opt. Soc. Am. B* **34** 601

[3] Koptev M Yu, Anashkina E A, Andrianov A V, Dorofeev V V, Kosolapov A F, Muravyev S V, and Kim A V 2015 *Opt. Lett.* **40** 4094

[4] Tang Y, Wright L G, Charan K, Wang T, Xu C, and Wise F W 2016 *Optica* **3** 948

[5] Anashkina E A, Andrianov A V, Koptev M Yu, Muravyev S V, and Kim A V 2014 *Opt. Lett.* **39** 2963

[6] Granzow N, Schmidt M A, Chang W, Wang L, Coulombier Q, Troles J, Toupin P, Hartl I, Lee K F, Fermann M E, Wondraczek L, and Russell P St J 2013 *Opt. Express* **21** 10969

[7] Anashkina E A, Shiryaev V S, Koptev M Y, Stepanov B S, and Muravyev S V 2018 *J. Non-Cryst. Solids* **480** 43

[8] Jackson S D 2012 *Nat. Photon.* **6** 423

[9] Zhu X, Zhu G, Wei C, Kotov L V, Wang J, Tong M, Norwood R A, and Peyghambarian N 2017 *J. Opt. Soc. Am. B* **34** A15.

[10] Yakovlev A I, Snetkov I L, Dorofeev V V, and Motorin S E 2018 *J. Non-Cryst. Solids* **480** 90

[11] Smayev M P, Dorofeev V V, Moiseev A N, and Okhrimchuk A G 2018 *J. Non-Cryst. Solids* **480** 100

[12] Kut'in A M, Plekhovich A D, Balueva K V, and Dorofeev V V 2018 *J. Non-Cryst. Solids* **480** 95

[13] Kotov L V, Koptev M Yu, Anashkina E A, Muravyev S V, Andrianov A V, Bubnov M M, Ignat'ev A D, Lipatov D S, Gur'yanov A N, Likhachev M E, and Kim A V 2014 *Quantum Electron.* **44** 458

[14] Koptev M Yu, Anashkina E A, Andrianov A V, Muravyev S V and Kim A V 2014 *Opt. Lett.* **39** 2008
[15] Dorofeev V, Moiseev A, Kraev I, Motorin S, Churbanov M, Kosolapov A F, Dianov E, Plotnichenko V, and Philippovskiy D 2012 Tungstate-tellurite glass fibers for spectral region up to 3 μm *in Advanced Photonics Congress, OSA Technical Digest (online)* paper JM5A.3
[16] Anashkina E A, Andrianov A V, Dorofeev V V, and Kim A V 2016 *Appl. Opt.* **55**, 4522
[17] Kosolapov A F, Dorofeev V, Moiseev A, Motorin S, Plotnichenko V, Koltashev V, and Filippovskiy D 2014 Er-doped Tungstate-Tellurite Glasses and Fibers *in Advanced Photonics*, paper JM5A.10
[18] Denker B I, Dorofeev V V, Galagan B I, Motorin S E, and S E Sverchkov 2017 *Laser Phys.* **27** 095801
[19] Merzliakov M A, Kouhar V V, Malashkevich G E, Pestryakov E V 2018 *Opt. Mater.* **75** 142
[20] Ketkova L A, Churbanov M F 2018 *J. Non-Cryst. Solids* **480** 18
[21] Wang W C, Yuan J, Li L X, Chen D D, Qian Q, and Zhang Q Y 2015 *Opt. Mater. Express* **5** 2964
[22] Ma Y, Guo Y, Huang F, Hu L, and Zhang J 2014 *J. Lumin.* **147** 372
[23] Gomes L, Oermann M, Ebendorff-Heidepriem H, Ottaway D, Monro T, Librantz A F H, and Jackson S D 2011 *J. Appl. Phys.* **110** 083111
[24] Lachheb R, Damak K, Assadi A A, Herrmann A, Yousef E, Rüssel C, Maâlej R Chara 2015 *J. Lumin* **161** 281
[25] Damak K, Maâlej R, Yousef El S, Qusti A H, Rüssel C 2012 *J. Non-Cryst. Solids* **358** 2974
[26] Gomes L, Lousteau J, Milanese D, Scarpignato G C, and Jackson S D 2012 *J. Appl. Phys* **111** 063105
[27] Moiseev A N, Chilyasov A V, Dorofeev V V, and Kraev I A 2007 Method for purifying tellurium dioxide, Patent RF 2301197, Bull. 17
[28] Dorofeev V V, Moiseev A N, Churbanov M F, Kotereva T V, Chilyasov A V, Kraev I A, Pimenov V G, Ketkova L A, Dianov E M, Plotnichenko V G, Kosolapov A F, Koltashev V V 2011 *J. Non-Cryst. Solids* **357** 2366
[29] Moiseev A N, Dorofeev V V, Chilyasov A V, Kraev I A, Churbanov M F, Kotereva T V, Pimenov V G, Snopatin G E, Pushkin A A, Gerasimenko V V, Kosolapov A F, Plotnichenko V G, Dianov E M 2011 *Opt. Mater.* **33** 1858
[30] Dorofeev V V, Moiseev A N, Churbanov M F, Snopatin G E, Chilyasov A V, Kraev I A, Lobanov A S, Kotereva T V, Ketkova L A, Pushkin A A, Gerasimenko V V, Plotnichenko V G, Kosolapov A F, Dianov E M 2011 *Opt. Mater.* **33** 1911
[31] Anashkina E A, Dorofeev V V, Koltashev V V, and Kim A V 2017 *Opt. Mater. Express* **7** 4337

IOP Conf. Series: Materials Science and Engineering **347** (2018) 012039 doi:10.1088/1757-899X/347/1/012039

Calcium phosphate composite cements based on simple mixture of brushite and apatite phases

A A Egorov[1], A Yu Fedotov[1], I S Pereloma[1], A Yu Teterina[1], N S Sergeeva[2], I K Sviridova[2], V A Kirsanova[2], S A Akhmedova[2], AV Nesterova[3], I V Reshetov[4], S M Barinov[1] and V S Komlev[1]

[1]Baikov Institute of Metallurgy and Materials Science, Russian Academy of Sciences, Moscow, 119334, Russia
[2]P.A. Herzen Moscow Oncology Research Institute—Branch of the National Medical Research Radiological Centre of the Ministry of Health Care of the Russian Federation, Moscow, 125284, Russia
[3]N.I. Pirogov Russian National Medical University of the Ministry of education and science of the Russian Federation, Moscow, 117997, Russia
[4]I.M. Sechenov First Moscow State Medical University, Ministry of Health of the Russian Federation, Moscow, 119991, Russian Federation

E-mail: komlev@mail.ru

Abstract. The composite cements based on simple mixtures brishite and apatite with ratio 70/30, 50/50, 30/70 were developed. The processes of phase formation, microstructure and mechanical properties were studied. The kinetics of degradation in simulated body fluid depending on the microstructure and the materials phase composition was carried out. The biological test *in vitro* were performed using the MTT-test on the human fibroblast immortalized (hFB) cell line and the human osteosarcoma cell line MG-63. The materials didn't have acute cytoxicity and possessed surface matrix properties. It was determined that the both line of cells actively proliferated, with viable cells values higher 20-60 % then control at all observation periods.

1. Introduction

Nowadays, the replacement of bone tissue is regarded as one of the most common surgical procedures. In fact a variety of techniques and synthetic substitutes have been developed in order to properly reconstruct tissue defects. CPCs represent a adequate synthetic substitute, which are widely used as filling materials in the field of orthopedic and dental medicine because of their good biocompatiblity, resorbability and a low toxicity [1]. The type of setting product formed by the reaction is affected by the pH value conditions in the cement paste. The two most commonly encountered end products are precipitated apatite or brushite. Brushite cements possess resorption rate quicker than apatite cements due to the higher solubility of brushite phase compared to that of calcium-deficient hydroxyapatite, which may be an advantage for bone regeneration [2].

Despite numerous advantages, it is widely accepted that CPCs need further improvements to their properties to broaden their potential clinical applications. The properties of CPCs are affected by many factors. Therefore, it is imaginable that all the factors, such as the chemical composition of the cement, the relative proportions of the reactants in the mixture, powder or liquid additives acting as accelerators or retarders, particle size, liquid / powder ratio, pressure applied during sample preparation and aging conditions, will affect its properties [3]. Many of the studies directed to improving CPC properties for clinical applications. Most of them were focused on improving next points: processing techniques [5], CPC mechanical properties [6], influence of CPC material properties on cell behavior [7], CPC materials as drug delivery systems [8-9], stem cell delivery by CPC [9]. The improvement of these properties is carried out by introducing a different polymer, fiber or ceramic additives in CPC based on brushite or apatite. Therefore, when optimizing any property of CPC, it is important to consider all the crucial properties. The main idea of this work, the development of composite materials by combining the advantage brushite and apatite cements. The study of these cements are poorly described in the literature.

On the other hand, the disadvantage of these materials is low porosity, do not contribute to the integration of the cement with the surrounding bone tissue. The absence of interconnected pores disturbs the circulation of the liquid inside the cement and the absence of fine pores reduces the adhesion of cells on the surface of the cement. The creation of porous cements with different pore-forming additives (PA) can solve this problem. PA introduced at the stage of mixing the, for example, crystals of soluble salts, particles of sugar, ice, water-soluble polymers cement and removed after hardening of the binding system. With the introduction of such additives can be achieved with a porosity of 50 %. Also, porosity in cements can be created and with the use of foaming agents, for example, ammonium carbonate or carbonate of calcium, the interaction of which with the cement liquid (usually phosphoric acid) is accompanied by release of carbon dioxide. It is possible to achieve higher porosity, but it is difficult to control the pore size [10, 11].

The purpose of this study was to prepare brushite/apatite composite cements and investigate the effect of PA on the phase formation, microstructure and mechanical properties, *in vitro* cytocompatibility and matrix (for cells) properties in dynamics of cultivation (up to 7 days) of the final cement composites.

2. Experimental

In this work the composition of CPCs were chosen: brushite cement (dicalcium phosphate dihydrate (DCPD)) based on reaction of the metastable hydrolysis of orthophosphate in the aquatic environment and apatite cement (apatite (HA) products) based on acid reaction with the formation of a neutral compound.

The cement paste was prepared by manually mixing the powder α-tricalcium phosphate (α - TCP) average size of 1-10 μm with hardening liquid (HL) comprising a magnesium phosphate solution in orthophosphoric acid with a spatula in a glass mortar, maintaining a liquid-to-powder ratio for brushite cement of 0.25 ml g^{-1} and apatite cement of 0.22 ml g^{-1}. Composite cement prepared by simple mechanical mixture apatite and brushite cement paste with 70 wt.%/ 30 wt.%, 50 wt.% / 50 wt.%, 30 wt.% / 70 wt.%.

Porous CPCs were prepared by the standard procedures reported in literature [12] using two types of PA: chemical and soluble additives. The introduction of chemical-reactive PA ($CaCO_3$, Na_2CO_3) to the CPC leads to chemical interaction with CPC. It was resulted from obtaining and releasing of CO_2 gas. The formation of the tunnel interconnected pores with a size about 30-40 μm and porosity up to 30 % was occurred. The porosity formation soluble PA (NaCl) was due to dissolution PA with contact simulated body fluid. PA were taken in a quantity of up to 5 wt.% for Na_2CO_3 and 15 wt. % for $CaCO_3$ and NaCl.

The cement setting time was determined with a Vicat apparatus based on the instant of disappearance of the Vicat needle trace on the sample surface. The resulting mass was put into 6 mm diameter × 12 mm height cylindrical Teflon moulds. After 5 – 7 minutes, the sample were allowed to set at 100 % relative humidity for 7 days at 37 °C.

The samples were investigated by scanning electron microscopic analysis (SEM, Tescan Vega II). The secondary and backscattered electron modes was used for microstructure studies of the ceramic samples. The samples were coated with a 25 nm-thick gold layer to impart electrical conductivity to the specimen surfaces for SEM.

The phase composition was analyzed by conventional X-ray diffraction (XRD) technique (Shimadzu XRD-6000 (Japan), Ni-filtered $CuK_{\alpha 1}$ target, λ=1.54183 Å).

The porosity, intergranular size and specific surface area of the investigated materials were determined by mercury porometry (TriStar 3000, Micromeritics, USA).

The mechanical tests were carry out by axial compression of cylindrical samples using Instron 5581 testing machine. The solubility testing was carried out according to GOST R ISO 10993-14-2001 by measuring of mass loss. The samples were kept in buffering solution up to 28 days at a constant volume of the liquid phase (closed system), pH 4.5, and 37°C.

The kinetics of degradation was evaluated by weight loss CPCs when exposed in SBF up to 28 days. The formation of calcium phosphate layer was investigated by scanning electron microscopy. In this study we used the SBF consist of composition (g/L): NaCl 6.457; $NaHCO_3$ 2.268; KCl 0.373; $Na_2HPO_4 \cdot 12H_2O$ 0.178; $MgCl_2 \cdot 6H_2O$ 0.305; Na_2SO_4 0.071; $CaCl_2 \cdot 2H_2O$ 0.368; Tris buffer 6.057.

The biological studies of CPCs in vitro were assessed on the hFB, clone 1608 hTERT (Engelhardt Institute of Molecular Biology of Russian Academy of Sciences) and the human osteosarcoma cell line MG-63 (Russian Cell Culture Collection, Institute of Cytology of The Russian Academy of Sciences). Before in vitro experiments, sterile CPC samples (γ irradiation of 15 kGy) were placed into 96 well plates for cultivation (Costar, United States) in triplicate and one plate per each incubation period (1, 3,7 days) and covered with complete growth medium (CGM) that contained DMEM medium (PanEko,Russia), 10% fetal calf serum (FCS, Austria), glutamine (0.65 mg/mL, PanEko,Russia), and gentamycin (50 μg/mL, PanEko,Russia). After neutral values of pH were established (1–3 days of the sample exposure to the CGM with a change of medium), the cell suspension (hFB at density of 2x104 cells per well or MG-63 at 1,5x104 cells per well) at 200 μL CGM was added to the plates with samples (experiment) and without them (control with cultural plastic, polystyrene). All the procedures were performed under sterile conditions in an atmosphere of moist air that contained 5% CO2 at 37°C. The viability of the hFB and MG-63 cell lines over time was assessed using an MTT test [13]. Briefly, after appropriate co- culture incubations, MTT (3-(4,5-dimethylthiazol-2-yl)-2,5-diphenyltetrazolium bromide (Sigma–Aldrich, USA,5 mg/mL) was added to each well and incubated for 3 h until coloured formazan compounds were visible. The supernatants were aspirated and isopropanol was added to dissolve formasan precipitates. Optical density at 540 nm was measured using an automatic reader for microtiter plates (Multiscan FC, Thermoscientific USA).

At cultivation stages, a pool viable cell (PVC) was calculated as a ratio of the optical density of the formazan solution in the experiment to this value in the control. The results were processed by conventional methods of variational statistics using Microsoft Excel 2000. The significance of differences was assessed using a parametric Student t-test; differences were considered statistically significant at $p < 0.05$.

3. Results and Discussion

First, the most promising formulation was selected to investigate the phase formation after the setting reaction. For this purpose, the effect of the brushite / apatite cement ratio in the mixture was investigated in order to find the reasonable values for phase composition, setting time and compressive strength. Table 1 gives the results obtained for these parameters.

Table 1 Characteristics of CPC cements as a function of composition.

Composition phase mixture (wt. %)	Final composition phase (wt%)	Setting time (min)	Compressive strength (MPa)
brushite	DCPD/HA/TCP 85/10/5	8	6
brushite/apatite 70/30	DCPD/HA/TCP 75/20/5	9	22
brushite/apatite 50/50	DCPD/HA/TCP 65/30/5	10	28
brushite/apatite 30/70	DCPD/HA/TCP 10/80/10	13	35
apatite	HA/TCP 90/10	15	55

It was shown that very small amount of the hydroxyapatite (HA) phase (10 wt. %) was detected for brushite cement, whereas about 20 wt. % of HA in the final reaction product was obtained for the 70/30 brushite/apatite composition. The (65 and 10 wt. %) DCPD, (30 and 80 wt. %) HA phases were detected for 50/50 and 30/70 brushite/apatite compositions. The setting time increases with increasing content of apatite phase. This the dependence of phase formation can be justified by a difference in solubility between DCPD and HA increasing upon the HA conversion. Recently, we reported the additional component (DCPD) influences significantly the structural behavior of HA, acting as the HA crystallization inhibition agent [14].

Further it was chosen the materials with the following composition: brushite, 50/50 and apatite. This choice of materials was made due to the following reasons: brushite/apatite 50/50 composition had optimal ratio of setting time and compressive strength. Brushite and apatite were chosen as extreme concentrations.

The XRD analysis had shown that apatite cements had an amorphous structure, with α-TCP particles up to 40 wt. %. The composite cements 50/50 had amorphous apatite phase (up to 30 wt. %), brushite phase (up to 30 wt. %) and crystalline α-TCP (up to 40 wt. %). The brushite cements had amorphous apatite phase (10 wt. %), brushite phase (up to 70 wt. %) and crystalline α-TCP (up to 20 wt. %). The addition of $CaCO_3$ in an amount of 5 wt. % brushite cement leads to the formation of apatite phase up to 10 wt. %. The introduction of additives in apatite cement had no effect on the phase composition (Figure 1).

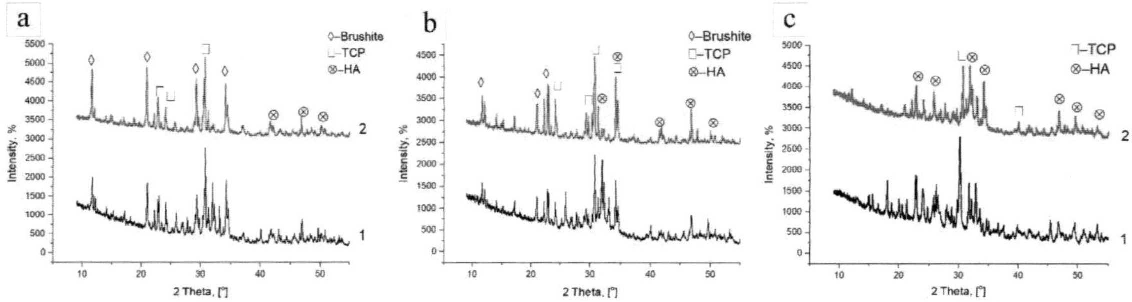

Figure 1. Comparative XRD pattern (1) dense and (2) porous CPC: (a) brushite; (b) 50/50, (c) apatite.

According to SEM observation the CPCs cement was predominantly homogeneous with α-TCP particles with average size of 1-10 μm embedded into the matrix (Figure 2). Brushite and 50/50 cements are composed of plate shape brushite crystals up to 2 μm thick and 5-20 μm length, with α-TCP particles and apatite phases.

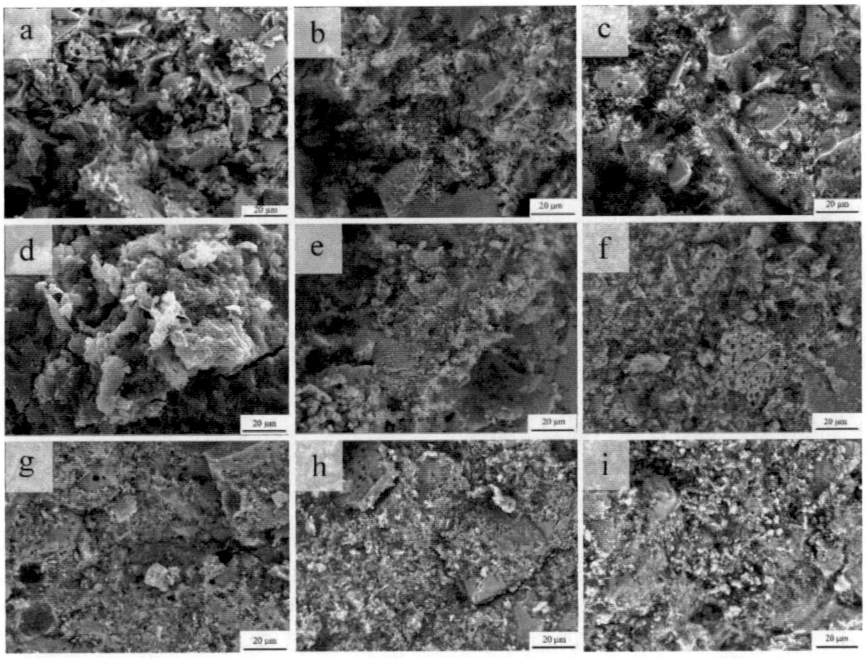

Figure 2. SEM images of the porous CPC samples with PA wt. 5 %: (a) – brushite with $CaCO_3$, (b) - brushite with Na_2CO_3, (c) - brushite with NaCl, (d) – 50/50 with $CaCO_3$, (e) – 50/50 with Na_2CO_3, (f) – 50/50 with NaCl, (g) – apatite with $CaCO_3$, (h) - apatite with Na_2CO_3, (i) - apatite with NaCl.

As shown in Figure 3, the interaction between brushite and apatite cement didn't occur. Uniform distribution of phases was observed.

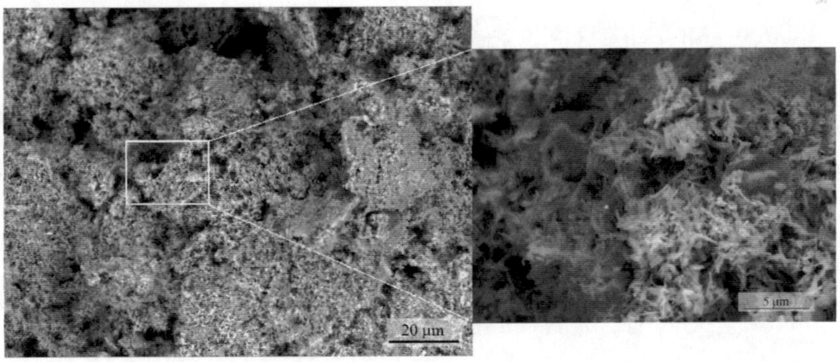

Figure. 3. Microstructure of the 50/50 composite material.

The open porosity dependence on PA (Figure 4) is linear and increases till the PA maximum concentration up to 15 wt. %. The samples obtained with the injection of the $CaCO_3$ and Na_2CO_3 have a high level of open porosity to 25 and 22 %, respectively, due to the release of CO_2 gas when the additive interacts with the hardening liquid. Such materials have interconnected pores of irregular shape with a size of 1-20 μm; wherein the gas realise posed by the initial reagents reaction had form of the tunnel porosity. The injection of NaCl leads to the formation of spherical pores with of 1 - 5 μm a size.

IOP Conf. Series: Materials Science and Engineering **347** (2018) 012039 doi:10.1088/1757-899X/347/1/012039

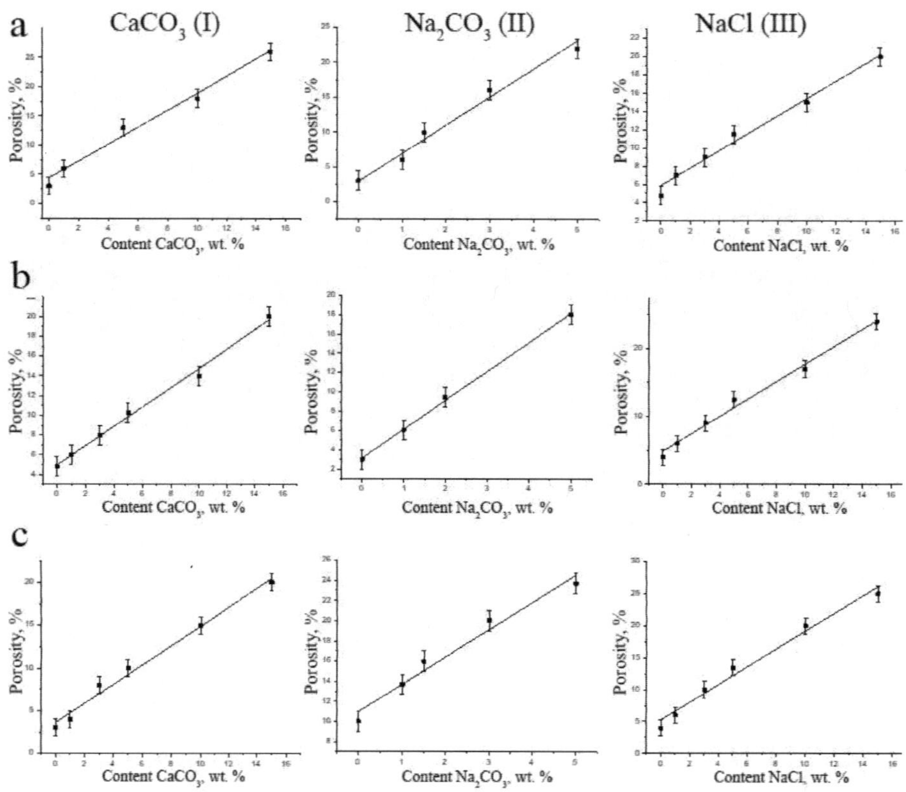

Figure 4. Dependence of porosity on the content and type of the additive: (a) – brushite, (b) - 50/50, (c) – apatite, (I) – $CaCO_3$, (II) – Na_2CO_3, (III) – NaCl.

The compressive strength of the CPCs increased from 6 ± 0,5 MPa for brushite, to 25 ± 3 MPa brushite/apatite (50/50) and 50 ± 3 MPa for apatite (Figure 5). It is established that the compressive strength values of CPCs monotonically decrease when increasing amount of PA, which is consistent with the results of open porosity. The highest compressive strength are for samples with $CaCO_3$ with 15 wt. %: 4, 16 and 27 MPa, respectively, for brushite, 50/50 and apatite cements. Based on the results CPC 50/50 with the additive $CaCO_3$ 5 wt. % has been chosen optimal combination of properties.

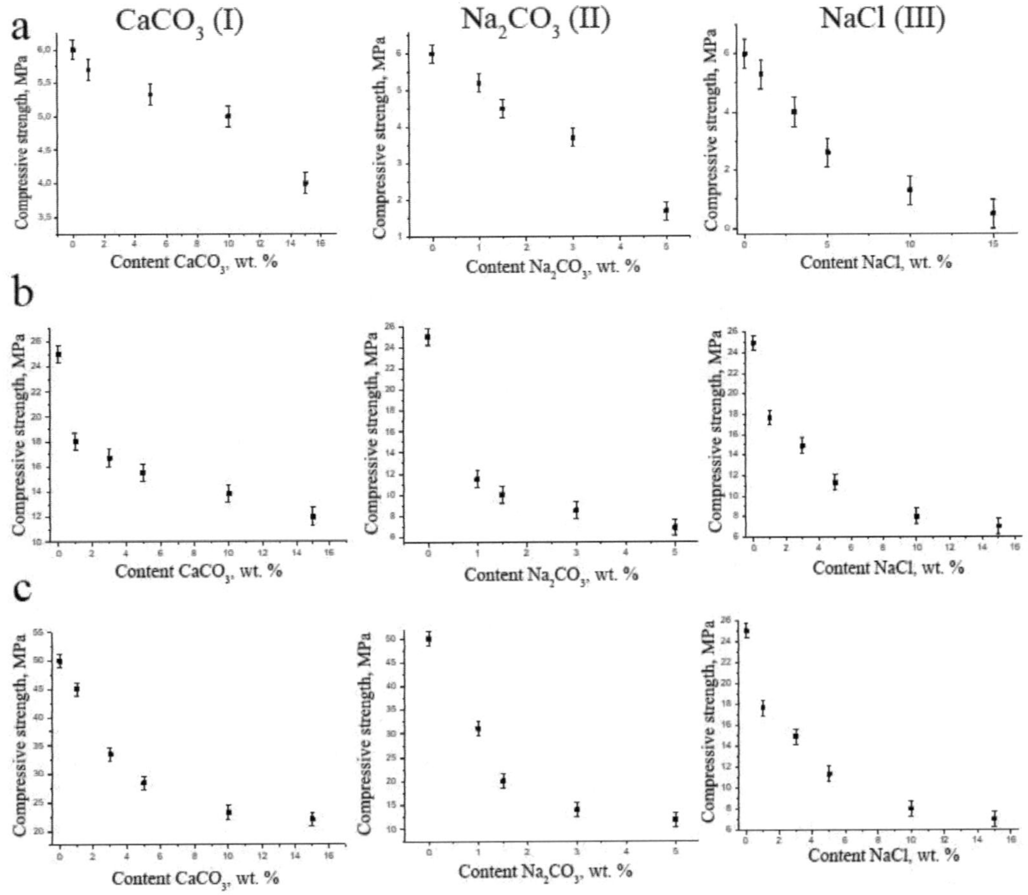

Figure 5. Dependence of compressive strength on the content and type of the additive: (a) – brushite, (b) - 50/50, (c) – apatite, (I) – $CaCO_3$, (II) – Na_2CO_3, (III) – NaCl.

The kinetics of dissolution of CPCs of all compositions at the initial stage can be described by an exponential function (Figure 6). The mass loss of CPCs was increased significantly by 5 to 7 days in a buffer solution, and after what dissolution rate slows down. The quantitative weight loss values of CPCs were different according to the composition and porosity of the material. Weight loss for CPC 50/50 in the buffer solution were 30 - 32 % for $CaCO_3$, 15-16 % for Na_2CO_3 and 10 -12 % for NaCl. For Apatite CPC weight loss were 16-17 % for $CaCO_3$, 20-21 % for Na_2CO_3 and 23 – 24 % for NaCl.

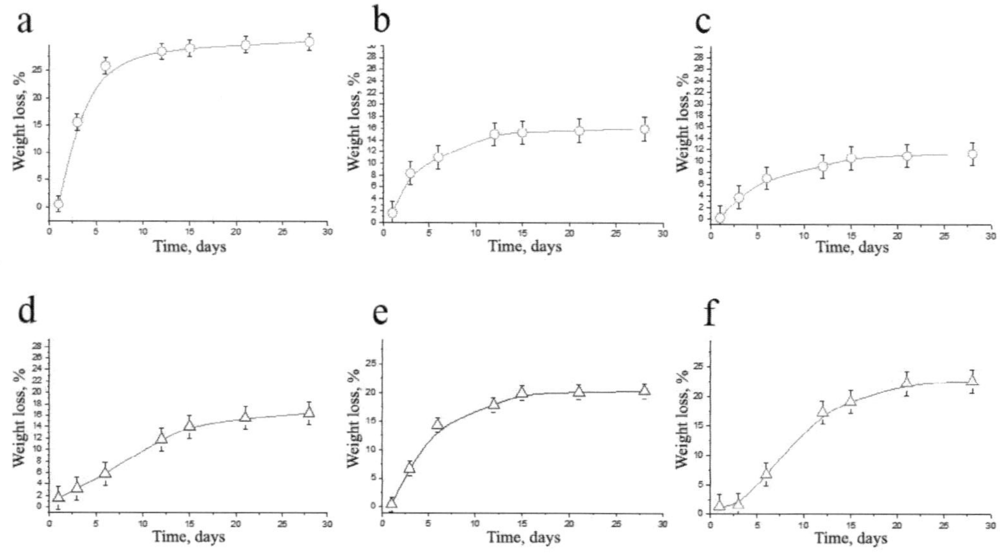

Figure 6. Kinetic of weight loss CPC in buffer solution: (a) 50/50 with $CaCO_3$, (b) – 50/50 with Na_2CO_3, (c) – 50/50 with NaCl, (d) – apatite with $CaCO_3$, (e) - apatite with Na_2CO_3, (f) - apatite with NaCl.

In vitro test of hFB and MG-63 on the CPC with 5 wt. % $CaCO_3$ demonstrated the absence of toxicity and the presence of surface matrix properties with varying degree (Figure 7). Thus, the results of the experiments indicate that the material with compositions brushite, 50/50, apatite were cytocompatibility with respect to the line of hFB and MG-63 cells. The both line cells after plating and adhesion/spreading to the CPC were actively proliferating and had viable cells values higher 20-60 % then control at all observation periods (Figure 8).

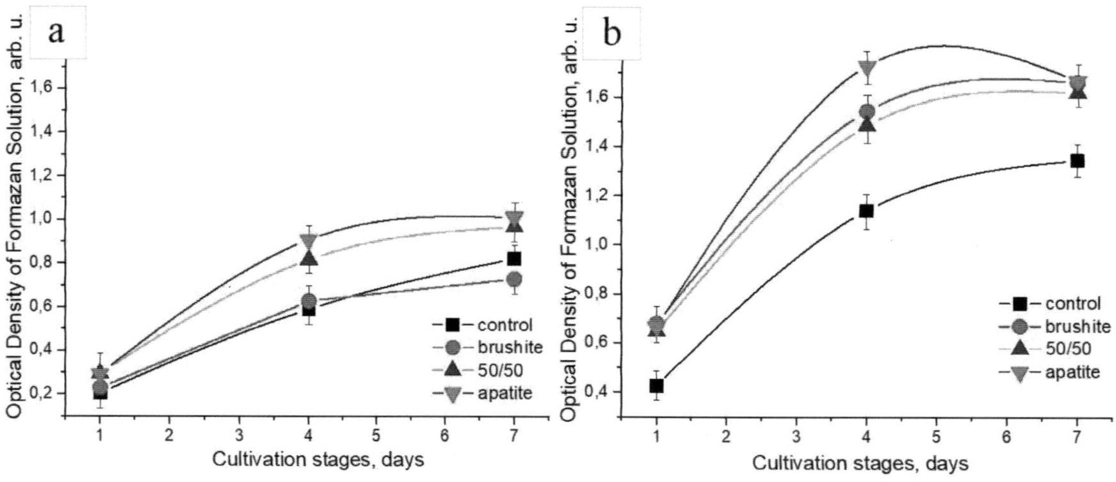

Figure 7. Increase in the population of (a) hFB and (b) MG-63 cells at cultivation stages.

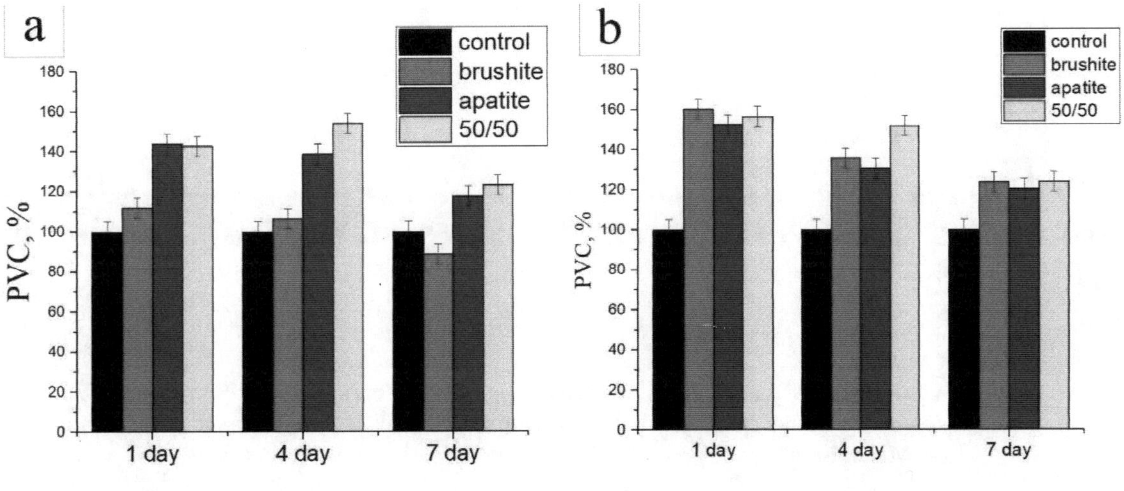

Figure 8. Value of VPC of (a) hFB and (b) MG-63 cells at cultivation stages.

4. Conclusion

The composite cements in the brushite-apatite system with different ratios were developed. The most promising composition was brushite/apatite 50/50 with the final phases DCPD/HA/TCP 65/30/5 wt. %. The pore-forming additives ($CaCO_3$, Na_2CO_3, NaCl) CPC were able to obtain a porosity from 10 to 27 %. The optimal properties CPC: setting time 10 min, porosity 15 %, compressive strength 18 MPa and weight loss 30-32% were obtained with $CaCO_3$ 5 wt. %. The absence of toxicity and the presence of appropriate surface matrix properties were demonstrated. The developed composite cements can be used for bone defects replacement.

Acknowledgments

This work was supported by the Russian Foundation for Basic Research, project no. 15-29-04795.

References

[1] Gemma M *et al* 2015 *Acta Biomaterialia* **11** 459-466
[2] Butscher A *et al* 2013 Acta Biomaterialia **9** 5369-5378
[3] Zhang J *et al* 2014 *Acta Biomaterialia* **10** 1035–1049
[4] Ginebra M *et al* 2010 *Acta Biomaterialia* **6** 2863–2873
[5] Zhang J *et al* 2014 *Acta Biomaterialia* **10** 1035–1049
[6] Samavedi S *et al* 2013 *Acta Biomaterialia* **9** 8037–8045
[7] Bose S *et al* 2012 *Acta Biomaterialia* **8** 1401–1421
[8] Ginebra M *et al* 2012 *Acta Biomaterialia* **64** 1090–1110
[9] Verron E *et al* 2012 *Acta Biomaterialia* **8** 3541–3551
[10] Dorozhkin S 2008 *J. Mater. Sci.* **43** 3028-3057
[11] Fedotov A *et al* 2008 *Powder Metallurgy Progress* **8** 351-357
[12] Teterina A *et al* 2015 *Doklady Chemistry* **461** 104-107
[13] Mossman T 1983 *J Immunol Methods* **65** 55-63
[14] Rau J *et al* 2010 *Cryst. Growth Des.* **10** 3824–3834

IOP Conf. Series: Materials Science and Engineering **347** (2018) 012040 doi:10.1088/1757-899X/347/1/012040

Construction ceramics from silicon nitride with calcium aluminates additives received by the sintering method in the SHS-reactor

A S Lysenkov[1], Yu F Kargin[1], D D Titov[1], N V Petrakova[1], S N Ivicheva[1], A I Zakharov[2], N A Popova[2], V V Zakorzhevskii[3], I P Borovinskaya[3] and I S Melnikova[4]

[1]Baykov Institute of Metallurgy & Material Science RAS, Moscow, Russia
[2]Mendeleyev University of Chemical Technology of Russia, Moscow, Russia
[3]Institute of Structural Macrokinetics and Material Science RAS, Chernogolovka, Russia
[4]Federal State Unitary Enterprise State Research Institute of Civil Aviation, Moscow, Russia

E-mail: toxa55@bk.ru

Abstract. The paper considers preparation of a strong and dense ceramics based on silicon nitride by sintering in a SHS reactor. As a sintering agent, an additive was used in the Al2O3-CaO system. The content of the additive varied from 5 to 15 wt. %. It is established that an increase in the amount of silicon introduced into the initial charge increases the strength characteristics of the material due to the formation of secondary silicon nitride during sintering.

1. Introduction

Silicon nitride is one of the most promising high-temperature materials. Silicon nitride and materials based on it have a high operating temperature up to 1400-1750°C, wear resistance, chemical inertness, high (especially silicon nitride) strength and crack resistance [1, 2]. Therefore, the appearance of these materials made it possible to raise the issue of replacing metal products with ceramic ones for operation in extreme thermomechanical conditions, for example, in the hot zone of gas turbine and piston engines. The replacement of carbide and other types of cutting tools with a more efficient ceramic tool continues, ceramics favorably differs from them, for example, in the field of processing of cast iron or superalloys. Currently research is being conducted to use nitride and silicon carbide products in virtually all industries in developed countries. They have already found wide application in aerospace, metallurgy, chemical, electronic and other industries [3, 4].

There are several methods for obtaining ceramics from silicon nitride: firing in a nitrogen medium without applying pressure, reaction sintering, hot pressing (GP), hot isostatic sintering (HIP). The listed methods are the most applicable in industry. High properties are obtained from the materials synthesized by the latter two methods, however, the high energy intensity and the impossibility of obtaining blanks of complex shapes, as well as the use of high-density and expensive graphite as a press tool, makes it possible to obtain Si_3N_4 ceramic. It is impossible to obtain dense materials with a porosity of less than 5 vol. % by the method of reaction sintering. The chemical deposition method from the gas phase leads to the practically theoretical density production. How ever their mass production is difficult because of technological difficulties [5].

In this work a different method of calcination was used. Ceramic blanks were sintered in the process of self-propagating high-temperature synthesis (SHS). The SHS method is based on the fact that the heat released by the exothermic reaction, due to energy transfer, it heats the substance neighboring layers, excites the reaction in them and leads to self-propagating combustion process occurrence. This method has not found wide application in the production of silicon nitride products yet, but it has prospects for implementation. Previously, ceramic materials based on silicon nitride in the firing mode of SHS were obtained [6]. Yttrium-aluminum garnet and mullite were used as sintering additives. In this paper, calcium aluminate was used as a sintering additive, which had already been used to produce nitride-silicon ceramics [7]. The aim of the work was obtaining a durable, strong silicon nitride ceramic with the calcium aluminate addition by firing in SHS reactor and studying the physicochemical and mechanical properties of sintered materials.

2. Materials and Method

Silicon nitride and aluminum nitride powders had been obtained by self-propagating high-temperature synthesis (SHS) (ISMAN RAS) and were used as starting materials. The well crystallized fibrous Si_3N_4 powder was obtained by the SHS method. An average size at a length was about 2 μm and a thickness up to 200 nm after grinding in a planetary mill (Figure 1). The content of α - silicon nitride was not less than 95%, the specific surface area of the powder was 8.2 m^2/g.

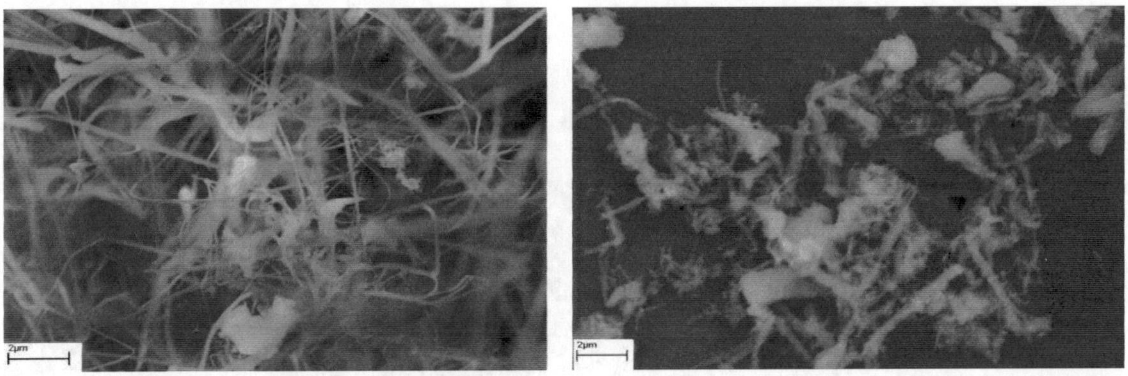

a) b)

Figure 1. SEM micrographs α-Si_3N_4 powder: a) befor milling;b) after milling.

The sintering additive (eutectic composition 33.5 wt.% CaO + 66.5 wt.% Al_2O_3) was selected on the basis of phase equilibrium data in the Al_2O_3-CaO system (Figure 2). It was used commercially poweders $Al(OH)_3$ («Labteh») and $CaCO_3$ («Labteh») to synthesise thesintering additive. The powders were mixed in a planetary mill with zirconium oxide balls in an alcohol medium (mass ratio material: balls = 1:5), the resulting powder was dried and wiped through 0.063 mm sieve cells). The mixture was calcined at 1200°C for two hours to obtain the desired phase composition additive. The additive was disaggregated in a planetary mill in acetone medium for 30 minutes after synthesis. According to X-ray the synthesis of calcium aluminate supplements does not go completely under these conditions. There are the $Ca_{12}Al_{11}O_{33}$ phase presented in addition to the main phases of $CaAl_2O_4$ and $CaAl_4O_7$. Therefore the additional heat treatment was carried out for 4 hours, it ensures the production of the required composition eutectic mixture (Figure 3). The commercially available technical silicon powder KR00 was used in the experiments (GOST 2169-69).

The silicon nitride, silicon and sintering additive (7, 10 and 15 wt.%) were mixing in a planetary mill. The 15 and 30 wt.% elemental silicon were introduced into the starting charge over the 100 wt.%. The introduction of elemental silicon is associated with partial reaction sintering due to the resulting material (secondary silicon nitride) increases in volume and fills internal pores [9].

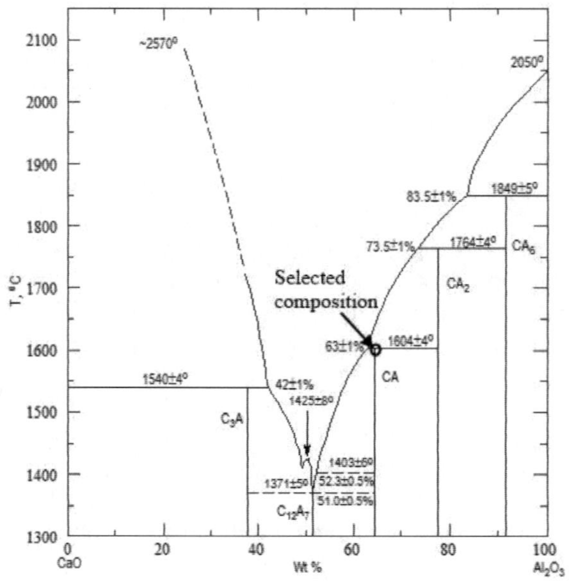

Figure 2. Diagram of the CaO-Al_2O_3 system.

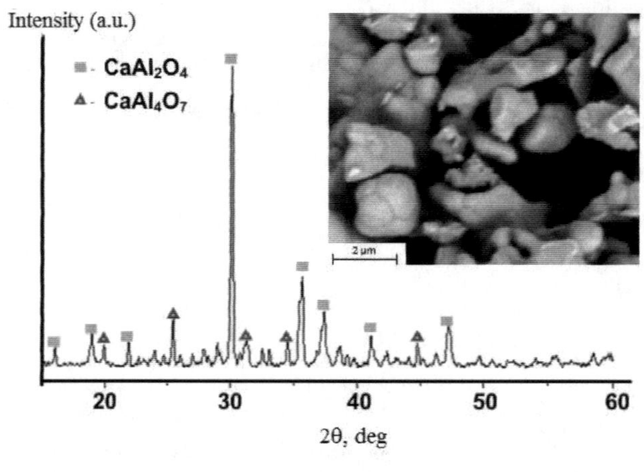

Figure 3. XRD patterns and SEM micrograph of CaO-Al_2O_3 powder

The samples formed in 50x5x5 mm balks and 70x20x5 mm plates by the semi-dry uniaxial two-sided pressing method. The balks were placed inside the charge in the SHS reactor along the combustion front (Figure 4). The ratio of the fill components was 45:55 (Si:β-Si_3N_4). The weight of the backfill was 5 kg. Samples were heated to 1900°C for 120 seconds in the SHS reactor, the holding time at this temperature was approximately 150 seconds, then the reactor was cooled. The entire process from samples loading to extraction was about 2 hours.

Figure 4. Location of samples in the SHS reactor

The X-ray diffraction analysis was used to identify a phase of the chemical composition (XRD 6000 SHIMADZU diffractometer, CuKα radiation, $\lambda = 1.5406$ Å, scanning speed $2\theta = 2$ deg/min). The phase was identified using the PDF-2 database, JCPDS-ICDD (Set 1-2002). The microstructure and phase composition were examined on an electron microscope (Supra 50 VP (LEO, Germany) with an INCA Energy + Oxford microanalysis system with a prefix for local X-ray spectroscopy). Density was measured by the Archimedes method using water. The flexural strength three bending test were studied by Instron 5581.

3. Results and discussion

The structure and ceramic properties of these samples were studied after sintering. Preliminary studies were carried out on silicon nitride with a different calcium aluminate concentration without silicon addition. The samples were porous and a low density (Table 1). The balks with 7 wt.% sintering additive composition had 26.5% open porosity. In view of this, up to 30 wt.% Si was added to the starting charge. It was noted above, addition 30 wt.% Si into the starting charge led to reduce the closed porosity and increased the grain size due to the secondary silicon nitride formation.

As can be seen in Table 1, the silicon addition reduces the number of closed pores, increased density and bending strength up to 540 MPa.

Table 1. The starting compositions, sintering parameters, density, open porosity and flexural strength of the Si_3N_4 ceramics

Samples	Composition, wt. %			Sintering parameters (°C/sec)	Density (g/cm^3)	Open porosity (%)	Flexural strength (MPa)
	Si_3N_4	Al_2O_3-CaO	*Si				
SHS-1	93	7	-	1900/150	2.49	26.58	123
SHS-2	90	10	-	1900/150	2.55	19.87	245
SHS-3	85	15	-	1900/150	2.62	10.75	366
SHS-Si-1	93	7	30	1900/150	2.93	5.97	195
SHS-Si-2	90	10	30	1900/150	2.99	3.26	510
SHS-Si-3	85	15	30	1900/150	3.05	2.78	540

*over 100 wt. %

4. Conclusion

Thus, a ceramics based on silicon nitride with a different calcium aluminate concentration (sintering additive) was obtained using the firing method in low-pressure reactor SHS (SHS-30). The ceramics properties with 15 wt.% sintering additive (without the introduction of silicon): flexural strength up to 350 MPa and density of 3.03 g/cm3 at room temperature. The best properties were flexural strength up to 540 MPa and density of 3.09 g/cm3 at room temperature ceramics based on silicon nitride with 15 wt.% calcium aluminates and by adding 30% Si with the following properties.

Acknowledgments

The reported study was funded by RFBR project No 18-38-00383. The methodical part of the work (X-ray, XRD, SEM, flexural strength, density) was carried out according to the state task No 007-00129-18-00.

References

[1] Evans A G and Langdon T G 1980 Structural Ceramics (Moscow: Metallurgiya) p 256
[2] Shevchenko V Ya and Barinov S M 1986 Technical ceramics (Moscow: Metallurgiya) p 277
[3] Garshin A P, Gropyanov V M, Zaytsev G P and Semyonov S S 2003 Ceramics for mechanical engineering (Moscow: Nauchtekhlitizdat) p 384
[4] Andrievsky R. A and Spivak I I 1984 Silicon nitride and materials on its basis (Moscow: Metallurgiya) p 136
[5] Perevislov S N and Chupov V D 2010 *Refractory materials and technical ceramics* **3** 17
[6] Lysenkov A S, Zakharov A I and Zakorzhevskii V V 2007 *Glass and Ceramics* **64** 86
[7] Kargin Yu F, Lysenkov A S, Ivicheva S N, Zakharov A I, Popova N A and Solntsev K A 2010 *Neorganicheskie Materialy* **46** 892
[8] A. Zherebtsov, S. A. Archugov, and G. G. Mikhailov, *Rasplavy*, 2, 63-65 (1999)
[9] Guzman I Ya 1996 Reactionary agglomeration and its use in technology of ceramics and refractory materials: Manual (Moscow: RHTU of D.I. Mendeleyev) p 55

IOP Conf. Series: Materials Science and Engineering **347** (2018) 012041 doi:10.1088/1757-899X/347/1/012041

Influence of synthetic calcium silicates on the strength properties of fine-grained concrete

S B Yarusova[1,2], P S Gordienko[1], A V Kozin[3], I G Zhevtun[1] and A V Perfilev[1]

[1] Institute of Chemistry of the Far-Eastern Branch of the Russian Academy of Sciences, 159 Prospekt 100-letya Vladivostoka, Vladivostok, Russian Federation, 690022
[2] Vladivostok State University of Economics and Service, 41 Gogolya street, Vladivostok, Russian Federation, 690014
[3] Far-Eastern Federal University, 8 Sukhanova street, Vladivostok, Russian Federation, 690950

E-mail: yarusova_10@mail.ru

Abstract. The effect of additives based on acicular calcium hydrosilicates (xonotlite and tobermorite) and wollastonite, obtained from boric acid production waste in autoclave synthesis at a temperature of 220 °C, on the strength of fine-grained concrete, has been studied in this paper. It was shown that when the calcium hydrosilicates and wollastonite are introduced, an increase in the strength characteristics of concrete is observed. After heat and moisture treatment, the maximum increase in strength is observed with the addition of 4% of mass content of calcium hydrosilicates and 6% of mass content of wollastonite. After 28 days of hardening under normal conditions, the maximum increase in strength of concrete is observed with the addition of 4% of mass content of both types of additives. It was shown that the water absorption of concrete decreases with a maximum when 4% of mass content is added, as in the case of the introduction of calcium hydrosilicates, and wollastonite. With a further increase in the number of additives, the amount of water absorption increases, but these values remain below the values for the control sample without additives.

1. Introduction

Calcium hydrosilicates $nCaO{\cdot}mSiO_2{\cdot}pH_2O$ and wollastonite $Ca_6Si_6O_{18}$ have found wide application in the production of building materials, paper, paints, plastics, composite polymeric and metal-ceramic materials, sorbents for water purification. Scientists from different countries are conducting research to develop optimal methods for obtaining calcium silicates from available natural and technogenic raw materials.

Analyzing the prospects for the use of calcium silicates in the building materials industry, it is necessary first of all to focus on the variety of assortment of building materials and products based on cement, other binding substances and solidifying dispersed systems [1, 2].

A number of Russian and foreign studies have been devoted to the study of the effect of natural and synthetic wollastonite on the functional properties of building materials such as cement and concrete

[3–13]. It is shown that when using wollastonite as a filler, the strength characteristics, frost resistance, durability of building materials increase, and in some cases their resistance to corrosion increases, too.

Since 2007, the Institute of Chemistry of the Far-Eastern Branch of the Russian Academy of Sciences, in conjunction with a number of Russian universities and Armenian scientists, has been actively involved in the development of the physico-chemical basis for the complex processing of boric acid production waste (borogypsum) to produce calcium hydrosilicates, wollastonite and potassium fertilizers [14]. The possibility of using an additive based on wollastonite and pseudo-wollastonite obtained from borogypsum during the production of concrete has been shown [15, 16]. It is established that the additive under study contributes to the increase in strength, to a decrease in water absorption and to an increase in the frost resistance of concrete. The work continued with the production of calcium hydrosilicates and wollastonite with needle-like shape of particles by autoclave method in the temperature range 180–220 °C [17, 18].

The aim of this work is to study the effect of additives based on acicular calcium hydrosilicates (xonotlite and tobermorite) and wollastonite, obtained from borogypsum in autoclave synthesis (220 °C), on the strength of fine-grained concrete.

2. Experimental part

For the production of calcium hydrosilicates, borogypsum was mixed with the potassium hydroxide solution *p.a.* in the stoichiometric ratio. The synthesis was carried out in an autoclave made by Parr Instrument (USA) 4848 at a temperature of 220 °C for 3 hours. The reaction rate was monitored for the residual concentration of potassium hydroxide in the solution. To obtain wollastonite, the precipitate after autoclave treatment was calcined in the temperature range 900–1000 ° C for 1 hour.

X-ray patterns of the samples were taken on an automatic D8 ADVANCE diffractometer with rotation of the sample in Cu K_α radiation. X-ray phase analysis was carried out using the EVA search program with the PDF-2 powder data bank.

The specific surface was determined by the method of low-temperature adsorption of nitrogen using Sorbtometr-M instrument.

The morphology of the samples was studied on a high resolution scanning electron microscope (SEM) Hitachi S 5500 (Japan).

For the production of concrete beams and cubes, the following components were used (in relation to the weight part of the cement taken as 1): superplasticizer C-3 – 0.01; sand – 3; additive based on calcium silicates – 0.02–0.08, water – 0.42. The water-cement ratio was maintained constant in all compositions (W / C = 0.42). To the aqueous solution of the superplasticizer C-3, a calcium silicate-based material was added as a powder and mixed with a mixer for 2 minutes until a homogeneous suspension was obtained. The cement was poured into a bowl of a laboratory mixer (type 1.0203.01 made by Testing), water was poured into it, then stirred for 30 seconds. The resultant mass was added to the prepared suspension and stirred for 30 seconds. The next step was gradually adding sand to the mixture and stirring at 140 rpm for 2 minutes and at 285 rpm for 30 seconds. The prepared mixture was layered manually into the 3FK-70 (3FB-40) molds and vibrated on the vibrating pad (model SMZh-539) for 10 seconds. The mold with the samples was covered with glass, and after 1 day, stripping was done. The samples were placed in a normal hardening chamber (model KPU-1M) on liners and stored for 27 days. The temperature in the chamber is 20 °C, relative air humidity is 95%. After 3, 7 and 28 days from the date of manufacture, some samples were taken out of the chamber. Within 4 hours the samples were in the natural conditions of the room, in which they were subsequently tested, that is, at an air temperature of 20 ± 5 °C and a relative humidity of at least 55%. With heat and moisture treatment (HMT), the samples in the molds were placed in the KPU-1M steaming chamber and held for 2 hours at a temperature of 20 °C. Then there was an even rise in temperature to 80 °C for 3 hours and isothermal heating at a temperature of 80 °C for 6 hours. After isothermal warming, the molds with samples were cooling for no less than 2 hours, then the chamber cover was opened, 24 hours after production samples were molded and immediately tested. Samples were tested for bending and compression on a combined machine (type 1.0244 made by Testing).

Samples were tested for water absorption in accordance with the guidelines described in [19].

3. Results and discussion

According to X-ray diffraction analysis, the following were found in the synthesized sample: crystalline phases of unreacted anhydrous calcium sulfate $CaSO_4$, those of tobermorite 11 Å $Ca_5(OH)_2Si_6O_{16}\cdot 4H_2O$ of orthorhombic modification (PDF-2, 00-019-1364) with crystal cell parameters: a–11.27000; b–7.35000; c–22.74000; $\alpha = 90.000$; ß = 90.000; $\gamma = 90.000$ and of those of xonotlite $Ca_6[Si_6O_{17}](OH)_2$ of monoclinic modification (PDF-2, 00-029-0379) with the parameters of the crystal cell: a–17.02900; b–7.35600; c–7.00700; $\alpha = 90.000$; ß = 90.340; $\gamma = 90.000$. The specific surface area of the obtained precipitate is 30.7 $m^2\cdot g^{-1}$. After annealing at 900 °C, crystalline phases of unreacted anhydrous calcium sulfate $CaSO_4$ and wollastonite of the triclinic modification (PDF-2, 01-084-0654) were found in the sample with crystal cell parameters: a–7.92580; b–7.32020; c–7.06530; $\alpha = 90.055$; ß = 95.217; $\gamma = 103.426$. The specific surface of the sample obtained is 4.5 $m^2\cdot g^{-1}$. As a result of annealing at 1000 °C, only the wollastonite phase of the triclinic modification with the same parameters of the crystalline cell is present in the sample composition (figure 1).

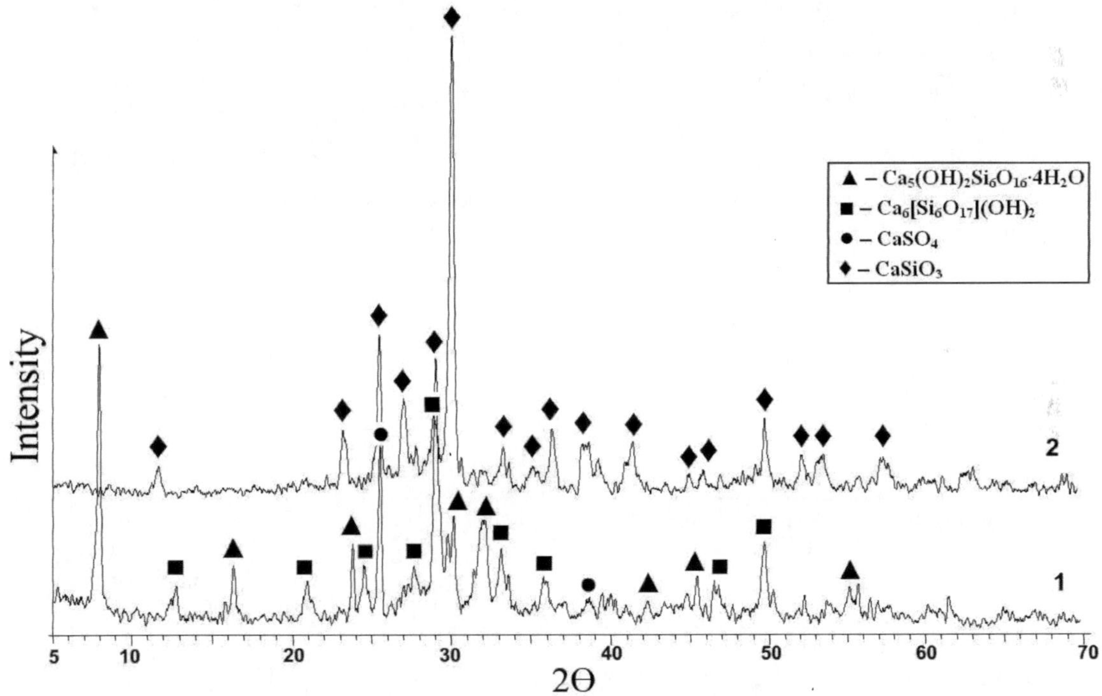

Figure 1. Diffractograms of samples obtained as a result of autoclave borogypsum treatment: 1 – initial sample (before annealing); 2 – after annealing at 1000 °C.

According to the data of scanning electron microscopy, the samples obtained, both initial and after annealing, consist of agglomerates of nano-sized particles having a predominantly acicular shape (figure 2). However, after annealing, the pinching of the needle particles is observed (figures 2c, d). After annealing at 900 and 1000 °C, no significant differences are observed in the particle morphology.

IOP Conf. Series: Materials Science and Engineering **347** (2018) 012041 doi:10.1088/1757-899X/347/1/012041

Figure 2. SEM images of microparticles of the sample obtained as a result of autoclave synthesis, after drying at 85 °C (a, b) and after annealing at 1000 °C for 1 hour (c, d).

The table 1 gives data on the strength characteristics of concrete obtained with the addition of calcium hydrosilicates and wollastonite in an amount of 2–8% by mass content.

Table 1. Strength characteristics of concrete with the addition of calcium hydrosilicates and wollastonite.

Additive	Mass content of additive (%)	HMT		Normal hardening conditions (28 days)	
		Bending (MPa)	Compression (MPa)	Bending (MPa)	Compression (MPa)
Based on calcium hydrosilicates (xonotlite and tobermorite)	No additive	3.4	21.7	5.0	27.4
	2	4.3	23.0	5.1	31.6
	4	4.4	25.9	5.2	33.5
	6	3.4	22.3	5.1	31.3
	8	3.3	20.4	5.0	28.5
Based on wollastonite obtained after calcining at 900 °C	No additive	3.3	21.2	5.07	28.3
	2	3.7	23.1	5.5	34.6
	4	4.3	24.9	5.6	36.6
	6	4.8	27.4	5.3	33.1
	8	4.3	22.9	5.07	31.8

As can be seen from the table, with the addition of calcium hydrosilicates and wollastonite, the strength characteristics of concrete are increased. After HMT, the maximum increase in strength is observed when 4% of mass content is introduced for calcium hydrosilicates and 6% of mass content for wollastonite. In this case, the strength increases with bending by almost 30% for both additives, with compression – by 19 and 24% for calcium hydrosilicates and wollastonite, respectively. After 28 days of hardening under normal conditions, the greatest increase in strength is observed with adding 4% of the mass content of both additives. When calcium hydrosilicates are introduced, the flexural strength increases insignificantly, when compressed – by 22%. In case of the addition of wollastonite, the strength characteristics of bending are increased by 25%, while compression is increased by 21%. With an increase in the amount of both types of additives up to 8% of mass content, there is a decrease in strength. However, in case of the addition of wollastonite in an amount of 8% by mass content, the values of bending strength and compression remain above the control values (without additives).

Figure 3 shows the dependences of the water absorption of concrete on the amount of the additive based on calcium hydrosilicates and wollastonite.

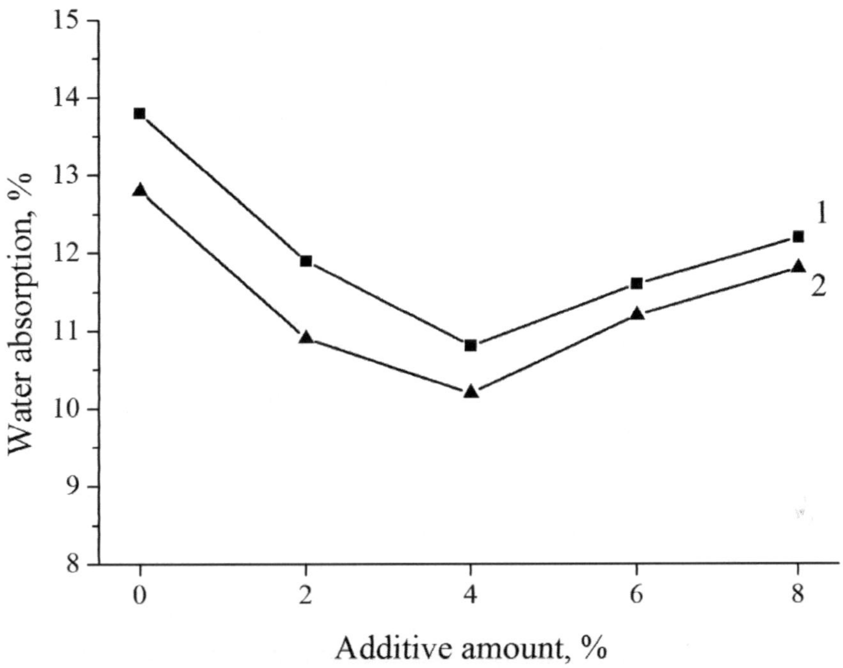

Figure 3. Dependences of the water absorption of concrete on the amount of the additives based on calcium hydrosilicates (xonotlite and tobermorite) (1) and wollastonite (2).

As can be seen from the above dependences, when the additives are introduced, a decrease in the water absorption is observed with a maximum of 4% of mass content. With a further increase in the amount of the additives, the amount of water absorption increases, but these values remain below the values for the control sample without additives.

The authors continue to actively study the relationship between the conditions for obtaining calcium silicates, their structure, the shape and size of the particles and the functional properties of concrete with a different amount of described additives.

4. Conclusions

Thus, studies on the effect of additives based on acicular calcium hydrosilicates (xonotlite and tobermorite) and wollastonite, obtained from boric acid production waste in autoclave synthesis at a temperature of 220 °C, have shown that when they are introduced, the strength characteristics of fine-grained concrete are increased. After heat and moisture treatment, the maximum increase in strength is observed when 4% of mass content is introduced for calcium hydrosilicates and 6% for wollastonite.

After 28 days of hardening under normal conditions, the greatest increase in strength is observed with the addition of 4% of mass content for both types of additives. It is shown that the water absorption of concrete decreases with a maximum when 4% of mass content is added, as in the case of the introduction of calcium hydrosilicates, and wollastonite. With a further increase in the number of additives, the amount of water absorption increases, but these values remain below the values for the control sample without additives.

Acknowledgements
The financial support from the government contract of Universities of the Ministry of Education and Science of the Russian Federation number 4.5913.2017/8.9 is gratefully acknowledged.

References

[1] Gladun V D, Akateva L V and Kholkin A I 2011 *Synthetic calcium silicates* (Moscow: Irisbuk).
[2] Tyulnin V A, Tkach V R, Eirikh V I and Starodubtsev N P 2003 *Wollastonite: a unique multi-purpose mineral raw material* (Moscow: Ruda i metally).
[3] Low N M P and Beaudoin J J 1993 The effect of wollastonite micro-fibre aspect ratio on reinforcement of Portland cement-based binders *Cement and Concrete Research* **23** pp 1467–79.
[4] Eirikh V I, Berezovsky SV, Tarantul N P, Ioramashvili I N and Konov G V 2002 On the application of wollastonite in the production of composite building materials and products based on cement *Building Materials* **1** pp 14–17.
[5] Sharifov A, Akramov A and Dzhabborov I S 2006 Modification of portland cement with additives for the use in production of special concretes *Reports of the Academy of Sciences of the Republic of Tajikistan* **49(5)** pp 458–463.
[6] Mathur R, Misra A K and Goel P 2007 Influence of wollastonite on mechanical properties of concrete *Journal of Scientific & Industrial Research* **66** pp 1029–34.
[7] Ransinchung G D, Kumar B and Kumar V 2009 Assessment of water absorption and chloride ion penetration of pavement quality concrete admixed with wollastonite and microsilica *Construction and Building Materials* **23** pp 1168–77.
[8] Panina A A, Samigullin R R, Tsyplakov D S, Kornilov A V and Tukhvatshina A I 2010 Wollastonite – an unconventional mineral additive – a filler in portland cement Bulletin of *Kazan Technological University* **8** pp 377–378.
[9] Kuldasheva A Kh 2011 Experimental studies of strength properties of concretes based on wollastonite raw materials *Vestnik MSCU* **7** pp 627–630.
[10] Kalla P, Misra A, Gupta R C, Csetenyi L, Gahlot V and Arora A 2013 Mechanical and durability studies on concrete containing wollastonite–fly ash combination *Construction and Building Materials* **40** pp 1142–50.
[11] Kalla P, Rana A, Chad Y B, Misra A and Csetenyi L 2015 Durability studies on concrete containing wollastonite *Journal of Cleaner Production* **87** pp 726–734.
[12] Bian H, Hannawi K, Takarli M, Molez L and Prince W 2016 Effects of thermal damage on physical properties and cracking behavior of ultrahigh-performance fiber-reinforced concrete *Journal of Materials Science* **51** pp 10066–76.
[13] Ismail S, Kwan W H and Ramli M 2017 Mechanical strength and durability properties of concrete containing treated recycled concrete aggregates under different curing conditions *Construction and Building Materials* **155** pp 296–306.
[14] Patent 2601608 the Russian Federation, IPC C01B 33/24. Method of complex borogypsum processing / Gordienko P S, Yarusova S B, Kozin A V, Stepanova V A, Shabalin I A, Grivanova O V – № 2015141651/05; application of 30.09.2015; published 10.11.16, Bulletin № 31.
[15] Gordienko P S, Kozin A V, Yarusova S B and Zgibly I G 2014 Complex processing of boric acid production wastes with obtaining materials for the construction industry *Architecture*

and construction of the Far East: Mining information and analytical bulletin (scientific and technical journal). Separate articles (special issue) **12** pp 60–66.

[16] Gordienko P S, Yarusova S B, Kozin A V, Ivin V V, Silantiev V Ye, Lizunova P Yu and Shornikov K O 2017 Material based on synthetic wollastonite and its influence on the functional properties of fine-grained concrete *Perspective materials* **9** pp 40–48.

[17] Patent 2595682 the Russian Federation, IPC C01B 33/24; C30B 7/10; C30B 29/34; C30B 29/62; B82B 3/00; B82Y 40/00. Method of obtaining wollastonite / Gordienko P S, Yarusova S B, Kozin A V, Stepanova V A, Shabalin I A and Zhevtun I G – № 2015141614/05; application of 30.09.2015; published 27.08.16, Bulletin № 24.

[18] Yarusova S B, Gordienko P S, Sharma Y C, Perfilev A V and Kozin A V 2017 Industrial waste as raw material for producing synthetic wollastonite in Russia *International Journal of Environmental Science and Development* **8** pp 1–5.

[19] State standard of the USSR GOST 12730.3–78. Concretes. Methods for determination of water absorption (Moscow: Gosstroy USSR).

IOP Conf. Series: Materials Science and Engineering **347** (2018) 012042 doi:10.1088/1757-899X/347/1/012042

Investigation of the influence of the composition on mechanical properties poly(glycolide-DL-lactide)

A S Baikin, M A Sevostyanov, E O Nasakina, K V Sergienko, M A Kaplan, S V Konushkin, A A Kolmakova, A D Yakubov, A G Kolmakov

A.A.Baikov Institute of Metallurgy and Materials Science, Russian Academy of Sciences

Email: baikinas@mail.ru

Abstract. In this paper we describe the creation of films from poly (glycolide-DL-lactide). Studied the mechanical properties of developed polymer films of poly (glycolide-DL-lactide). The effect of the molecular weight of poly (glycolide-DL-lactide) on the mechanical properties of the resulting polymer films is shown. The dependence of the mechanical properties of poly (glycolide-DL-lactide) films on the polymer concentration in chloroform was studied. The possibility of creating biodegradable films with specified mechanical properties is shown.

1. Introduction

At the end of the 20th century, a special device was invented and put into practice, called a "cava filter" (a medical device implanted in the lumen of the inferior vena cava to catch clots carried by the bloodstream). Currently, in medical practice, operations involving the use of a "cava filter" are widespread, and active development of new products continues, since in addition to the merits they all also have significant drawbacks. Together with the establishment of "cava-filters", drugs are treated. Most drugs used for systemic prevention can cause bleeding. Therefore, now work is underway to create "cava-filters" with a superficial drug layer that provides a local effect of the drug. However, there is a rapid leaching of the drug agent from the surface of the "cava filter". At the same time, the required duration of exposure to the drug is not achieved and its desired concentration in the blood is not maintained. This problem can be solved by administering the drug to the biopolymer layer to prevent rapid elution and to provide a controlled release of the drug into the bloodstream. [1-4]

When choosing a material for the development of controlled drug delivery systems, the following considerations are guided: the material must be chemically inert, do not contain leachable impurities, be easily synthesized, and in the case of hydrolysis (or resorption), do not form toxic compounds. They are silicone, polyurethanes, polyethylene, polyethylene vinyl acetate. Long-used materials are being used in medicine, new polymers are being developed that meet the requirements that are imposed on the rate of disintegration and the physical characteristics of delivery systems. The need to use biodegradable materials for this purpose led to the active development of biodegradable polymers of synthetic and natural origin, among which the greatest development was obtained by polyesters: polylactides, polylactide glycolides, polyethylene glycol, polyanhydrides, polyorthoesters, polysaccharides (starch, dextran, chitosan). [5-8]

In this paper, films based on poly (glycolide-lactide) of different molecular weights are considered. Polyglycolide is the simplest polyester in which, due to the close arrangement of ester groups, intermolecular interactions are strongly pronounced, has a high degree of crystallinity, a high melting point (~ 300 ° C), and extreme hydrolytic instability. Polylactide is an aliphatic polyester whose monomer is lactic acid. It is biodegradable, biocompatible and thermoplastic. Using polylactide / polyglycolide copolymers, the rate of bioresorption varies depending on the type of polymer and its molecular weight [9].

The physico-chemical properties of these polymers are determined by the molar ratio and the sequential arrangement of lactic and glycolic acids. Copolymers can be obtained with different molecular weights and the structure of macromolecules, which makes it possible to vary the degree of interaction between macromolecules. Based on lactide and glycolide, it is possible to obtain a whole family of copolyesters whose properties will differ within a certain range. The more the ratio between lactide and glycolide, the more hydrophobic the substance will be and the more soluble it will be. The decomposition times are shorter for polymers with a lower molecular weight, greater hydrophilicity and a larger content of the amorphous portion, as well as a higher content of glycolide in the copolymers. [10]

Poly (glycolide-D, L-lactide) is soluble in most solvents, including chlorinated solvents, tetrahydrofuran, acetone, ethyl acetate. In the work, chloroform was chosen as the solvent. [11]

2. Materials and methods

To create the films, we prepared samples of polymers of mass 2; 6 and 10 g (± 0.01 g). Chloroform, 200 ml in volume, was placed in a 500 ml flask and heated to 80^0 C with a magnetic stirrer.

Further, the polymer samples weighed were dissolved to a homogeneous state in chloroform at 80° C for 1 hour with constant stirring by means of an electronic overhead stirrer.

At the end of dissolution of the polymer, the volume of the solution was adjusted with chloroform to 200 ml. The resulting solution was aged for 5 minutes at 80^0 C and spread over glass pallets.

Drying was carried out for 2 days in air at 37^0 C in a thermostat.

At the end of the drying, the resulting films were removed.

The obtained samples were assigned ciphers used for independent research in the future (Table 1):

Table 1. Code samples of films

Code	Polymer	Molecular weight, kDa	Concentration, g per 100 ml of chloroform
1.1	Poly (glycolide-DL-lactide)	45	1
1.2	Poly (glycolide-DL-lactide)	45	3
1.3	Poly (glycolide-DL-lactide)	45	5
2.1	Poly (glycolide-DL-lactide)	90	1
2.2	Poly (glycolide-DL-lactide)	90	3
2.3	Poly (glycolide-DL-lactide)	90	5
3.1	Poly (glycolide-DL-lactide)	180	1
3.2	Poly (glycolide-DL-lactide)	180	3
3.3	Poly (glycolide-DL-lactide)	180	5

Tensile strength studies of polymeric films from polylactide were carried out on a universal test machine INSTRON 3382 at a loading rate of 10 mm / min. Samples of polymer films for testing were made in accordance with GOST 14236-81, in the form of a double blade. The sample was fixed in the

IOP Conf. Series: Materials Science and Engineering **347** (2018) 012042 doi:10.1088/1757-899X/347/1/012042

grippers of the testing machine, which was tightened evenly to ensure that the sample did not slip during the test. Tests of polymer films with determination of the relative elongation, yield strength and tensile strength were carried out in accordance with GOST 14236-81. The processing of the test results in determining the characteristics of mechanical properties was carried out using the software INSTRON Bluehill 2.0. The measurement error of the test machine is less than 1%. Five samples were tested for one experimental point. The values of yield strength σ_y, tensile strength σ_B and deformation δ were determined.

3. Results and discussion

Table 2 presents the averaged results of mechanical tests of polymer films for each composition based on polyglycolide-lactide (PGLA) with different molecular weights from different dissolution concentrations in chloroform.

From the results obtained, graphs of the dependence were plotted, along which it is possible to trace the tendency of the change in mechanical characteristics (Figures 1-3).

As can be seen from the diagrams shown in Figures 1-3, the thickness of the poly (glycolide-lactide) films, the mechanical properties of the films (strength, ductility) depend on the conditions for their production. By varying the molecular weight and concentration of the polymer in the chloroform solution, films with the desired characteristics can be obtained.

Increasing the concentration of the polymer in the solution promotes thickening of the polymer layer, thereby increasing plasticity, but decreasing the strength of the polymer film.

Table 2. Mechanical properties of polymer films.

Code	Polymer and concentration	Deformation, %	Yield strength, MPa	Ultimate Strength, MPa	Thickness, μm
1.1	PGLA 45kDa 1x100	139,52	3,96	5,51	66
1.2	PGLA 45kDa 3x100	303,48	1,88	3,63	155
1.3	PGLA 45kDa 5x100	447,17	0,37	1,82	224
2.1	PGLA 90kDa 1x100	262,21	2,42	5,17	63
2.2	PGLA 90kDa 3x100	369,34	0,47	4,18	148
2.3	PGLA 90kDa 5x100	432,73	0,27	2,35	258
3.1	PGLA 180kDa 1x100	198,32	5,59	7,62	32
3.2	PGLA 180kDa 3x100	376,73	0,48	4,47	154
3.3	PGLA 180kDa 5x100	554,91	0,27	1,83	252

IOP Conf. Series: Materials Science and Engineering **347** (2018) 012042 doi:10.1088/1757-899X/347/1/012042

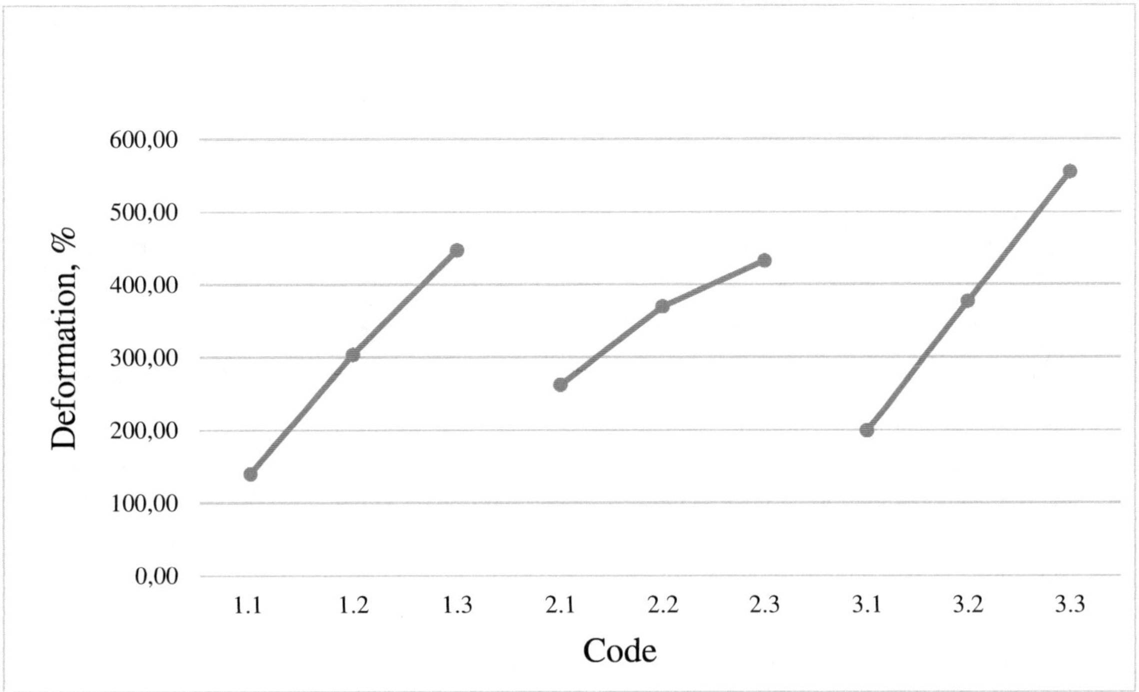

Figure 1. Diagram of deformation depending on the concentration of the solution for PGLA

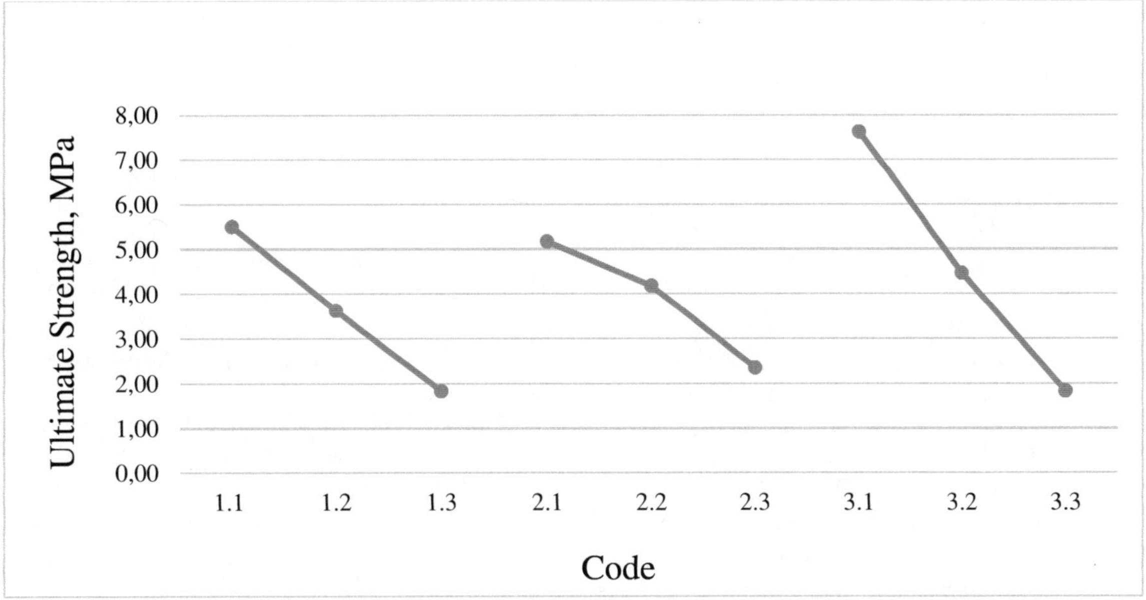

Figure 2. Diagram of the dependence of tensile strength on the concentration of solution for PGLA

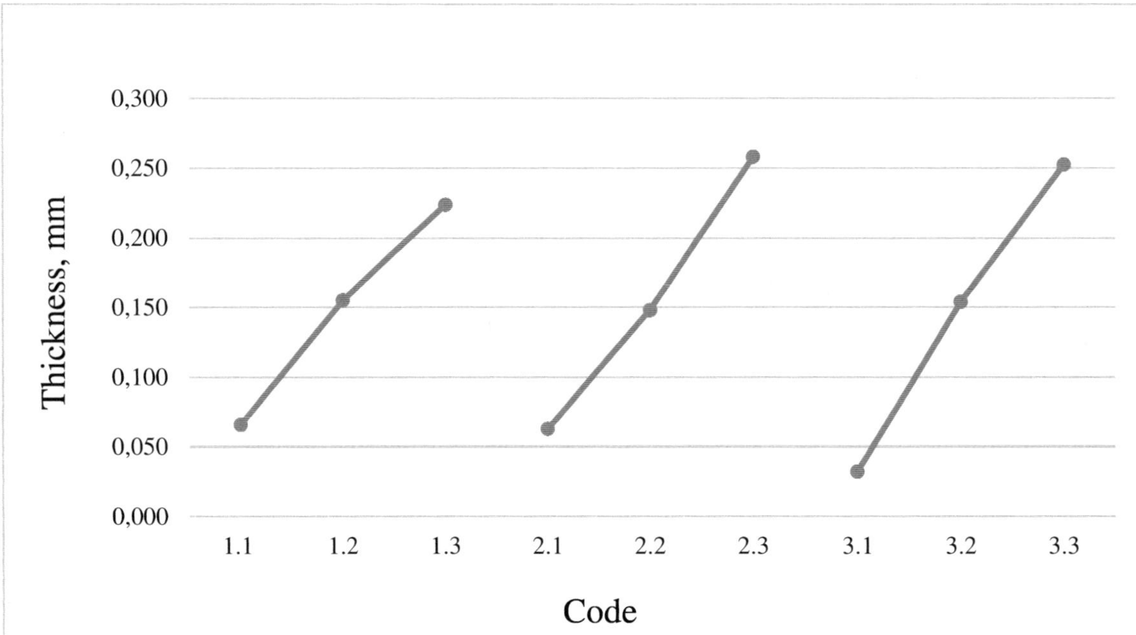

Figure 3. The diagram of the dependence of the thickness of polymer films on the concentration of solution for PGLA

4. Conclusions

Studies of the mechanical properties of the developed polymer films of poly (glycolide-lactide) were carried out. The dependences of the mechanical properties of the resulting polymer films on the molecular weight of poly (glycolide-lactide) and on its concentration in the chloroform solution are revealed. The possibility of creating biodegradable films with specified mechanical properties is shown.

Acknowledgements

The work was supported by the Ministry of Education and Science of Russia (grant identifier RFMEFI60417X0196).

References

[1] Giessen WV, Lincoff A, Schwartz R, Beusekom HV, Serruys P, Holmes D, et al. Marked inflammatory sequelae to implantation of biodegradable and nonbiodegradable polymers in porcine coronary arteries. *Circulation* 1996;**94(7)**:1690–7.

[2] Nasakina E.O., Baikin A.S., Sevost'yanov M.A., Kolmakov A.G., Zabolotnyi V.T., Solntsev K.A. Properties of nanostructured titanium nickelide and composite based on it *Theoretical Foundations of Chemical Engineering*, 2014. - **V.48. №.4.** - P.477–486.

[3] Nasakina E.O., Sevost'yanov M.A., Mikhailova A.B., Gol'dberg M.A., Demin K.Yu., Kolmakov A.G., Zabolotnyi V.T. Preparation of a nanostructured shape memory composite material for biomedical applications *Inorganic Materials*, 2015, **Vol. 51, No. 4**, pp. 400-404

[4] Cooper SL, Visser SA, Hergenrother RW, Lamba NMK. Polymers.In: Ratner BD, Hoffman AS, Schoen FJ, Lemons JE, editors. *Biomaterials science an introduction to materials in medicine*. 2nd ed. San Diego: Elsevier Academic Press; 2004. p. 67–79.

[5] Stack R, Califf R, Phillips H, Pryor D, Quigley P, Bauman R, et al. *Interventional cardiac catheterization at Duke medical center*. Am J Cardiol 1988;**62**:3F–24F.

IOP Conf. Series: Materials Science and Engineering **347** (2018) 012042 doi:10.1088/1757-899X/347/1/012042

[6] Lincoff AM, Schwartz RS, Giessen WJVD, Beusekom HMMV, Serruys PW, Holmes DR, et al. Biodegradable polymers can evoke a unique inflammatory response when implanted in the coronary artery. *Circulation* 1992;**86(Suppl.4)**:

[7] Supper S., Anton N., Seidel N., Riemenschnitter M., Curdy C., Vandamme T. Thermosensitive chitosan/glycerophosphate-based hydrogel and its derivatives in pharmaceutical and biomedical applications. *Expert Opin. Drug Deliv.* 2014;**11**:249–267. doi: 10.1517/17425247.2014.867326.

[8] Sevost'yanov M.A., Fedotov A.Yu., Nasakina E.O., Teterina A.Yu., Baikin A.S., Sergienko K.V., Kolmakov A.G., Komlev V.S., Ivanov V.E., Karp O.E., Gudkov S.V., Barinov S.M. Kinetics of the release of antibiotics from chitosan-based biodegradable biopolymer membranes *Doklady Chemistry*, 2015. **V.465. Part.1.** P.278-280.

[9] Timchenko TV, Shcherbakova LI, Kompantsev VA Poly-D, L-lactide-co-glycolide: methods of preparation, properties and use for the development of drugs with micro- and nanodelivery means. *Modern problems of science and education*. 2015. **№ 4**. P.559-570 (in russian)

[10] Kedik SA, Zhavoronok ES, Sedishev IP, Panov AV, Polymers for delayed drug delivery systems (review) (May 3, 2013) (in russian)

[11] Glotova VN, Novikov VT, Izhenbina TN, Titova NG Solubility of lactide and glycolide in organic solvents *Polzunovskii vestnik*. 2014. **No. 3.** P. 145-147. (in russian)

IOP Conf. Series: Materials Science and Engineering **347** (2018) 012043 doi:10.1088/1757-899X/347/1/012043

Preparation and mechanical properties of SiC-TiN composite

A V Leonov, M A Sevostyanov, A S Lysenkov, A S Baikin, M G Frolova and A M Tsareva

Baikov Institute of Metallurgy & Materials Science (IMET RAS), Moscow, Leninskiy prospect 49, Russia

Email: a_leonov.imet@list.ru

Abstract. A method has been developed for the production of a SiC-TiN composite material based on combining the processes of hot pressing and reaction sintering. The main difference of the method is the use of pure titanium as initial components. The mechanical properties and microstructure of the obtained samples were studied. X-ray phase analysis showed the absence of pure titanium and carbotitanium in the samples. Investigations of the influence of the type of sintering additive on the structure and properties of the resulting material were carried out.

1. Introduction

Recently a lot of attention in materials given to new superhard composite materials. Materials based on the silicon carbide (SiC) is widely used in this area. Silicon carbide has unique combination of chemical, physical, electrical and magnetic properties. Because of these properties SiC is widely used in many industry areas and products such as: heating elements, nuclear power plant details, turbine and pump details, etc. One of the most important advantages of silicon carbide is high wear resistance and heat resistant. [1] However, it is not possible to obtain the SiC material with theoretical density without additives. It has been reported that some additives can enhance the SiC material and improve its properties. Composite material based on silicon carbide with titanium nitride (TiN) is perspective. TiN is a material with good chemical and thermal stability, high hardness and thermal conductivity. TiN additives in SiC materials also reduce the electrical resistance of the material, which will make possible to apply the method of electro-erosion cutting for processing process. [2]

Obtaining dense, high-quality materials based on refractory compounds requires a long exposure at high temperatures (> 2000C). To reduce the firing temperature of the SiC-TiN material, sintering additives of various systems, such as Y2O3-Al2O3-ZrO2 and Y2O3-Al2O3, can be used. In the current literature there are data on composite SiC-TiN materials [3,4], however for the production of such materials an expensive powder of titanium nitride powder is used. It is possible to simplify the production of the SiC-TiN composite material by carrying out the nitriding of pure titanium during the sintering of the ceramic material.

2. Experimental procedure

SiC obtained by the SHS synthesis method (<25 μm), Ti powder obtained by sputtering in a plasma torch (<25 μm) were used. As a sintering additive, the addition of the Y2O3-Al2O3-ZrO2 system, obtained by the plasmachemical method (YAlZr), and also Y2O3-Al2O3 (3: 5) (YAG), obtained by solid-phase synthesis, were used. 7 different compositions were prepared, differing in the amount of Ti and also in the form of a sintering additive.

The compositions are listed in Table 1. Components were mixed in a Pulverisette 6 planetary mill in teflon beakers for 1 hour in isopropyl alcohol medium with a material: balls ratio of 1:10. As grinding bodies were balls of zirconia. After mixing, the resulting powders were dried in a drying oven at a temperature of 80 ° C, and then sieved through a sieve having a cell size of 63 μm and then hot-pressed. The blank was a disc d = 25 mm, the weight of the sample was 10 g. The firing was carried out in a nitrogen medium in a graphite mold. The firing was carried out stepwise. The first exposure was carried out at a temperature of 1600 ° C for 30 minutes. Then the temperature was raised to 1850 ° C and a second exposure was carried out for 30 minutes. The pressure was 30 MPa. The samples were then removed from the mold and the samples obtained were mechanically processed before testing and measurement.

Table 1. Compositions of different samples and processing conditions

Material	mass %				Processing conditions
	SiC	Ti	YAG	YAlZr	
SiC-Ti-0-1	93	0	7	0	T1=1600, 30 min T2=1850, 30 min
SiC-Ti-10-1	83	10	7	0	
SiC-Ti-20-1	73	20	7	0	
SiC-Ti-0-2	93	0	0	7	
SiC-Ti-10-2	83	10	0	7	
SiC-Ti-20-2	73	20	0	7	
SiC-Ti-40-2	53	40	0	7	

The density of the material was determined by hydrostatic weighing in water, while the true density was calculated by the additivity rule. For bending tests, specimens were produced. The tests were carried out on an Instron 3382 unit with a load cell of 100 kg. For microhardness measurements and microscope studies, metallographic samples were made. Microhardness was measured on a Wolpert Wilson Instruments 401/402-MVD unit with a load of 0.98 N and. Microstructure studies were carried out on a Scanning Electron Microscope (SEM) of the Tescan Vega II SBU (Tescan, Czech Republic). For X-ray phase analysis, the Ultima IV (Rigaku, Japan) was used.

3. Results and discussion

The main task of the work was to carry out the nitration of the initial titanium in the billet during sintering. The firing regimes of the SiC-TiN composite were selected [5]. According to X-ray analysis, the material obtained is represented by two main phases of SiC and TiN (Figure 1). Titanium in its pure form, as well as carbotitan, which could have formed during firing in a graphite mold, was not detected.

Investigations of the influence of the type of sintering additive and the amount of titanium in the composition on the mechanical properties of the material and the microstructure were carried out. Data are presented in Table 2.

Table 2. Mechanical properties of materials

Material	Density, g cm^{-3}	Relative density	Bending stress, MPa	Microhardness, GPa
SiC-Ti-0-1	3,11	0,85	225,46	9,45
SiC-Ti-10-1	2,98	0,87	238,00	10,73
SiC-Ti-20-1	2,89	0,8	263,14	12,80
SiC-Ti-0-2	3,07	0,83	190,73	10,11
SiC-Ti-10-2	2.95	0,86	175,38	11,14
SiC-Ti-20-2	2,81	0,77	166,15	13,27
SiC-Ti-40-2	3,25	0,81	153,91	21,83

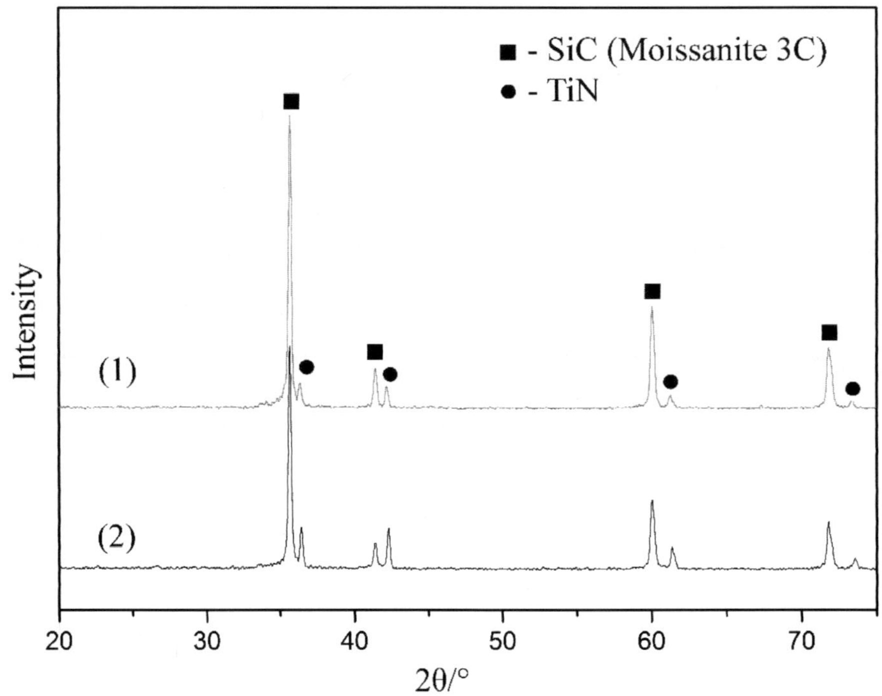

Figure 1. XRD patterns of SiC-Ti-10-1 (1) and SiC-Ti-20-1 (2)

It can be seen from the studies carried out that the use of the sintering additive YAG leads to a greater strength value than the use of the YAlZr additive. The bending strength of the material using the YAG additive is in the range of 230-270 MPa, while for the YAlZr additive, within the range of 150-170 MPa. The relative density of the resulting materials is not high and on the average is 0.82. As can be seen from the microphotographs of the microstructure presented (Figure 2), the increased bending strength of the samples with the addition of YAG is associated, first of all, with the shape of inclusions of titanium nitride (light areas in the figure). Image (a) shows that TiN in the material is distributed in the form of elongated inclusions, while in the images (b) and (c) TiN has a basically spherical shape.

An increase in the amount in the initial components of titanium, and as a consequence, an increase in the amount of titanium nitride leads to an increase in the microhardness of the samples. A particularly sharp increase in microhardness was found in the SiC-Ti-40-2 sample.

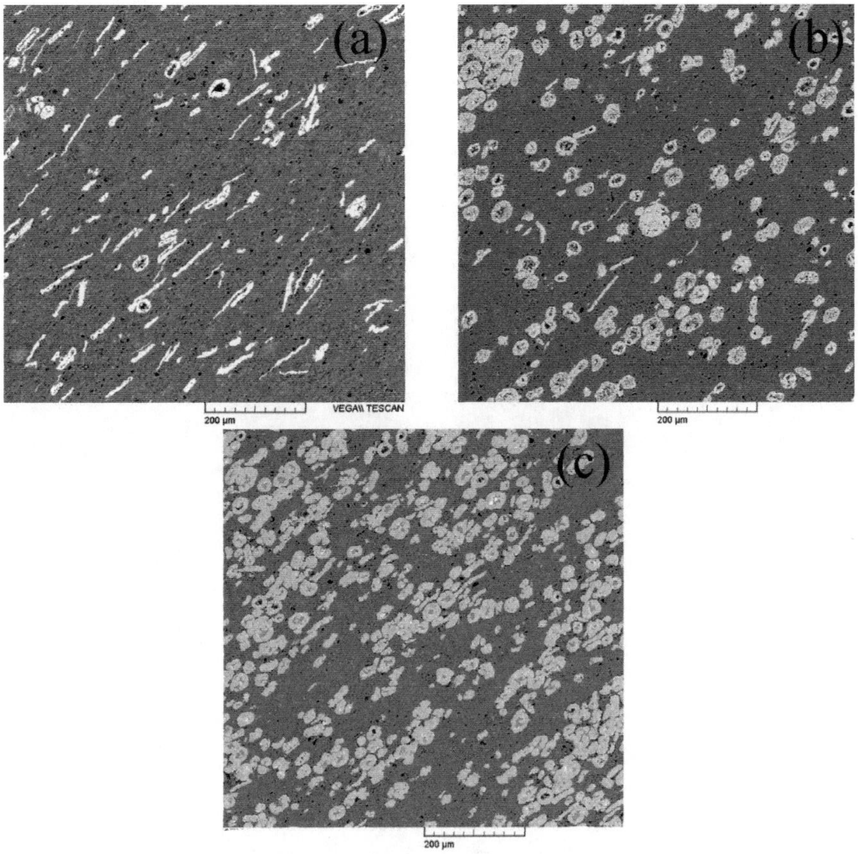

Figure 2. Microstructure of SiC-based composites: (a) SiC-Ti-20-1; (b) SiC-Ti-20-2; (c) SiC-Ti-40-2

4. Conclusion
The samples of the composite ceramic material SiC-TiN were obtained. The prospect of introducing titanium nitride into a carbide-silicon matrix was shown. The introduction of TiN led to increase the microhardness of the material. The technology of obtaining the material by combining the processes of hot pressing and reaction sintering has also been studied. For the production of titanium nitride, hot pressing was carried out in a nitrogen atmosphere, which resulted in the simplification of manufacturing technology and the cheaper material due to the use of cheaper components. The use of the YAG additive is more promising for obtaining a stronger material.

Acknowledgments
The work was carried out according to the state task No. 007-00129-18-00

References
[1] Pavel Istomin, Elena Istomina, Aleksandr Nadutkin, Vladislav Grass, Aleksandr Leonov , Mikhail Kaplan , Mikhail Presniakov 2017 Fabrication of Ti3SiC2 and Ti4SiC3 MAX phase ceramics through reduction of TiO2 with SiC *Ceramics International* **43** 16128–16135
[2] Kwang Joo Kim, Kun Mo Kim and Young-Wook Kim 2014 Highly conductive SiC ceramics containing Ti2CN *Journal of the European Ceramic Society* **34** 1149–1154
[3] Xingzhong Guo, Hui Yang, Lingjie Zhang, Xiaoyi Zhu 2010 Sintering behavior, microstructure and mechanical properties of silicon carbide ceramics containing different nano-TiN additive

Ceramics International **36** 161–165

[4] Lingjie Zhang, Hui Yang, Xingzhong Guo, Jianchao Shen and Xiaoyi Zhu 2011 Preparation and properties of silicon carbide ceramics enhanced by TiN nanoparticles and SiC whiskers *Scripta Materialia* **65** 186–189

[5] K B Kuznetsov, A P Stecovskiy, A S Chernyavskiy, K A Solntsev 2008 *Perspektivnie materiali* **1** 56-59 (Rus)

Hydrogels based on polysaccharide-calcium phosphate with antibacterial / antitumor activity for 3D printing

A Yu Teterina[1], A Yu Fedotov[1], Yu V Zobkov[1], N S Sergeeva[2], I K Sviridova[2], V A Kirsanova[2], P A Karalkin[2] and V S Komlev[1]

[1]A.A. Baikov Institute of Metallurgy and Material Sciences of Russian Academy of Science, Moscow, 119334 Russia
[2] P.A. Herzen Moscow Oncology Research Institute—Branch of the National Medical Research Radiological Centre of the Ministry of Health Care of the Russian Federation, Moscow, 125284 Russia

Email: komlev@mail.ru

Abstract. The purpose of this study was to develop hydrogels for 3D printing of sodium alginate /gelatin /octacalcium phosphate-based constructs with antibacterial and antitumor activity intended for bone defects replacement in patients with malignant diseases. In this work, we evaluated the drug release kinetic and physico-chemical characteristics of constructs, as well as their specific activity, biocompatibility and osteoplastic properties by means of *in vitro* and *in vivo* tests. The principal possibility of creating the biocompatible bone substitutes with antibacterial /antitumor activity and osteoconductive-retaining properties of 3D printing method was demonstrated.

1. Introduction

Many resent concepts for osteoplastic materials investigations are inspired by the structure of natural bone tissue which extracellular matrix comprises a perfect nanocomposite consisted of collagen fibrils chemically linked to carbonated apatite. In this term, different biomimetic approaches involving natural polymers and calcium phosphate (CP) ceramic were developed in last decades [1, 2]. In our study, we used biocompatible polymers—sodium alginate and gelatin in combination with octacalcium phosphate (OCP)-based ceramics. Being a precursor of biological apatite OCP promotes a mineralization of new bone tissue and also demonstrates convenient solubility in simulating body fluids [3, 4].

At the present time, a number of requirements have been postulated for biomaterial: biocompatibility, high surface area, high degree of porosity, optimal rate of resorption (biodegradation) *in vivo*, corresponding to the rate of new bone formation, osteoconductive and, more preferable, osteoinductive properties. Recent technologies of rapid prototyping, e.g. 3D printing, are capable for creation of personalized grafts with precise geometry of defects according to patient's CT/MRI data [5]. On the other hand, the functionalization of three-dimensional printed structures for local delivery of antitumor drugs or antibiotics can theoretically increase the effectiveness of such therapy by improving the accumulation of the drug in bone tissues, reducing its dosage and toxicity without significantly increasing the duration of bone regeneration [6].

The study was devoted to hydrogels based on polysaccharide-calcium phosphate with antibacterial / antitumor activity for 3D printing of constructs for bone defects replacement. Thus, we studied effects of the hydrogel composition on resulting structure, porosity and mechanical properties of 3D printed constructs. A complex composition of constructs indicated to the possibility of their multipurpose use: not only as a "simple" implant material, but also as a bioconstruction containing biologically active molecules (growth factors) or drugs.

2. Materials and methods

An experimental three-dimensional constructs (3DC) was produced according to original technology (Patent RU2606041C2) using a custom-designed injection 3D printer. The hydrogel ("ink") for 3D printing consisted of sodium alginate, gelatin, and octacalcium phosphate in proportion of 56/14/30 wt. %. The functionalization of constructs with vancomycin (3DCVanc) or doxorubicin (3DCDox) we performed by addition of drugs to the initial dry components of hydrogel. The process of 3D printing was as follows. The hydrogel having viscosity 10–29 Pa was placed in disposable syringe of 3D printer and then injected through a needle onto the precooled (–20°C) surface with alcohol and calcium chloride water solution (70/30) filled bath. The printed semi-solid constructs were freeze-dried at –50°C and 6 × 10–5 atm. for 10–12 h. Solidified 3DCs were washed in distilled water and dried at 37°C for 24h to finalize the synthesis of nonorganic components.

Scanning electron microscope (SEM) Vega II (Tescan, Czech Republic), working in secondary and backscattered electron modes, was used for microstructure studies. The samples were sputter-coated with a 25 nm thick gold layer prior imaging to impart electrical conductivity of the surfaces. FTIR spectroscopy was carried out with Nicolet Avatar system (Thermo Fisher Scientific, USA) using the powder mixtures of KBr samples.

Compression testing was carried out using an Instron Electro Puls E3000 testing machine (Bucks, UK) operating at a crosshead speed of 1 mm min^{-1}. The 3D printed cylindrical samples for testing were about 12 mm in height and 6 mm in diameter.

Biocompatibility of three-dimensional constructs was evaluated on the model of their subcutaneous implantation of male BDF1 mice (Scientific Center for Biomedical technologies of FMBA of Russia). The experiment implied 4 groups of 12 animals formed according to the number of functionalized compositions of implantable 3D constructs and 1 control group. Thus, 4 types of 3DCs saturated with 30 or 70 wt. % vancomycin (3DCVanc30 and 3DCVanc70) and 3 or 6 wt. % doxorubicin (3DCDox3 and 3DCDox6) were tested. Animals were carried for 2, 4, 8, 12 weeks after surgery. At the indicated time, two animals from each group were withdrawn from the experiment. Samples of materials were extracted and observed with stereomicroscope equipped with a digital video camera (Olympus, Japan), then fixed in a 10% formalin solution and embedded in paraffin. Histological samples (a thickness of 4 um) were stained with hematoxylin-eosin and documented with Eclipse Ti microscope and digital camera DS-Fi1c (Nikon, Japan).

3. Results

Basically, our approach for the production of 3D constructs based on polysaccharide-calcium phosphate involved the prepeartion of the original hydrogels into which the required concentrations of antibacterial (vancomycin) and antitumor (doxorubicin) drugs were added at the stage of dry components mixing. Using capillary viscosimetry, it was found that the administration of antibiotic in concentrations of 30 and 70 wt. % and antitumor preparation in concentrations of 3 and 6 wt. % did not significantly affect the values of viscosity and hydrogel fluidity, allowing the providing of 3D printing. According to SEM data, 3DCVanc maintained its porosity and tabular structure in the presence of 30 wt. % vancomycin. Increasing of concentration of vancomycin in the hydrogel slightly transformed the structure: pore size increased to 200–300 μm; there was a uniform spreading of drug on the surface of the specimens, which thickness increases proportionally to concentration of vancomycin (figure 1). Addition of doxorubicin led to serious changes in the inner structure of 3DCDox.

IOP Conf. Series: Materials Science and Engineering **347** (2018) 012044 doi:10.1088/1757-899X/347/1/012044

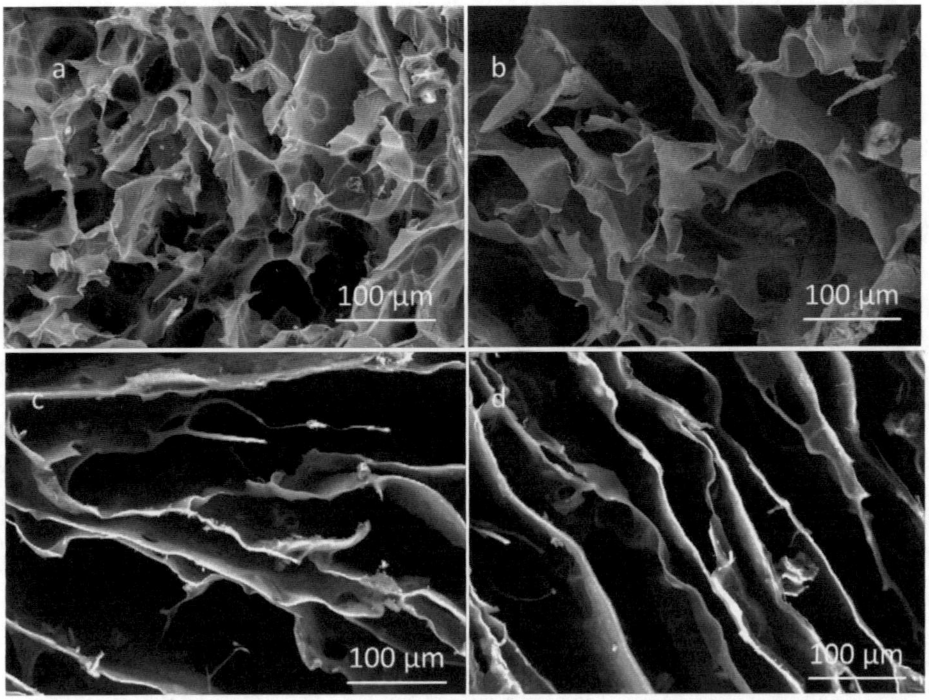

Figure 1. SEM images of microstructure of the 3DCs with 30% vancomycin (a), with 70% vancomycin (b) and 3% doxorubicin (c), 3% with 6 % doxorubicin (d).

The compressive strength of 3DCs was decreased after addition of higher drug concentrations to the hydrogel for 3D printing. The compressive strength (5.5 MPa) did not differ between 3DCVanc30 and control 3DCs. Meanwhile, its value decreased to 4.4 MPa after addition of 50 wt. % vancomycin and to 3.7 in case of 70 wt. %. For the materials comprising 3 and 6 wt % of doxorubicin the value of strength was changed more dramatically and decreased 2.5-fold compared to empty 3DCs (1.8 and 2.2 MPa, respectively). The determination of the effects on the component composition after the introduction of drugs to 3DC was determined by IR spectroscopy. In all investigated constructs, IR spectra contain bands corresponding to the ν_1, ν_2 and ν_4 modes of the phosphate groups in the range of waves 960, 1020-1120 cm^{-1} and 560-660 cm^{-1}, respectively (figure 2). The characteristic bands of vancomycin 1650, 1506, 1230 and 1047 cm^{-1} showed that antibiotic administration to the hydrogel did not lead to a change in the chemical composition of vancomycin and did not cause any noticeable change in the composition of the 3DC. The IR spectra of composite materials based on the hydrogel with introduced doxorubicin, according to the corresponding bands 1616, 1583, 1207 cm^{-1}, also confirmed the absence of a change in the chemical composition.

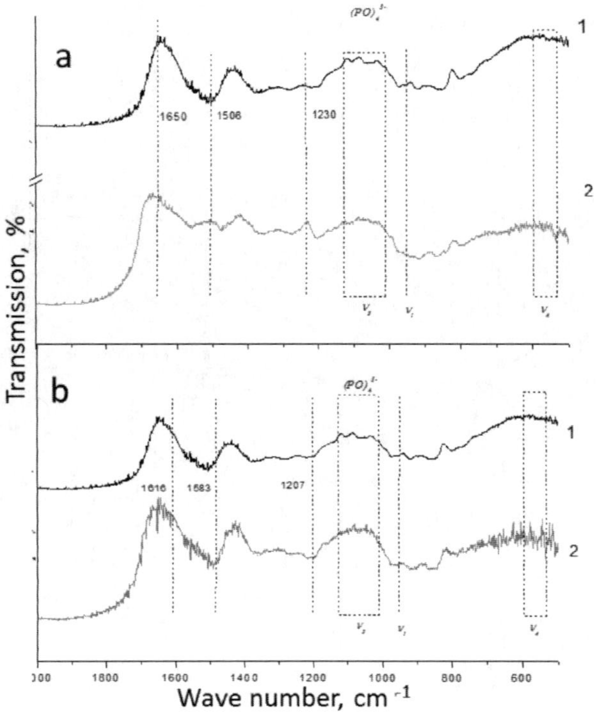

Figure 2. IR absorption spectra of 3DC based on polysaccharide-calcium phosphate – line 1 (a, b); with vancomycin – line 2 (a); with doxorubicin – line 2 (b).

In case of control printed 3DC based on sodium alginate / gelatin / OCP without medicinal drugs a thin fibrous capsule comprised connective tissue with a regular orientation of the fibers, moderate neutrophil infiltration and active vascularization could be visible in 2 weeks after subcutaneous implantation (Figure 3). Aseptic inflammation in the implantation zone substantially ameliorated in 4 weeks after the surgery. This indicated that composition of 3DC promoted the "transition" of surround tissue to the regenerative state.

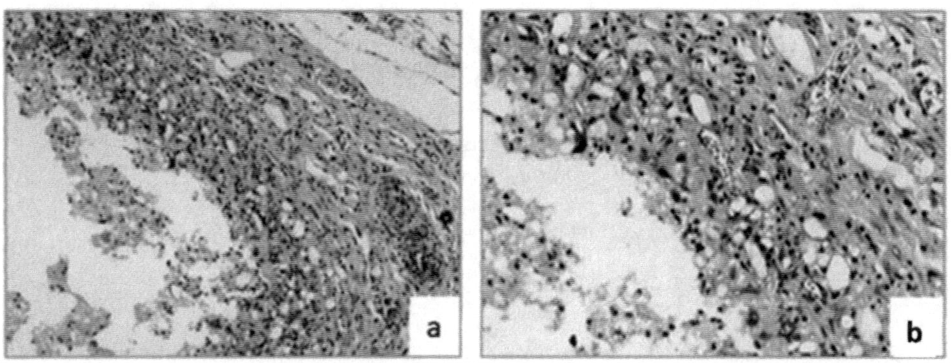

Figure 3. Images of the 3D printed samples in 2 weeks after subcutaneous implantation in mice. HE staining. Magnification: a - x100, b - x200.

After subcutaneous implantation the 3DCs saturated with antibiotic vancomycin (3DCVanc30 or 3DCVanc70) demonstrated well biocompatibility similar to those for control constructs. The signs of inflammation or necrosis were absent, and the formation of a loose connective tissue capsule even thinner in comparison with the capsule around control samples could be observed (Figure 4). Numerous foreign-body giant cells indicated the beginning of resorption processes. A rather different picture was observed when biocompatibility studying of 3DCs with antitumor drug doxorubicin was performed. In the early stages (2 weeks) after implantation the strong inflammatory reactions accompanied with intensive leukocyte infiltration developed in surrounding tissues (figure 4). There were no bioresorption signs in both polymeric and CP components of the 3DCDox composite material. The spreading of connective tissue in the inner part of implants and new blood vessels formation was hampered by the formation of thick connective tissue capsule. The dermal cells from internal layers the skin possessed cytoplasm vacuolization and apoptotic morphology. Thus, the antitumor drug released from the constructs rendered cytotoxic effect on surrounding tissues, which led to the development of aseptic inflammation and gradual incapsulation of 3DCDox. Nevertheless, the severity of the these processes allowed us to expect a powerful local antitumour effect after functionalized 3DC implantation, that can be useful in filling bone defects after surgery for primary malignant bone tumors or metastases.

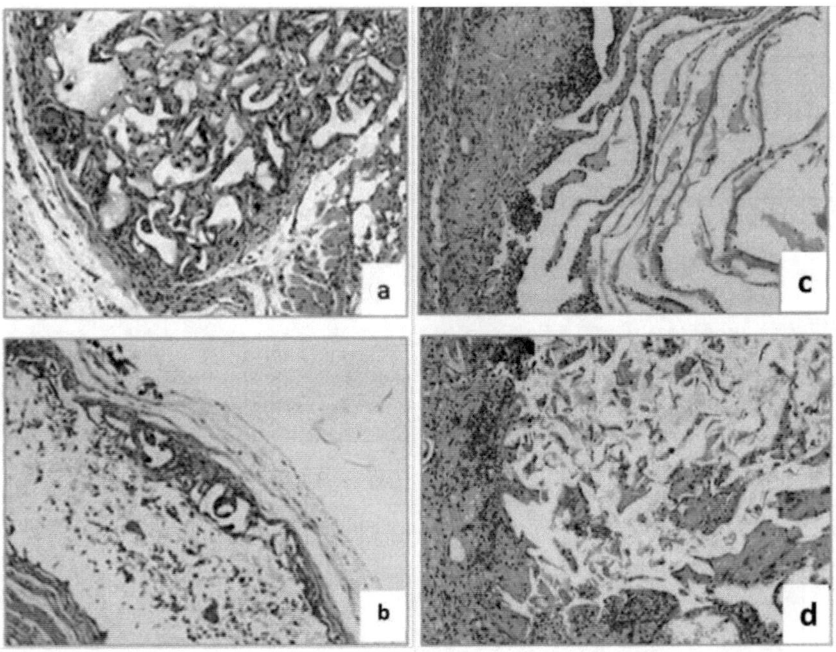

Figure 4. Images of functionalized 3DC in 2 weeks after subcutaneous implantation in mice. 30% vancomycin (a), 70% vancomycin (b) 3% doxorubicin (c), 6 % doxorubicin (d). HE staining. Magnification: x100.

4. Conclusion

The approach for 3D printing with low viscosity biopolymers and calcium phosphate components containing drugs and intended for bone defects replacement was created. Also the influence of the polymer - drug ratio on fluidity of hydrogel for 3D printing was estimated. Physicochemical characteristics of compositions based on sodium alginate, gelatin and octacalcium phosphate with different concentrations of drugs were studied as well. It was found, that components of the system were stable during 3D printing. The functionalization of printed 3DCs did not cause dramatic deterioration of their biocompatibility properties. As a result, a complex study of physicochemical and biological features of functionalized constructs revealed that 3DCVanc and 3DCDox retain valuable features of composite bone grafts, such as porosity, phase compound and mechanical strength, but also demonstrated functional activity in vivo.

Acknowledgments

This work was supported by the Russian Science Foundation, project no. 15-13-00108.

References

[1] Generosi A et al 2010 *J. Phys. Chem.* **114** 973-979
[2] Butscher A *et al* 2013 *Acta Biomaterialia* **9** 5369-5378
[3] Dorozhkin S 2008 *J. Mater. Sci.* **43** 3028-3057
[4] Driessens F *et al* 1994 *J. Mater. Sci. Mater.Med.* **5** 164-170
[5] Fedotov A et al 2008 *Powder Metallurgy Progress* **8** 351-357
[6] Teterina A *et al* 2015 *Doklady Chemistry* **461** 104-107

IOP Conf. Series: Materials Science and Engineering **347** (2018) 012045 doi:10.1088/1757-899X/347/1/012045

Limitation of biocompatibility of hydrated nanocrystalline hydroxyapatite

V V Minaychev[1,2], A T Teleshev[3], V N Gorshenev[4], M A Yakovleva[3], V A Fomichev[3], A S Pankratov[5], K A Menshikh[6], R S Fadeev[1,2], I S Fadeeva[1,2], A S Senotov[1], M I Kobyakova[1,2], Yu B Yurasova[7] and V S Akatov[1,2]

[1]Institute of Theoretical and Experimental Biophysics RAS, Pushchino, Moscow region, Russia
[2] Pushchino State Institute of Natural Science, Pushchino, Moscow region, Russia
[3]Moscow State Pedagogical University, Moscow, Russia
[4]Emanuel Institute of Biochemical Physics RAS, Moscow, Russia
[5]Sechenov University, Moscow, Russia
[6]Lomonosov Moscow State University, Moscow, Russia
[7]Central Institute of Traumatology and Orthopedia, Moscow, Russia

E-mail: vminaychev@gmail.com, akatov.vladimir@gmail.com

Abstract. Nanostructured hydroxyapatite (HA) in the form of hydrated paste is considered to be a promising material for a minor-invasive surgical curing of bone tissue injure. However questions about adhesion of cells on this material and its biocompatibility still remain. In this study biocompatibility of paste-formed nanosized HA (nano-HA) by *in vitro* methods is investigated. Nano-HA (particles sized about 20 nm) was synthesized under conditions of mechano-acoustic activation of an aqueous reaction mixture of ammonium hydrophosphate and calcium nitrate. It was ascertained that nanocrystalline paste was not cytotoxic although limitation of adhesion, spreading and growth of the cells on its surface was revealed. The results obtained point on the need of modification of hydrated nano-HA in the aims of increasing its biocompatibility and osteoplastic potential.

1. Introduction

Artificial constructions made of composite materials combining organic and mineral phase – hydroxyapatite evoke a high interest in creation of biomaterials out of them. These materials could provide regeneration of bone tissue in the area of its damage. A plenty of results showed high biocompatibility potential of the composites made from HA and organic matrix for instance collagen. Stimulation of osteogenesis was also demonstrated as the material played role of matrix for a newly forming bone [1-4].

However, in the creation of artificial biocomposites there remain unsolved problems connected with controlling of forming of material structure and its composition which are able to provide efficient regeneration of bone tissue. Excessive ratio of HA in the material can inhibit the process of osteogenesis [5-7]. Significant influence can be caused by level of dispersion of composite's mineral component [8, 9].

IOP Conf. Series: Materials Science and Engineering **347** (2018) 012045 doi:10.1088/1757-899X/347/1/012045

Several investigations point that synthetic HA which particles have high dispersion level looks more like a mineral constituent of bone tissue and is more likely to activate osteoblastic differentiation of mesenchymal cells than HA with bigger particles [10]. Paste-formed nano-HA is of practical interest as a material for less-invasive surgical intervention in the treating of bone defects but from time to time it appears that surgeons have contrasting views on the possibilities of this material [11]. Some of them think consider it to be a beneficial material but others point on its insufficient biocompatibility lack of its resorption and regeneration of bone tissue in the site of implantation.

Objective reasons of this situation remain unclear. In the current study we attempted to estimate whether these reasons could be adhesive properties and biocompatibility of nano-HA in relation to cells of mesenchymal origin.

2. Materials and methods

2.1 Synthesis of nano-HA

The process of receiving HA was held in the rotary pulsation apparatus (RPA) (Scientific production enterprise 'Aviatekhnika', LLC) [12] using a well-known reaction [13]

$$6(NH_4)_2HPO_4 + 10Ca(NO_3)_2{\times}4H_2O + 8NH_4OH \rightarrow Ca_{10}(PO_4)_6(OH)_2 + 20NH_4NO_3 + 46H_2O$$

For the ion exchange reaction and receiving HA, two solutions were used: (1) calcium nitrate tetrahydrate (450.6 g were dissolved in 5 l of distilled water and the pH was adjusted to pH 10.1 by adding a concentrated ammonia solution) and (2) diammonium phosphate (151 g were dissolved in 2 l of distilled water and the pH adjusted to 10 by adding a concentrated ammonia solution). Solution 1 in cooling conditions was placed in the loading chamber of the RPA, the mechano-acoustic activation mode (3000 min^{-1}) was switched on and then solution 2 was added into the chamber at a rate of 0.5 l/min. The mixture was treated in this regime for 2 minutes. Purification of HA in the resulting pulp from ammonium nitrate was carried out by washing with distilled water after centrifugation (2500 g, 15 min, 4 cycles). The content of mineralized calcium in the paste-formed material after the final washing was 100 mg/g of paste weight.

2.2. Analysis of nano-HA

The particle size of the dispersed phase was evaluated by dynamic light scattering using device Zetasizer Nano «Malvern».

IR spectra were recorded in the reflection mode on a Nicolet 380 spectrometer, ZnSe glass.

The ζ-potential of particles dispersed in an aqueous medium was measured by Zetasizer Nano device.

Elemental analysis was carried out on a CNE analyzer Eager 300.

Calorimetric curves were obtained on a Netzsch instrument, model DSC-204 F1.

X-ray analysis was performed on an Emma diffractometer, GBC Scientific Equipment Pty Ltd.

2.3. Cell culture

Mouse embryonic fibroblasts NIH/3T3 were obtained from the Russian Cell Culture Collection (Institute of Cytology RAS, St. Petersburg). The cells were cultured in a DMEM/F12 culture medium (Sigma-Aldrich, USA) supplemented with 10% fetal bovine serum (Gibco, USA) and 80 μg/ml gentamicin sulphate (Sigma-Aldrich, USA) at 37°C under conditions 5% CO_2 in the air.

2.4. Investigation of biocompatibility of hydrated HA in vitro

To investigate the biocompatibility of nanocrystalline paste-formed hydroxyapatite *in vitro*, a 0.5-mm HA paste layer was placed on one half of the cover glass and the other half was uncovered. The glass was placed in a 35 mm diameter Petri dish, and a cell suspension (1×10^4 cells/cm^2) was added. After 24 hours and 72 hours after seeding a cytotoxic effect, mitotic activity and cell microphotography were performed. As a control sample commercial ReproBone novo paste (Ceramisys, England) was used.

2.5 *Cytotoxicity analysis.* For the cytotoxicity analysis the number of live and dead cells was determined by fluorescence microscopy. Cells were stained for 20 minutes at 37°C in a culture medium with adding of 1 μg/ml calcein AM and 2 μg/ml propidium iodide (Sigma-Aldrich, USA). Using a confocal fluorescence microscope TCS SP5 (Leica, Germany) at least 500 cells were analyzed.

2.6. Analysis of mitotic activity. The analysis of mitotic activity was performed by fluorescence microscopy using a vital staining with fluorescent nuclear dye Hoechst 33342 (Sigma-Aldrich, USA) at a concentration of 1 μg/ml and 2 μg/ml of propidium iodide. Mitotic cells were detected by the chromatin distribution characteristic of prophase, metaphase, anaphase, and telophase using a fluorescent edmicroscope DM 6000 (Leica, Germany). At least 500 cells were analysed. The mitotic index (MI) was calculated from the formula $MI = (P+M+A+T)/N * 100\%$ where (P+M+A+T) is the sum of cells in the prophase, metaphase, anaphase and telophase, respectively, and N is the total number of cells analyzed.

2.7. Statistical analysis
The results of the experiments were presented as the mean ± standard error (M±SEM). The experiments were performed in at least three replicates (n≥3). The statistical significance of the difference was determined using the Mann-Whitney test.

3. Results and discussion

Mechano-acoustic activation of the exchange reaction between diammonium phosphate and calcium nitrate allows the formation of nanosized particles of HA. Reduction of the feed rate of the solution of diammonium phosphate to the reactor and increase in the mixing speed of the reactants lead to increase in dispersion. A HA dispersion characterized by particles having an average diameter of 20 nm was obtained at a rotation RPA rotor speed of 3000 min^{-1} and the rate of introduction of 0.5 l/min solution of diammonium phosphate into the reactor. The mass concentration of such particles is ≈90% in the reaction medium (Figure 1).

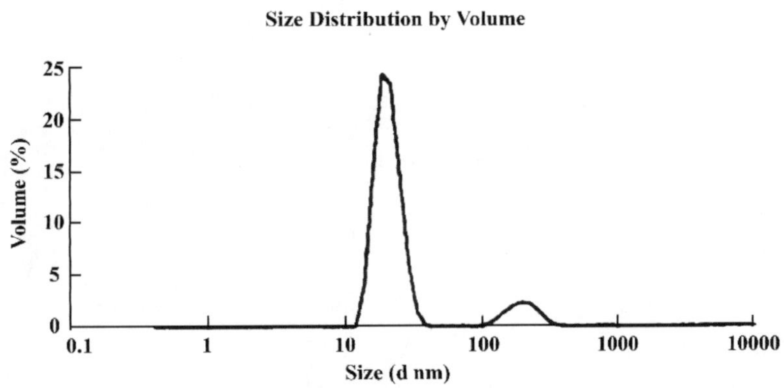

Figure 1. The distribution of particle sizes of HA obtained under mechano-acoustic treatment conditions. RPA rotor rotation speed is 3000 min^{-1}. The rate of introduction of a solution of diammonium phosphate into the reactor is 0.5 l/min

An important stage of the HA synthesis is its purification from the ammonium nitrate. Centrifugation can significantly increase the speed of the cleaning process in comparison with decantation. Thus filtering the pulp, washing the retentate once with distilled water, drying it at 120°C for 3 hours leads to a sample of technical HA contaminated with ammonium nitrate (absorption bands

1330, 820 cm^{-1}, Figure 2b). Four-time washing with centrifugation and subsequent drying (120°C for 3 hours) successfully removes ammonium nitrate as well as heat treatment of HA (350°C for 30 min). After these treatments these absorption bands are not detected in the samples (Figure 2a).

The elemental analysis of purified HA confirms the complete absence of nitrogen-containing impurities in four-time washed with centrifugation as well as in heat-treated HA (reference-ammonium nitrate).

Additional evidence of the absence of ammonium nitrate in purified samples of HA was the results of calorimetric analysis (Figure 3). Calorimetric curves after 4 times washing and after heat treatment (300°C) coincide.

The ζ-potential of the particles of the non-washed hydroxyapatite dispersed in distilled water was +30 mV (Figure 4a) but it was reduced to about +5 mV (Figure 4b) after 4 times washing.

Figure 5 shows the diffractogram of a heat-treated purified HA sample. The crystalline phase of HA has been identified, which has an elementary hexagonal cell with dimensions (Å): a=9.4257, c=6.8853, space group P63/m. The experimental characteristics of the X-ray spectrum of HA are given in Table 1. The results indicate the presence of phases of not only crystalline but also amorphous HA.

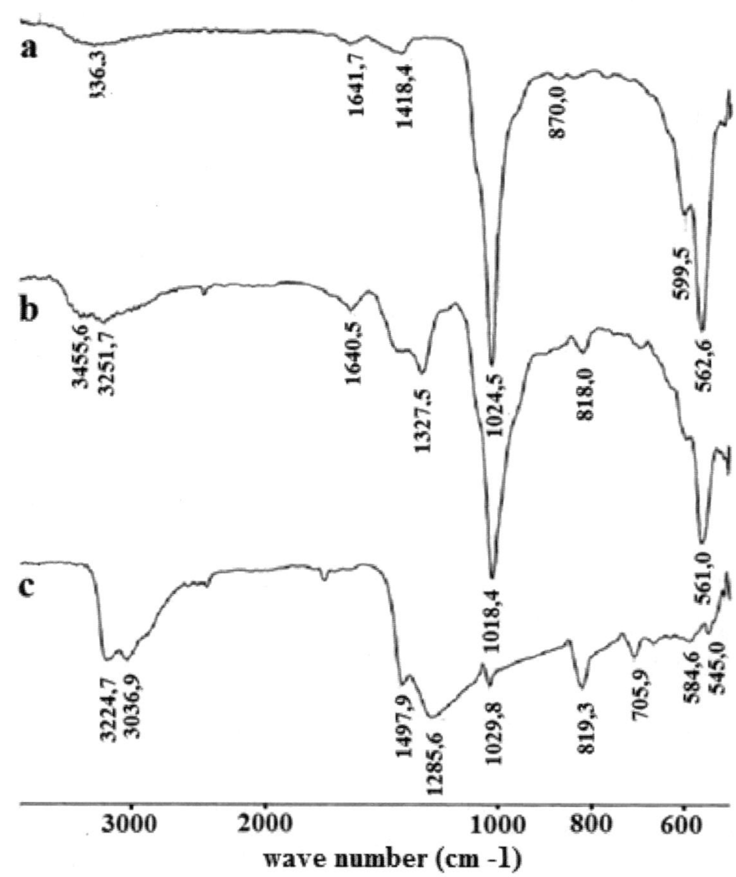

Figure 2. IR spectra: a – HA after heat treatment at 300 °C, b – technical HA, c – ammonium nitrate

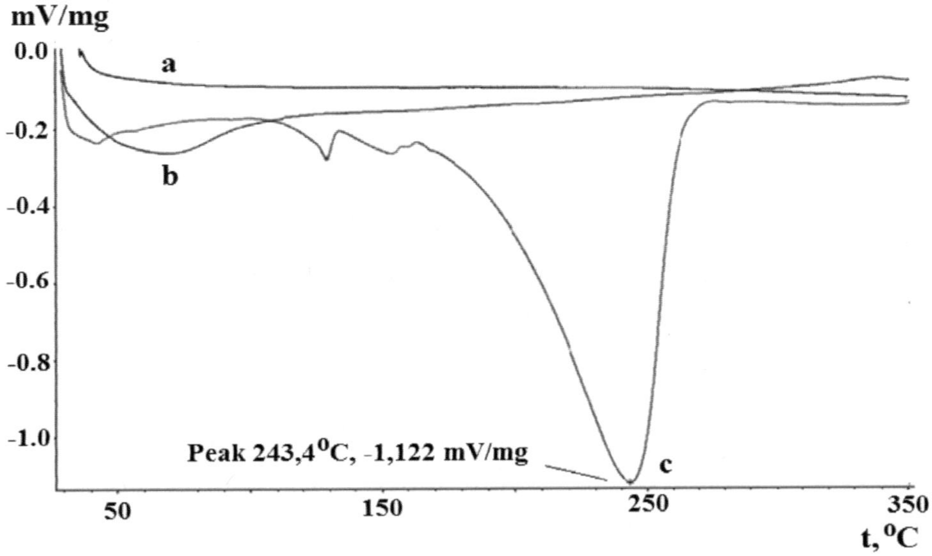

Figure 3. Calorimetric curves: a – HA heated at 300 °C, b – technical HA after a single wash, c – ammonium nitrate

a

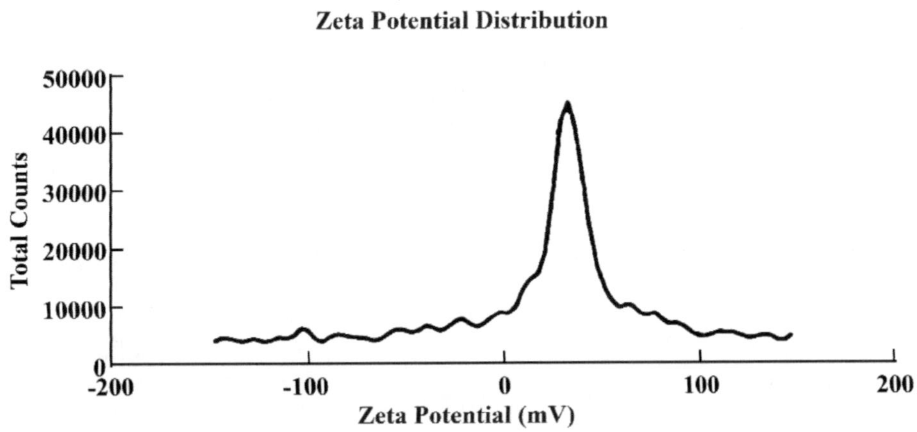

b

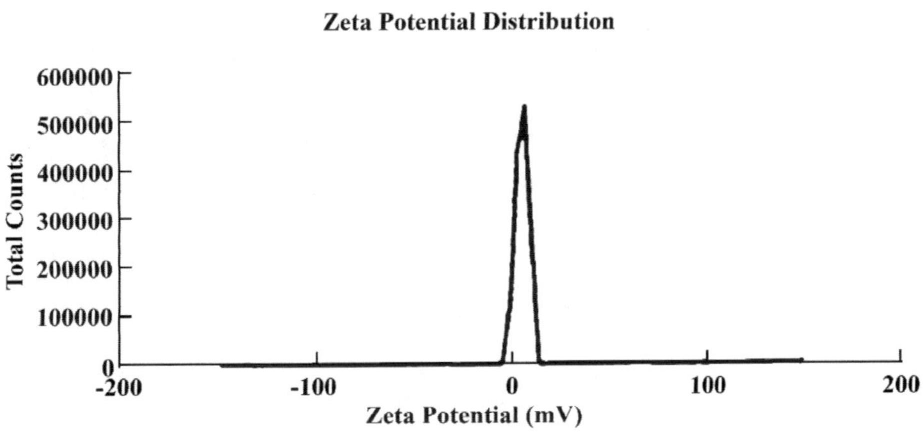

Figure 4. ζ-potential of dispersed HA, crude (a) and purified (b) from impurities

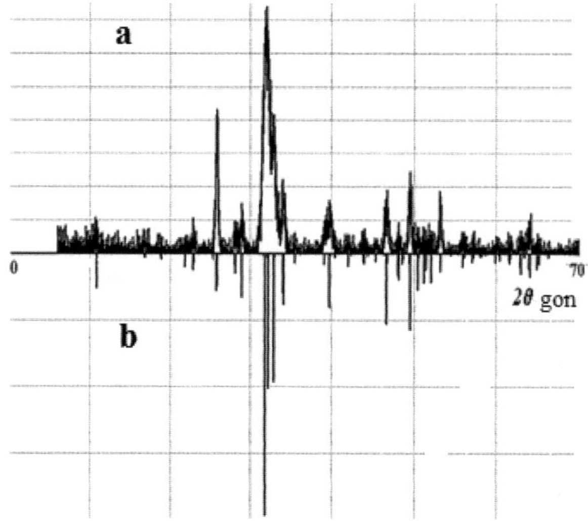

Figure 5. Diffractogram of purified and heat-treated HA (a). For comparison, the diffractogram of HA presented in [14] is shown (b)

Table 1. The experimental values of the angles (2θ), the relative intensities (I_o) and the interplanar distances (d/n) for the diffractogram of the purified and heat-treated HA

№ reflection	2θ gon	I_o	d/n, Å
1	25.8541	60	3.443
2	31.73	96	2.818
3	31.8108	100	2.811
4	31.88	91	2.805
5	34.20	21	2.62
6	39.4575	9	2.282
7	46.64	27	1.946
8	49.42	32	1.843
9	51.82	18	1.763

A study of the cytotoxicity of HA paste was carried out 24 and 72 hours after cell seeding. It was established that the HA paste and the ReproBone novo paste didn't have a cytotoxic effect. Figure 6 shows the results of studying cytotoxicity 72 hours after cell seeding. The number of dead cells on these samples of HA didn't exceed 5% and didn't differ from the number of dead cells when cultivated on culture dish surface.

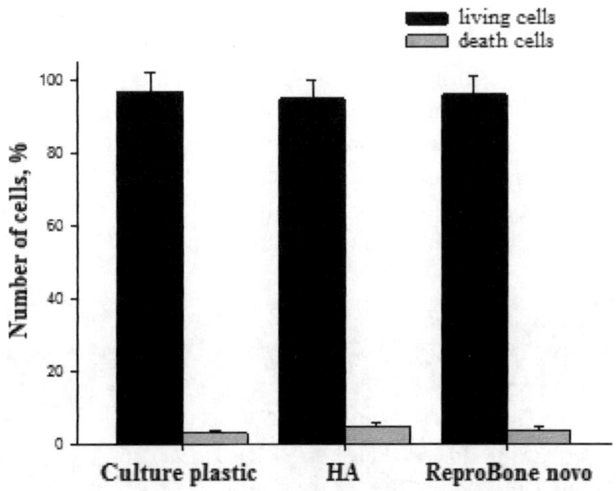

Figure 6. Absence of cytotoxic effect of HA paste and ReproBone novo paste after seeding NIH 3T3 cell culture on paste

The mitotic activity of cells cultivated on glasses coated with HA paste was studied 72 hours after the start of the cultivation. The data presented in Figure 7 show that the cells on the surface of the HA paste and the ReproBone novo paste are incapable of dividing. The mitotic index after 72 hours of cultivation was 4.1±0.8% when the cells were cultured on culture plastic.

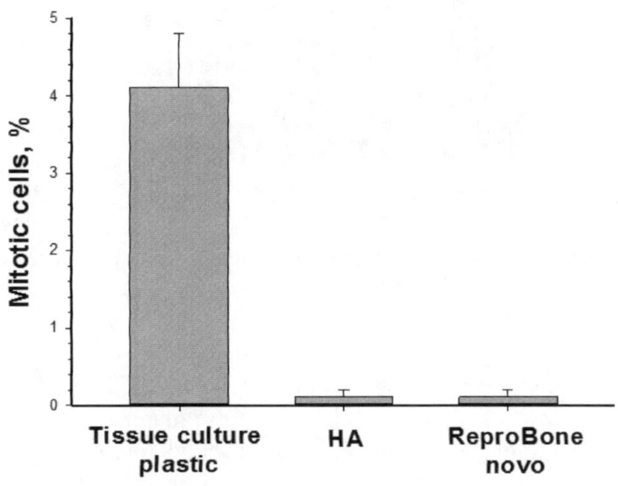

Figure 7. Mitotic activity of cells cultured on the surface of HA and ReproBone novo pastes

Taking into account the obtained data on the absence of the cytotoxic effect of the test paste and the suppression of the mitotic activity of the cells we assumed a violation of adhesion and cell spreading on their surface. We carried out a morphological analysis of the adhesion and cell spreading on the surface of nanocrystalline HA and ReproBone novo pastes to verify this assumption.

IOP Conf. Series: Materials Science and Engineering **347** (2018) 012045 doi:10.1088/1757-899X/347/1/012045

It has been established that the cells attach to the paste surface however the spreading of cells on the pastes is limited in comparison with cells on the culture plastic. Only 35±5% of the cells were spread on the HA while on the culture plastic 90±6% of the cells were spread. In Figure 8 photomicrographs are presented illustrating this effect. On micrographs the cells spread on the HA had a spindle-shaped elongated shape and were more "compressed" than on the plastic (Figure 8A). On HA and ReproBone novo pastes were observed many attached but not spreaded cells having a rounded shape (indicated with rows on Figure 8A and B). On culture plastic almost all cells were spreaded and had a spindle-shaped or elongated triangular shape (Figure 8C).

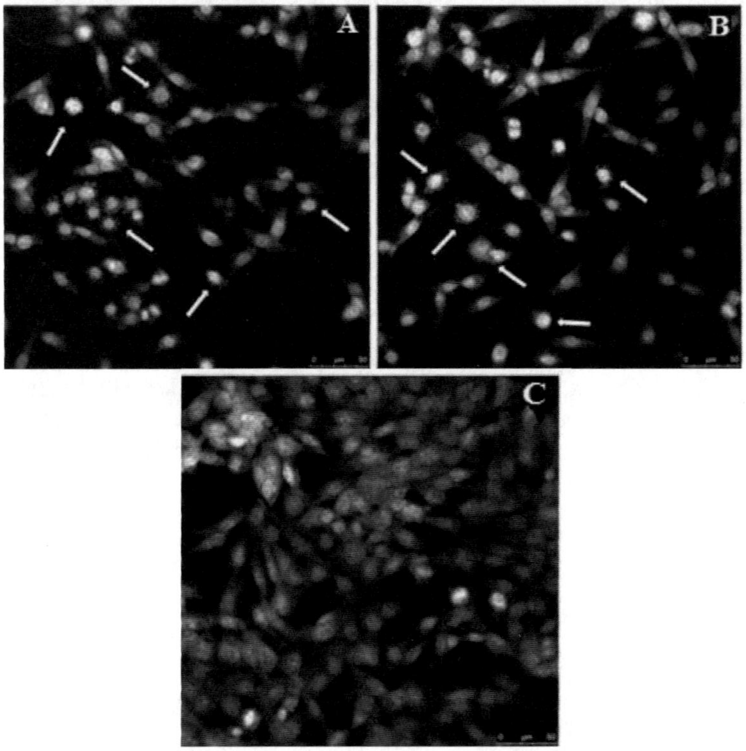

Figure 8. Microphotographs of NIH/3T3 cells after 72 hours of cultivation on a layer of nanocrystalline HA paste (A), ReproBone novo (B) paste and culture plastic (C). Fluorescence microscopy, coloration of acridine orange. The arrows indicate the cells attached but not spread on the paste. Magnification: objective x20, ocular x10

4. Conclusion

The obtained results make it possible to suggest limiting biocompatibility of hydrated nanocrystalline hydroxyapatite and disordering of cell adhesion to this material. Cell adhesion disorder to nanocrystalline HA will inevitably be a limiting factor for the migration of cells into it and as a consequence will limit its osteoconductive properties. A modification of the hydrated HA paste with osteoconductive components which ensures the adhesion of cells and their migration into the material is necessary to overcome this difficulty. Another possible reason for limiting cell adhesion to hydrated nanocrystalline HA can be the sorption of the components of the intercellular fluid (the culture medium in our studies) and as a result the ionic, amino acid and other compositions in the material zone is disrupted. In order to determine which of these reasons can be decisive and how to overcome them additional studies of cell response to hydrated nanocrystalline HA modified by osteoconductive and osteoinductive activities in accordance with the proposed operational hypotheses are needed to be carried out *in vitro* and *in vivo*. This work is actual since the creation of hydrated HA paste which has a high osteoplastic potential and possesses osteoconductive properties is urgently needed for minimally invasive bone tissue repair surgery.

Acknowledgments

The research was carried out with the support of the Government grant of the Russian Federation № 14.Z50.31.0028 using the instruments of the centre of collective usage of ITEB RAS.

References

[1] Gorshenev VN, Teleshev AT, Ershov YuA, Kaziev GZ, Kolesov VV, Sklyanchuk ED 2013 *Patent RF № 2482880*
[2] Gorshenev VN, Ershov YuA, Teleshev AT, Sklyanchuk ED, Prosvirin AA, Grigorie SA 2014 *Biomedical Engineering* **48** 33-6
[3] Huang J, Best S, Bonfield W, Brooks RA, Rushton N, Jayasinghe SN, Edirisinghe MJ 2004 *J. Mater Sci Mater Med* **15** 441–5
[4] Sklyanchuk ED, Prosvirin AA, Malyshev IYu, Fadeeva IS, Fesenko N.I., Gorshenev VN, Teleshev AT, Ershov YuA., Guryev VV 2014 *Journal Technologies of Living* Systems **6** 20-35
[5] Müller KH, Motskin M, Philpott AJ, Routh AF, Shanahan CM, Duer MJ, Skepper JN 2014 *J Biomaterials* **35** 1074–88
[6] Shi Z, Huang X, Cai Y, Tang R, Yang D 2009 *Acta Biomater* **5** 338-45
[7] Theiszova M, Jantová S, Dragúnová J, Grznárová P, Palou M 2005 *Biomed Pap Med Fac Univ Palacky Olomouc Czech Repub* **149** 393-6
[8] Mestres G, Espanol M, Xia W, Persson C, Ginebra MP, Ott MK 2015 *PLoS One* 10(4):e0120381
[9] Lebre F, Sridharan R, Sawkins MJ, Kelly DJ, O'Brien FJ, Lavelle EC 2017 *Sci Rep.* **7** 2922
[10] Pezzatini S, Solito R, Morbidelli L, Lamponi S, Boanini E, Bigi A, Ziche M *J Biomed Mater Res A* 2006 **76** 656–63
[11] Habraken W, Habibovic P, Epple M, Bohner M 2016 *J Materials Today* **19** 69-87
[12] Kesel BA, Fedorov AD, Gimushin IF Kesel BA; Fedorov AD; Gimushin IF; Volkov GA; Gataullin RSh; Voskoboinikov DV; Veseliev DA 2006 *Patent RF № 2166986*
[13] Kovaleva E.S., Filippo Y.Yu., Putlyaev VI, Tretyakov YD, Ivanov VK, Silkin NI, Galiullina LF, Rodionov AA, Mamin G B.V., Orlinsky S.B., Salahov M.H. 2010 *J Uchenye Zapiski Kazanskogo Universiteta. Seriya Estestvennye Nauki* **152** 79-98
[14] Saenger AT, Kuhs WF. *Golden Book of Phase Transitions* Wroclaw 2002 **1** 1-123

IOP Conf. Series: Materials Science and Engineering **347** (2018) 012046 doi:10.1088/1757-899X/347/1/012046

Features of the phase composition and morphology of the particles of sialon synthesized from silicon and aluminum nitrides

S N Ivicheva, A S Lysenkov, N A Ovsyannikov, D D Titov and Yu F Kargin

A.A. Baikov Institute of Metallurgy and Materials Science of Russian Academy of Sciences, 119334, Moscow, Leninskii pr., 49, Russia

E-mail: ivitcheva@mail.ru

Abstract. The phase composition and morphological features of sialons were studied under the same conditions of firing (duration, temperature) using different initial components, silicon nitride, aluminum nitride, and a mixture of silicon nitrides and aluminum with the application of nitrides of the corresponding oxide (aluminum or silicon) sol-gel method. The effect of the initial reagents composition on the phase composition of the final product and the morphological features of the sialon powders obtained in a single firing step in a nitrogen atmosphere is shown.

1. Introduction

The development of reliable methods for the synthesis of ceramic and composite materials based on oxonitrides of silicon and aluminum, capable of withstanding high temperatures and operating in corrosive environments is an urgent task of modern materials science. Sialons or oxonitrides of silicon-aluminum (which are oxonitridoaluminosilicates and in literature mostly called oxynitrides) are sufficiently widely studied ceramic materials [1-6], which combines high thermomechanical properties and the ability to work under aggressive conditions of metallurgical processes [5]. In the four-component Si_3N_4-SiO_2-AlN-Al_2O_3 system in addition to the β-sialon $Si_{6-x}Al_xO_xN_{8-x}$ ($0 \leq x \leq 4{,}2$, the solid solution of isomorphic β-Si_3N_4, the space group $P6_3/m$), O'-sialon $Si_{2-x}Al_xO_{1+x}N_{2-x}$ ($0 \leq x \leq 0{,}3$), X-phase of $Si_{12}Al_{18}O_{39}N_8$ composition, and a number of ordered polytypes based on AlN (8H, 15R, 12H, 21R, 27R, $2H^{\delta}$) with wurtzite-type structure are formed. The so-called α-sialons have the formula $M_mSi_{12-(pm+n)}Al_{(pm+n)}O_nN_{16-n}$ realized for M^{p+} = Li, Mg, Ca, Sr, Y and rare-earth metals.

Because of its high strength, chemical stability and heat resistance sialon-based products are used in conditions of high-speed mechanical loading, thermal shock, exposure to aggressive chemical agents and abrasive. Depending on the additives sialon-based materials are used not only as high-temperature structural ceramics and corrosion-resistant materials [6], but when doping sialons with rare-earth elements as luminophores [7-9].

The main difficulty in obtaining sialons is the need for high temperatures to form a liquid phase that facilitates the sintering of sialonic ceramics and the tendency of sialonic materials to decompose at high temperatures. Therefore, it is important to carry out research to develop effective and reproducible methods for the synthesis of sialon from the initial components of various chemical nature (nitrides, oxides, hydroxides, salts) in order to obtain the target powder or ceramic product and reduce the temperature of obtaining composite materials based on it. In this respect, it is of interest to

IOP Conf. Series: Materials Science and Engineering **347** (2018) 012046 doi:10.1088/1757-899X/347/1/012046

use methods that make it possible to obtain substances that are initial for synthesis with an increased reactivity and high dispersity by applying to the crystal nitrides of silicon or aluminum the hydrolysis products of silicon and aluminum alkoxides based on the stoichiometry of the required sialon composition.

2. Results and discussion

A comparative analysis of structural and morphological features of sialons obtained under identical conditions from different starting components was carried out using three different initial reaction mixtures, the calculated composition of which corresponded to the phase of $Si_3Al_3O_3N_5$: 1 - 3AlN + $3SiO_2$ (reference designation SiAlON-I); 2 - $2Si_3N_4$ + $3Al_2O_3$ (reference designation SiAlON-II); 3 - Si_3N_4 + AlN + Al_2O_3 (reference designation SiAlON-III). Calculated amounts of silicon or aluminum oxides were introduced into the reaction mixtures as neutral or acid sols prepared from alkoxides of silicon or aluminum (tetraethoxysilane or aluminum sec-butoxide), simple or polyhydric alcohols with the addition of mineral acids.

Interest in aluminum and silicon alkoxides during the synthesis of sialons is primarily due to the use of their hydrolysis products to obtain pure (in terms of chemical and phase composition) and active highly dispersed oxides of silicon and aluminum and more uniform distribution of the initial nitride materials over the grain surface. The alumina and silicon oxides obtained with the use of alkoxy technologies differ in structural and texture characteristics from traditional crystalline varieties and amorphous hydroxides obtained by coprecipitation, and varying the conditions of hydrolysis can affect the main characteristics of the final product, including its phase composition.

Powders of silicon nitride and aluminum nitride synthesized by the self-propagating high-temperature synthesis (SHS) at the Institute of Structural Macrokinetics and Materials Science, Russian Academy of Sciences (ISMAN, Chernogolovka) were used as the starting nitrides. Si_3N_4 powder (N = 37.2%, basic impurities: O - 1.64%, Fe - 0.023%) was represented by particles of silicon nitride with an α-Si_3N_4> 95% content of predominantly a fibrous structure with a length of about 2 μm and a thickness of up to 200 nm, the specific surface area of the powder - 8.2 m^2/g. AlN aluminum nitride at $\geq$ 96% consisted of the main substance (the content of N-33.0 wt.%) contained some amounts of impurities (O - $\leq$ 1.2%, Fe - $\leq$ 0.1%, C - $\leq$ 0.05%), specific surface of the powder - 5.0 m^2/g. According to the microscopy data aluminum nitride powder is represented by bulk particles of detrital form with particle sizes from 200 nm to 2 μm. For preparing the sols, silicon and aluminum alkoxides were used: tetraethoxysilane (TEOS), aluminum sec-butoxide, ethyl and isopropyl alcohols, 30% hydrochloric acid or nitric acid.

Gelation of the sol containing amorphous hydrated $SiO_2 \cdot nH_2O$ and $Al_2O_3 \cdot nH_2O$ particles on AlN and Si_3N_4 powders, respectively, was carried out in air, aging of the sols in the resulting mixtures occurred at room temperature. Preheating and heat treatment to remove water, alcohols, acids and products of hydrolysis of tetraethoxysilane and aluminum sec-butoxide were not carried out. The firing of the powders of the obtained composites from nitrides and xerogels or alumogels was carried out without preliminary pressing in corundum crucibles in a furnace with a graphite heater under a nitrogen atmosphere at 1600-1650°C for 2-3 hours.

To identify the phase and chemical composition of the initial nitride powders of Si_3N_4 and AlN and the firing products after the application of silicasols and aluminosols X-ray diffraction analysis was used (XRD 6000 SHIMADZU diffractometer, CuKα radiation, λ = 1.5406 Å, scanning speed 2θ = 2 deg/min). The phase composition of the samples was identified using the PDF-2 database, JCPDS-ICDD (Set 1-2002). Morphology and structural features of the samples were studied by scanning electron microscopy (electron microscope NVision 40, Carl Zeiss).

IOP Conf. Series: Materials Science and Engineering **347** (2018) 012046 doi:10.1088/1757-899X/347/1/012046

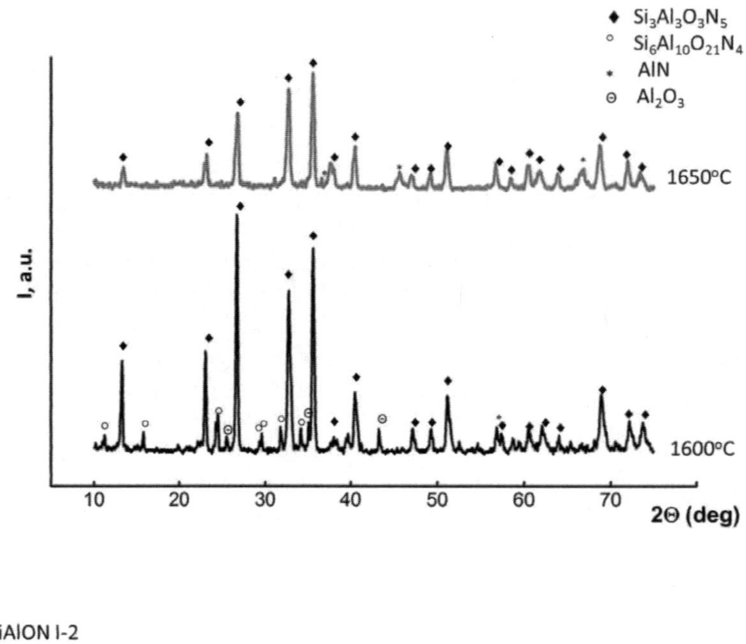

Figure 1. Change in the phase composition of SiAlON-I powders prepared using acidic silica sols from the calcination temperature.

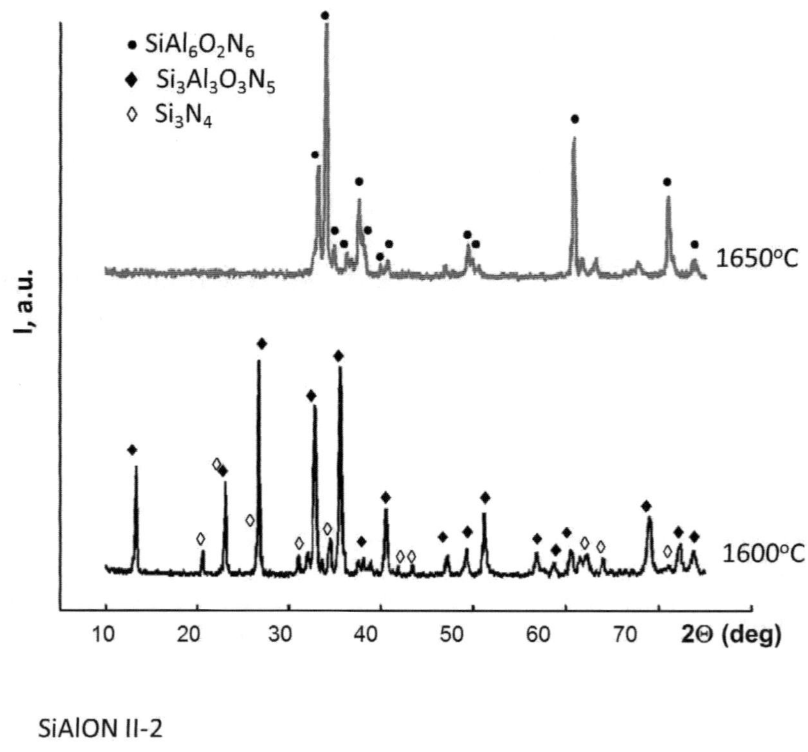

Figure 2. Change in the phase composition of SiAlON-II powders prepared using acidic silica sols from the calcination temperature.

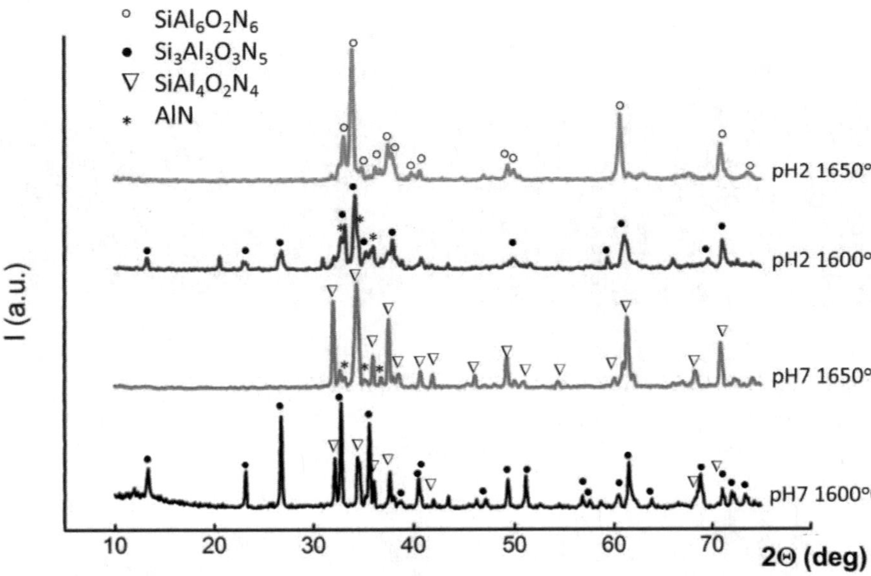

Figure 3. Change in the phase composition of SiAlON-III powders prepared using neutral and acidic silica sols from the calcination temperature.

According to X-ray diffraction analysis, all samples obtained as a result of firing are mainly β-SiAlON powders (Card No. 36-1333 for the composition $Si_3Al_3O_3N_5$) with a hexagonal crystal structure (figures 1-3). It should be noted that the samples of powders are not always single-phase, and can consist of a mixture of sialons of various compositions or contain a certain amount of impurities of the initial components. The phase composition of sialon can vary not only depending on the composition of the initial mixture, but also on the conditions of preparation and heat treatment of the samples.

For example, in samples of SiAlON-I (prepared using neutral silica sols at pH 7) an increase in the duration of heat treatment at 1600°C for 1 to 3 hours results in the appearance of a phase of X-sialon, which is belongs to the structural type of mullite and have an elongated form of crystallites. In the case SiAlON-I (prepared with the use of acid silica sols at pH 1÷2) along with β-SiAlON γ--phase ($Si_6Al_{10}O_{21}N_4$, PDF no. 36-0832) isostructural AlN is formed (figure 1). In samples of SiAlON-II (prepared from silicon nitride powders and neutral and acid aluminosols) after firing for 2 hours at 1600°C traces of the initial nitrides are present, but with increasing duration of heat treatment the samples of $Si_3Al_3O_3N_5$ obtained are single-phase. With an increase in the firing temperature of SiAlON-II samples to 1650°C is observed the formation a polytype 15R isostructural AlN (figure 2). In SiAlON-III samples, as the firing time increases, traces of the initial nitrides also disappear and the composition of β-SiAlON changes (figure 3).

Figure 4 shows micrographs of SiAlON-II powders obtained on the basis of silicon nitride and aluminosol, which illustrating the morphological features of crystalline particles at different processing temperatures of 1600°C and 1650°C . A change in the morphology of the crystallites of SiAlON-III samples with an increase in the firing time is illustrated in figure 5.

a b

Figure 4. SEM micrographs of SiAlON-II powders illustrating the morphological features of crystallites at different processing temperatures of 1600°C (a) and 1650°C (b) samples obtained by depositing neutral (pH 7) -a and acidic (pH 2) aluminosols on Si_3N_4.

a b

Figure 5. SEM micrographs of crystallites of SiAlON-III samples with increasing firing time (a-1hour, b – 3 hours).

The obtained samples of the phase $Si_3Al_3O_3N_5$ with the structure of β-SiAlON were used to study thermoexoelectronic emission (TEE) [10]. The choice of the composition of the samples was not accidental and is due to the effective atomic number Z_{ef} = 11.7, which is close to the effective atomic number of bone tissue, which is important for individual dosimetry. The positions of the peaks of the TEE peaks at 190, 225, 282 (the main peak of the TEE) and at 359 and 416°C were determined by decomposition of the experimental TEE curves into components. Studies have shown that sialons (in particular $Si_3Al_3O_3N_5$) may be of interest as sensors for thermoexoelectronic dosimetry of electron radiation.

Acknowledgments
The work was supported by the Russian Foundation for Basic Research, project No. 17-03-00630a.

References

[1] Jack K H, Wilson W I 1972 Sintered Bodies of the Silicon Nitride-Alumina System *Nature*

Phys. Sci. **238** pp 28–29.
[2] Oyama Y, Kamigaito O 1972 Sintered body of silicon nitride and alumina system *Yogyo Kyokai* Shi (Japan) **80** pp 327-336.
[3] Brown I W M, Barris G C, Bowden M E, Kenneth MacKenzie, Sheppard C M, White G Vaughan 2003 Synthesis, Densification and Properties of SiAlON Bodies and Composites *Key Engineering Materials* **237** pp 95-104
[4] Izhevskiya V A, Genova L A, Bressiani J C, Aldinger F 2000 Progress in SiAlON ceramics Journal of the European Ceramic Society **20** pp 2275-2295
[5] Suvorov S A, Dolgushev N V, Ponikarovskii A I, Domakova A V, Prokhorov K D 2006 Sintered heat-resistant sialon materials Refractories & technical ceramics No 3 pp 2-5
[6] Ganesh I 2011 Development of β-SiAlON based ceramics for radome applications . Review paper Processing and Application of Ceramics **5** pp 113–138
[7] Van Krevel J W H, Van Rutten J W T, Mandal H, Hintzen H T, Metselaar R 2002 Luminescence properties of terbium-, cerium, or europium-doped α-sialon materials *J Solid* State Chem. **165** pp 19-24
[8] Xie R-J and Hirosaki N 2007 Silicon-based oxynitride phosphors for white LED's *Sci.* Technol. Adv. Mater. **8** p 588
[9] Kargin Yu F, Akhmadullina N S, Lysenkov A S, Ashmarin A A, Ishchenko A V *et al* 2012 Synthesis and cathodoluminescence characteristics of europium-doped Ca-sialons *Inorganic Materials* **48** pp 827-831
[10] Slesarev A I, Ivicheva S N, Lysenkov A S, Kargin Yu F, Solntsev K A, Ishchenko A V, Yagodin V V, Shul'gin B V 2017 Thermostymulated exoelectron emission of sialon $Si_3Al_3O_3N_5$ *Problemy spektroskopii and spektrometrii* **Iss 38** pp 3-7

IOP Conf. Series: Materials Science and Engineering **347** (2018) 012047 doi:10.1088/1757-899X/347/1/012047

Physicochemical and osteoplastic characteristics of 3D printed bone grafts based on synthetic calcium phosphates and natural polymers

E K Nezhurina[1], P A Karalkin[1], V S Komlev[2], I K Sviridova[1], V A Kirsanova[1], S A Akhmedova[1], Ya D Shanskiy[1], A Yu Fedotov[2], S M Barinov[2] and N S Sergeeva[1]

[1]P.A. Hertsen Moscow Oncology Research Institute – branch of National Medical Research Radiological Center of Ministry of Health of Russian Federation, Russia, Moscow
[2]Baikov Institute of Metallurgy and Materials Science, Russia, Moscow

E-mail: eliznezhurina@gmail.com

Abstract. A creation of personalized implants for regeneration of bone tissue seems to be a very promising biomedical technological approach. We have studied the physicochemical characteristics, cyto- and biocompatibility of three-dimensional constructs based on sodium alginate and gelatin in combination with 2 types of calcium phosphate (tricalcium phosphate or octacalcium phosphate) obtained by inkjet 3D printing. In our experiments, we have studied the physical and chemical properties of the constructs – their porosity, chemical composition, microarchitecture of the surface and mechanical elasticity. The cytocompatibility of 3D constructs and matrix-for-cell properties were investigated in vitro on a model of human osteosarcoma MG-63 cell line by means of MTT assay. The biocompatibility of 3D constructs was studied on the model of subcutaneous implantation in mice up to 12 weeks. All types of 3D constructs were cytocompatible in vitro, demonstrated good matrix-for-cells properties, and had supported cell proliferation for 2 weeks. In results of subcutaneous in vivo test all constructs demonstrated biocompatibility with slow bioresorption of organic and inorganic components. Osteogenesis proceeded more actively in rat tibia model defects (marginal excision), substituted by 3D printed 3-component implants based on alginate, gelatin and octacalcium phosphate.

1. Introduction

The development of biomedical materials intended for bone defects replacement is still an important medical and social issue. At the present time, a number of requirements have been postulated for "ideal" biomaterial: biocompatibility, high surface area, high degree of porosity (at least 60-85%) with interconnected pores with diameters within the range of 100-1000 um, optimal rate of resorption (biodegradation) *in vivo*, corresponding to the rate of new bone formation, osteoconductive and, more preferable, osteoinductive properties. Finally, a biomaterial for reconstructive plastic surgery of bone tissue must provide a mechanical stiffness at the site of implantation and hold its shape during the reparative phase [1]. Many current concepts for osteoplastic materials are inspired by the structure of natural bone tissue which extracellular matrix comprises a perfect nanocomposite consisted of collagen fibrils chemically linked to carbonated apatite. In this term a different biomimetic approaches involving natural polymers and calcium phosphate (CP) ceramic were developed in last decades [2,3]. In this study we investigated the structural features, biocompatibility and osteoplastic properties of

3D-constructs based on sodium alginate, gelatin, and two types of calcium phosphates (tricalcium phosphate and octacalcium phosphate) obtained by three-dimensional printing. Being a precursor of biological apatite OCP promotes a mineralization of new bone tissue and also demonstrates convenient solubility in simulating body fluids [4,5]. Recent technologies of rapid prototyping, e.g. 3D printing, are capable of creating personalized grafts with precise geometry of the defects according to patient's CT/MRI data [6]. This was the method we chose to construct bone replacement implants.

2. Materials and Methods

2.1. Fabrication of bone grafts using 3D printing technology

The production of an experimental three-dimensional (3D) constructs was performed according to original technology (patent RF No. RU2606041C2) by using custom-designed extrusion-based 3D printer. The hydrogel ("ink") for 3D printing contained 2 % sodium alginate (CAS 9005-38-3, Acros), 10 % gelatin solutions (142060.1211, Blooms) and granules of TCP or OCP. The process of 3D printing was as follows. The hydrogel having viscosity 10-29 Pa·s was placed in disposable syringe of 3D printer and then injected through a needle onto the precooled (-20°C) surface with 10% alcohol water (70/30) calcium chloride solution. The printed semi-solid constructs were freeze-dried at –50°C and 6×10^{-5} atm. for 10–12 h. Solidified 3D constructs were washed in distilled water and dried at 37°C for 24h to finalize the synthesis of nonorganic components. The 3D constructs were sterilized by using γ-ray irradiation (15 kGy). The 3D constructs in our tests were represented by cylinders with the 4,0-4,5 diameter and the height of 3-4 mm while weighing 3,5-4,4 mg.

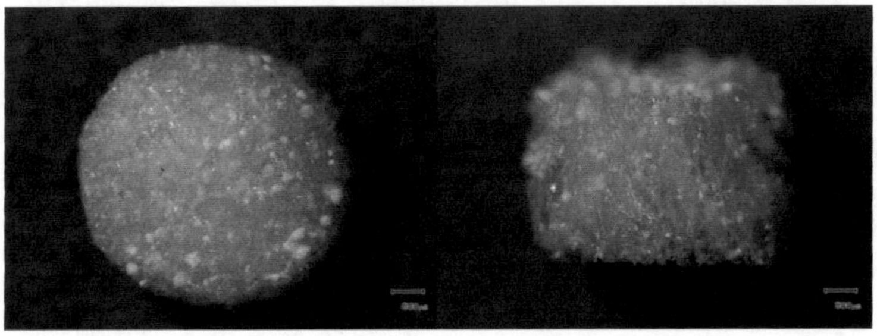

Figure 1. The external view of 3D construct based on Alginate-TCP.

2.2. Physicochemical analysis

Scanning electron microscope (SEM) Vega II (Tescan, Czech Republic), working in secondary and backscattered electron modes, was used for microstructure studies. The 3D constructs were sputter-coated with a 25 nm thick gold layer prior imaging to impart electrical conductivity of the surfaces.

FTIR spectroscopy was carried out with Nicolet Avatar system (Thermo Fisher Scientific, USA) using the powder mixtures of KBr. Compression testing was carried out using an Instron 4082 testing machine (Bucks, UK) operating at a crosshead speed of 1 mm × min−1.

2.3. Assessment of cytocompatibility and matrix-for-cells property in vitro

Cytocompatibility was assessed on a model of human breast osteosarcoma MG-63 cell line (Institute of Cytology of the Russian Academy of Sciences, St Petersburg). The sterilized 3D constructs (sodium alginate + TCP; sodium alginate + OCP) were placed in 96-well culture plate (Costar, USA) in DMEM (PanEco, Russia) supplemented with 10% fetal bovine serum (HyClone, USA) for 3-4 hours till pH achieved 7,0-7,2. After that medium was removed and MG-63 cells at an initial density of 2 × 104 cells/cm2 in 20,0 µl of medium were seeded on the surface of the constructs and in the control-wells without constructs , then plates were placed into CO2-incubator for 3 hours (37°C, 5 % CO2). After that extra 180 µl of medium were added in every well and the plates were incubated during 1-14 days. The medium was changed twice a week during the incubation period.

The cellular survival was assessed by using MTT (3-(4,5-dimethylthiazol-2-yl)-2,5-diphenyltetrazolium bromide) assay performed after 1, 3, 7, 10 and 14 days of culture. The amount of formazan produced by metabolically active viable cells was measured at 550 nm wavelength using a Multiscan microplate reader (ThermoFisher Scientific, USA).

2.4. Biocompatibility assessment of 3D constructs in vivo

Biocompatibility was estimated on the model of subcutaneous implantation in male BDF1 mice. The grafts were soaked in physiologic saline solution immediately before the surgical procedure. All mice were divided into 3 groups (10 mice per a group) according to the 2 types of grafts: 1st group -sodium alginate -TCP; 2nd group - sodium alginate-OCP. The animals were anesthetized by intraperitoneal injection of mixture ketamine/relanium (1:1) and sterilized 3D constructs were implanted subcutaneously in the back at the level of thoracic spine. The mice were sacrificed at 2, 4, 8 and 12 weeks after implantation (2 mice at a time). The specimens of implanted 3D constructs with adjacent tissues were incised and fixed in 10 % formalin solution, followed by paraffin embedding. Sections (3–4 μm) were prepared and stained with hematoxylin and eosin (HE) for histological examination. The slides were observed using an Axiovert 40C microscope, and the images were captured by an Axiocam MR3 Camera (both Zeiss, Germany).

2.5. Assessment of osteoplastic properties of 3D constructs

Osteoplastic properties were investigated on a model of rat tibia fracture. Forty male Wistar rats (180-200 g in weight) were used for this study. The Ethical Committee of P.A. Hertsen Moscow Oncology Research Institute approved the protocol of experiments. The principles of laboratory animal care were followed as in the national laws. The surgical procedures (marginal excision of tibia) were carried out under general anesthesia. The parameters of fracture: 6-8 mm in length, 1,5-2,0 mm in width, 1,5-2,0 in depth. The area of fracture was connected with bone canal which was previously separated from the bone marrow. The 3D constructs of four following types were used: alginate –TCP, alginate –gelatin – TCP, alginate –OCP and alginate –gelatin – OCP. After 3, 6, 9 and 12 weeks after the surgery two rats at a time from each group were sacrificed. The specimen containing the bone graft was fixed with 10% formalin solution for 7 days and then removed calcium using the EDTA solution (0,3 M; 37°C; 25-30 days) after rinsing with distilled water. The paraffinized specimen block was sectioned longitudinally in the center of each defect, and then slides were HE stained for histological examination.

2.6. Statistical analysis

The results of this study were analysed using standard Student's t-test in Microsoft Excel 2013 Software. Statistical significance was determined at a $p < 0.05$.

3. Results

3.1. Scanning electron microscopy (SEM)

The analysis of SEM-images showed that all four types of 3D constructs had irregular plate-type structure which consisted of alginate with 2-3 mm wide plate, with 500 μm pores between them and with the round shape ceramic granules regardless the type of calcium phosphates. The addition of gelatin to the composition of "ink" did not change the microstructure as well as arrangement and sizes of inner pores (figure 2).

IOP Conf. Series: Materials Science and Engineering **347** (2018) 012047 doi:10.1088/1757-899X/347/1/012047

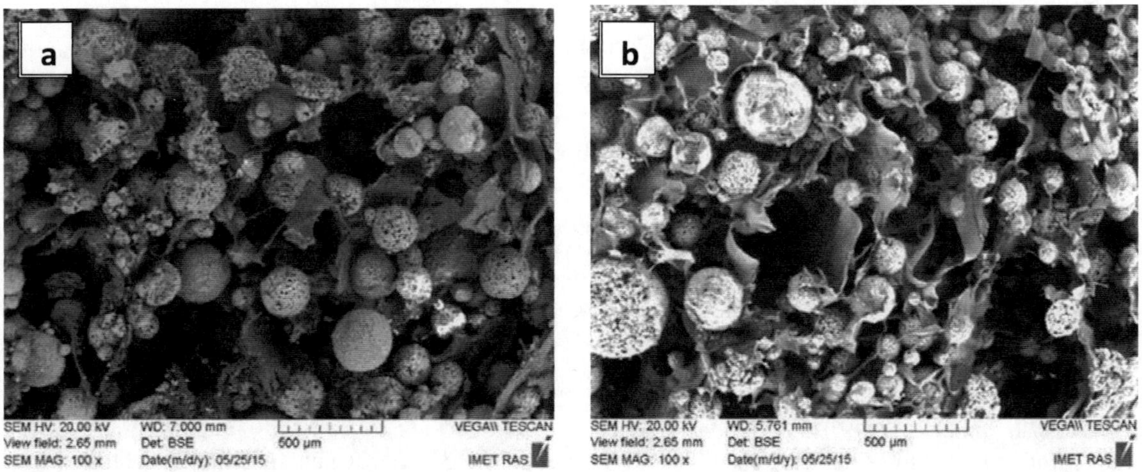

Figure 2. SEM images of (**a**) Alginate-TCP 30 mass %; (**b**) Alginate-Gelatin(90/10)-TCP 30 mass % (magnification x100).

3.2. Component analysis

X-ray phase analysis showed that phase composition of calcium phosphates did not change after addition to the hydrogel and printing of 3D construct. Insignificant decrease of intensity of their peaks determined by alginate content (figure 3).

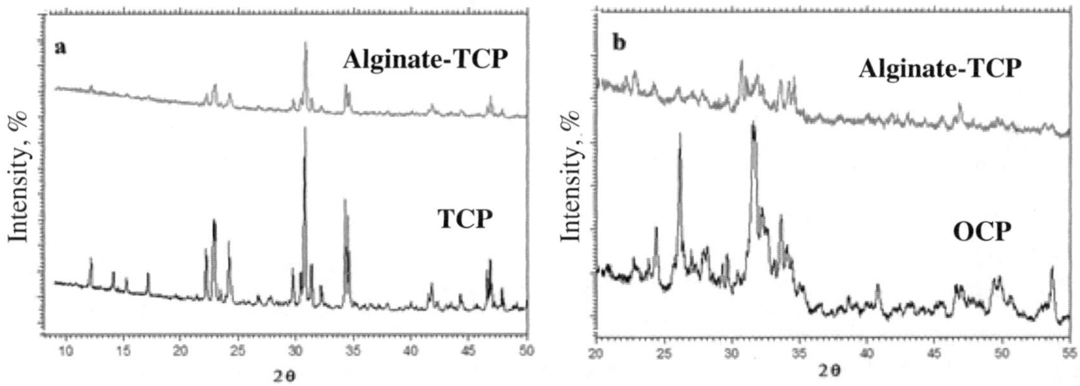

Figure 3. The results of X-ray phase analysis of (**a**) Alginate-TCP composition in comparison to TCP; (**b**) Alginate-OCP composition in comparison to OCP.

3.3. The structure of the 3D constructs

Having taken into the account the fact that following aspects such as efficiency of cells seeding, nutrients and gases delivery, speed of blood vessels growth depend on the porosity of the biomaterial after being implanted in vivo, we paid more attention to the porosity of different compositions. As the result, we could made the conclusion that the decrease of porosity correlated with the increase of calcium phosphate content, 81, 98 % and 63, 92 % for 10 and 30 mass % correspondingly (table 1). Statistically significant increase in pores sizes of 3D constructs based on alginate and 30 mass % TCP was registered after the addition of gelatin in the composition (table 1).

IOP Conf. Series: Materials Science and Engineering **347** (2018) 012047 doi:10.1088/1757-899X/347/1/012047

Table 1. The structure of 3D printed alginate-TCP-based constructs.

Sample composition	Total porosity of the sample (%), M±s.e.m	Size of inner pores, WxL, μm
Alginate-TCP – 10 mass %	81.98±0.28	400x500
Alginate-TCP – 20 mass %	75.06±0.62	400x500
Alginate-TCP – 30 mass %	63.92±2.33	300x600
Alginate-Gelatin-TCP – 30 mass %	68.23±0.94	600x800

3.4. The mechanical properties of 3D constructs

The comparative assessment of mechanical properties of 3D constructs revealed the following: resistance to deformation under compression steadily increased, while the elasticity decreased together with increase of TCP content. The addition of gelatin to the material improved its resistance to deformation under compression (from 3,53±0,39 to 5,28±0,020 MPa) with no practical impact on the elasticity (table 2).

Table 2. Mechanical properties of 3D constructs of different composition.

Sample composition	σ^a, MPa, M±s.e.m	δ^b, %, M±s.e.m
Alginate-TCP – 10 mass %	2.71±0.12	15.4±0.16
Alginate-TCP – 20 mass %	3.01±0.13	13.8±0.11
Alginate-TCP – 30 mass %	3.53±0.39	12.4±0.35
Alginate-Gelatin-TCP – 30 mass %	5.28±0.020	12.2±0.16

[a] – resistance to deformation under compression

[b] – elasticity

3.5. The results of cytocompatibility assessment

During 2 weeks of culturing an optical density of formazan increased in all experimental groups but the speed of MG-63 cell line proliferation on the different types of 3D constructs differed from each other: for alginate-TCP-based constructs no statistically significant difference from a control group was observed while on the alginate-OCP-based construct cells proliferated more slowly than in a control group.

3.6. Analysis of biocompatibility and osteoplastic properties

The biocompatibility properties of 3D constructs we analysed on a model of subcutaneous implantation by means of tissue reactions as well as toxicity and immune cells activity correlated with chemical composition of material. Thus, we assessed the following factors: firstly, the degree of integration of connective-tissue capsule and surrounding tissues as well as its vascularization, secondly, presence/absence of inflammation. There were additional characteristics of biocompatibility such as number of cell layers in the capsule, biodegradation qualities of biomaterial and morphological characteristic of tissue replacing the implanted 3D constructs.

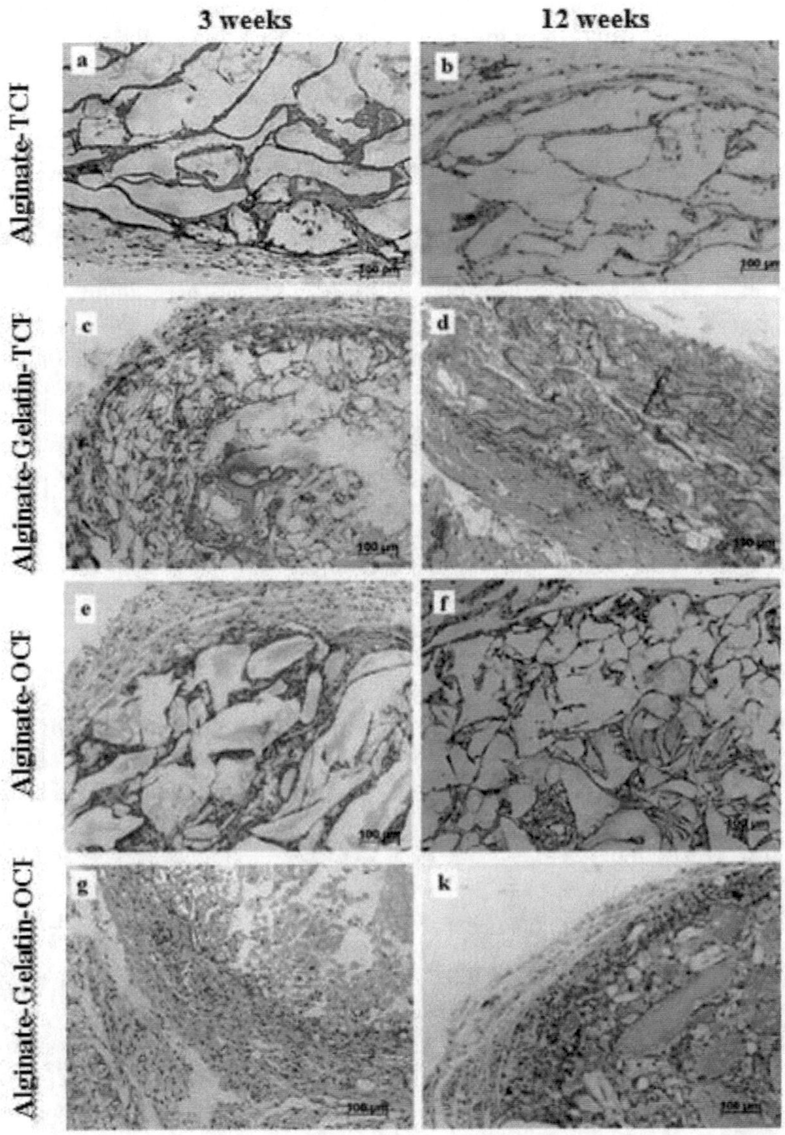

Figure 4. Histological sections of the subcutaneously implanted 3D constructs at 2 weeks (**a,c,e,g**) and 12 weeks (**b,d,f,k**) after surgery. **a,b** – Alginate-TCP-based 3D construct; **c,d** - Alginate-Gelatin-TCP-based 3D construct; **e,f** -Alginate-OCP-based 3D construct; **g,k** - Alginate-Gelatin-OCP-based 3D construct. HE, magnification x100.

After subcutaneous implantation of 3D constructs based on Alginate-TCP at early stages (2 weeks after surgery) we observed thin fibrous capsule (up to 8-10 cell layers) with fibers of connective-tissue oriented concentrically and with sprouting small blood capillaries. After the first weeks of implantation we registered active fragmentation of alginate content, and fibroblasts migration to the area of construct. Inner part of constructs was filled with immature connective tissue. 8-12 weeks later the amount of connective tissue inside the implant increased dramatically (figure 4a,b) We observed the appearance of foreign-body giant cells in the newly formed connective tissue together with the reduction of TCP granules which indicated the beginning of biodegradation of the implants. The addition of gelatin to the constructs accelerated these processes which was noticeable at early stages (2 weeks) (figure 4c, d).

The different picture was observed in the situation with OCP-based grafts. At early stages capsule consisted of loose connective tissue with normally oriented fibers, neutrophilic infiltration and active vascularization. 2 weeks after the surgical procedure aseptic inflammation in the area of implantation slowed down. This observation indicated that applied material stimulated the tissue transition from the

IOP Conf. Series: Materials Science and Engineering **347** (2018) 012047 doi:10.1088/1757-899X/347/1/012047

inflammation to regeneration. Steady involution of 3D construct with its active replacement by connective tissue with fibroblast-type cells and foreign-body giant cells took place. Relatively small number of blood vessels could be observed (figure 4e, f). It is important to mention that the speed of connective tissue replacement was the maximal when alginate-gelatin-OCP constructs were used (figure 4g, k).

Taken together, experiments in vivo demonstrated good biocompatibility for all types of examined 3D constructs. A higher speed of OCP degradation in comparison to TCP was registered as well as faster alginate fragmentation when gelatin was added. All this led to more intensive connective tissue replacement.

Assessment of bone-replacement properties was measured on a model of rat tibia fracture. 3 weeks after the implantation we observed the beginning of alginate layers resorption with its further replacement by fibrous structures with fibroblast-type cells. Sporadic periosteal osteogenesis formations were visualized on the border with healthy bone tissue. Numerous blood vessels were observed throughout the whole area of the cross sections (figure 5a).

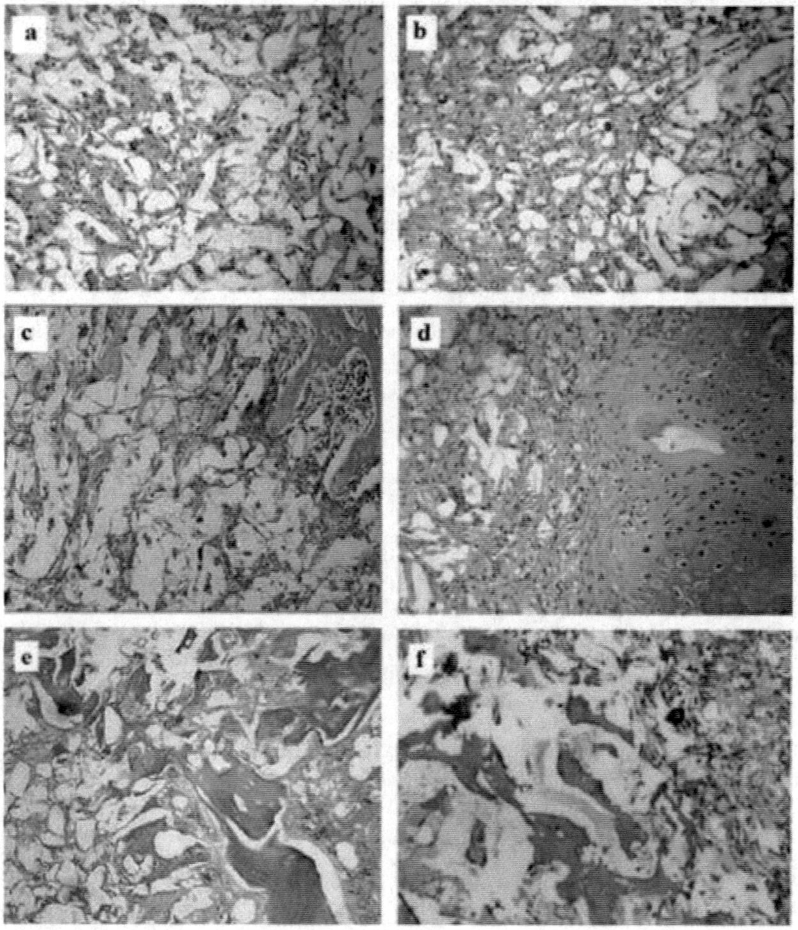

Figure 5. Histological sections of Alginate-TCP-based 3D construct (**a,b,c**) and Alginate-Gelatin-TCP-based 3D construct (**d,e,f**) in the area of rat tibia fracture.

a- 3 weeks after implantation; **b** – 6 weeks after implantation; **c** – 9 weeks after implantation; **d** – 3 weeks after implantation; **e** – 6 weeks after implantation; **f** – 12 weeks after implantation. HE staining, magnification x 100.

After 6 weeks of implantation the fibrous hyperplastic periosteum was observed on the frontier of the fracture. Inside the implant, periosteal osteogenesis was accompanied by endochondral osteogenesis. At the same time, in the area of fracture above thin layer of bone tissue newly formed trabeculae began to appear (figure 5b).

The process of regeneration continued at the late stages of observation. On the one hand, the area of defect was filled with connective tissue, which substituted with spongy bone tissue; on the other hand, granules of TCP were steadily replaced by bone trabecula with hematopoietic loci (fig. 5c). The addition of gelatin to the composition of 3D constructs did not significantly change the terms and the way of reparative osteogenesis (figure 5 d, e, f).

In case of Alginate-OCP-based 3D constructs implantation, processes of reparative osteogenesis were similar but more intensified: 3 weeks later in the area of defect multicellular fibrous periosteum appeared, the area of fracture was filled with cellular connective tissue, as well as this, on the border with compact bone lower cortical layer we observed newly formed bone tissue trabecula (figure 6a). The newly formed bone tissue trabecula continued to grow together with active bioresorption of OCP granules and osteoclasts accumulation (figure 6b). Direct osteogenesis was also accompanied by endochondral osteogenesis (figure 6c).

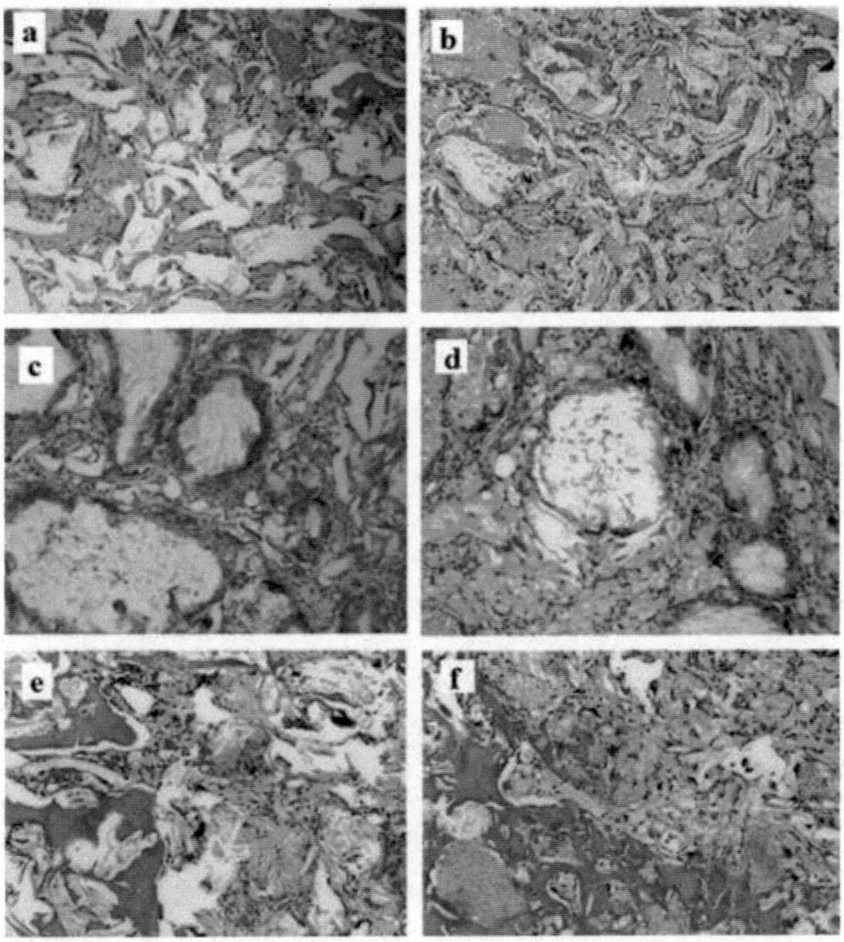

Figure 6. Histological sections of Alginate-OCP-based 3D construct (**a,b,c**) and Alginate-Gelatin-OCP- based 3D construct (**d,e,f**) in the area of rat tibia fracture.
a- 3 weeks after implantation;
b – 6 weeks after implantation;
c – 9 weeks after implantation;
d – 3 weeks after implantation;
e – 6 weeks after implantation;
f – 12 weeks after implantation. HE staining, magnification x 100.

The most intensified osteogenesis was observed when the 3-component 3D constructs were used (Alginate-Gelatin-OCP). 3 weeks after the implantation, partly resorbed alginate and gelatin contents were replaced by newly formed connective tissue. The bioresorption of OCP granules also took place which was indicated by the presence of numerous osteoclasts (figure 6 d,e,f). Up to the 12 weeks of observation the newly formed bone tissue trabecula continued to grow (on the frontier of the fracture, as well as, inside the implant) together with the formations of hematopoiesis. The periosteal osteogenesis, as well as, endochondral osteogenesis was observed.

4. Conclusion

In this study, we investigated the mechanical properties, cyto- and biocompatibility, as well as, bone replacement properties of two- and three-component grafts, which were prepared by using 3D printing[8]. Two-component 3D constructs consisted of alginate and one of the two types of calcium

IOP Conf. Series: Materials Science and Engineering **347** (2018) 012047 doi:10.1088/1757-899X/347/1/012047

phosphate (TCP or OCP) while three-component 3D constructs additionally included gelatin, which increased their resistance to deformation under compression as well as porosity. The experiments on the model of subcutaneous implantation in mice demonstrated well biocompatibility of all types of 3D constructs. In case of application of OCP-based grafts the highest speed of biodegradation and well-structured connective tissue were observed. The results of implantation of 3D constructs in the rat tibia defects showed the most intensified osteogenesis for the 3-component alginate-gelatin-OCP-based 3D constructs. Thus the results of this study show the feasibility and long-term benefits of applying three-component mineral-polymer composites based on alginate, gelatin and calcium phosphates as osteoplastic materials for printing 3D constructs intended for bone defect replacement.

References

[1] Oryan A, Alidadi S *et al.* 2014 *J Orthop Surg Res.* **9** pp 18-42
[2] Kolk A, Handschel J *et al.* 2012 *J Craniomaxillofac Surg.* **40** pp 706-18
[3] Pertici G, Carinci F *et al.* 2015 *J Biol Regul Homeost Agents.* **29** (Suppl. 1) pp 136-48
[4] Komlev V, Barinov S, Bozo I *et al.* 2014 *ACS Appl. Mater. Interfaces.* **6** 16610-20
[5] Suzuki O, Imaizumi H *et al.* 2008 *Curr. Med. Chem.* **15** 305-13
[6] Bose S, Vahabzadeh S *et al.* 2013 *Materials Today* **16** pp 496-504

IOP Conf. Series: Materials Science and Engineering **347** (2018) 012048 doi:10.1088/1757-899X/347/1/012048

Porous silicon nanoparticles for drug delivery

E N Abramova, A M Khort, A G Yakovenko, A I Lvovsky, E A Slipchenko, D I Prokhorov, V I Shvets

Department of material science and technology of functional materials and structures, Moscow technological university, Institute of fine chemical technologies, prospect Vernadscogo, 86, Russia

anavenko@yandex.ru

Abstract: Based on main principles of porous silicon (por-Si) formation the research of producing containers for drug delivery systems was launched. Nanoparticles for loading medicine were obtained.

1. Introduction

Targeted drug delivery to pathologic focuses of a human body is a way to increase the effectiveness of medicines. An advanced approach for directional drug delivery is using nanostructured containers that should be biocompatible, bioavailable, biodegradable and nontoxic.

Porous silicon (por-Si) is a very promising material for such containers, as long as it corresponds to the conditions mentioned above [1-3]. To use porous silicon as nanocontainers one should take into account its characteristics, including radial and axial pore dimensions, pore partition width, porosity, surface extension measure, chemical composition of por-Si surface, electronic state of por-Si surface. The chemical composition and electronic state of por-Si surface are important for researching the mechanisms of adsorption of a medicine, possible chemical interactions, and peculiarities of drug loading.

These parameters depend on the silicon wafer properties (the conductivity type, dopant concentration, structure imperfections, etc.) as well as the porous etching conditions (the electrolyte solutions composition, current density, etching time, etc.). Simultaneous taking into consideration all the factors is extremely difficult. In this regard, it is more reasonable to determine general principles of por-Si formation and dependences of por-Si parameters on the etching conditions to get por-Si with required characteristics.

2. Experimental

We used n- and p-type silicon wafers with the resistivity in the range ρ = (0.01 – 100) ohm·cm, oriented long the (100) and (111) crystallographic planes. The wafers were etched with hydrofluoric acid solutions (45% HF) in H_2O or C_2H_5OH (96%) in a galvanostatic mode. The etching cell, made of Teflon, allowed for one-sided horizontal etching. The anode used was a polished massive copper plate, and the cathode had the form of a thin platinum plate. The electric field applied during etching was varied by diaphragms of various sizes and changing the electrode separation. The current density were 1-50 mA/cm^2.

The nanopore size, shape, and orientation and the thickness of nanoporous layers were determined using a JEOL electron microscope and a POLARM R optical microscope.

The chemical composition of por-Si surface was investigated with infrared spectroscopy on the spectrophotometer Equinox 55 (Bruker).

Por-Si nanoparticles were made with ultrasonic treatment of por-Si layers after their preparation. It lasted 4 h. The ultrasonic frequency was 35 khz (capacity was 80 W).

Por-Si nanoparticles sizes were investigated with photon correlation spectroscopy and electrophoretic light scattering on the particle-size analyser Delsa^{TM}Nano and with polarization intensity differential scattering (PIDS) on the device LS13320.

3. Experimental results and discussion

Based on experimental data, some of which are described in our previous articles [4-6], it is necessary to focus on the most important principles of por-Si formation that crucially influence on the por-Si characteristics:

1. The pore frequency function increases while the electric-field intensity decreases.
2. Pores, generated on (100) wafers, are perpendicular to the substrate. As to pores on (111) wafers they are at some angle the substance surface, that results in branched structures and a larger internal surface.
3. Under other conditions being equal, the thickness of n-type porous layers significantly overtops the thickness of p-type porous layers. The thickness of por-Si layers, obtained on silicon wafers with P dopant (n-type) $8 \cdot 10^{14}$ cm^{-3} (ρ= 5 ohm·cm), $4 \cdot 10^{14}$ cm^{-3} (ρ= 10 ohm·cm), j=25 mA/cm^2, t=30 minutes, was 180-220 micron. The thickness of por-Si layers, obtained on silicon wafers with B dopant (p-type) $1.2 \cdot 10^{15}$ cm^{-3} (ρ= 10 ohm·cm) with other conditions being equal, did not exceed 50-60 micron.
4. Porous layers on p-type of Si are more fragile than on n-type.
5. The correlation between radial pore dimensions and Si resistivity (ρ) (dopant concentration in Si substrate) is near-linear in the range from 0.01 to 10 ohm·cm. According to SEM data, the radial pore dimensions varies from 12 to 600 nm on n-Si and from 10 to 30 nm on p-Si.
6. .The dopant type does not reveal any significant influence.
7. Figure 1 shows the results of infrared spectroscopy. There are mainly Si-H, Si-OH and Si-O-Si bonds on por-Si surface.

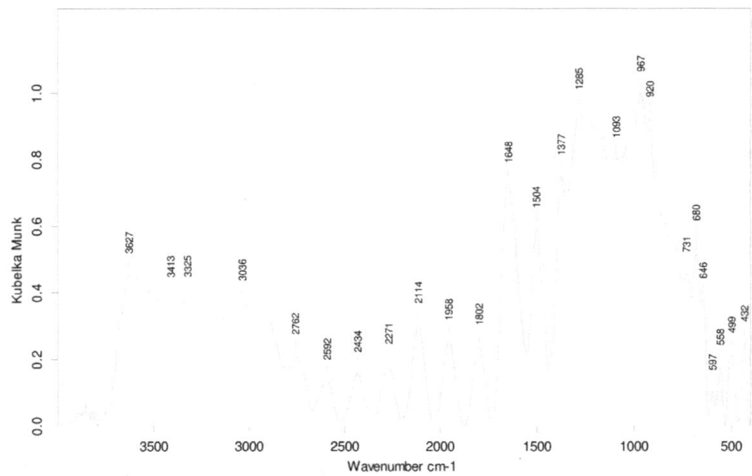

Figure 1. The dependence of por-Si layers thickness on the etching time

IOP Conf. Series: Materials Science and Engineering **347** (2018) 012048 doi:10.1088/1757-899X/347/1/012048

Table 1. Assignment of the infrared observed in anodically oxidized por-Si

ω_{IR} [cm^{-1}]	mode
460	Si-O-Si bending
640	Si-H bending
835	Si-H_2 wagging
870	Si-H_2 scissors
1040-1160	Si-O-Si stretching
2100	Si-H stretching
2250	O-Si-H stretching
3420-3580	O-H stretching

We have revealed [4-6] that porous etching of Si is stipulated not F^- ion, but $(HF_2)^-$ ion, which is contained in HF solutions > 1mol/l.

Preferable parameters of por-Si containers for the drug delivery of psychoneurological medicine include the pore size about 20 nm, porosity near 50%, and the container size around 50 nm to avoid embolism. Based on the proposed mechanism of por-Si formation in our previous articles [4-6] electrochemical etching conditions of Si were determined. The conditions stipulate the formation of por-Si with the parameters required to solve the task of the principle possibility of producing nanocontainers to load carbamazepine. We chose basic silicon wafers of p-type with ρ = (10 – 50) ohm·cm. Etching time was 30 minutes, current density was 5-50 mA/cm^2. It led to por-Si layers with the depth about 50 μm and the radial pore size around 20 nm.

Immediately after por-Si forming porous layers were processed with ultrasonic treatment (frequency is 35 khz) in H_2O or C_2H_5OH. It is observed, that the ultrasonic treatment of por-Si immediately after its preparation results in nanoparticles of various sizes, including the required ones less than 100 nm.

The results of por-Si nanoparticles sizes measurements with photon correlation spectroscopy, electrophoretic light scattering, polarization intensity differential scattering (PIDS) and scanning electron microscopy are on figures 2.

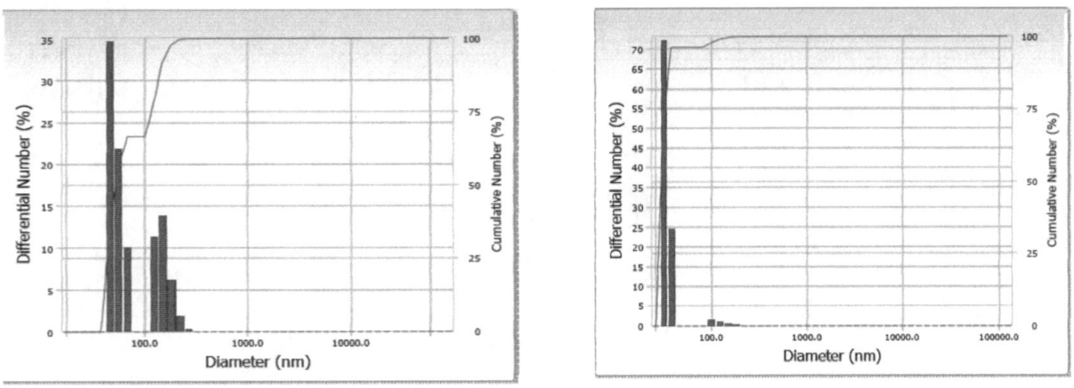

Figure 2. The measurements of por-Si nanoparticles size on Delsa™Nano

The nanoparticles of preferable sizes to make nanocontainers were separated with centrifugation and filtration.

To prevent coagulation and sedimentation por-Si layers were processed with organic substances. After such a treatment, and taking into account peculiarities of the material and the specifics of the research method, sizes of nanoparticles were also measured with PIDS on LS13320 to confirm pervious results.

It was determined that the ultrasonic treatment of porous layers let obtain nanoparticles with sizes in the required range (figure 3).

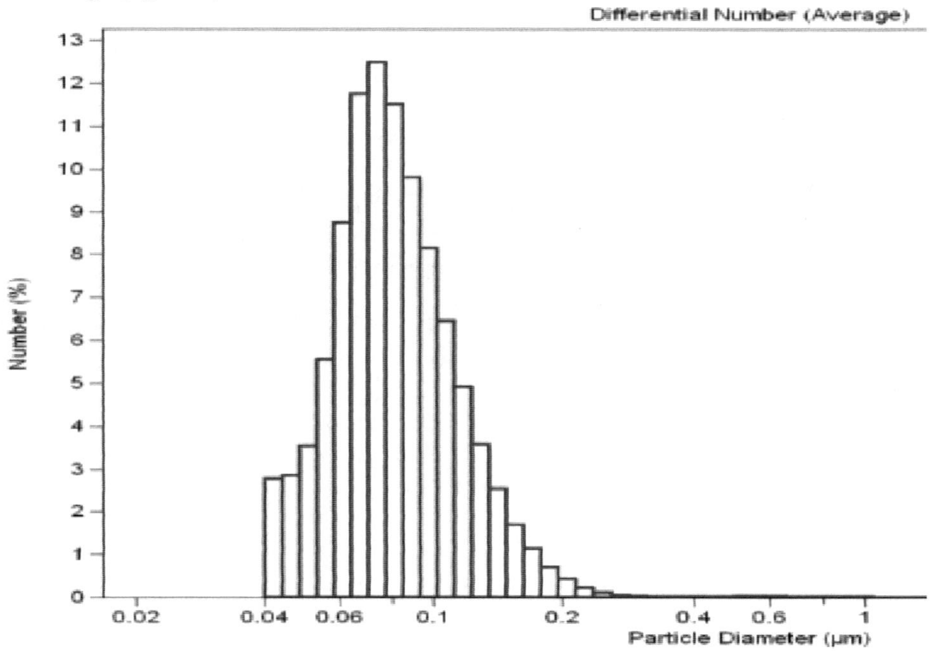

a)

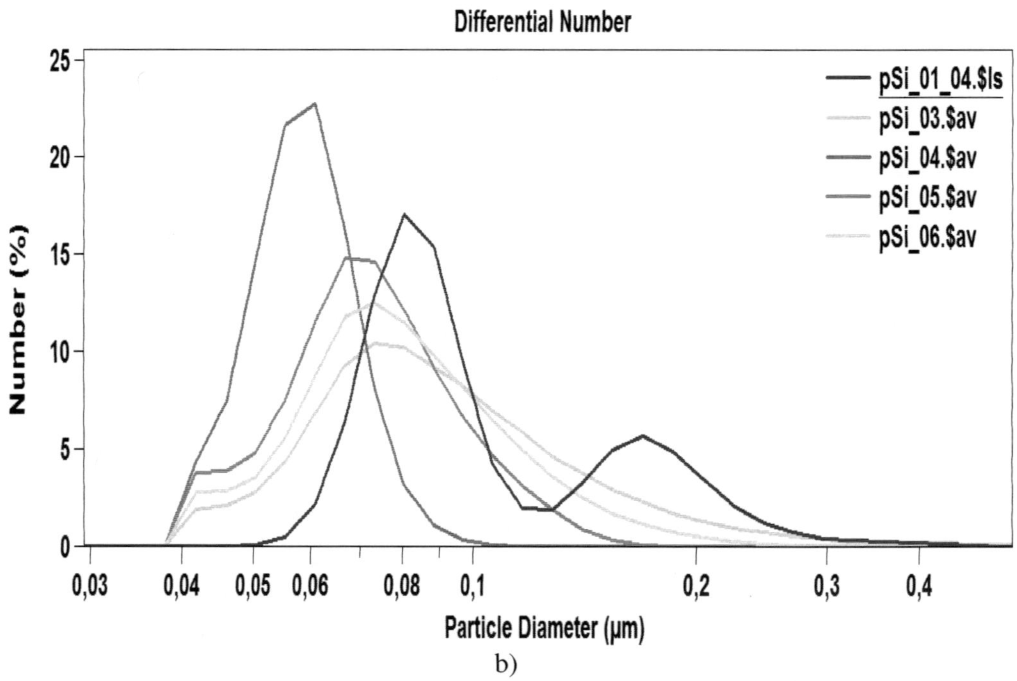

b)

Figure 3. The measurements of por-Si nanoparticles size with PIDS on LS13320: a) average distribution, b) distribution during discrete measurements

IOP Conf. Series: Materials Science and Engineering **347** (2018) 012048 doi:10.1088/1757-899X/347/1/012048

When solving the problems of loading medicine in por-Si, one should take into consideration drugs electronic state and por-Si surface electronic state, which was covered in our articles about the mechanism of pores formation.

4. Conclusions

Based on key regularities of por-Si formation the porous etching conditions were chosen to obtain por-Si with the required characteristics. The nanoparticles for loading psychoneurotical medicine were produced. The research of the possibility of loading drugs were started.

References

[1] Ji-Ho Park, Luo Gu, Geoffrey von Maltzahn , Erkki Ruoslahti , Sangeeta N. Bhatia, Michael J. Sailor 2009 *J. Nature Materials* **8** pp 331-6
[2] Nalin H. Maniya, Sanjaykumar R. Patel and Z. V. P. Murthy 2016 *J. Rev.Adv.Mater. Sci.* **44** pp 257-72
[3] Tushar Kumeria , Steven J. P. McInnes , Shaheer Maher, Abel Santos 2017 *J. Expert opinion on drug delivery* **14 (12)** pp 1-16
[4] Abramova E N, Gvelesiani A A, Khort A M, Yakovenko A G, Shvets V I 2016 *J. Doklady chemistry* **470(1)** pp 252-4
[5] Abramova E N, Khort A M, Yakovenko A G, Shvets V I 2015 *J. Neorganicheskie marterialy* **51(8)** pp 815-22
[6] Abramova E N, Khort A M, Yakovenko A G, Shvets V I 2015 *J. Doklady chemistry* **462(1)** pp 48-51

IOP Conf. Series: Materials Science and Engineering **347** (2018) 012049 doi:10.1088/1757-899X/347/1/012049

Investigation of static properties of medical alloys Ti-(20-30)Nb-(10-13)Ta-5Zr

K V Sergienko [1], M A Sevost'yanov [1], S V Konushkin [1], E O Nasakina [1], A S Baikin[1], L A Shatova [2] and A G Kolmakov [1]

[1] Baikov Institute of Metallurgy & Materials Science (IMET RAS), Leninsky Prospekt 49, Moscow, Russia
[2] Voronezh State Technical University, street 20 let Oktyabrya 84, Voronezh, Russia

Email: shulf@yandex.ru

Abstract. In the work, static properties of TiNbTaZr titanium alloy were carried out. The search for a NiTi alloy replacement is necessary for medical products to eliminate the negative effects of nickel on the body. Conclusions are drawn about the adequacy of the mechanical properties of the test alloy for use in stent implants.

1. Introduction

Materials used in medicine for the manufacture of implants, should exhibit a set of properties that ensure the compatibility of medical products and the body, such as superelasticity, low values of the Young's modulus, high corrosion resistance and biocompatibility.

Low values of the Young's modulus (50-80 GPa) and superelastic (SU) behavior [4], which are close to the behavior of bone tissue, exhibit alloys with shape memory (SPF), in particular Ti-Ni alloys. This led to their widespread use in medicine. However, the toxic properties of nickel limit their further medical use in a number of developed countries (Japan, Germany) [5-10].

To satisfy the requirements of biochemical compatibility, the alloy should not cause inflammatory processes and allergic reactions in the body. The basis for this is the use of only safe elements as alloy components, which include: Ti, Nb, Ta, Zr.

Great attention was attracted to studies on the effect of tantalum on the functional properties of Ti-Nb-Ta-Zr alloys [1]. Tantalum alloying provides the maximum range of α-martensite concentration and the highest barrier for the formation of the ω phase by other β-stabilizers. In the immediate vicinity of the α "+ β / β boundary in quenched Ti-Nb-based alloys, a decrease in the Young's modulus and manifestation EhP associated with reversible β↔α "martensitic transformation.

In this case, alloys containing 10 and less% Ta consist of a β-phase and an ω-phase, while alloys containing more than 10% Ta consist of only β-phase. At the same time, the most perfect superelastic properties are Ti-30Nb- (5-10) Ta-5Zr alloys, while alloys with higher tantalum content lose their tendency to superelastic behavior. Also, doping Ti- (Nb, Ta) of zirconium alloys promotes hardening of the solid solution and reduces the probability of brittle ω-phase formation, thereby improving the functional properties of metastable Ti-Nb-based memory alloys.

Zirconium, usually in all titanium alloys, acts as a neutral hardener, however, as shown in [1] in β-titanium alloys, zirconium can also have a β-stabilizing effect by lowering the martensitic transformation temperature, and also suppressing the formation of the athermal ω phase. In addition, titanium, niobium

IOP Conf. Series: Materials Science and Engineering **347** (2018) 012049 doi:10.1088/1757-899X/347/1/012049

and tantalum have similar atomic radii (0.145 ... 0.146 nm), while zirconium has a large atomic radius (0.160 nm). Therefore, while zirconium doping with titanium should promote an increase in the interatomic distance in the alloy, a decrease in the bond strength between the atoms, and, consequently, a decrease in the modulus of elasticity, while titanium doping with niobium and tantalum the period of the β-phase lattice should at least not decrease.

It was shown in [2] that the Zr content in the range of 5% corresponds to the Young's modulus minimum for the Ti- (13-35.5) Nb- (5-22) Ta- (4-7,2) Zr (mass%) system alloys. .

In [3] the alloy Ti-29Nb-13Ta-4,6 Zr was studied, exhibiting high characteristics important for biomedical alloys: low Young's modulus, good mechanical properties and excellent fatigue properties.

Thus, it is possible to consider alloys of the composition Ti- (20-30) Nb- (10-13) Ta-5Zr as promising materials for the manufacture of medical products such as "stent" and "cava filter".

2. Materials and methods

For research in this work, thin wires of alloys of the following compositions were chosen: Ti-20Nb-10Ta-5Zr; Ti-20Nb-13Ta-5Zr; Ti-25Nb-10Ta-5Zr; Ti-25Nb-13Ta-5Zr; Ti-30Nb-10Ta-5Zr, Ti-30Nb-13Ta-5Zr, selected as optimal for use in the manufacture of medical devices.

Iodide titanium, iodide zirconium, technically pure niobium and technically pure tantalum were used (Table 1).

Table 1. Content of impurities in charge materials.

Material	O, %	N, %	C, %	H, %
Titanium	0,001	0,001	0,001	0,0001
Zirconium	0,005	0,005	0,008	0,0001
Niobium	0,0033	0,015	0,002	0,0002
Tantalum	0,0017	0,011	0,006	0,0004

Mechanical tests of wire tensile samples were carried out on a universal testing machine INSTRON 3382 with a loading rate of 1 mm / min, with accuracy of the crosshead speed ± 0.2% of the value of the set speed. Accuracy of load measurement: ± 0.5% of the measured value and up to 1/100 of the maximum value of the load cell, according to the data sheet of the instrument. The test data are given in Table. 2.

The length of the working part of the samples is 28 mm. At one experimental point, 3 samples are tested, samples were polished before testing. The installation and fixing of samples in the testing machine does not cause additional stresses from misalignment of the specimens and grippers. All samples of the intended series are tested on the same machine, which satisfies metrological requirements. The total error in measuring the samples during the test does not exceed 0.5% of the measured stretching tension, according to the data sheet of the instrument.

3. Results and discussion

Table 2. Results of mechanical testing of samples.

Deformation (%)	Yield strength (MPa)	Ultimate strength (MPa)	Young's modulus (GPa)	Alloy composition
0,5	503,0	547,0	40,8	Ti-20Nb-13Ta-5Zr
0,7	627,7	743,3	48,2	Ti-25Nb-13Ta-5Zr
0,6	414,3	550,3	37	Ti-25Nb-10Ta-5Zr
0,5	504,0	586,7	38,6	Ti-20Nb-10Ta-5Zr
0,6	576,0	680,7	42,5	Ti-30Nb-13Ta-5Zr

0,8	566,7	715,3	46,9	Ti-30Nb-10Ta-5Zr

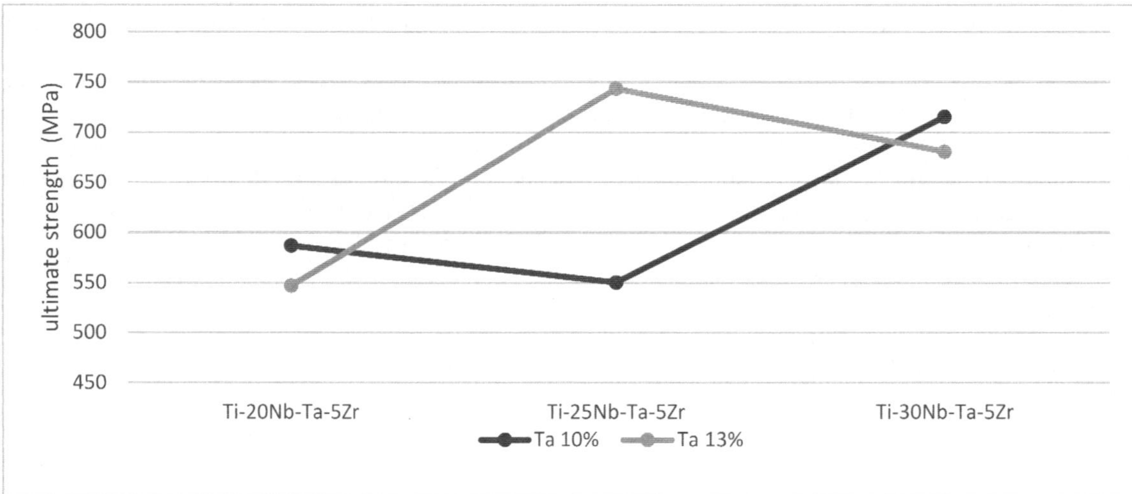

Figure 1. Dependence of the ultimate strength on the composition of the alloy.

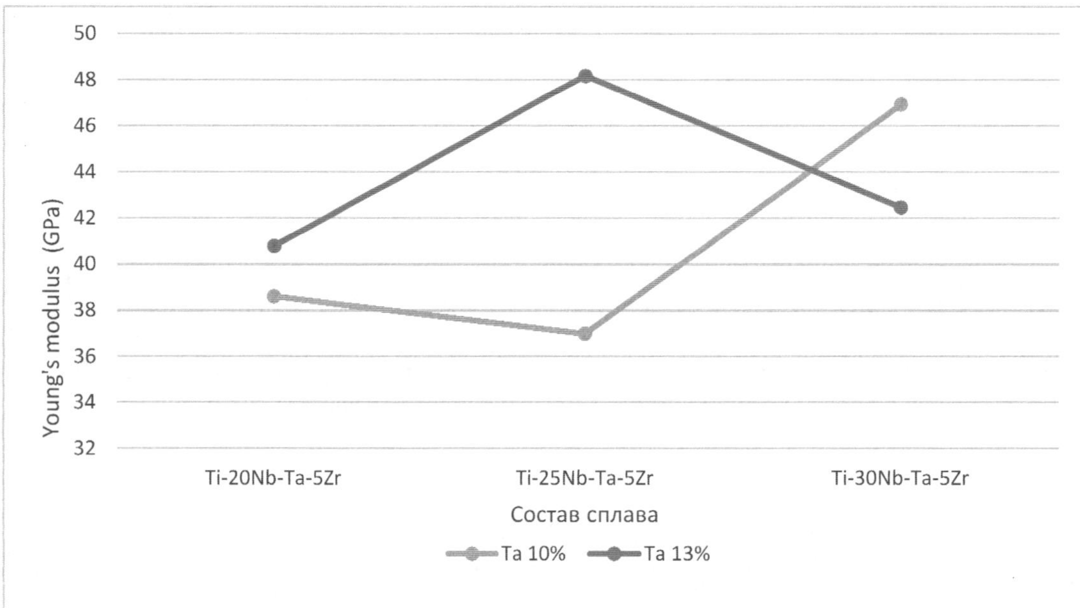

Figure 2. Dependence of the Young's modulus on the composition of the alloy.

IOP Conf. Series: Materials Science and Engineering **347** (2018) 012049 doi:10.1088/1757-899X/347/1/012049

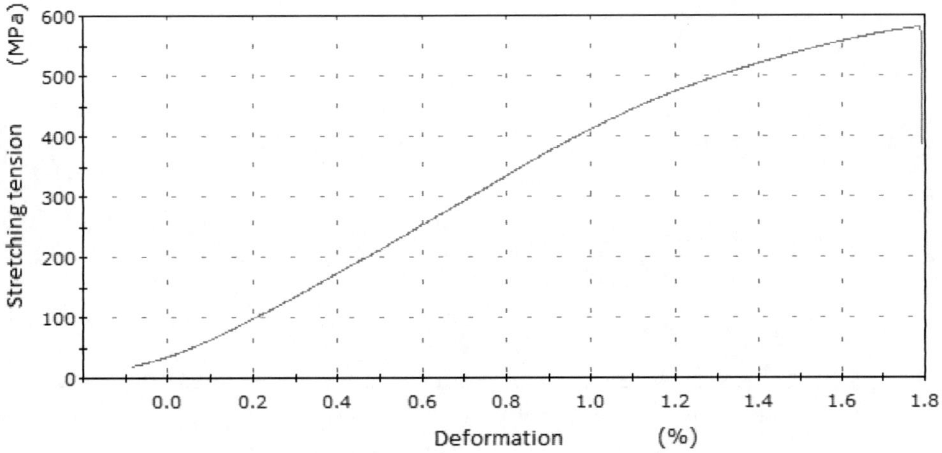

Figure 3. Diagram of stretching tension of Ti-20Nb-13Ta-5Zr.

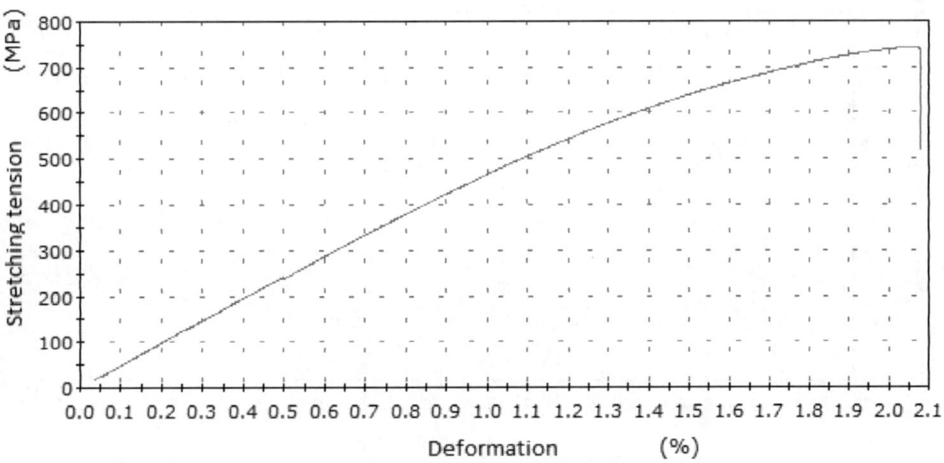

Figure 4. Diagram of stretching tension of Ti-25Nb-13Ta-5Zr.

IOP Conf. Series: Materials Science and Engineering **347** (2018) 012049 doi:10.1088/1757-899X/347/1/012049

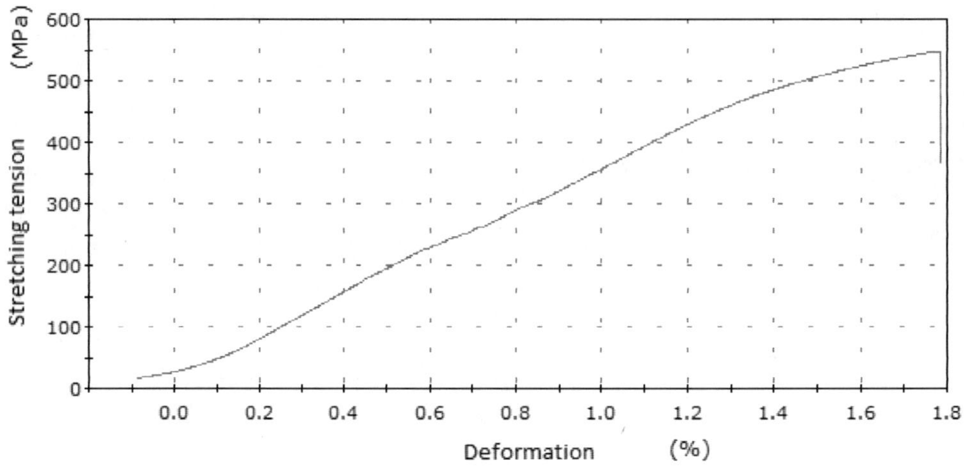

Figure 5. Diagram of stretching tension of Ti-25Nb-10Ta-5Zr.

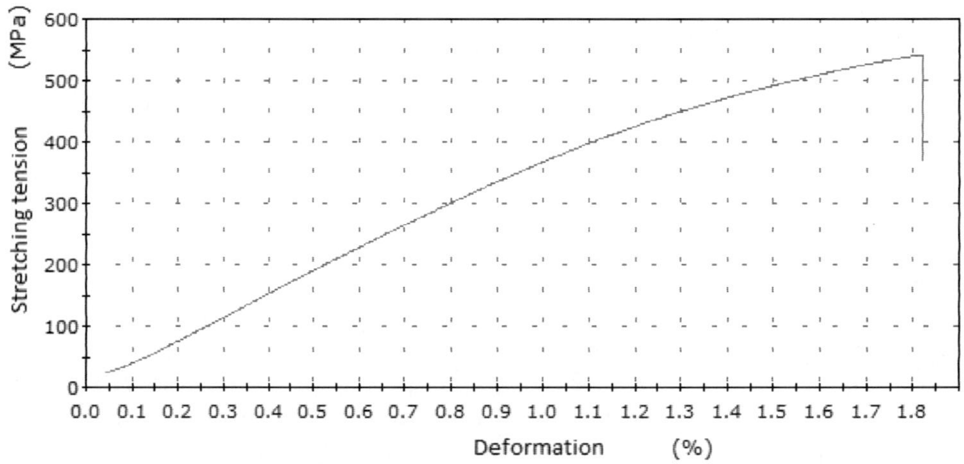

Figure 6. Diagram of stretching tension of Ti-20Nb-10Ta-5Zr.

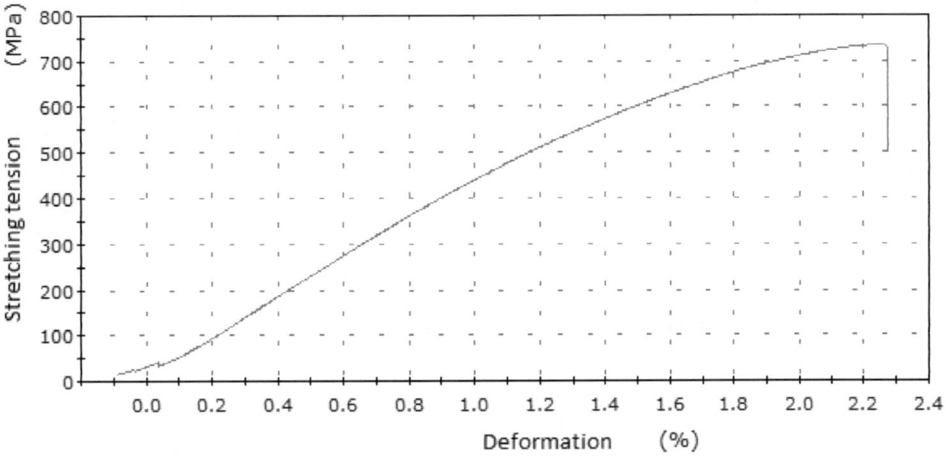

Figure 7. Diagram of stretching tension of Ti-30Nb-13Ta-5Zr.

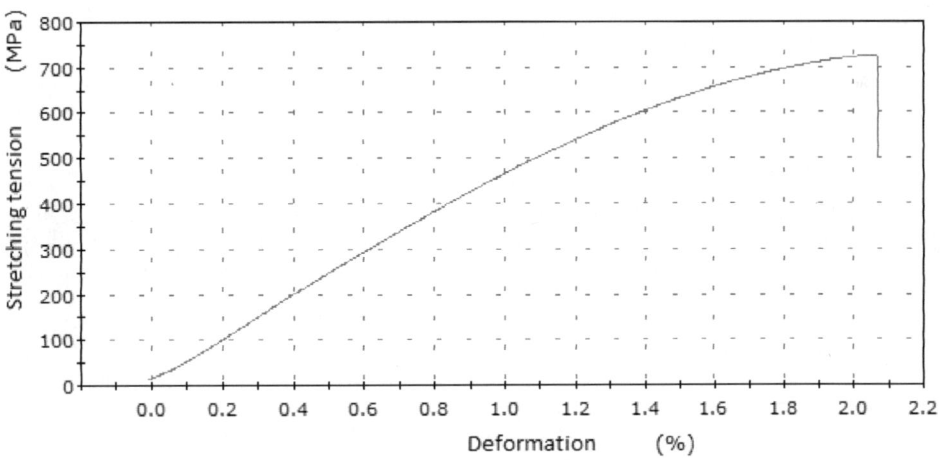

Figure 8. Diagram of stretching tension of Ti-30Nb-10Ta-5Zr.

According to the tests carried out, it can be concluded that the wire from biocompatible Ti-Nb-Ta-Zr alloys satisfies Jung's modulus as a material for making Cava filters and stents. It can also be noted that the composition of Ti-30Nb-10Ta-5Zr differs in the best complex of properties, in relation to other tested ones.

IOP Conf. Series: Materials Science and Engineering **347** (2018) 012049 doi:10.1088/1757-899X/347/1/012049

4. Conclusions

The use of stents and Kava filters from materials with shape memory effect in medical practice is a relatively new direction of the method of treatment. In this regard, the problem of negative effects of the implants on the body remains an important task. The search for nickel-free alloys that satisfy the necessary mechanical and functional properties is the most important task for the further development of this direction in medicine. The studied Ti-Nb-Ta-Zr alloy satisfies Jung's modulus and ultimate strength as a material for the manufacture of Kava filters and stents.

Acknowledgements

Russian Foundation for Basic Research supported this work (grant identifier 16-08-01132 A).

References

[1] Le Gu´ehennec L, Soueidan A, Layrolle P, Amouriq Y 2007 Surface treatments of titanium dental implants for rapid osseointegration *Dental Materials* **23** 844–854

[2] Tang X, Ahmed T, Rack H J 2000 Phase transformations in Ti-Nb-Ta and Ti-Nb-TaZr alloys *Journal of Materials Science* **35** 1805-1811

[3] Yilmazer H, Niinomi M, Nakai M 2012 et al. Heterogeneous structure and mechanical hardness of biomedical β-type Ti–29Nb–13Ta–4.6Zr subjected to high-pressure torsion *Journal of the Mechanical Behavior of Biomedical Materials* **10** 235–245

[4] Nasakina E O, Sevost'yanov M A, Mikhailova A B, Gol'dberg M A , Demin K Yu , Kolmakov A G , Zabolotnyi V T 2015 Preparation of a nanostructured shape memory composite material for biomedical applications *Inorganic Materials* **51(4)** 400-404

[5] Nasakina E O, Baikin A S, Sergienko K V, Sevost'yanov M A, Kolmakov A G, Goncharenko B A, Zabolotnyi V T, Fadeev R S, Fadeeva I S, Gudkov S V, Solntsev K A2015 Biocompatibility of nanostructured nitinol with titanium or tantalum surface composite layers formed by magnetron sputtering *Doklady Chemistry* **461(1)** 86–88

[6] Shabolovskaya S A 1995 Biological aspects of TiNi alloy surfaces, J. Phys. IV, Colloque C8, *supplement au J. Phys.* **III** 1199–1204

[7] Vasilevsky R Dj 1979 Shape memory effect in Ti–Ni based alloys is one of several aspects of stress-induced martensitic transformation, Shape Memory Effects in Alloys, *Metallurgy* 205–230

[8] Savada K, Okita H, Daja A 1983 The covered method for Ti–Ni devices, Patent 60-21374, Japan C22C 14/22 C 22 G 1/10, received: 16.07.83, No. 58-130019

[9] Kossiy G G, Trusov G N, Goncharenko B A, Micheev V S Corrosive and electrochemical characteristics of Ti–Ni intermetallic compounds in acid solutions, *Zashch.* **14** 662

IOP Conf. Series: Materials Science and Engineering **347** (2018) 012050 doi:10.1088/1757-899X/347/1/012050

A method for introduction of Al_2O_3 nanofiber into aluminum alloy

Yijin Chen, He Qin, Y A Kurganova and V K Gaaze

Department of Engineering Technology, Bauman Moscow State Technical University, 5/1, ul.Baumanskaya 2-ya, Moscow, 105005, The Russian Federation

E-mail: chenyijin@yandex.ru

Abstract. Experimental samples of aluminum matrix composites (AMCs) reinforced with Al_2O_3 nanofibers were obtained. In order to increase the wettability in liquid phase combination conditions, it is proposed to use copper powder as a carrier of nanofiber. When studying the structure of the samples, a modifying effect of the introduction of reinforcement, demonstrated by grinding the grain, was revealed.

1. Introduction

The development of modern science and technology requires the optimization of technical characteristics of machines and mechanisms, which can be provided on condition that new materials are created and fully utilized [1,2]. Current researches have proved that aluminum matrix composites (AMCs) are very promising new generation materials, and AMCs have aroused considerable interest among researchers. AMCs combine the best properties of their two constituents, such as ductility and toughness of the aluminum matrix, high modulus and strength of the reinforcements [3]. Important AMCs applications in the ground transportation (auto and rail), thermal management, aerospace, industrial, recreational and infra-structure industries have been enabled by functional properties that include high structural efficiency, excellent wear resistance, and attractive thermal and electrical characteristics [4,5]. AMCs can reduce the curb weight of vehicles, thereby improving fuel efficiency [6]. 10% reduction in the weight of cars and light-duty trucks can reduce CO_2 emissions by 72 million metric tons. A market report from Technavio Research in 2017 says that the global metal matrix composites market size is projected to reach 10.8 kilotons by 2021, growing at a CAGR of over 6% during the forecast period. According to market report from Grand View Research, the largest product segment was the aluminum-based metal matrix. It accounted for 30% of the demand. The need for lightweight and high tensile strength parts are the driving factors for the demand of aluminum to soar [7]. The North America market is expected to remain the largest market for 2022 (figure 1).

So far, various efforts have been made to exploit aluminum matrix composites with different reinforcement materials. In particular, since nanosized materials own a series of excellent performances, many studies have been done to develop AMCs with nanosized reinforcements (CNTs, graphene, SiC nanoparticles, Al_2O_3 nanoparticles, etc.) [3,8,9]. On the other hand, a few results have been reported on alumina nanofibers reinforced aluminum matrix composites.

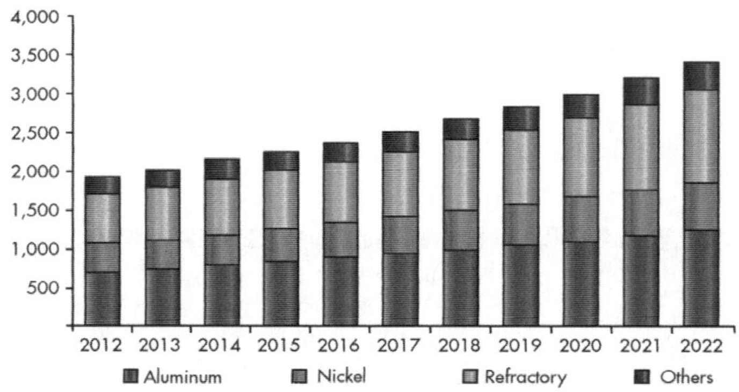

Figure 1. North America metal matrix composite market volume by product, 2012-2020 (Tons) [7]

Alumina nanofiber is one of the most important ceramic nanomaterials, which finds applications in many areas such as high-temperature insulation, reinforcement components, microelectronics and optics, fire protection and catalysis [10,11]. Alumina nanofibers (Al_2O_3-NFs) have very high elastic modulus and high tensile strength. Compared to other nanosized reinforcement materials, aluminum oxide nanofibers have high temperature properties. Therefore, they are suitable as reinforcement materials of large scale aluminum matrix composites.

There are various methods to synthesize AMCs [2,3,12]. Generally, these methods are classified into: solid state processing, liquid state processing and vapour deposition [12]. In liquid state methods, the aluminum matrix is in the state of iquid. As a useful method of liquid state processing to produce AMCs, casting is a more industrial compatible technique to produce composites in large scales [13].

However, the fabrication of composites with uniformly distributed Al_2O_3 nanofibers by casting is very difficult. The problem is that, due to the physicochemical features, the aluminum matrix and alumina nanofibers reinforcement are quite difficult to connect. Nanomaterials with a large specific area (up to 1000 m^2/g) usually tend to easily agglomerate due to the presence of van der Waals forces. In addition, a significant difference in surface tension and density between nanosized reinforcements and matrices complicates the connection. Thus, the main technological problems of introducing discrete Al_2O_3 nanofibers into aluminum alloys are to overcome the surface tension forces and ensure homogeneity of their distribution.

The analysis of the special literature showed that an effective way to improve wettability is alloying the matrix alloy and applying special coatings to the reinforcing phase. The hypothesis of the experimental part of the studies is the use of copper as a component that ensures an improvement in wettability [14,15]. Copper is proposed to be introduced in the form of powder, which simultaneously acts as a carrier of nanofibers.

2. Experimental procedure

2.1: Materials

In the current research, AK6 aluminum alloy was chosen as the matrix alloy. The chemical composition of AK6 aluminum alloy according to GOST 4784-97 [16] is shown in Table 1. Table 2 shows the properties of AK6 aluminum alloy.

Table 1. Chemical composition of AK6 aluminum alloy (wt.%).

Composition	Al	Cu	Si	Fe	Mn	Mg	Zn	Ti	Ni
wt.%	93.3-96.7	1.8-2.6	0.7-1.2	<0.7	0.4-0.8	0.4-0.8	<0.3	<0.1	<0.1

IOP Conf. Series: Materials Science and Engineering **347** (2018) 012050 doi:10.1088/1757-899X/347/1/012050

Table 2. Properties of AK6 aluminum alloy.

Material	Density (g/cm^3)	Vickers Hardness (MPa)	Young's modulus (GPa)	Tensile strength (MPa)
AK6	2.75	95-100	72	400

Discontinuous nanosized alumina fibers were used as the modifying additive. "Nafen" brand Al_2O_3-NFs with an average diameter of 10-20 nm and an average length>100 nm were offered by ANF Technology Ltd (Estonia). The properties of Al_2O_3-NFs were shown in Table 3.

Table 3. Properties of Al_2O_3 nanofibers.

Material	Density (g/cm^3)	Thermal resistance (°C)	Young's modulus (GPa)	Tensile strength (GPa)
Al_2O_3-NFs	3.98	1200	400	12

2.2: Fabrication of composites

To combine the matrix alloy with discrete Al_2O_3 nanofibers placed on copper powder, a casting method of mechanical stirring was used. As illustrated in figure 2, the introduction of Al_2O_3 nanofibers into the aluminum melt was carried out at a temperature of 800°C with constant stirring at a rate of 600 rpm for 10 min. Figure 3 shows the agitated Al_2O_3 nanofibers/Al composites.

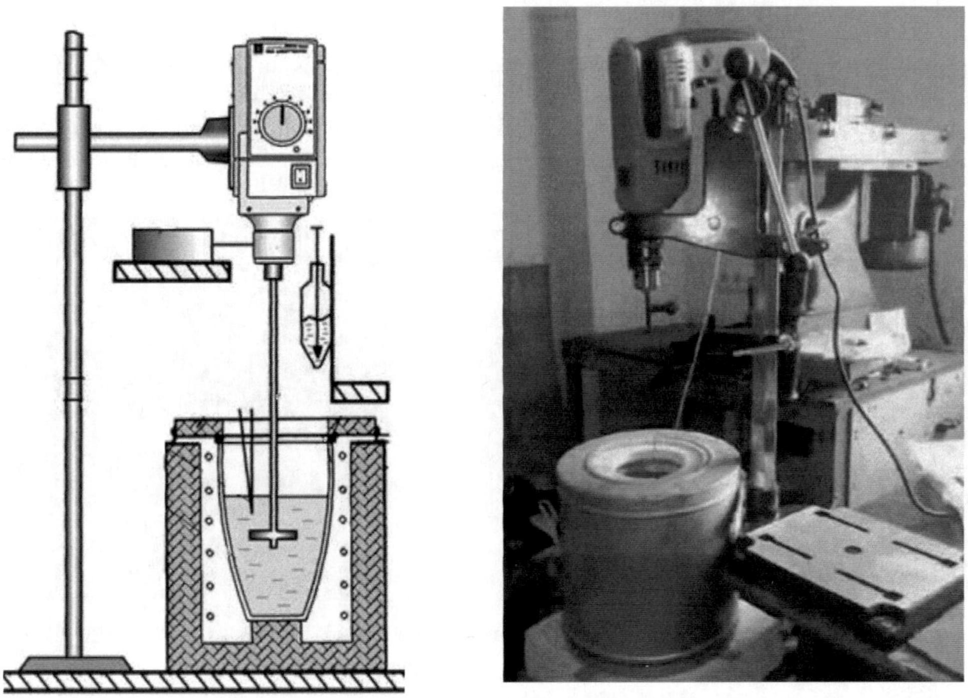

a b

Figure 2. Schematic diagram (a) and experiment device (b) of the liquid process for composite manufacturing by mechanical stirring

IOP Conf. Series: Materials Science and Engineering **347** (2018) 012050 doi:10.1088/1757-899X/347/1/012050

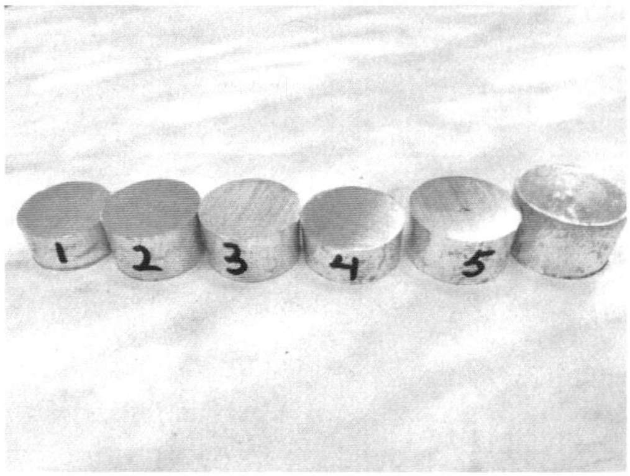

Figure 3. Al_2O_3 nanofibers/Al composite

2.3: Composite characterizations
The macrostructure of the aluminum alloy and the as-fabricated composites was characterized by Optical Microscopy (Neophot 21), and the microstructure of the aluminum alloy and the as-fabricated composites was characterized by Optical Microscopy (Olympus GX51).

3. Results and discussion

3.1 Comparative analysis of the macrostructure
The results of comparative studies of experimental initial and filled samples, carried out on the Neophot 21 microscope with an attachment for macroscopic studies with zoom of 20x, are shown in Figure 4. It indicates that the addition of Al_2O_3 nanofibers has reduced the grain size of aluminum matrix. Besides, the porosities are visible in the composite, which could be attributed to gas entrapment during the stirring.

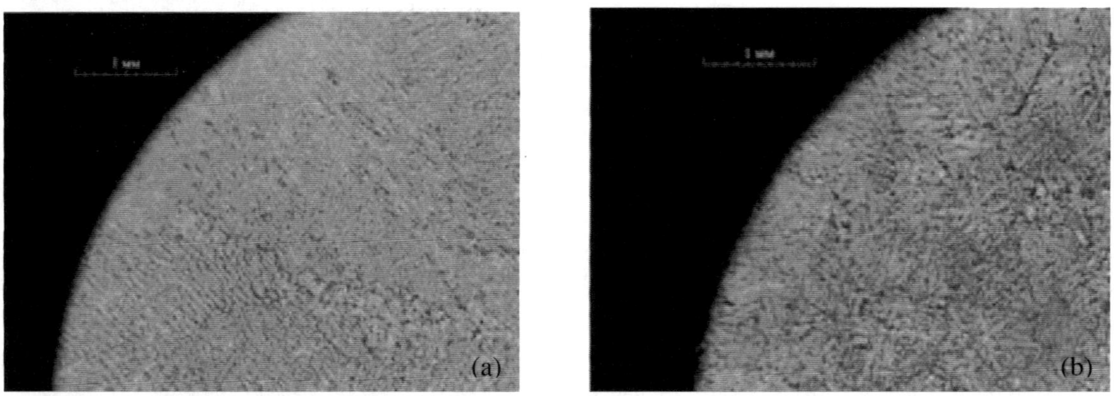

Figure 4. Optical macrostructures of cast materials: (a) AK6 alloy matrix; (b) Al_2O_3 nanofibers/Al composite

3.2 Comparative analysis of the microstructure
The results of comparative studies of the microstructure performed on the Olympus GX51 microscope after etching in 0.5 % aqueous HF solution with zoom of 100x and 500x are shown in figure 5.

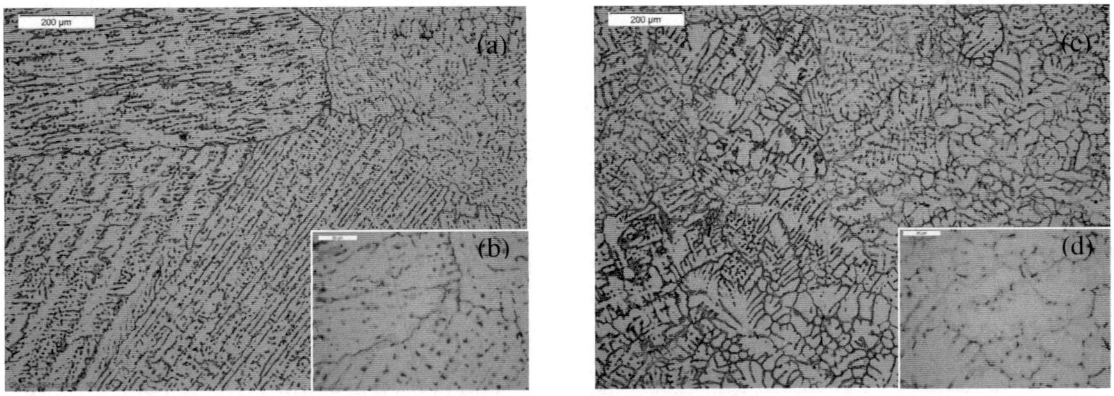

Figure 5. Optical microstructures of cast materials: (a,b) AK6 alloy matrix; (c,d) Al_2O_3 nanofibers/Al composite

Analysis of the structure at various magnifications, demonstrated the presence of a modifying effect, expressed in the grinding of structural components. So, if the average grain size is taken as the quantitative index, then the average grain area, estimated by the method of GOST 5639-82 [16], is equal to 1.1 mm^2 and 0.015 mm^2 for the initial and modified samples, respectively.

4. Conclusion

Based on the present study, the following conclusions can be drawn:
(1) As a result of the work, by mechanical mixing, samples of a composite material with a filler in the form of a thin-discrete alumina with a diameter of 10-20 nm were obtained.
(2) In the course of the study, the effect of introduction of nanofibers on structure refinement is established.

Thus, the analysis of the results of the study showed the presence of a modifying effect in aluminum alloy from the introduction of Al_2O_3 nanofibers .

Acknowledgments

The authors are grateful to the employee of the company ANF Technology Ltd M. V. Lobanov for providing Al_2O_3 nanofibers. The authors would like to thank S. P. Scherbakov for the preparation of the composite casting and useful discussion. The lead and second authors (Yijin Chen and He Qin) gratefully appreciate the financial support of China Scholarship Council (CSC).

References

[1] Kurganova Y A and Kolmakov A G 2015 *Structural Metal Matrix Composites* (Moscow: Bauman Moscow State Technical University Press) p 141
[2] Chen Itszin 2017 Metal-matrix composite materials. Prospect of effective application and production (review) *Metal technology* **10** 25-30
[3] Tjong S C 2013 Recent progress in the development and properties of novel metal matrix nanocomposites reinforced with carbon nanotubes and graphene nanosheets *Materials Science and Engineering R* **74** 281–350
[4] Miracle D B 2005 Metal matrix composites – From science to technological significance *Composites Science and Technology* **65** 2526–40
[5] Kurganova Y A, Chernyshova T A, Kobeleva L I, and Kurganov S V 2011 Service properties of aluminum□-matrix precipitation-hardened composite materials and the prospects of their use on the modern structural material market *Russian Metallurgy (Metally)* **7** 663-666
[6] 2017 Global Metal Matrix Composites Market 2017-2021 TechNavio - Infiniti Research Ltd.
[7] 2015 Metal Matrix Composite (MMC) Market Analysis By Product (Aluminum, Nickel, Refractory) By End-use (Ground Transportation, Electronics/Thermal Management, Aerospace) And Segment Forecasts To 2022 Grand View Research

IOP Conf. Series: Materials Science and Engineering **347** (2018) 012050 doi:10.1088/1757-899X/347/1/012050

[8] Kollo L, Leparoux M, Bradbury C R, Jaggi C, Carreno-Morelli E and Rodriguez-Arbaizar M 2010 Investigation of planetary milling for nano-silicon carbide reinforced aluminium metal matrix composites *J. Alloys Comp.* **489** 394–400

[9] Han Q, Setchi R and Evans S L 2016 Synthesis and characterisation of advanced ball-milled Al-Al_2O_3 nanocomposites for selective laser melting *Powder Technology* **297** 183–192

[10] Hao Yu, Jian Guo, Shuqi Zhu, Yaogang Li, Qinghong Zhang and Meifang Zhu 2012 Preparation of continuous alumina nanofibers via electrospinning of PAN/DMF solution *Materials Letters* **74** 247–249

[11] Hongqin Dai, Jian Gong, Hakyong Kim and Doukrae Lee 2002 A novel method for preparing ultra-fine alumina-borate oxide fibres via an electrospinning technique *Nanotechnology* **13** 674–677

[12] Moghadam A D, Omrani E, Menezes P L and Rohatgi P K 2015 Mechanical and tribological properties of self-lubricating metal matrix nanocomposites reinforced by carbon nanotubes (CNTs) and grapheme – A review *Composites Part B* **77** 402–420

[13] Kurganova Y A and Scherbakov S P 2017 Influence of a discrete additive of aluminum oxide on structure and properties of aluminium alloy *Journal of mining institute* **228** 717–721

[14] Serpova V M, Shavnev A A, Grishina O I, Krasnov E I, Solyayev Y O 2014 Wetability and interfacial interaction in MKM on an aluminum matrix reinforced with Al_2O_3 *Materials Science* **12** 29–35.

[15] Leon C A, Mendoza-Suarez G and Drew R A L 2006 Wettability and spreading kinetics of molten aluminum on copper-coated ceramics *Journal of Material Science* **41** 5081–87

[16] GOST 4784-97 *Aluminium and wrought aluminium alloys. Grades* (Moscow: Standardinform) p 15

[17] GOST 5639-82 *Steels and Alloys. Methods for detection and determination of grain size* (Moscow: Standardinform) p 16

IOP Conf. Series: Materials Science and Engineering **347** (2018) 012051 doi:10.1088/1757-899X/347/1/012051

Influence of the silicon carbide deposit on the thermal resistance of fire protection

K A Kim and D O Lemeshev

Mendeleev University of Chemical Technology of Russia, Moscow, Russia

Const552@gmail.com

Abstract. The ceramics samples with structure of SiC-Al_2O_3-Fireclay having good thermal resistance were received. As materials were used: black α-SiC F-120, corundum α-Al_2O_3 F-1000 and Kudinovsky fire-clay. As a temporary technological bundle used polyvinyl alcohol (PVA). Thermal stability was determined by method of heat changes.

1. Introduction

Nowadays the main method of creating new unique in properties materials is combining of different materials – composites. Properties of materials in common with each other acquire properties of the new material and qualities of formed composites are unique.

Composites compete with common construction materials, for example aluminum, stainless steel, titanium. They are used in material producing relating to aviation, aerospace, chemical manufacture, automotive industry and medicine. Value of composites is high as known; this is due to the fact that complex producing, technologies and raw materials are really expansive. But in this case there is a saving resources in complex details producing due to reducing the number of components [1-2].

In recent decades the considerable attention was paid to the SiC-Al_2O_3 composite. The combination of silicon carbide and aluminum oxide will give us the certain ceramics composite materials with high durable and high chemical and temperature stability. This composite is characterized with multifunctional in usage, availability of raw materials, corrosion resistance and non-power-intensive production technology. Aluminum oxide and silicon carbide have unique complex of physic-mechanical properties such as high solidity, low coefficient of thermal expansion and also possibility of use in extreme conditions.

There are different types of composites: based on carbide-silicon (80-90 % SiC), carbide-silicone containing materials (10-30 % SiC), it depends on content quantity of SiC as matrix or reinforcing material.

It is not possible to obtain dense material from pure silicon carbide due to oxidation of SiC at high temperature up to 2000 °C in the air. For this reason the most used ceramics are contain different binders, as well as most carbide-silicon refractories are made with addition of α-Al_2O_3 and fire-clay because they do not oxidize in high temperature with air and they form the mechanism of liquid sintering. This way does not require high energy consumption.

Products with SiC-Al_2O_3-fireclay composition are characterized by high thermal conductivity, crack resistance, wear resistance and high temperature resistance. Ceramics of composition SiC-Al_2O_3 are widely used in ferrous and non-ferrous metallurgy due to high thermal properties [3].

IOP Conf. Series: Materials Science and Engineering **347** (2018) 012051 doi:10.1088/1757-899X/347/1/012051

Basic refractory materials in the metallurgical industry are dinas refractories, chamotte, periclase and also concrete materials. Dinas refractory contains more 93 % of silicon carbide SiO_2. It has high durable and high operating temperature but heat resistance does not exceed on average 1 – 2 water heat changes.

Chamotte refractories are one of the most used refractory materials and available. Chamotte contains SiO_2 and Al_2O_3 up to 45 %. In one hand it has high heat resistance on average 10 – 20 water heat changes but in other hand the strength is low, that is the main lack.

Periclase refractories are able to operate at high temperature about 1700 °C and contain no less than 85 % MgO. Heat resistance of periclase is low. It is on average 1 – 2 water heat changes.

In the metallurgical industry, the main part of rafractories are used in induction furnaces. Metal is actively mixed in these furnaces and it is possible to lead to destruction. Because the depth of the lining has to be about 10 – 15 cm, durable and chemical resistance ought to be high also.

2. Materials and Method

To obtaining refractory materials with high durable the mixture is added: SiC (F-120) with a content 20, 30, 35 mass %, Al_2O_3 (F-1000) in quantity of 55, 45,40 mass % and Kudinovsky field fire-clay content was the same in all of the mixtures 25 mass %. It is known that with increasing oxide phases initially, the porosity decreases and mechanical properties grow simultaneously.

Variation of the components content allowed us to study in detail the temperature resistance dependence of content SiC reinforcing component.

It is known that silicon carbide has high mechanical properties and high thermal conductivity. Because of this presence of coarse particles SiC allowed to obtain reinforced structure with high crack resistance. Using of fireclay with ceramic mixture allowed us to compensate low diffusion activity of SiC and Al_2O_3.

Use of the liquid-phase sintering principle allowed to achieve mutual dissolution of particles, which eventually led to activation of atoms diffusion process on the phase boundaries [4].

In this way the liquid-phase sintered material structure with particles of SiC, Al_2O_3 and silicate interfacial was formed.

Initial components were mixed in a roller mixer with addition of corundum balls with a diameter of 15 mm in proportion of 3:1 by weight during 2 hours. Using a greater number of grinding balls leads to fragmentation of particles.

As a temporary technological bundle was used the 8 % aqueous solution of polyvinyl alcohol in quantity of 7 mass %. For uniform distribution of particles and bundle the mixture was sowed. With addition of the temporary technological bundle it occurs wetting and lubrication of particles surfaces. It allows to reduce friction between particles and improves their movement at high pressure. Due to this it was possible to achieve the closest packing at pressure. Also an important factor was using fireclay powder and aluminum oxide with the particular size smaller than particles of silicon carbide. It allowed us to obtain high density of samples.

The samples were formed by semi-dry pressing from the obtained mass. Pressing pressure was 100 MPa and cylindrical samples sizes were d=h=35 mm. The samples were dried at a temperature of 23-25 °C during 24 hours and burned in air at a temperature of 1400 °C with 4 hours of isothermal aging. With this temperature it occurs formation of liquid phase with more fusible phase of fireclay.

It begins the process of liquid sintering at which a liquid phase spreads along grains boundaries of aluminum oxide and silicon carbide and pulls them together with filling the pores. Simultaneously the grains and structure become larger.

The main advantage of liquid sintering mechanism is impossibility of pores formation but it is possible the formation of large grains in the structure influencing on the physic-mechanical properties.

IOP Conf. Series: Materials Science and Engineering **347** (2018) 012051 doi:10.1088/1757-899X/347/1/012051

3. Results and discussion

The obtained testing results of average density, open porosity and water absorption are shown in the table 1. The results of density and porosity were determined by the hydrostatic weighing.

Table 1. Properties of ceramic samples

Content Al_2O_3;SiC;fire-clay, mass. %	ρ, g/cm³	P, %	W, %
50;25;25	2,72	12	5
45;30;25	2,55	13	6
40;35;25	2,53	15	6

The content of pores and defects in the volume of the material affects the material strength and heat-resistance. There is an increase in the stress gradient at thermal shock, eventually there are breaking cracks. Cracks propagation occurs if the energy of crack spread is more than the restraining energy [5].

Obtained material's heat resistance was measured by heat-changes method. The samples were heated to 1000 °C in the silicate furnace with carbide-silicon heaters then cooled instantly in water at a temperature of 22-25 °C. This method allows to obtain results about thermal resistance on a par with actual operating conditions of refractories use. With instant cooling there is a stress gradient on the surface and inside of structures. By the number of heat-changes it is possible to judge the cycles of use in production. Results of the experiments are shown in table 2. The samples with the greatest content of SiC (35 mass %) were destroyed firstly.

Table 2. Heat changes results

Material	Quantity of heat changes before	
	crack appearance	destruction
25 mass % SiC	4	47
30 mass % SiC	5	28
35 mass % SiC	5	24

This is due to the fact that the presence of an oxide phase and reinforcing components allows to reduce the spread of main cracks in the ceramic structure. Coefficient of thermal expansion difference is weakens the adhesion between particles of SiC and oxide phase. This mechanism allows to restrain the cracks spread along the SiC boundaries, increasing the spread way. Eventually it requires more energy consumptions. Cracks in this material spread and pass around the grains of silicon carbide with intergranular mechanism at high temperature and thermal stresses [3-5].

An important factor is equal distribution of SiC particles in the Al_2O_3 matrix, it was a dominating role. With the increase SiC concentration there is a local congestion and agglomeration particles that leads to formation of unreinforced areas. These areas contribute to passing of break cracks through the material structure.

On the pages 1, 2 and 3 showed polished surfaces and character of SiC particles distribution. It is obviously that with increasing of SiC content occurs a local congestion as indicated earlier. Type cracks spreading (in the images below) characteristic for thermal shock and contracting-compressing thermal stresses.

IOP Conf. Series: Materials Science and Engineering **347** (2018) 012051 doi:10.1088/1757-899X/347/1/012051

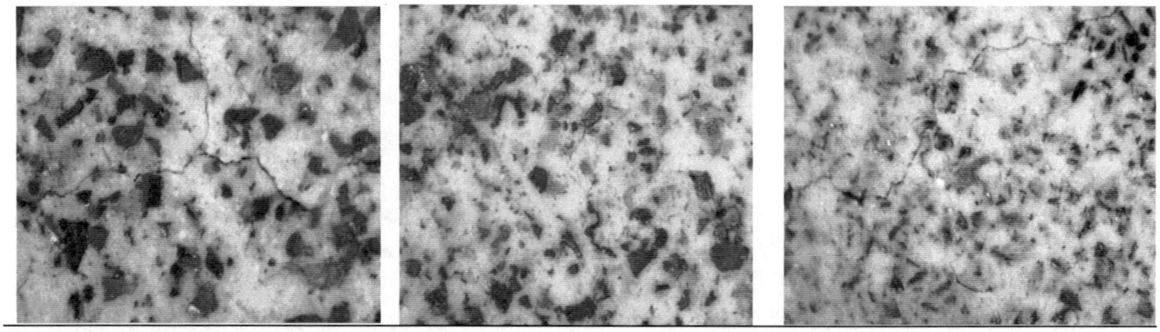

a b c

Figure 1. SiC particles spread (a - 25 mass %, b – 30 mass %, c – 35 mass %)

4. Conclusions

A study was conducted to determine ceramics SiC-Al_2O_3-fire-clay properties of heat resistance and results allow us to make conclusions:

- Ceramic properties are investigated: density, porosity and water absorption. Porous samples with high density were obtained.
- By the method of heat changes the values of heat resistance with different SiC content were determined. As a result of experiments it was found that the samples with the content of 25 % have the best thermo mechanical properties due to SiC particles spread in the ceramic structure. This affected the increase in crack resistance.
- Developed material containing 25 mass % of SiC has high thermo mechanical properties, about 47 heat changes. This exceeds the standard chamotte and high-alumina refractories which are used nowadays in metallurgy.

References

[1] Pilipovskiy L and Grudkina G 1990 Composite materials in mechanical engineering (Moscow: Science) p 139
[2] Inamura Y 1984 Refractories and their application (Moscow: Metallurgy) p 448
[3] Perevislov S N and Nesmelov D D 2014 Refractories and Industrial Ceramics **4-5** 3
[4] Sciti D and Bellosi A 2000 Journal of Materials Science **35** 3849
[5] Perevislov S N 2018 Doctoral Dissertation

IOP Conf. Series: Materials Science and Engineering **347** (2018) 012052 doi:10.1088/1757-899X/347/1/012052

Concentration dependence of sodium alloys based on tin surface tension

B B Alchagirov, O Kh Kyasova, A N Uzdenova and A Kh Khibiev

Kabardino-Balkarian State University, Chernyshevskogo Street 173, Nalchik, Kabardino-Balkaria 360004, Russia

E-mail: oksikyas160993@gmail.com

Abstract. The concentration dependence of the surface tension (ST) for alloys of the Sn-Na system in the range of compositions with a content of 0.06 to 5.00 at.% Na is studied by the large droplet method using high-purity components and a corresponding ST isotherm for T = 573 K is constructed. It has been established that small additions of sodium to tin significantly reduce ST of the studied melts. Calculations of sodium adsorption in alloys have shown that there is a maximum on the adsorption curve corresponding to alloys with a content of about 1.5 at.% Na in tin.

1. Introduction

Great interest in the study of alloys with the participation of alkali metals due to the possibility of their use in many areas of new technology. In particular, tin-sodium melts are considered as promising coolants for nuclear reactors, for the development and creation of new power-intensive chemical current sources, etc [1]. However, the thermophysical properties of alloys of the system Na-Sn is poorly studied [2, 3], in this connection it is necessary to obtain precise data on the thermophysical properties of this system. The present work is devoted to the study of the concentration dependence of the surface tension (ST) $\sigma(x)$ of sodium alloys on the basis of tin.

In the literature there are only a few papers devoted to the investigation of the ST of alloys of the Na-Sn system [4, 5]. The influence of sodium (from 0.03 to 4.6 at.% Na) on ST Sn was studied in [4]. But it has several disadvantages: a) in measurements of ST, the authors of [4] used tin of technical purity for the preparation of alloys; b) due to the lack of information about the density of Sn-Na alloys in the literature, which are necessary for the calculations of ST, the authors of [4] made a rough assumption: since the Na concentrations in the alloys with Sn are small, the density of alloys should not differ significantly from the density pure tin. In this approximation, the ST of all 10 sodium alloys studied with tin was calculated.

In the framework of a simple statistical model based on the formation of complexes and on the concept of the presence of a layered structure near the phase boundary, the isotherm of the Sn-Na alloys is plotted in [5] in the total composition interval, in which both the minimum and the maximum where found, figure 1. The data $\sigma(x)$ obtained in [5] for the concentration regions of the reciprocal dilute solutions of the Sn-Na system are quite strictly subject to the criteria for the surface activity of the components in binary metallic melts [6] and, in this respect, they should be considered correct. However, it is not possible to verify the reliability of the presence of a maximum and a minimum on the isotherm $\sigma(x)$, unfortunately, due to the lack of experimental data on $\sigma(x)$ of the Sn-Na system in the complete concentration range of the compositions. In the present paper, the task is set to refine the isotherm of

IOP Conf. Series: Materials Science and Engineering **347** (2018) 012052 doi:10.1088/1757-899X/347/1/012052

σ(x) Na alloys based on Sn using the components of improved purity and modern techniques and experimental technology.

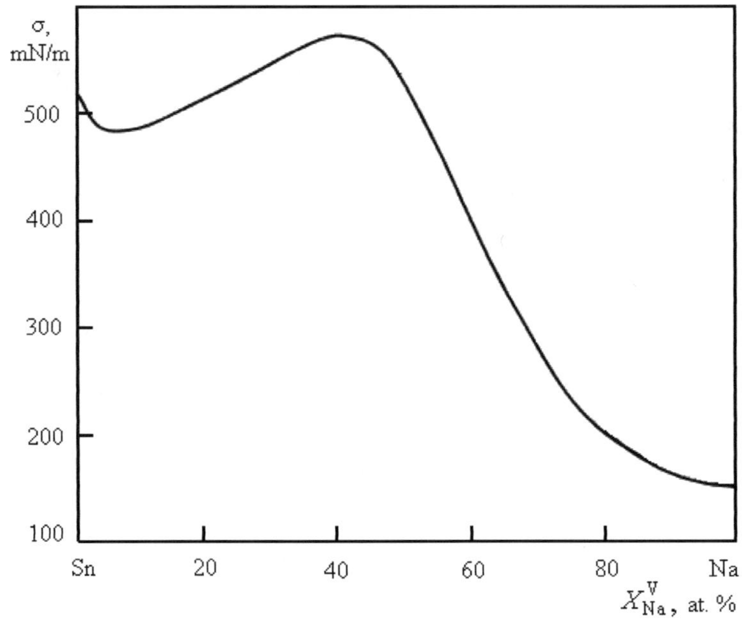

Figure 1. Concentration dependence of the surface tension (873 K) of alloys of the Sn-Na system [5].

The surface tension is determined by the method of the lying drop with an error of about 2% on samples of alloys from high-purity components (Sn-000 and 99.993% Na) in a vacuum of 10^{-4} Pa. The whole-soldered measuring cell used by us makes it possible to prepare alloys and measure ST in the same conditions of static vacuum without opening and depressurization in the course of experiments, and, most importantly, it ensures the thermodynamic equilibrium of the surfaces of alloys with its own saturated vapor. The features of the experimental setup and the measurement procedure are described in our papers [7, 8].

2. Results of the study ST sodium alloys based on tin

The results of measuring the ST of sodium based on tin alloys in the range of compositions studied by us graphically presented in figure 2. It shows that the obtained isotherm of the ST σ (x) is a smooth falling curve, i.e. additions of sodium to tin lead to a significant decrease in the ST of alloys. In particular, the addition of 1 at.% Na to Sn reduces the ST of tin by 80 mN/m, i.e. by 15 %, which indicates a rather high surface activity, manifested by sodium at a low content in alloys with tin.

As noted above, there is only one work [4], devoted to the experimental study of the ST alloys of the Sn-Na system, whose data are also shown in figure 2 in the form of light circles. In figure 2 that our data on the isotherm σ(x) are in qualitative agreement with the results of [4], but are located systematically somewhat lower than the σ (x) isotherm obtained in [4]. The maximum deviation of the compared ST isotherms reaches 20 mN/m, which exceeds the experimental error by several times.

Using the ST σ (x) isotherm obtained in the experiments, we estimated the limiting surface activity of sodium α_{Na} in alloys with tin at a temperature of 573 K:

$$\alpha_{\mathrm{Na}} = \left(\lim_{\Delta x_{\mathrm{Na}} \to 0} \frac{\Delta\sigma}{\Delta x_{\mathrm{Na}}} \right)_T ,$$

where Δσ is the change in ST depending on the composition of the melt, and x is the concentration of sodium in atomic fractions.

It turned out that α_{Na} in alloys with tin it amounted to about 15 N/(m×at.of a share), which indicates a higher surface activity exhibited by sodium with its small content in tin than the limiting surface activity α_{Na} calculated from the data of [4].

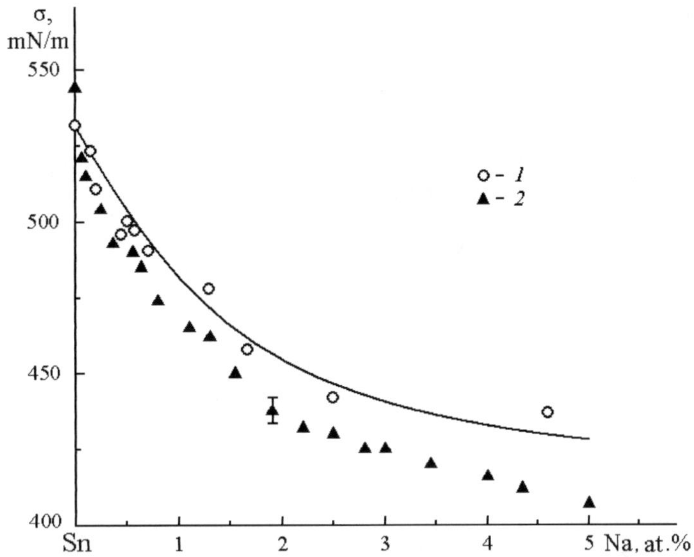

Figure 2. Concentration dependences of the surface tension of sodium alloys based on tin at a temperature of 573 K: 1 - data [4], 2 - our data.

Calculations of sodium adsorption in alloys with tin are carried out by the known Guggenheim-Adam formula in the N-variant by the method described in [9-11], using for this purpose the experimental data obtained on the isotherm of the ST σ(*x*) of sodium alloys based on tin at 573 K:

$$\Gamma_2^{(N)} = -\frac{x_2(1-x_2)}{RT}\left(\frac{\partial\sigma}{\partial x_2}\right)_T$$

in which x_2 – are the atomic fractions of the second component (sodium) under the normalization condition $x_1 + x_2 = 1$, *R* is the universal gas constant, and *T* is the temperature in K.

We note that the derivatives $(\partial\sigma/\partial x_2)_T$ are determined until recently by "manual" graphical differentiation of the isotherm σ (x), adhering to the "first-derivative step" principle and substituting them later into the Guggenheim-Adam equation. The main drawback of such calculations is the considerable error in the calculations of adsorption, which usually reaches 15% or more [11]. To reduce the error in calculating the adsorption of the components of alloys, we have used the method of calculating the adsorption of components, which we improved, by processing the experimental data of the concentration dependence of the ST σ(x) by the method of least squares and then differentiating it in the MathCAD package [12]. In this case, a general regression is used representing a linear combination of power functions x^m, $m = \overline{1, 3}$ and an isobaric hyperbola $1/(x+\alpha)$, where the numerical parameter α is determined from the condition of regression convergence to the data obtained experimentally. As a result, the error in calculating the adsorption of components in metallic melts decreased several times by a new method and, importantly, it is possible to eliminate the timeconsuming and inefficient procedure for graphically differentiating the isotherm of the ST σ(x).

The data of a large number of computational experiments make it possible to assert that the proposed method and procedure of approximation make it possible not only to minimize the errors in determining the adsorption of the components of the solutions under study, but also to check the informativeness and accuracy of data processing.

From the data obtained, it follows that sodium in the region of its small content in alloys with tin exhibits a high surface activity, the maximum value of which is achieved in the composition range of ~ 1.5 at. % Na in tin, figure 3. The greatest value of adsorption obtained in this work is 10% higher than the adsorption maximum found earlier by Pokrovsky et al. at [4].

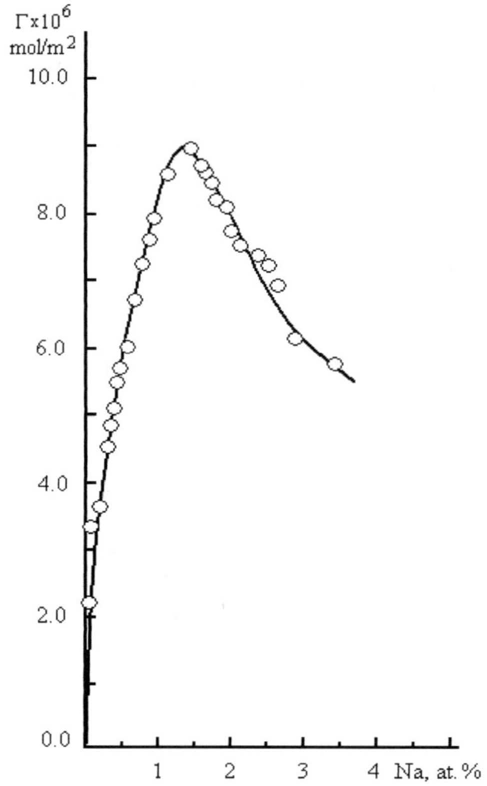

Figure 3. Results of calculations of sodium adsorption in alloys based on tin.

3. Conclusions

Under conditions of a high static vacuum, the surface tension of high purity tin, as well as of its 19 alloys with a content of 0.06 to 5.0 at.% Na, was measured and a corresponding isotherm ST $\sigma(x)$ was constructed for a temperature T = 573 K. It has been shown that the limiting surface activity of sodium in alloys with tin is about 15 N/(m×at. of a share). Sodium in the region of its small content in alloys with tin exhibits a high surface activity, the maximum value of which is achieved in the composition region with a content of about 1.5 at.% Na in Sn, which is in agreement with the rule of generalized moments in Semenchenko and Zadumkin, as well as with other criteria for the surface activity of components in binary metallic systems.

References

[1] Drits M E and Zusman L L 1986 *Alloys of Alkali and Alkaline Earth Metals: A Reference Book* (Moscow: Metallurgiya) [in Russian] p 248

[2] Nizhenko V I and Floka L I 1981 *The Surface Tension of Liquid Metals and Alloys: A Reference Book* (Moscow: Metallurgiya) [in Russian] p 208

[3] Alchagirov B B, Arkhestov R Kh, Dyshekova F F and Taova T M 2013*High Temp.* **51** (2) 183

[4] Pokrovsky N L and Galanina N D 1949 *J. of Phys. Chem.* **23** (3) 324

[5] Anusionwu B C 2003 *J. Alloys Comp* **359** (1-2) 172
[6] Zadumkin S N, Khokonov Kh B and Karamurzov B S 2014 *Physics of Interfacial Phenomena in Condensed Media* (Nalchik: KBSU) [in Russian] p 246
[7] Alchagirov B B, Dyshekova F F and Arkhestov R Kh 2016. *J. of Phys. Chem.* **90** (11) 1710
[8] Alchagirov B B, Dyshekova F F and Kokov Z A 2017 *Pribory* **2** (200) 18 [in Russian]
[9] Guggenheim E A and Prue J E 1955 *Physicochemical calculations* (Interscience publishers inc. New York) p 491
[10] Dadashev R Kh 2008 *Thermodynamics of surface phenomena* (Cambridge Int Sci Publishing) p 280
[11] Alchagirov B B, Khokonov Kh B and Chochaeva A M 2004 *Calculation of Component Adsorption, Structure and Thickness of Surface Layers of Binary Metallic Solutions* (Nalchik: KBSU) [in Russian] p 58
[12] Alchagirov B B, Lesev V N, Afaunova L Kh, Arkhestov R Kh, Dyshekova F F and Kegadueva Z A 2009 *Doklady AMAN* **11** (1) 125 [in Russian]

IOP Conf. Series: Materials Science and Engineering **347** (2018) 012053 doi:10.1088/1757-899X/347/1/012053

Magnetic properties of powder hard magnetic Fe–27Cr–10Co–0.5Mo and Fe–27Cr–10Co–2Mo alloys

I M Milyaev[1], M I Alymov[1,2], I N Bouryakov[2], V S Yusupov[1] and D M Abashev[3]

[1]Baikov Institute of Metallurgy and Materials Science, Russian Academy of Sciences, Leninskii 49, Moscow 119336, Russia
[2]Merzhanov Institute of Structural Macrokinetics and Materials Science, Russian Academy of Sciences, Akademika Osipyana 8, Chernogolovka, Moscow Region 142432, Russia
[2]JCS Spetsmagnit, Dmitrovskoe 58, Moscow 127238, Russia

E-mail: imilyaev@mail.ru; yusupov@2000aport.ru; bouriakov@mail.ru

Abstract. Magnetic hysteresis properties of powder hard magnetic Fe–27Cr–10Co alloys with 0.5 and 2.0 wt % molybdenum were studied. For Fe–27Cr–10Co–0.5Mo alloy: B_r = 1.35 T, H_{cB} = 45.6 kA/m and $(BH)_{max}$ = 41 kJ/m^3. For Fe–27Cr–10Co–2Mo alloy: B_r = 1.29 T, H_{cB} = 41 kA/m, $(BH)_{max}$ = 34.8 kJ/m^3. It was shown that an alloying with 2 wt. % molybdenum of the low-cobalt powder hard magnetic Fe–27Cr–10Co alloy unlike high-cobalt hard magnetic FeCrCo alloys did not increase its magnetic hysteresis properties, reduced the camber of a demagnetization curve in a second quadrant.

1. Introduction

The problem of research of economically alloyed and effective hard magnetic materials (HMMs) with the use of new processes of their production is of undoubted interest in the context of the constant growth of raw material resources on the global market and requirements of reducing the labor and energy expenses for greater their competitiveness. Only deformable hard magnetic alloys based on the Fe–Cr–Co system combine high magnetic hysteresis properties with high mechanical (first of all strength) ones. The Russian metallurgical industry has mastered the mass production of sheet and rod from hard magnetic FeCrCo alloys containing a moderate amount of cobalt (15 wt. %). Further development of these alloys was aimed at the study of the alloys both with lower cobalt content with a view to reducing their cost and with its high content for greater magnetic hysteresis properties owing to an increase in the saturation magnetization.

Alloys containing more than 9–10% by weight of cobalt possess rather high values of residual induction B_r (above 1.3 *T*) and maximum energy product $(BH)_{max}$ (> 40 kJ/m^3) at relatively low values of coercive force H_{cB} (40–44 kA/m) [1].

To prepare the low-cobalt FeCrCo alloy with higher values of coercive force, the Fe–27Cr–10Co alloy was alloyed with 0.5 and 2 wt. % molybdenum. Molybdenum was chosen as the additional alloying element according to [2]. In [3, 4], it was shown that an additional alloying with molybdenum of high-cobalt FeCrCo alloys increases the value of H_{cB} to 80 kA/m.

In this work, the samples of the investigated alloys were prepared by powder metallurgy methods, which are cost-effective for manufacturing of permanent magnets.

IOP Conf. Series: Materials Science and Engineering **347** (2018) 012053 doi:10.1088/1757-899X/347/1/012053

2. Experimental

The alloys for the study were obtained by method of powder metallurgy using element powders of metals of industrial purity: carbonyl iron of the VS brand, PHS-1 chrome, PK-1 cobalt, powdered direct-reduction molybdenum with a size of particles less than 40 μm. Homogeneous powder mixtures were obtained in a C2.0 turbulent mixer. The powders were mixed in a glass vessel with capacity of 200 ml for 80 min without intermediate stops. The prepared powder mixes were refined using a Pulverizett-7 mill for 15 min with a rate of 600 rpm. In some cases, we added surfactants (stearin and oleic acids). The mixture was subjected to uniaxial pressing using a KNUTH-130042 manual press in demountable matrixes with a diameter of 13.6 mm under a pressure of 600 MPa. As a result, the cylindrical samples 20 mm in height were obtained. After molding, the samples were sintered in an SShV-1.2.3/25I1 mine vacuum electric furnace in vacuum no worse than 10^{-2} Pa. The heat treatment of the sintered powder samples was carried out in standard muffle furnaces in the container with a diameter of 60 mm, which was filled with metal ballast in order to simulate real operating conditions. The temperature and time were controlled using a TPM251 programmable PID-regulators. The thermomagnetic processing was carried out in a laboratory installation with the armor-clad electromagnet providing magnetic field intensity $H = 320$ kA/m (> 4000 Oe). Magnetic hysteresis properties were measured using a Permagraph L hysteresisgraph.

The magnetic hysteresis properties of the alloys were evaluated and the heat treatment regimes were optimized both by the method of one-factorial experiment and by the method of design of experiment with plotting central composite 2^4 designs + star points. The choice of the central composite (consecutive) designs applied to the description of almost stationary area, which cannot be described by means of linear approach, was caused by the fact that the preliminary experiments on evaluation of near-zero magnetic properties of the alloys with the use of heat treatment regimes did not show their sharp fluctuations. The chosen designs were rotatable (information contained in the regression equations is distributed uniformly over a sphere on which points of carrying out experiment get out, and the researcher do not know the area of the response surface where the optimum region of interest is located) and randomized (the order of carrying out experiments is defined in a random way).

Based on the existing concept on the mechanism of formation of a high-coercive state in hard magnetic Fe–Cr–Co alloys [5,6], we consider that generally the magnetic hysteresis properties (residual induction of B_r, coercive force H_c and maximum energy product $(BH)_{max}$) depend on the following: (1) tempering temperature of for α-solution; (2) temperature of isothermal thermomagnetic processing (ITMP); (3) time of ITMP; (4) cooling rate $V1$ in the temperature range from temperature of ITMP to 600–580°C; (5) tempering time at 600 (580°C), (6) cooling rate $V2$ in the temperature range of 600 (580)–500°C.

Other factors, such as, for example, a rate of heating to the temperature of ITMP, were stabilized as much as possible. In particular, the tempering of samples of the studied alloys was carried out from 1100°C in water with a holding at this temperature for 15 min; heating of the container to the temperature of ITMP was carried out for 40 min. The processing of experimental results was carried out using Statgraphics XVI and Statistica 10 software packages.

3. Results and discussion

The zero regime of the heat treatment of the hard magnetic Fe–27Cr–10Co–0.5Mo alloy involved the following: tempering at 1100°C (15 min) + heating to 670°C (10 min) + cooling in magnetic field to 580°C with a rate $V1 = 12$ °C/h + cooling without field to 500°C with $V2 = 7$ °C/h with variation of the temperature of the beginning of thermomagnetic processing (TMP) ± 10°C (factor A), cooling rates $V1$ ± 2°C/h (factor B) and cooling rates $V2$ ± 2°C/h (factor C). The experimental results and design matrix for anisotropic samples are given in table 1.

The statistical analysis of data in table 1 gives analytical dependences in the form of the regression equations for residual induction Br, coercive force Hc and maximum energy product (BH)max as a function of chosen variable factors.

IOP Conf. Series: Materials Science and Engineering **347** (2018) 012053 doi:10.1088/1757-899X/347/1/012053

Table 1. Magnetic hysteresis properties of anisotropic samples of the Fe–27Cr–10Co–0.5Mo alloy.

	Temperature of the beginning of TMP, °C	Cooling rate $V1$ to 580°C, °C/h	Cooling rate $V2$ to 500°C, °C/h	B_r, T	H_{cB}, kA/m	$(BH)_{max}$, kJ/m^3
1	680 (1)	10 (–1)	9 (1)	1.34	44.75	40.3
2	670 (0)	12 (0)	10.4 (1.68)	1.36	45.4	42.2
3	687 (1.68)	12 (0)	7 (0)	1.35	45.7	42.1
4	660 (–1)	14 (1)	5 (–1)	1.35	45.9	43.9
5	680 (1)	14 (1)	5 (–1)	1.37	46.0	43.5
6	670 (0)	12 (0)	7 (0)	1.34	45.45	40.5
7	670 (0)	8.64(–1.68)	7 (0)	1.33	45.2	40.0
8	670 (0)	12 (0)	3.64 (–1.68)	1.35	46.0	43.1
9	660 (–1)	10 (–1)	5 (–1)	1.34	45.9	40.7
10	670 (0)	12 (0)	7 (0)	1.36	46.3	43.1
11	670 (0)	12 (0)	7 (0)	1.35	45.0	40.4
12	670 (0)	15.4 (1.68)	7 (0)	1.37	44.75	40.9
13	653 (–1.68)	12 (0)	7 (0)	1.35	45.9	42.9
14	680 (1)	10 (–1)	5 (–1)	1.36	45.4	42.1
15	660 (–1)	14 (1)	9 (1)	1.36	44.65	42.1
16	660 (–1)	10 (–1)	9 (1)	1.325	43.9	38.3
17	680 (1)	14 (1)	9 (1)	1.36	45.2	42.2

However, it can be seen in figure 1 from the standardized Pareto charts, which make it possible to estimate the statistical significance of the coefficients of the regression equations (coefficients are statistically significant when the vertical line on the chart crosses the column image of the corresponding coefficient), that it is reliably possible to speak only about average values of B_r, H_{cB} and $(BH)_{max}$, which are 1.35 T, 45.6 kA/m and 41 kJ/m^3, respectively.

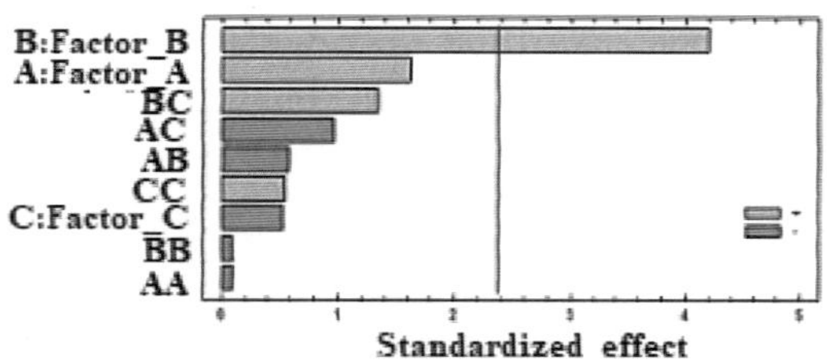

(a)

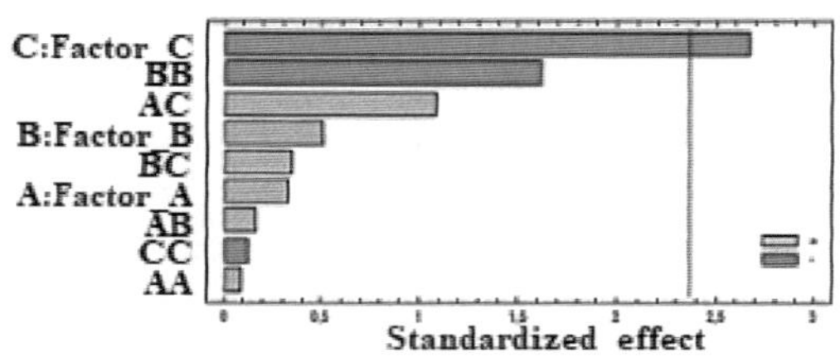

(b)

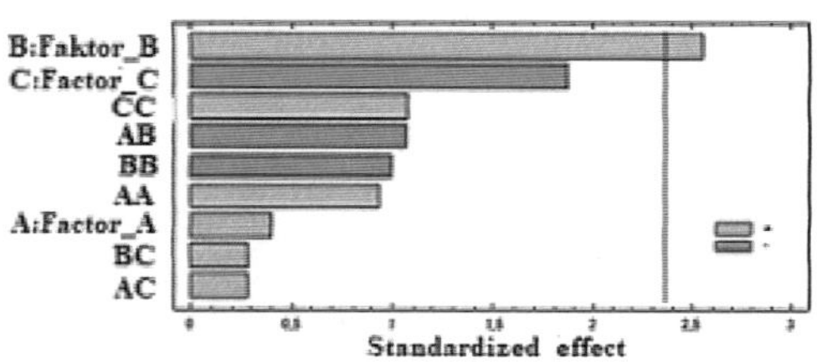

(c)

Figure 1. The standardized Pareto charts of alloy Fe-27Cr-10Co-0,5Mo for residual induction B_r (a), coercive force H_{cB} (b) and maximum energy product of $(BH)_{max}$ (c) in the anisotropic state.

According to the Statgraphics program, the following values can be obtained: $B_r = 1.38$ T, $H_{cB} = 46.8$ kA/m and $(BH)_{max} = 46$ kJ/m^3.

IOP Conf. Series: Materials Science and Engineering **347** (2018) 012053 doi:10.1088/1757-899X/347/1/012053

Upon study of the magnetic hysteresis properties of powder Fe–27Cr–10Co–2Mo alloy after tempering from 1100°C, a zero (initial) point was the regime of ITMP at 640°C (40 min) + cooling with $V1$ = 15 °C/h to 580°C + cooling with $V2$ = 7 °C/h to 500°C. The variation was ± 10°C for ITMP temperature (factor A), ± 10 min for time of ITMP (factor B), ± 2 °C/h for cooling rates $V1$ (factor C) and ± 2 °C/h for cooling rates $V2$ (factor D). The results of the carried out experiments and design matrix for anisotropic samples are given in table 2.

Table 2. Magnetic hysteresis properties of anisotropic samples of the Fe–27Cr–10Co–2Mo alloy.

Temperature of ITMP, °C	Time of ITMP, min	Cooling rate $V1$, °C/h	Cooling rate $V2$, °C/h	B_r, T	H_{cB}, kA/m	H_{cM}, kA/m	$(BH)_{max}$, kJ/m^3
1 670	1 50	1 17	1 9	1.29	39.9	40.0	34.2
–1 650	–1 30	–1 13	1 9	1.29	40.1	40.3	33.5
–2 640	0 40	0 15	0 7	1.31	41.0	41.1	32.9
–1 650	–1 30	–1 13	–1 5	1.30	42.1	42.3	36.3
2 680	0 40	0 15	0 7	1.22	41.6	41.7	31.2
–1 650	1 50	–1 13	–1 5	1.30	42.1	42.3	35.4
0 660	0 40	0 15	0 7	1.29	40.2	40.3	33.6
1 670	–1 30	–1 13	–1 5	1.27	42.9	43.1	34.0
–1 650	1 50	1 17	–1 5	1.31	41.1	41.9	34.1
–1 650	–1 30	1 17	1 9	1.30	38.9	39.1	34.0
–1 650	1 50	–1 13	1 9	1.285	39.9	40.2	32.6
0 660	0 40	0 15	0 7	1.30	41.5	41.8	35.1
0 660	0 40	0 15	0 7	1.30	41.9	42.1	35.6
–1 650	1 50	1 17	1 9	1,30	39.2	39.4	34.0
0 660	0 40	–2 11	0 7	1.26	42.6	42.8	32.7
1 670	1 50	–1 13	–1 5	1.29	42.1	42.2	36.2
0 660	0 40	0 15	2 11	1.30	39.6	39.8	33.8
1 670	–1 30	–1 13	1 9	1.265	40.4	40.5	32.3
0 660	2 60	0 15	0 7	1.34	41.6	42.0	35.8
–1 650	–1 30	1 17	–1 5	1.29	46.8	41.0	34.8
1 670	–1 30	1 17	–1 5	1.30	42.1	42.2	36.2
0 660	0 40	2 19	0 7	1.26	40.35	40.6	32.4
1 670	–1 30	1 17	1 9	1.30	42.0	42.2	36.9
0 660	0 40	0 15	–2 3	1.29	41.3	41.6	36.2
0 660	0 40	0 15	0 7	1.26	40.85	41.1	35.2
1 670	1 50	–1 13	1 9	1.27	40.9	41.0	32.8
1 670	1 50	1 17	–1 5	1.30	41.05	41.3	36.0

The statistical analysis of data in table 2 gives analytical dependences in the form of the regression equations for residual induction B_r, coercive force H_c and maximum energy product of $(BH)_{max}$ as a function of chosen variable factors.

The standardized Pareto charts, which make it possible to estimate the statistical significance of coefficients of the regression equations, are presented in figure 2 (coefficients are statistically significant when the vertical line on the chart crosses the column image of the corresponding coefficient).

As can be seen from presented charts, the residual induction B_r is described by the nonlinear equation:

$$B_r = 1.29 + 0.007BB - 0.01A - 0.007CC \qquad (1)$$

in which the free term gives the average value of Br in the anisotropic state in the chosen ranges of variation of the factors.

In the regression equation for coercive force H_{cB}, as can be seen from Pareto chart, the linear terms of F, D and C are statistically significant:

$$H_{cB} = 41 + 0.3A - 0.7D - 0.4C \qquad (2)$$

IOP Conf. Series: Materials Science and Engineering **347** (2018) 012053 doi:10.1088/1757-899X/347/1/012053

The Pareto chart for maximum energy product $(BH)_{max}$ shows that statistically significant coefficients are linear D and square AA; the regression equation for $(BH)_{max}$ has the following form:

$$(BH)_{max} = 34{,}8 - 0{,}7D - 0{,}55AA \quad (3)$$

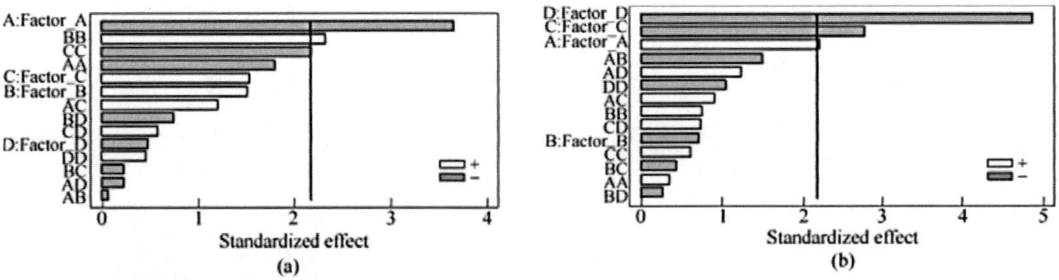

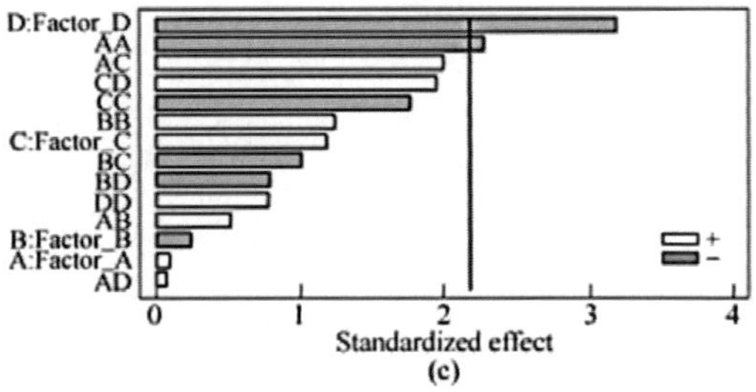

Figure 2. The standardized Pareto charts of alloy Fe-27Cr-10Co-2Mo for (*a*) residual induction B_r, (*b*) coercive force H_{cB} and (*c*) maximum energy product $(BH)_{max}$ in the anisotropic state.

The appearance of response surface in the phase space of the varied factors and especially their sections contains the useful and obvious information from the technological point of view. Figure 3 shows these response surfaces for B_r, H_{cB} and $(BH)_{max}$ with the corresponding sections for anisotropic state.

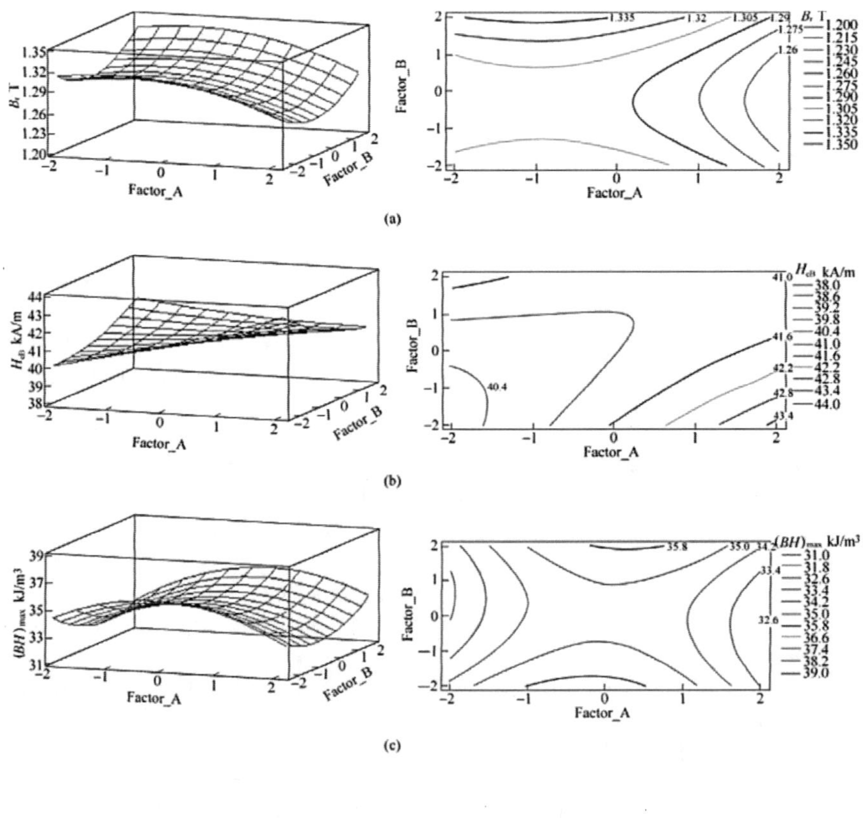

Figure 3. Estimated response surfaces and their sections for (*a*) residual induction B_r, (*b*) coercive force H_{cB} and (*c*) maximum energy product $(BH)_{max}$ (factor $C = 0$ and factor $D = 0$) for anisotropic state.

According to the data of table 2, for Fe–27Cr–10Co–2Mo alloy the maximum received value of coercive force $H_{cB} = 43$ kA/m (the average value of $H_{cB} = 41$ kA/m) that is confirmed by a type of the response surface for H_{cB} in figure 3.

Thus, the addition of 2 wt % molybdenum into the ternary Fe–27Cr–10Co alloy does not lead to an expected increase in the coercive force. However, we showed in [7] that the value of H_{cB} of a similar cast Fe–27Cr–10Co–0.4Si–0.7Ti–0.6Mo alloy reached 48–50 kA/m.

In this connection, it should be noted that if the optimum content of molybdenum in FeCrCo alloys containing more than 15–20 wt % cobalt is 2–3 wt %, in low-cobalt alloys (cobalt content of 10 wt % or less), it is apparently less than 1 wt %.

Lastly, note that molybdenum in hard magnetic FeCrCo alloys reduces the camber of the demagnetization curve in the second quadrant (see figure 4). The camber coefficient $\eta = (BH)_{max}/B_rH_{cB}$ in the ternary Fe–27Cr–10Co–0.5Mo alloy is 72% and only 68.6% in the Fe–27Cr–10Co–2Mo alloy.

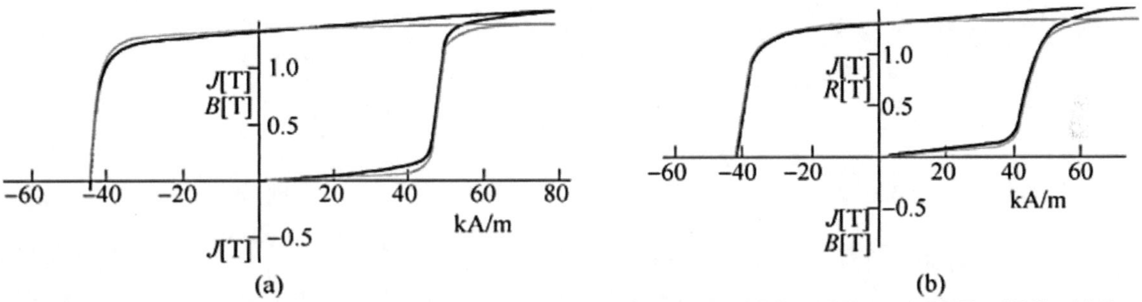

Figure 4. Curves of magnetization and demagnetization for (*a*) Fe–27Cr–10Co and (*b*) Fe–27Cr–10Co–2Mo alloys.

4. Conclusions

The addition of a small amount of molybdenum (0.5 wt. %) into the powder hard magnetic Fe–27Cr–10Co alloy leads to an increase in its magnetic hysteresis properties.

The additional alloying with molybdenum of up to 2 wt. % of the ternary alloy does not lead to an increase in the coercive force, there is a decrease in the residual induction B_r and maximum energy product $(BH)_{max}$.

The introduction of molybdenum into the low-cobalt hard magnetic FeCrCo alloys, just as in the alloys with moderate amount of cobalt and in the high-cobalt hard magnetic alloys, reduces the camber of the demagnetization curve in the second quadrant.

Acknowledgements

Authors would like to express their sincere gratitude to the leading researcher V. A. Zelenskii and research associate A. B. Ankudinov for the help in the receiving powder samples of the studied alloys.

Work performed on the state task No. 007-00129-18-00 with financial support of FTsNTP-2017 under the state contract of 26.09.2017 No. 14.579.21.0149.

References

[1] Jin S and Gayle N W 1980 *IEEE Trans. Magnetic.* **3** 526–529.
[2] Kaneko H and Inoue K 1974 Patent USA 3 806 336 Int. Cl. C22c 39/16 H01f 1/00. U.S. Cl. 75–122.

[3] Milyaev I M, Yusupov V S, Stelmashok S I and Milyaev A I 2016 *Innovatsionnye metallicheskie materialy* ed V M Kolokoltseva (Magnitogorsk, Nosov Magnitogorsk State Technical University) chapter 3 pp 59–75.
[4] Milyaev I M, Yusupov V S, Milyaev A I and Laysheva N V 2009 *Perspektivnye materialy* **6** 359–361.
[5] Milyaev I M, Milyaev A I and Yusupov V S 2008 *Perspektivnye materialy* 277–280.
[6] Milyaev I M, Milyaev A I and Yusupov V S 2009 *Russian Metallurgy (Metally)* **3** 250–252.
[7] Milyaev A I, Kovneristiy Yu K, Efimenko S P and Korznikova G F 2004 *Trudy Nizhegorodskogo gosudarstvennogo tekhnicheskogo universiteta Materialovedenie i metallurgiya* **42** 127–132.

IOP Conf. Series: Materials Science and Engineering **347** (2018) 012054 doi:10.1088/1757-899X/347/1/012054

Effect of cryogenic treatment on the fracture toughness of aircraft aluminum alloy 7075

V A Ermishkin and Y B Soloveva

Baikov Institute of Metallurgy & Materials Science (IMET RAS), Moscow, Leninskiy prospect 49, Russia

Email: jylsol1403@yandex.ru, eva@imet.ac.ru

Abstract. Influence of three types of the treatment on fracture toughness of the Al-7075 alloy was investigated in this study. Commercial Al-7075 alloy in the solid solution heat-treated condition was processed by hardening with post-cryogenic deformation treatment and PVD deposition titanium and copper coatings. The fracture toughness was estimated with using macroscopic and microscopic approaches. The conditions for the coincidence of the fracture toughness estimates between brittle fracture mechanics and the photometric analysis of structural images (PHASI) methods were achieved. The highest fracture toughness was obtained by applying hardening, cryogenic compression, ageing and deposition of the Ti-coating, leading to dispersion particles precipitation.

1. Introduction

The development of strengthening treatments for high-strength aluminum alloys is a big challenge of the Aircraft Industry. Modern strengthening treatments of the high-strength alloys include combination of the several strengthening mechanisms at the same time: a) alloying by chemical elements which are able to provide solid-solution and dispersion hardening effects; b) heat treatment and deformation in a wide range of temperatures [1-8].

There is limitation for the high-strength materials associated with increasing possibility of the brittle fracture in the result of complex multi-stage strengthening treatments. Among many mechanical properties of the high-strength aluminum alloys the fracture toughness is one of the most important due to exceptional structurally sensitiveness. The fracture toughness is able to be a criterion limiting high-strength materials application [9].

The fracture toughness is determined by the methods of fracture mechanics as special fracture criteria for three modes of destruction: the opening mode (Mode I), the in plane shear mode (Mode II) and out-of-plane shear mode (Mode III) [10,11]. These criteria are called the stress intensity factors (SIF) and are denoted as K_{Ic}, K_{IIc} and K_{IIIc}. (The SIF depends significantly on the sample thickness). In case of plane strain minimum SIF is typical for large sample thicknesses. For plane stress minimum SIF is typical for small sample thicknesses. In real constructions the high-strength materials are used to produce thin-walled shells, therefore, the KIc is of more interest for the evaluation of fracture toughness.

It should be noted that fracture mechanics determine the limiting conditions for the equilibrium of bodies with macroscopic cracks. The conditions of microcracks nucleation and mechanisms of their growth up to macroscopic dimensions are not possible for fracture mechanics methods. However, understanding the structural mechanisms of cracks nucleation and growing is a necessary requirement to solve the problems of materials strengthening. Thus, the study of elementary mechanisms of destruction at the micro-level opens new possibilities for correct estimation of SIF.

2. Experimental procedure

Experiments were performed on plates (6-mm thick, 20 × 20 mm) of cold rolled 7075 aluminum alloy, which was produced using standard industrial practices. The chemical composition of this alloy is shown in Table 1.

Table 1 Chemical composition of 7075 aluminum alloy (wt. %)

Alloy	Al	Si	Fe	Cu	Mn	Mg	Cr	Zn	Ti	Other Each	Others Total
7075	Rem	0.40	0.50	2.0	0.30	2.9	0.28	6.1	0.20	0.05	0.15

Every sample had electric spark notch with 4.0 mm length and 0.4 mm width cut along the vertical axis of sample symmetry.

Specimens were treated by using three types of technological operations, which are shown in Table 2.

Table 2 Technological operations of the Al-7075 samples

№	Treatment	Applied operations	Treatment parameters			
			D (mm)	Δh (μm)	Th / t, (°C/min)	Ta / t, (°C/min)
1	Original state	Cold rolled condition	-	-	-	-
2	Cryogenic strengthening + ageing	Hardening + cryogenic compression + ageing	-	-	-	45/30
3	Cryogenic strengthening + Cu-coating	Hardening + cryogenic compression + ageing+ Cu-coating	11	0,18	450/30	45/30
4	Cryogenic strengthening + Ti-coating	Hardening + cryogenic compression + ageing+ Ti-coating	11	0,18	450/30	45/30

D – strengthening zone diameter; Δh – depth of the pressure in strengthening zone; T_h- hardening temperature; T_a - ageing temperature; t- operation duration.

Hardening and ageing processes were performed in vacuum atmosphere with using quartz ampoules. Cryogenic compression was conducted in liquid nitrogen fluid in T= - 196 °C. Titanium and copper coatings were deposited on the one surface of the samples in the vacuum universal system VUP-5 (USSR). Electric spark notches were cut after completion of treatment operations for each sample.

Specimens were loaded by inserting a wedge into the notch according to Hillman technique [12] with 1 mm/ min rate on an Instron-3382 (USA) test machine. The loading scheme of the samples is shown in Figure 1

IOP Conf. Series: Materials Science and Engineering **347** (2018) 012054 doi:10.1088/1757-899X/347/1/012054

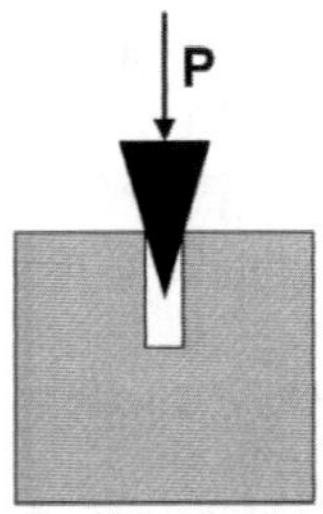

Figure 1. Hillman loading technique

Figure 2 shows the loading curve of the ordinary state sample in the coordinates load (P) – time (t).

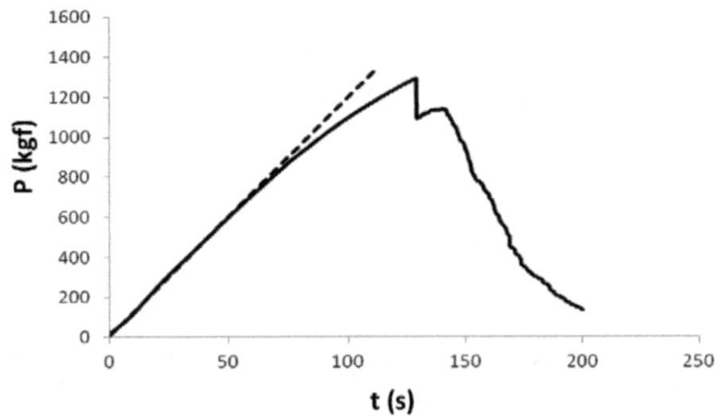

Figure 2. Load – time curve of ordinary state sample Al-7075

During loading samples surfaces images were recorded using video camera Panasonic HDC-HS60. The images were selected with 10s, 5s and 2s time steps and then analyzed using photometric analysis of structural images (PHASI) [13], which had been developed by high-voltage microscopy laboratory team at Institute of Metallurgy and Material Science.

PHASI is a method based on the comparative analysis of the digital images using spectrums brightness of visible light reflection from the surface before loading and in the process of loading. PHASI is a program-analytical platform that allows to dye the chosen parts of the spectrum in different colors, to move these colors to the digital image and determine spectrum density of any part of the image. Each spectrum parameter of images matches with physical parameter of changing, which were observed during experiment. On the basis of this data, calibration graph was constructed, where spectrum parameters transform to 1. local physical characteristics of materials, 2. level of degradation material as result of physical influence, including kinetics of this process.

Coordinates of PHASI comparative analysis: p (I) - spectrums brightness density of the reflection - intensity of reflection (I). Spectrums brightness density of the reflection p (I) was calculated using the formula:

$$p(I) = \frac{n\,(I_i)}{N} \tag{1}$$

n(Ii) – images pixels with intensity I_i; *N*- overall amount of the pixels.

Intensity of reflection (I) was measured in the conventional units of the linear scale, where 0 – surface condition of the total visible light absorption; 1 – surface condition of the total visible light reflection.

Figure 3 shows comparison of the sample surface structures and spectra of the visible light reflection before start loading (a) and in 10s loading (b).

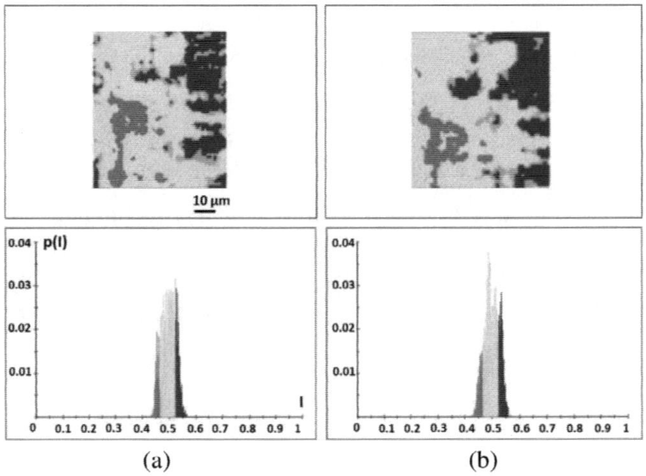

Figure 3. Sample surface structures and spectra of the visible light reflection before start loading (a) and in 10s loading (b), obtained by PHASI method

The PHASI allows to estimate the degree of the material structural modification (Ds) and describe it as:

$$D_s = \frac{S_i(t) - S_i(0)}{S_{max} - S_{min}} \quad (1')$$

$S_i(t)$ - spectral curve area of the i fragment captured at moment t; $S_i(0)$ - spectral curve area of the i fragment captured at start of loading; S_{max} and S_{min} - maximum and minimum spectral curve areas at the time interval [0, t].

Using this criterion, the value of the load corresponding to the beginning of the crack propagation was determined. After that the stress intensity factor (SIF) was determined. For the SIF definition, the standard methods of the fracture mechanic and PHASI method were applied. According to fracture mechanic method SIF was determined using finite-element method [14] as:

$$K_{Ic} = \frac{P}{tW^{1/2}} \times f(a) \quad (2)$$

P – load of the crack initiation moment; t – sample thickness; W – geometry parameter; $f(\alpha)$ – coefficient, defined as:

$$f(\alpha) = (2+\alpha)\cdot(0,8072+8,858\alpha-30,23\alpha^2+41,088\alpha^3-24,15\alpha^4+4,951\alpha^5)\cdot(1-\alpha)^{-3/2} \quad (2')$$

$\alpha = a/W$, a - The length of a crack from tip of a crack to the loading axis.

PHASI allows to estimate energy of the light reflection from the sample surface (E_s) in conventional units. This energy is numerically equal to the spectral curve area. According to [15], the body emission energy is related to its internal energy (U) as:

$$E_s = AU \quad (3)$$

A – coefficient describing the possibility of the spontaneous body light emission;

Coefficient A was determined from the equation (3) by plugging into it elastic energy value for the set moment of time and the elastic energy value in the conventional units defined by PHASI method.

This energy balance equation can be described as:

$$\Delta U = U_\sigma - U_o = \frac{\sigma^2}{2E} \quad (4)$$

Where U_σ - internal energy for the sample under the load; U_0 - internal energy for the sample before the load; E - Young's modulus (modulus of elasticity). σ – stress

Thus, strain can be calculated as:

$$\sigma = \sqrt{2\text{A}E\Delta U} \quad (5)$$

To estimate strain in the tip of the crack the notch area of the sample was divided into three fragments bordered to each other (figure 4)

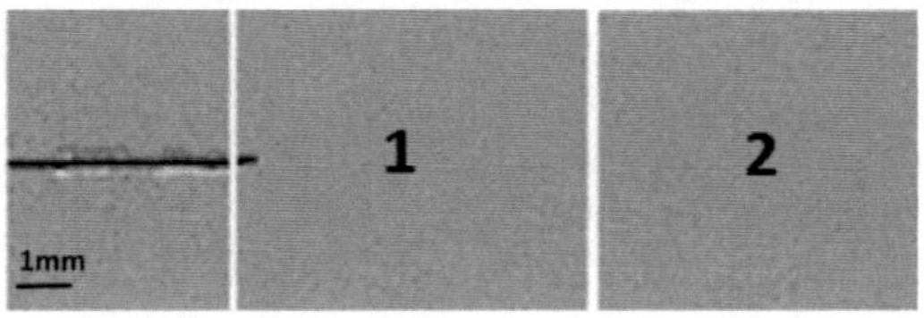

Figure 4. Sample fragments with notch for PHASI analysis: 1 - the microfragment closest to the notch, 2 - the fragment at the edge of the sample, without a notch.

The stress in the center of the fragments was determined from the equation (5). Fracture toughness values (K_{Ic}) were calculated from [11]:

$$K_I = \sigma_y\sqrt{2\pi r} \quad (6)$$

Where r - distance between tip of the crack and center of the fragment

3. Results and discussion

Figure 5 shows Internal energy – time (t) curve of the 7075 alloy after technological operation №2 (hardening + cryogenic compression + ageing).

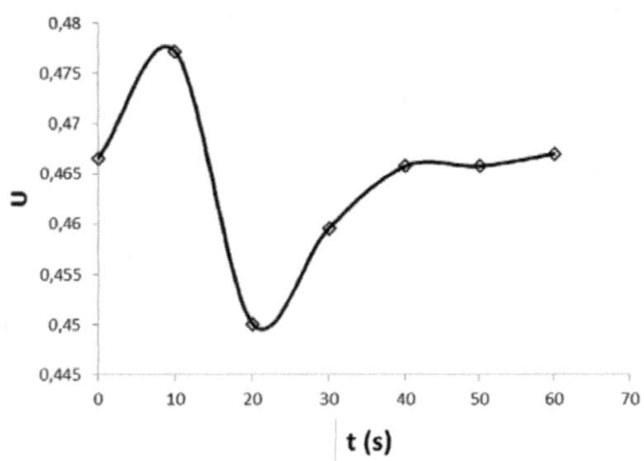

Figure 5. Internal energy – time curve of the 7075 alloy after technological operation №2

Figure 6 shows material structural modification (Ds) – time (t) curve of the 7075 alloy after technological operation №2 (hardening + cryogenic compression + ageing).

IOP Conf. Series: Materials Science and Engineering **347** (2018) 012054 doi:10.1088/1757-899X/347/1/012054

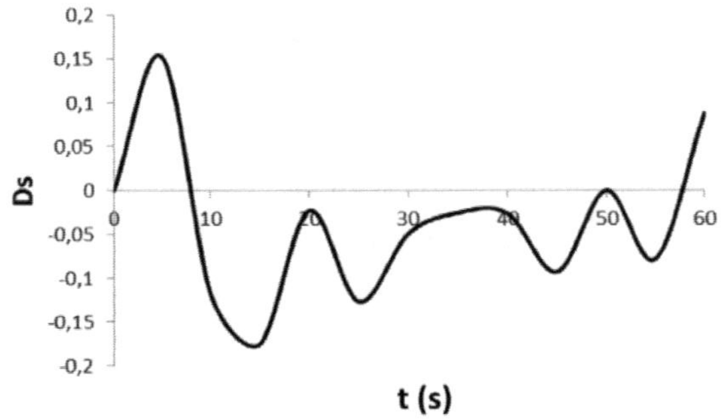

Figure 6 Material structural modifying– time curve of the 7075 alloy after technological operation №2

PHASI of the sample surface showed that crack grows by consistent change of the three main stages: the first - stage of the plastic zones forming in the fronts, the second - the stage of the notch opening, the third – the stage of the crack propagation. Figure 7 shows material structural modifying (Ds) – distance from crack tip (r) curve of the 7075 alloy after technological operation 2. The maximum Ds value is fixed at some distance from the crack notch r = 1.65 mm. It means that microstructure damage happens when three modes of destruction are in action (Mode I+ Mode II + Mode III) more precisely, under plane strain conditions [12, 16].

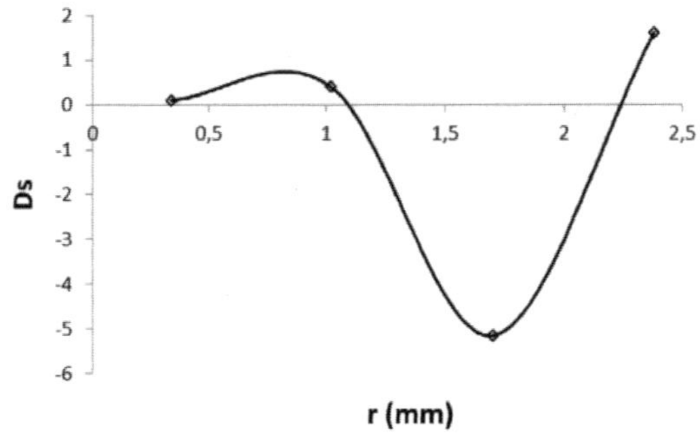

Figure 7 Material structural modifying (Ds) – distance from crack tip (r) curve of the 7075 alloy after technological operation 2

Figure 8 shows the stress intensity factor (K_{Ic}) – time (t) curves. The K_{Ic} criteria are calculated by using fracture mechanic method (equation 2) and PHASI method (equation 6). The calculated dependencies shown in Figure 8 are approximated by analytical formulas, which are given in Table 3.

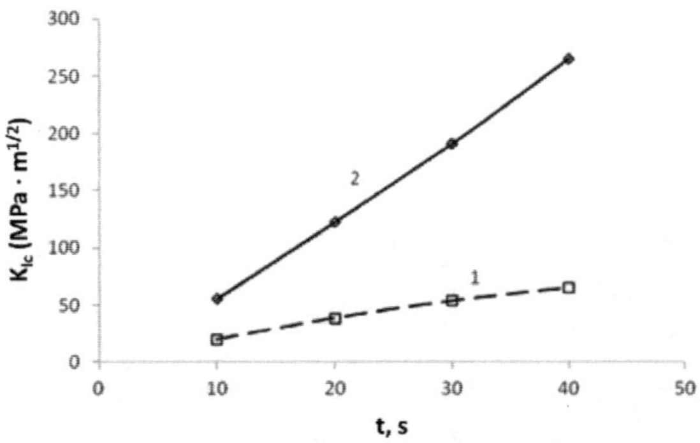

Figure 8. Stress intensity factor (K_{Ic}) – time (t) curves of the 7075 alloy after technological operation 2, r=1,165 mm. 1 – K_{Ic} were calculated using fracture mechanic method; 2 – K_{Ic} were calculated using PHASI method.

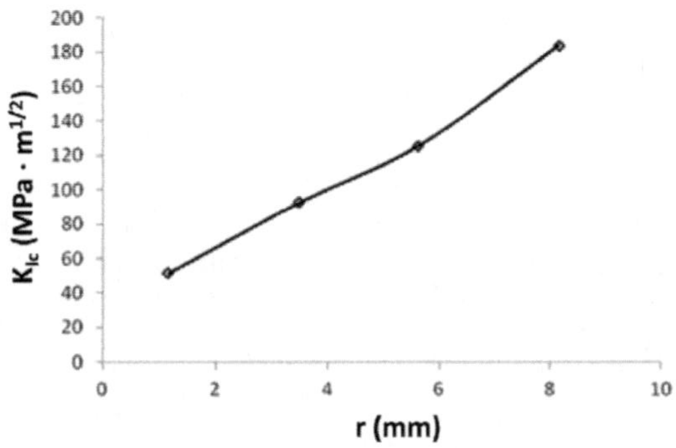

Figure 9. Stress intensity factor (K_{Ic}) – distance from crack tip (r) curves of the 7075 alloy after technological operation 2, t=40s.

Table 3 Analytical formulas approximating calculated dependencies $K_{Ic} = f(t)$

Equation for K_{Ic}	Approximating equation	Significance test
(2)	$K_{Ic} = 2{,}7277 \cdot t^{0{,}8702}$	0,9943
(6)	$K_{Ic} = 6{,}9875 \cdot t^{-16{,}063}$	0,9995

In this research [17] the condition of equality of the K_{Ic} obtained by using fracture mechanic and PHASI methods was determined. These values for alloy 7075 are the following: K_{Ic} = 49.1MPa for fracture mechanic method and K_{Ic} = 51.3 MPa for PHASI method. The experiment showed that K_{Ic} = 49.1MPa is reached within 20 s of the load. the same value of the fracture toughness determined using PHASI method was reached within 9s of the load. Both dependences K_{Ic} – t are practically linear dependences on time. The results of experiments and K_{Ic} calculating for all technological operations are presented in table 4.

According to the data in Table 4 the highest fracture toughness was obtained by applying technological operation №4, which includes hardening, cryogenic compression further ageing and deposition of the Ti-coating, which lead to dispersion particles precipitation.

Table 4 Results of the calculating K_{Ic} for all technological operations

№	Treatment	t (s)	P (kg)	Δl (mm)	Klc (MPa · $m^{1/2}$)	
					fracture mecanic	PHASI
1	Original state	13,3	165,6	0,67	14,7	40,4
2	Cryogenic strengthening + ageing	29,2	207,84	1,77	18,5	115,8
3	Cryogenic strengthening + Cu-coating	27,2	208,32	1,22	18,5	108,3
4	Cryogenic strengthening + Ti-coating	50	357,22	0,23	32,8	143,1

4. Conclusions

Investigation of the influence of the three types of treatment on the fracture toughness of the aluminum alloy 7075 process has been performed in this study. The results show that most perspective treatment for fracture toughness increasing includes hardening, cryogenic compression further ageing and deposition of the Ti-coating. This technology allows to increase fracture toughness by 2-4 fold (depending on the evaluation method) compared to the original state of 7075 alloy. The developed microscopic PHASI method is able to estimate fracture toughness using compact samples and to reach appropriate correlation with macroscopic approach of the fracture mechanic method. PHASI is a promising technique with potential for industrialization and therefore needs to be further investigated in any aspect.

References

[1] Rashed H, Bazlur Rashid A 2017 *J. Materials Science and Material Engineering* **2** 337
[2] Kannan C, Ramanujam R 2017 *Journal of Advanced Research* **8** 309
[3] Shaeri M 2015 *Progress in Natural Science: Materials International* **25** 159
[4] Xuetal X 2015 *Materials Science & Engineering* **A648** 367
[5] Kurmanaeva L, Topping T 2015 *Journal of Alloys and Compounds* **632** 591
[6] Rao P, Singh D, Jayaganthan R 2014 *J Material Design* **56** 97
[7] Abd El-Aty, Xu Guo Y A 2017 *Journal of Advanced Research* in Press.
[8] Anderson T L 1995 *Fracture mechanics: Fundamentals and applications* (New York: CRC Press LLC) p 688
[9] Oskouei R, Ibrahim R 2011 *J Procedia Engineering* **10** 1936
[10] Knott J F1973 *Fundamentals of fracture mechanics. Department of Metallurgy and Materials Science, Cambridge University* (London: Butterworths) p 289
[11] Cherepanov G P 1974 *Mechanics of brittle fracture* (Moscow: Nauka) p 640
[12] Gilman J J 1959 *Fracture* (New York: Wiley) p 708
[13] Ermishkin V A, Murat D P 2007*J Information Technologies* **11** 65
[14] Murakami Y 1990 *Stress intensity factors guide* (Moscow: Mir) p 448
[15] Astapenko V A 2010 *Radiation interaction with atoms and nanoparticles* (Dolgoprudniy: Intellect) p 492
[16] Novikov I I, Ermishkin VA 1991 *Metal damage micromechanisms* (Moscow: Nauka) p 366
[17] Ermishkin V, Soloveva Y 2015 Proc. *Int. Conf. Mechanical engineering and technosphere of the XXI century* (Donetsk: MSM press) p 120

IOP Conf. Series: Materials Science and Engineering **347** (2018) 012055 doi:10.1088/1757-899X/347/1/012055

Methods of obtaining and the prospect of using powders based on aluminum magnesium spinel

D O Lemeshev[1], M O Senina[2] and M S Pedchenko[3]

Dmitry Mendeleev University of Chemical Technology of Russia, Miusskaya sq. 9, 125047 Moscow, Russia

E-mail: [1]diolem@muctr.ru, [2]snnmarina@rambler.ru, [3]pesenkatalor@gmail.com,

Abstract. The prospects of use optically transparent ceramics from aluminum magnesium spinel are considered. Technologies of synthesis of spinel powders f for the subsequent application when receiving transparent ceramics are described.

Development of materials with the increased physical and mechanical properties is one of the perspective directions of the modern equipment. Thanks to a number of valuable properties, such as high hardness, durability, thermal stability and high light transmittance, transparent ceramic materials are of special interest optically. Such materials find broad application as arc lamps of high pressure, details of optical systems of navigation, fairings of rockets, etc. The problem of development of new transparent armored materials is relevant.

Effective protection against high-energy weapons of destruction with high penetration – armor-piercing rifle bullets with the heat-strengthened cores – is inconceivable without its use as an element of protective structure of ceramics – highly rigid, but fragile material [1]. And in the field of an individual armored protection and protection of the easy equipment against bullets of small arms, the use of ceramics for the last 20 years has sharply increased in connection with distribution of bullets with highly rigid steel and hard-alloy cores [2].

Transparent antipulse and splinterproof armor is used for production the bullet-proof of windows of cars, planes and helicopters, banks and offices. In the means of an individual armored protection (MIAP) the transparent ceramics is applied to production of visors for helmets and observation ports of armored shields.

The protecting ability of a barrier in relation to action of the shock getting weapons of destruction usually is characterized by its thickness or area density, which provides the standard nature of environmental damage [1].

The mechanism of interaction of bullets and ceramic plate comes down to the fact that for some time the ceramics at the expense of the high hardness does not allow the drummer to get into itself. At the same time the drummer is forced to be deformed or collapsed for barrier surfaces, spending own kinetic energy. It is accepted to call this time penetration delay time. After this time the rest of the core of a bullet gets into a ceramic crumb into which the ceramics in a point of influence manages to turn [2].

Armor, as a rule, has as a part of several layers.

IOP Conf. Series: Materials Science and Engineering **347** (2018) 012055 doi:10.1088/1757-899X/347/1/012055

The ceramic layer of multilayered structures is produced from separate ceramic elements with a 50 - 100 mm party size, 10 - 15 mm thick, which paste to an organoplastics or metal substrate. To prevent transfer of cracks to the next elements between them leave a small gap (about 0.5 mm) which is filled with elastomer [1].

Thus, the role of a ceramic layer of a multilayered barrier comes down to destruction of a head part of a bullet, increase in the area of impact on the subsequent layers of a barrier and absorption of a part of kinetic energy of the remains of a bullet in the course of its braking in already destroyed ceramics [1].

Ceramics from aluminum magnesium spinel $MgAl_2O_4$ (AMSh) is held a specific place among optically transparent ceramic materials. In comparison with other ceramic transparent armored material (ALON, polycrystalline Al_2O_3) spinel has advantage in the form of bigger availability of input products and smaller temperatures of synthesis.

The optical ceramics from AMSh represents material with an exclusive mechanical durability, wear resistance on attrition, erosion resistance, resistance to unilateral aerodynamic blow, chemical inertness. The material is mechanically stable up to the temperature of 1250 °C (to 1500 °C at short-term influence) [3], is characterized by transparency in the wide spectral range (from ultra-violet (UF) to infrared (IK) area of range).

As a rule, the ceramics from aluminum magnesium spinel is received in three stages: spinel powder synthesis, formation of semi-finished products, sintering in hydrogen or vacuum.

The phenomenon of inheritance is observed by the subsequent phase of the structure of the previous one in ceramics.

Properties optically of transparent ceramics strongly depend on structure, structure and dispersion of powdery precursors therefore when receiving optically transparent ceramic materials much attention should be paid to synthesis of initial powders.

Many researchers meet that for receiving dense transparent ceramics it is necessary to use the nanodimensional crystal powders consisting of weak agglomerates or separate small monodisperse particles with the form close to spherical. Also special attention need to be paid to chemical and phase purity [5].

At production of transparent oxide materials rather low-temperature ways of preparation of initial compositions first of all such as cryochemistry, coprecipifation, heterogeneous methods which provide high degree of uniformity of distribution of components [6] are the most widespread.

Cryochemical method. Four operations are the cornerstone of this method: preparation of the mixed solution of salts, freezing, sublimation drying and heat treatment of powdery products. Physical and chemical bases of a cryochemical method of receiving high-homogeneous powdery materials are based on charts of a state «water-salt» [6]. The cryochemical method allows to receive 0.2 – 0.4 microns united in strong clump [7]. Need of the special equipment, labor input of technological process and also the possible aggregation of particles which is negatively reflected in a ceramics light transmittance belong to the main lack of this method.

Hydrolysis of salts. Hydrolysis of the isopropoxide of $MgAl_2$ (OPri) 8 magnesium-aluminum is most often used for receiving transparent ceramics. The similar method is considered by authors of works [8, 9, 10] and consists in dissolution of magnesium and aluminum in isopropyl alcohol in the presence of activators (chlorides of tin and ammonium). Received thus isopropoxide is exposed to vacuum distillation for additional cleaning. The mix of hydroxyls of aluminum and magnesium formed as a result of hydrolysis is calcinated before formation of complex oxide. In a research [11] it is shown what applications of this way has allowed to reduce spinel formation temperature: the formation of a phase of spinel will read at 360 °C, and full crystallization happens at 900 °C. The samples made by this method do not differ in the high level of light transmittance (figure 1) [10]. The high cost of initial materials and complexity of execution belongs to the main shortcomings of this way.

Thermal synthesis from initial salts. The method is rather simple performed by and consists in calcinating of mix of the initial components taken in a certain ratio. The similar way is considered by authors of work [12]. Use of such way leads to emergence of defects in material and to receiving spinel with various degree of a frontage that can adversely affect transparency of ceramics.

Spray drying. Raspylitelny drying. The method consists in dispergating of initial solution in a heat carrier gas stream. As a rule, as the heat carrier air is used. Speed of evaporation of solvent and extent of supersaturation of solution depend on temperature of the heat carrier and speed of his giving. The main shortcomings of a method are considerable losses of material and the violation of its chemical homogeneity connected with redistribution of components in particles of a synthesizable powdery sample during process [7].

Sol-gel process. The principle sol-gel of process consists in transition of liquid solutions of alkoxide, esters or halogenides and also organic compounds to gel at hydrolysis and polycondensation which then in the course of heating at rather low temperatures is powdered [13]. The sol-gel technology has a number of advantages such as: a possibility of receiving dense amorphous bodies at lower temperatures; a possibility of receiving materials with adjustable distribution of particles by the sizes; possibility of obtaining high dispersion (0.1-0.05 microns) and purity; achievement of high uniformity of material [14]. This technology was used by authors of works [14] and [15]. Synthesis of spinel and particles up to 1 micron size happened to a faultless crystal lattice at 950-1000 °C, however, the ceramics on the basis of such powders had no high values of mechanical durability [15].

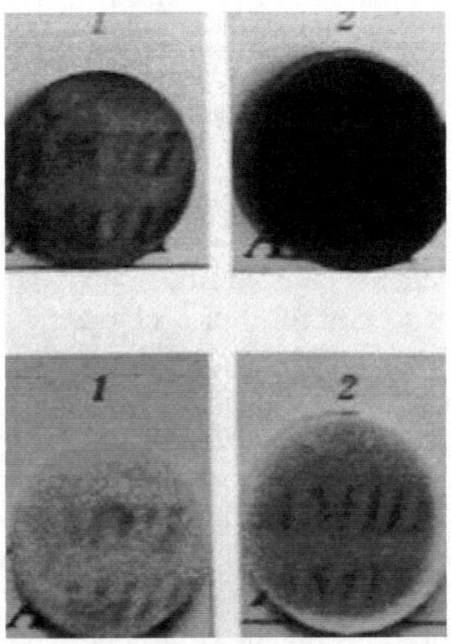

Figure 1. Ceramic samples before (upper) and after (lower) annealing, synthesized by hydrolysis of isopropoxides in the slow (1) and rapid heating mode (2) [10].

The Pechini method is one of modifications of a sol-gel way of receiving powders. The method is tolerant to water presence, does not demand the special atmosphere and careful control of the course of reaction. At the same time gel with ideally homogeneous distribution of atoms of metals turns out. Initial reagents at synthesis of complex oxides are served by water solutions of nitrates of metals which mix in a stoichiometric ratio. Organic compounds (lemon, wine, amino acetic, ethylene diaminetetra-acetic acids) are added to the received solution. Initially in the Pechini method as a complexing compound citric acid was applied. Synthesis by Pechini method can take place in two directions. Citric acid (saturated water solution) can be added to solution of nitrates and after to evaporate. There is an ignition of mix under heat treatment and amorphous carbon-bearing primary particles of compound oxides are formed. After cooling mix is frayed and annealed at more high temperature. In other case after formation of metal complexes polyatomic alcohol (ethylene glycol) into system is added. Solution is evaporate before formation of polymeric gel, which is subjected to heat treatment. In the course of synthesis ions of metals and carbonic acid form chelate complex. These complexes have free hydroxyl groups at the expense of which there is a polyetherification of

IOP Conf. Series: Materials Science and Engineering **347** (2018) 012055 doi:10.1088/1757-899X/347/1/012055

chelates to polyatomic alcohol. At the same time, ions of various metals will be evenly distributed in polymeric gel. At polymer decomposition formation of primary particles of plain oxides and their reaction with the subsequent formation of a single-phase high-disperse difficult oxidic product at the same time proceeds [16].

The main problem from the point of view of transparent ceramics production is that the particles as a result form strongly agglomerated powders. It is quite problematic to make a high density pore-free product of such powders. Besides, owing to a high non-equilibrium of process it is difficult to receive powders with strictly set parameters of an elementary cell and geometry of particles.

Heterophase methods. The method consists in receiving the low-hydrated slightly soluble connection by interaction of solid soluble salt of the main component with solution of the basis or acid. The amorphous hydroxides or oxalates received as a result of chemical interaction have high ability to absorption of ions from solutions in the "firm phase solution" system. Features of this method are that at transfer of solid salt of any element to slightly soluble connection an inheritance of structure of initial salt happens [6].

Coprecipitation method. Coprecipitation of the components which are a part of ceramics is reached in the form of slightly soluble hydroxides, carbonates or oxalates. Use of the concentrated solutions of salts and solution of the precipitator with various temperatures provides high dispersion of particles and also possibilities of regulation of the size of particles in very wide limits. Application of the return order of a coprecipitation provides course of all process at considerable excess of solution of the precipitator that allows to carry out simultaneous crystallization of components with formation of solid solutions of connections [6]. Often as the precipitator ammonia solution is used (an intermediate product in case of synthesis of oxides) the smaller size as the precipitator apply the bases which are at the same time well adsorbed on the surface of the stabilizer to receiving particles of a deposit. For example, applying tetraalkylammonia hydroxide as the precipitator, it is possible to receive particles of hydroxides of metals about 10-nanometer size [16].

Regulation of pH value allows to control phase composition of the received powders also. At coprecipitation of salts of magnesium and aluminum difficulties because of an essential difference in pH precipitation of hydroxides of magnesium (pH = 12) and aluminum (pH = 7,5) can arise water solution of ammonia, the probability of violation of the set stoichiometry of synthesizable connection is high [17]. In a research [12] it is shown that increase pH from 9 to 11 at coprecipitation with the subsequent heat treatment already at 400 °C is formed a phase of spinel and completely phases of hydroxides disappear. Also, according to data of work [18] sedimentation begins at pH = 4 and comes to an end at pH = 9, and at pH = 9 the stoichiometric ratio of cations in draft is reached [19].

Precipitation from water solutions is the most productive and simple method which does not demand the increased temperatures and pressure, organic solvents. Also advantage is rather short time of reaction [16].

The method of coprecipitation is most often applied at synthesis of powders of aluminum magnesium spinel. As initial components various salts of aluminum and magnesium are applied to synthesis of a precursor of AMSh. Are most often used $Mg(NO_3)_2{\cdot}6H_2O$, $Al(NO_3)_3{\cdot}9H_2O$ [20, 21, 22, 23, 24], $Mg(CO_3)_3(OH)_2{\cdot}4H_2O$, $Al(OH)_3$ [25], $NH_4AlSO_4{\cdot}12H_2O$, $MgSO_4{\cdot}7H_2O$ [12].

Important technological parameter for receiving spinel is synthesis temperature.

With temperature increase above 1000 °C the compound of spinels are significantly changed. There is a so-called γ- nonstoichiometry: aluminum magnesium spinel is beneficiates with quite considerable surplus of Al_2O_3. Any violation of a stoichiometry is followed by emergence of a disorder in a crystal – emergence of vacancies in knots of a lattice or settling of interstices [26] that can have an adverse effect on transparency and strength characteristics of material. It is established that process of a spinel formation depends on many factors: dispersions of initial components, their nature and a type of impurity, specially entered additives, annealing conditions, etc. Formation of spinel begins at 397-497 °C. However, the formed spinel is at the same time in extremely high-disperse, almost amorphous state. Good crystallization of spinel is observed at a temperature of synthesis of 697-897 °C [27].

For the purpose of decrease in temperature of agglomeration and also improvement of physical and mechanical characteristics of transparent ceramics from spinel use such method as introduction of the cementing additives LiF, B_2O_3, Y_2O_3.act as such connections.

For the purpose of increase in transparency of ceramics lithium fluoride into powders is added. Additive of LiF in quantities 0,5 – 3,0 % by weight promotes receiving ceramics with the porosity close to zero. At the same time presence of LiF leads to the intensive growth of crystals of spinel in the course of agglomeration that causes moderate sizes of microhardness and durability on a bend and negatively affects a light transmittance [8].

The similar effect is caused by spinel powder processing by boric acid (figure 2) [28].

Oxides of rare-earth elements often apply for receiving ceramic materials with the set properties as activators of process of synthesis of the main matrix. Along with B_2O_3 oxide of yttrium is considered to be one of the best mineralizers of a spinel formation. In a research [29] it is established that yttrium oxide introduction in number of 2 and 4% exerts strong impact on intensity of reaction of a spinel formation by method of solid phase synthesis in the air atmosphere is optimum for a combination of processes of synthesis and sintering. A spinel yield in this case at 1300 °C makes 100 and 99,7% respectively.

Perspective is an introduction of the additives promoting hardening of a matrix of spinel because of increased requirements to mechanical properties of armored ceramics [30].

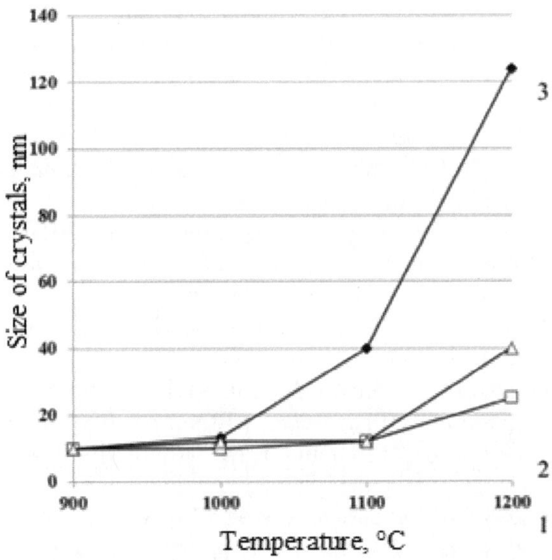

Figure 2. The effect of temperature on the crystal size for various powders: 1 - spinel powder, 2 - powder with the addition of lithium fluoride (1% by weight), 3 - powder with the addition of boric acid [28].

Using boron as oxide modifier is effective. Experimental results have shown that presence of B_2O_3 not only reduces precursor powder sintering temperature, but also suppresses ceramics grain growth that provides high optical transparency. Optimum concentration of B_2O_3 at which the high coefficient of transmission (80% is reached at a thickness of 1 mm) makes 0.15 % by weight [31]. It should be noted that the way of introduction of B_2O_3 to powder exerts strong impact on the ceramics grain size. So, during the use of solution of boric acid and removal of powder from solution before drying the intensive growth of grain [28, 32] while drying and sintering of the powder which is not separated from solution allows to receive fine-crystalline structure [31] is observed.

The positive effect gives introduction to a nanocomposite spinel matrix of nanoparticles of Si_3N_4. The received samples have high value of a light transmittance (>70%), the value of impact strength has increased up to 85% [33].

At sintering of transparent ceramics such methods as the hot pressing (HP), isostatic pressing in the hydro / gasostats – cold (HIP) are often applied (GIP), the spark plasma sintering (SPS) [34] more hotly.

These methods provide good caking of material and allow to receive ceramics with high value of a light transmittance and mechanical properties. However, difficulty is represented by the complex equipment and high cost of processes. The method of vacuum sintering is the easiest way when receiving transparent ceramics and also allows to receive samples with high operational properties [22, 33].

The ceramics on the basis of aluminum magnesium spinel meets requirements imposed to modern means of an individual armored protection and is perspective material for production of transparent armor.

Now when receiving ceramics from spinel the special attention is given to synthesis of initial powder. There is rather large number of the methods allowing to synthesize fine-crystalline powders with the set form and the size of particles. The universal method of receiving spinel does not exist. At the choice of a route of synthesis it is necessary to consider all technological chain in general and especially the mode of the subsequent heat treatment. However, preference to low-temperature heterophase methods is given.

References

[1] Kobylkin I F 2014 *Materials and structures of light armor protection* MSTU them. NE Bauman
[2] Saybotalov G M 2014 *Use of Armored Ceramics for protection aggregates of missile systems* **1** p. 333-35
[3] Sharypin V V 2016 *Optical Journal* **83** p 60-65
[4] Lukin E S Popova N A V S Glazachev L T Pavlyukova N A Kulikov 2015 *Constructions from composite materials* **3** p 24-36
[5] Golieva E V M D Mikhailov A A Dunayev B A Ignatenkov 2016 *Optical Journal* **83** p 67-72
[6] Lukin E S 1982 *Theoretical foundations of production and technology of optically transparent ceramics* MHTI them. DI. Mendeleev University
[7] Geraskin A A 2014 *Dis. Ph.D.* IONH them. N.S. Kurnakov Institute of Russian Academy of Sciences
[8] Alekseev M K Kulikova G I Rusin M Yu Savinina N N Balabanov S S Belyaev A V Gavrishchuk E M Ivanov A V Rizakhanov R N 2016 *Inorganic materials* **52** p 367-37
[9] Balabanov S S Vaganov V E Gavrishchuk E M. Drobotenko V V 2014 *Inorganic Materials.* **50** p 895 - 901
[10] Tolstikova D V Golieva E V Lebanin V S Mikhailov M D *2014 Optical Journal* **81** p 69 - 73
[11] Novikova A V 2014 *Functional Materials: Properties, Applications. Sat. theses* St. Petersburg. Lem. P 191 - 92
[12] Belykh G I Gritsynv V T Udalov L V 2004 *Questions of atomic science and techn.* **3**. P 101-107
[13] Martyukhova D A Skroznikova V V Popovich N V 2014 *X Int. Cong. of Young Scientists in Chemistry and Chemical* p 65 - 67
[14] Fikov P P *Dis. Ph.D.* 2007 RCTU them. DI. Mendeleev University
[15] Borshchishin I D Lutsyuk I V Vakhula Ya I Kochubei V V 2009 *Refractories and technical ceramics* **10** p 32-34
[16] Mamonova D V 2015 *Dis .... kand. Ph.D.* SPbSU. St. Petersburg
[17] Morozova L V Kalinina M V Drozdova I A Polyakova I G 2014 *Physics and chemistry of glass* **21** p 879 - 86
[18] Kuznetsov S V Fedorov P P Voronov V V Osiko. V V 2011 *Inorganic materials* **47** p 988 - 91
[19] Sokol V A Rohlenko D A Konovalova L I Bromberg A V 1981 *Inorganic materials* **17** p 896 - 901
[20] Ledovskaya E G Gabelkov S V Litvinenko L M Logvinkov D S 2006 *Questions of atomic science and techn* 2006 **1** p 160-63

IOP Conf. Series: Materials Science and Engineering **347** (2018) 012055 doi:10.1088/1757-899X/347/1/012055

[21] Gabelkov S V Tarasov R V Poltavtsev N S Kurilo Yu P 2007 *Inorganic materials* **4** p 462 - 70
[22] Li J Ikegami T Lee J Mori T 2000 J. Am. Ceram. Soc **83** p 2866–68
[23] Kobra N 2016 *Preparation, characterization, and catalytic application of metallic nanocrystalline MgAl2O4 in the synthesis of 3-hydroxy-3-indolyl-indolin-2-ones, symmetrical and unsymmetrical 3,30-bis (indolyl) indolin-2-ones, and 3,30-bis (indolyl) Arabian Journal of Chemistry* (electronic resource) URL: http://www.sciencedirect.com/science/article/pii/S1878535216000241
[24] Shuvendu T Deepash S S Debasis B 2016 *Journal of Asian Ceramic Societies* **4** p 149 – 54
[25] Galimov G G Sidorov A Yu Nikiforov A A 2014 *Refractories and technical ceramics* **9** p 21 – 26
[26] Kovtunenko P V 1997 *Glass and ceramics* **8** p 12 - 17
[27] Vydrik G A 1980 *Transparent ceramics*
[28] Evstropyev S K Sharypin V V Smirnov A N. Levit L G 2012 *Collected Works of the 10th Int Conf «Applied Optics»* p 235 - 37
[29] Belousova O L Domanskiy A I Ugolkov V L Dikaya L F 2008 *Physics and chemistry of glass* **34** p 616-21
[30] Wang S F Zhang J Luo D W Gu F 2013 *Progress in Solid State Chemistry* **41** p 20 – 54
[31] Tsukuma K 2006 *Journal of the Society of Japan* **114** p 802 – 806
[32] Evstrropiev S K Smirnov A N Sharypin V V 2014 *Glass and ceramics* **7** p 16-20
[33] Gledhill A Li D Mroz T. Goldman L Nitin Padture P 2012 *Acta Materialia* **60** p 1570 – 75
[34] Kachaev A A Graschenkov D V Lebedeva Yu E Solntsev S S Hasanov O L 2016 *Glass and ceramics* **4** p 3-10